Journal of Applied Logics - IfCoLog Journal of Logics and their Applications

Volume 5, Number 1

February 2018

Disclaimer
Statements of fact and opinion in the articles in Journal of Applied Logics - IfCoLog Journal of Logics and their Applications (JAL-FLAP) are those of the respective authors and contributors and not of the JAL-FLAP. Neither College Publications nor the JAL-FLAP make any representation, express or implied, in respect of the accuracy of the material in this journal and cannot accept any legal responsibility or liability for any errors or omissions that may be made. The reader should make his/her own evaluation as to the appropriateness or otherwise of any experimental technique described.

© Individual authors and College Publications 2018
All rights reserved.

ISBN 978-1-84890-274-9
ISSN (E) 2055-3714
ISSN (P) 2055-3706

College Publications
Scientific Director: Dov Gabbay
Managing Director: Jane Spurr

http://www.collegepublications.co.uk

Printed by Lightning Source, Milton Keynes, UK

All rights reserved. No part of this publication may be reproduced, stored in a retrieval system or transmitted in any form, or by any means, electronic, mechanical, photocopying, recording or otherwise without prior permission, in writing, from the publisher.

EDITORIAL BOARD

Editors-in-Chief
Dov M. Gabbay and Jörg Siekmann

Marcello D'Agostino	Melvin Fitting	Henri Prade
Natasha Alechina	Michael Gabbay	David Pym
Sandra Alves	Murdoch Gabbay	Ruy de Queiroz
Arnon Avron	Thomas F. Gordon	Ram Ramanujam
Jan Broersen	Wesley H. Holliday	Chrtian Retoré
Martin Caminada	Sara Kalvala	Ulrike Sattler
Balder ten Cate	Shalom Lappin	Jörg Siekmann
Agata Ciabttoni	Beishui Liao	Jane Spurr
Robin Cooper	David Makinson	Kaile Su
Luis Farinas del Cerro	George Metcalfe	Leon van der Torre
Esther David	Claudia Nalon	Yde Venema
Didier Dubois	Valeria de Paiva	Rineke Verbrugge
PM Dung	Jeff Paris	Heinrich Wansing
Amy Felty	David Pearce	Jef Wijsen
David Fernandez Duque	Brigitte Pientka	John Woods
Jan van Eijck	Elaine Pimentel	Michael Wooldridge
		Anna Zamansky

Area Scientific Editors

Philosophical Logic
Johan van Benthem
Lou Goble
Stefano Predelli
Gabriel Sandu

New Applied Logics
Walter Carnielli
David Makinson
Robin Milner
Heinrich Wansing

Logic and category Theory
Samson Abramsky
Joe Goguen
Martin Hyland
Jim Lambek

Proof Theory
Sam Buss
Wolfram Pohlers

Logic and Rewriting
Claude Kirchner
Jose Meseguer

Human Reasoning
Peter Bruza
John Woods

Modal and Temporal Logic
Carlos Areces
Melvin Fitting
Victor Marek
Mark Reynolds.
Frank Wolter
Michael Zakharyaschev

Automated Inference Systems and Model Checking
Ed Clarke
Ulrich Furbach
Hans Juergen Ohlbach
Volker Sorge
Andrei Voronkov
Toby Walsh

Formal Methods: Specification and Verification
Howard Barringer
David Basin
Dines Bjorner
Kokichi Futatsugi
Yuri Gurevich

Logic and Software Engineering
Manfred Broy
John Fitzgerald
Kung-Kiu Lau
Tom Maibaum
German Puebla

Logic and Constraint Logic Programming
Manuel Hermenegildo
Antonis Kakas
Francesca Rossi
Gert Smolka

Logic and Databases
Jan Chomicki
Enrico Franconi
Georg Gottlob
Leonid Libkin
Franz Wotawa

Logic and Physics (space time. relativity and quantum theory)
Hajnal Andreka
Kurt Engesser
Daniel Lehmann
lstvan Nemeti
Victor Pambuccian

Logic for Knowledge Representation and the Semantic Web
Franz Baader
Anthony Cohn
Pat Hayes
Ian Horrocks
Maurizio Lenzerini
Bernhard Nebel

Tactical Theorem Proving and Proof Planning
Alan Bundy
Amy Felty
Jacques Fleuriot
Dieter Hutter
Manfred Kerber
Christoph Kreitz

Logic and Algebraic Programming
Jan Bergstra
John Tucker

Logic in Mechanical and Electrical Engineering
Rudolf Kruse
Ebrahaim Mamdani

Logic and Law
Jose Carmo
Lars Lindahl
Marek Sergot

Applied Non-classical Logic
Luis Farinas del Cerro
Nicola Olivetti

Mathematical Logic
Wilfrid Hodges
Janos Makowsky

Cognitive Robotics: Actions and Causation
Gerhard Lakemeyer
Michael Thielscher

Type Theory for Theorem Proving Systems
Peter Andrews
Chris Benzmüller
Chad Brown
Dale Miller
Carsten Schlirmann

Logic Applied in Mathematics (including e-Learning Tools for Mathematics and Logic)
Bruno Buchberger
Fairouz Kamareddine
Michael Kohlhase

Logic and Computational Models of Scientific Reasoning
Lorenzo Magnani
Luis Moniz Pereira
Paul Thagard

Logic and Multi-Agent Systems
Michael Fisher
Nick Jennings
Mike Wooldridge

Logic and Neural Networks
Artur d'Avila Garcez
Steffen Holldobler
John G. Taylor

Logic and Planning
Susanne Biundo
Patrick Doherty
Henry Kautz
Paolo Traverso

Algebraic Methods in Logic
Miklos Ferenczi
Rob Goldblatt
Robin Hirsch
Idiko Sain

Non-monotonic Logics and Logics of Change
Jurgen Dix
Vladimir Lifschitz
Donald Nute
David Pearce

Logic and Learning
Luc de Raedt
John Lloyd
Steven Muggleton

Logic and Natural Language Processing
Wojciech Buszkowski
Hans Kamp
Marcus Kracht
Johanna Moore
Michael Moortgat
Manfred Pinkal
Hans Uszkoreit

Fuzzy Logic Uncertainty and Probability
Didier Dubois
Petr Hajek
Jeff Paris
Henri Prade
George Metcalfe
Jon Williamson

Scope and Submissions

This journal considers submission in all areas of pure and applied logic, including:

pure logical systems	dynamic logic
proof theory	quantum logic
constructive logic	algebraic logic
categorical logic	logic and cognition
modal and temporal logic	probabilistic logic
model theory	logic and networks
recursion theory	neuro-logical systems
type theory	complexity
nominal theory	argumentation theory
nonclassical logics	logic and computation
nonmonotonic logic	logic and language
numerical and uncertainty reasoning	logic engineering
logic and AI	knowledge-based systems
foundations of logic programming	automated reasoning
belief revision	knowledge representation
systems of knowledge and belief	logic in hardware and VLSI
logics and semantics of programming	natural language
specification and verification	concurrent computation
agent theory	planning
databases	

This journal will also consider papers on the application of logic in other subject areas: philosophy, cognitive science, physics etc. provided they have some formal content.

Submissions should be sent to Jane Spurr (jane.spurr@kcl.ac.uk) as a pdf file, preferably compiled in LaTeX using the IFCoLog class file.

CONTENTS

ARTICLES

Editorial 1
  Dov Gabbay and Jörg Siekmann

Possibilistic Reasoning from Partially Ordered Belief Bases with the Sure
  Thing Principle 5
  Claudette Cayrol, Didier Dubois and Fayçal Touazi

Characterization of a New Subquasivariety of Residuated Lattice 41
  Saeed Rasouli, Zeinab Zarin and Abass Hasankhan

Tuning the Program Transformers from CC to PDL 71
  *Pere Pardo, Enrique Sarión-Morrillo, Fernando Soler-Toscano and Fernando R.
  Velázquez-Quesada*

Lighthouse Principle for Diffusion in Social Networks 97
  Sanaz Azimipour and Pavel Naumov

A Labelled Sequent Calculus for Half-order Modal Logic 121
  Romas Alonderis and Jūratė Sakauskaitė

On Epicomplete MV-algebras 165
  Anatolij Dvurečenskij and Omid Zahiri

Paraconsistent Rule-based Reasoning with Graded Truth Values 185
  Francesco Luca De Angelis, Giovanna Di Marzo Serugendo and Andrzej Szałas

Paracomplete Logic K1 — Natural Deduction, its Automation, Complexity
  and Applications 221
  Alexander Bolotov, Danil Kozhemiachenko and Vasilyi Shagin

Suzumura Consistency, an Alternative Approach 263
  Peter Schuster and Daniel Wessel

Maximum Entropy Models for Σ_1 Sentences 287
  Soroush Rafiee Rad

Elementary Unification in Modal Logic $KD45$ 301
  Philippe Balbiani and Tinko Tinchev

Probabilistic Formal Verification of Communication Network-based Fault
  Detection, Isolation and Service Restoration System in Smart Grid . . . 319
  Syed Atif Naseem, Riaz Uddin, Osman Hasan and Diaa E. Fawzy

Elementary-base Cirquent Calculus I: Parallel and Choice Connectives . . . 367
  Giorgi Japaridze

Boolean-valued Models as a Foundation for Locally L^0-Convex Analysis
  and Conditional Set Theory 389
  Antonio Avilés and José Miguel Zapata

About Relationships Between two Individuals 421
  Robert Demolombe

On the Lattice of the Subvarieties of Monadic $MV(C)$-algebras 437
  Antonio Di Nola, Revaz Grigolia and Giacomo Lenzi

Editorial

1 Introduction

We are happy to introduce the first issue of the combined journal *Journal of Applied Logics - IfCoLog Journal of Logics and their Applications* .

This journal continues the publication of the Elsevier *Journal of Applied Logic* (JAL) together with our very successful *IfCoLog Journal of Logics and their Applications* (FLAP), as a free open access journal.

The Elsevier JAL was established by IfCoLog in 2002. In 2018 Elsevier discontinued the title and allowed IfCoLog to continue the Journal as a free open access journal.

In their communication to us dated 19th July 2017 Elsevier said among others:

- We agreed that for at least the first year of the new journal we would be able to add the line that it is supported by Elsevier. We agree with you that for PR and continuity this is a good option. We can review it after that

- We will also inform Scopus to ensure that they are aware of this for indexing purposes there too.

- Naturally towards the end of the year a press release, email announcement and a note on our JAL homepages can be arranged to ensure the community are aware this is a continuation of an old Journal and that Elsevier are fully supporting the Society in this endeavor.

We are grateful to Elsevier for their generosity and support of our Logic UK Charity IfCoLog and the idea of free open access.

The IfCoLog Journal of Logics and their Applications (JAL/FLAP) covers all areas of pure and applied logic, broadly construed. All papers published are free open access, and available via the College Publications website. This Journal is open access, puts no limit on the number of pages of any article, puts no limit on the number of papers in an issue and puts no limit on the number of issues per year. We insist only on a very high academic standard, and will publish issues as they come.

For example for the year 2017 we published 11 issues, containing about 4000 pages. Issue 4 for example, an issue dedicated to the Memory of Grigory Mints, was about 800 pages. No commercial publisher will ever do this.

The issues are available in both printed and electronic formats. It is published by College Publications, on behalf of the UK logic charity IfCoLog (www.ifcolog.net).

2 Background

The International Federation of Computational Logic (IfCoLog) sponsors THREE logic journals, two published by OUP (The Logic Journal of the IGPL and the Journal of Logic and Computation) and one published by Elsevier (Journal of Applied Logic). All three Journals are highly successful (attracting many submissions and high impact factor). The community has expressed a desire for some form of open access, and no limit on size of issues, thus giving immediate free access and also avoiding years of backlog in publications.

Publishers' current open access arrangement demands a hefty payment from authors, and they seem to be resisting any form of concession or compromise on size of issues.

In order to overcome this issue and set an example for other publishers, we proposed to start our own independent practically open access journal under the title "The (open access) IfCoLog Journal of Logics and their Applications".

This move is in the spirit of a recent call by the community to publish our own Journals. See this article `https://www.nature.com/news/mathematicians-aim-to-take-publishers-out-of-publishing-1.12243` in *Nature*.

From 2018, the Elsevier *Journal of Applied Logic* is no longer being published by Elsevier and is now being continued by the UK Charity IfCoLog.

We are making the *Journal of Applied Logics* free open access and amalgamating it with our current journal (FLAP), giving the unified journal the name *Journal of Applied Logics: IfCoLog Journal of Logics and their Applications*.

The impact factor of the *Journal of Applied Logic* is currently as follows

- CiteScore: 0.73

- More about CiteScore

- Impact Factor: 0.838

- 5-Year Impact Factor: 0.839

- Source Normalized Impact per Paper (SNIP): 0.936

- SCImago Journal Rank (SJR): 0.401

Editorial

We expect that the impact factor of the unified journals will continue and grow even stronger. We are also pleased that DBLP will index the new unified Journal.

We have amalgamated the lists of editors and area editors of both journals, and we are maintaining the high standard shared by all our Journals as sponsored by IfCoLog.

We are happy to present to you first issue of 2018 of the combined journal

Dov Gabbay
Jörg Siekmann
February 2018

POSSIBILISTIC REASONING FROM PARTIALLY ORDERED BELIEF BASES WITH THE SURE THING PRINCIPLE

CLAUDETTE CAYROL, DIDIER DUBOIS
IRIT, CNRS and Université de Toulouse, France.
`{claudette.cayrol,didier.dubois}@irit.fr`

FAYÇAL TOUAZI
University M'hamed Bougara, Independence Avenue, 35000 Boumerdes, Algeria.
`Faycal.touazi@univ-boumerdes.dz`

Abstract

We consider the problem of reasoning from logical bases equipped with a partial order expressing relative certainty, with a view to construct a partially ordered deductive closure via syntactic inference. At the syntactic level we use a language expressing pairs of related formulas and axioms describing the properties of the order. Reasoning about uncertainty using possibility theory relies on the idea that if an agent believes each among two propositions to some extent, then this agent should believe their conjunction to the same extent. This principle is known as adjunction. Adjunction is often accepted in epistemic logic but fails with probabilistic reasoning. In the latter, another principle prevails, namely the sure thing principle, that claims that the certainty ordering between propositions should be invariant to the addition or deletion of possible worlds common to both sets of models of these propositions. Pursuing our work on relative certainty logic based on possibility theory, we propose a qualitative likelihood logic that respects the sure thing principle, albeit using a likelihood relation that preserves adjunction.

Keywords : partially ordered bases, possibility theory, adjunction rule, comparative probability

1 Introduction

The representation of partial belief often uses a numerical setting, prominently the one of probability theory, but also weaker non-additive settings such as belief functions or imprecise probabilities (see [18] for a survey). However, this kind of approach requires the use of elicitation procedures so as to force agents to provide degrees of belief through a given protocol (for instance, using the betting metaphor, assigning prices to gambles or risky events, or using analogy with the frequentist setting of drawing balls from a known urn). Inevitably, the resulting numbers will not have infinite precision, which leads either to consider precise figures as suitable idealization, or to take into account the imprecision of assessments, which may lead to more complex computations.

Reasoning with uncertain knowledge often consists of attaching belief weights to propositions of interest and computing belief weights of other propositions of interest, using some appropriate inference methods. This approach was early considered by De Finetti [10] (see [29] for a translation), and then taken over by many other scholars (Adams and Levine[1], Coletti and Scozzafava [9], Nilsson [34], etc.).

In this paper, we deliberately give up assigning belief weights to propositions. We assume that uncertain knowledge is based on stating that some propositions are more believed than others. This is the least we can expect from agents expressing their beliefs. In the case of probability theory, it comes down to studying properties of the relation "more probable than" first introduced by De Finetti [10], and later by Ramsey, and Savage, among others (see Fishburn[21] for an early survey). Comparative probabilities are total orders on propositions, that obey a special case of the so-called sure thing principle of Savage [35], stating that the fact that a proposition is more probable than another one is not affected by the probabilities of their common models. We call this property *preadditivity*, to highlight the known fact that on finite settings this property is not sufficient to ensure the existence of a probability measure representing the ordering between propositions [26]. There is not a long tradition on logics for comparative probability that do not refer to a numerical underpinning. This point is discussed in detail by Walley and Fine [37] who provide an early overview on modal, conditional and comparative probability logics.

Another kind of uncertainty relation, originally introduced by Lewis [32], are comparative possibility relations, independently introduced by Dubois [11] along with their dual called necessity relations. While Lewis introduced these concepts in connection with the logical representation of counterfactuals, Dubois viewed possibility relations as the ordinal counterpart of Zadeh's possibility measures [39]. These relations are weak orders that do not obey the sure thing principle, but they are instrumental in non-monotonic reasoning and belief revision [17, 2] (where necessity relations are called epistemic entrenchments). This setting also captures the notion of accepted beliefs [15]: the agent reasons with such beliefs as if they were true ones, so that the condition that the conjunction of accepted beliefs is an accepted belief is adopted, like in epistemic logic. For the sake of clarity, we call qualitative plausibility and certainty relations the generalisation of possibility and necessity relations to the partially ordered setting. The key property for such relations is called *qualitativeness* [22], which encodes the idea that a possible world is always more likely than the disjunction of less likely worlds.

Logics for reasoning with totally ordered comparative possibility statements have been first studied by Lewis [31]. Possibilistic logic [19] is an alternative setting where a total order on a subset of propositions is encoded by means of weighted formulas, where weights attached to formulas are taken from a totally ordered symbolic scale. In this paper we focus on partial orders, as we consider that agents may only have a lacunary knowledge of the relative beliefs of propositions. Approaches to reasoning from logical bases equipped with a partial order expressing relative certainty have been proposed by Halpern [25] using a modal logic framework inspired by Lewis works, which means a very rich language. A simpler framework, called relative certainty logic and focusing on strict partial orders, yet adopting similar axioms as Halpern, is presented in [36], where the purpose is to construct a partially ordered deductive closure. The idea is to interpret a partially ordered base as a partial necessity ordering. At the syntactic level the language expresses pairs of related formulas; axioms and inference rules describe the properties of the partial certainty order. The semantics consists in assuming that the partial order on formulas stems from a partial order between the corresponding sets of models (and not between models as in possibilistic logic).

Moving from the totally ordered to the partially ordered setting is non-trivial. The difficult points are twofold: (i) equivalent definitions in the totally ordered case are no longer equivalent in the partially ordered one, and (ii) a partial possibility order on subsets of a set cannot be represented by a partial order between elements of this set. This point is especially explained in [36].

In this paper, we pursue the work initiated in [36] with a view to study how the preadditivity of comparative probability can be used to refine the relative certainty logic. In the totally ordered case, qualitativeness is almost incompatible with preadditivity [15]. In the partially ordered setting, we get a qualitative likelihood logic that is adjunctive, but respects the sure thing principle, that we compare with the qualitative certainty logic of [36]. Moreover we show that the latter logic can be used to facilitate inference in the former.

The paper is structured as follows: in the next section we provide an overview of confidence relations between sets of states, including comparative possibility and probability. Then we provide characteristic properties of qualitative plausibility, certainty and (preadditive) likelihood relations, in the partially ordered setting. We show that there is a bijection between qualitative plausibility and qualitative likelihood relations. Based on these new results we propose in section 3 a general setting for reasoning about uncertainty using confidence relations, which extends the methodology introduced in [36] for qualitative certainty logic. Then, in sections 4 and 5 we respectively focus on the qualitative likelihood logic and on its connection with relative certainty logic.

2 Qualitative confidence relations comparing subsets

In a non-numerical setting, it is natural to represent confidence in propositions by means of a partial preorder $\succeq$ on subsets $A, B, C, \ldots$ of a set of states of affairs S. This idea goes back to De Finetti's [10] comparative probabilities, and is presented in more details in Fine's book [20]. Other proposals are comparative possibilities of Lewis [32] later independently proposed, along with their dual necessity relations by one of the authors [11] in contrast with comparative probabilities. These are examples of complete preorders (reflexive, complete and transitive relations) on the power set $\wp(S)$. Various examples of confidence relations have been discussed by Halpern [23, 22] in connection with non-monotonic reasoning. They are called acceptance relations in [15]. In some cases, confidence orderings stem from a total or partial plausibility ordering on S. This is the case for comparative possibility relations and their refinements [14, 12], and also for relations built from a partial order on elements, studied by Halpern [25]. In this section we review such relations and their properties.

Given a reflexive relation $\succeq$ on $\wp(S)$ we can derive three companion relations:

- The strict part of $\succeq$: $A \succ B$ iff $A \succeq B$, but not $B \succeq A$
- The indifference relation $A \sim B$ iff $A \succeq B$ and $B \succeq A$
- The incomparability relation: $A \pm B$ iff neither $A \succeq B$ nor $B \succeq A$

Moreover, we can also define the *dual* $\succeq^d$ of a relation $\succeq$ on $\wp(S)$ as:

$$A \succeq^d B \text{ iff } \overline{B} \succeq \overline{A}$$

There are minimal requirements a confidence relation should satisfy in order to justify this name.

1. Compatibility with Inclusion (CI) If $B \subseteq A$ then $A \succeq B$

Indeed if B implies A there is no point for B to be more likely than A.[1]

2. Orderliness (O) If $A \succ B$, $A \subseteq A'$, and $B' \subseteq B$, then $A' \succ B'$

This property, already mentioned by Walley and Fine [37], and also used by Friedman and Halpern [22], is a variant of, but not equivalent to, the former. It also reflects compatibility with logical deduction.

3. Quasi-Transitivity (QT) If $A \succ B$, and $B \succ C$, then $A \succ C$

---

[1] Friedman and Halpern [23] call "Plausibility measure" a partial relation that satisfies (CI); however this name may be judged misleading, since plausibility is a notion dual to belief, as often used in evidence theory.

Should this property be false for an agent, one may question her rationality. These are the three minimal properties we can expect from a partial confidence relation.

Definition 1. *A relation on $\wp(S)$ is called a* confidence relation *if it satisfies (CI, O, QT). Its strict part is called a* strict confidence relation.

This terminology was proposed in [12]. It is clear that a confidence relation is reflexive and consistent in the sense that $S \succeq A \succeq \emptyset$ for all subsets A of S. Note that we can do away with the two monotonicity conditions (CI) and (O) if we modify the latter by requiring it for $\succeq$ instead of $\succ$. Moreover, a strict confidence relation is a strict partial order satisfying (O). Finally, it can be easily verified that the dual of a confidence relation is again a confidence relation.

2.1 Complete and transitive confidence relations

It is quite often the case that partial belief is represented numerically via a set-function $f : \wp(S) \to [0, 1]$, for instance a probability measure. A set-function f is said to *represent* a confidence relation $\succeq$ provided that for all subsets A, B of S, $A \succeq B$ if and only if $f(A) \geq f(B), f(\emptyset) = 0, f(S) = 1$.

Of course, if this is so, the confidence relation $\succeq$ should be transitive and complete (hence reflexive):

- *Transitivity*: If $A \succeq B$, and $B \succeq C$, then $A \succeq C$

- *Completeness*: $A \succeq B$ or $B \succeq A$

It is easy to see that complete and transitive confidence relations are represented by *capacities*, which are monotonic set-functions, such that if $A \subseteq B$ then $f(A) \leq f(B)$, which expresses (CI) (for instance, [18]). In fact, for transitive and complete relations, (CI) implies (O). Important examples of complete and transitive confidence relations are

- *Comparative probabilities* [10, 20]: They are complete and transitive confidence relations that obey the preadditivity property:

    Preadditivity (P) If $A \cap (B \cup C) = \emptyset$ then ($B \succeq C$ iff $A \cup B \succeq A \cup C$)

- *Comparative possibilities* [32, 11]: They are complete and transitive confidence relations that satisfy a property that is a variant of the former:

    Stability for Union (SU) If $A \succeq B$ then $A \cup C \succeq B \cup C$

Comparative possibility relations, denoted by $\succeq_\Pi$, can be represented by and only by possibility measures [11]. They are set-functions $\Pi : \wp(S) \to [0, 1]$ such that $\Pi(A \cup B) = \max(\Pi(A), \Pi(B))$ [39, 16]. This is because the (SU) axiom for complete and transitive confidence relations is equivalent to: If $A \succeq_\Pi B$ then $A \sim_\Pi A \cup B$. Comparative possibility relations on finite sets are completely characterised by the restriction $\geq_\pi$ of $\succeq_\Pi$ to singletons on S. Namely [11]:

$$A \succeq_\Pi B \iff \forall s_2 \in B, \exists s_1 \in A : s_1 \geq_\pi s_2 \qquad (1)$$
$$\iff \exists s_1 \in A, \forall s_2 \in B : s_1 \geq_\pi s_2 \qquad (2)$$

This property, which shows the simplicity of this approach, reflects the fact that a possibility measure Π derives from a possibility distribution $\pi : S \to [0, 1]$, in the sense that $\Pi(A) = \max_{s \in A} \pi(s)$. In the scope of uncertainty modeling, $\pi(s)$ can be viewed as a degree of plausibility of s, and the condition $\max_{s \in S} \pi(s) = 1$ must be satisfied. The possibility degree $\Pi(A)$ can be interpreted as a degree of unsurprizingness of A, i.e., the degree to which there is no reason not to believe A (which does not imply a reason for believing it).

The conjugate functions $N(A) = 1 - \Pi(\overline{A})$, called necessity measures [16], express the idea that A is certain to some extent, that is, A is true in all situations that are plausible enough. The corresponding necessity relations $\succeq_N$ have a characteristic axiom called

Stability for intersection (SI): If $A \succeq_N B$ then $A \cap C \succeq_N B \cap C$

It is easy to check [11] that necessity relations can be defined from possibility relations by duality: $A \succeq_N B$ if and only if $\overline{B} \succeq_\Pi \overline{A}$, so that

$$A \succeq_N B \iff \forall s_2 \in \overline{A}, \exists s_1 \in \overline{B} : s_1 \geq_\pi s_2 \qquad (3)$$
$$\iff \exists s_1 \in \overline{B}, \forall s_2 \in \overline{A} : s_1 \geq_\pi s_2 \qquad (4)$$

Comparative possibility relations satisfy properties that indicate their qualitative nature:

Qualitativeness (Q) If $A \cup B \succ_\Pi C$ and $A \cup C \succ_\Pi B$, then $A \succ_\Pi B \cup C$
Negligibility (N) If $A \succ_\Pi B$ and $A \succ_\Pi C$, then $A \succ_\Pi B \cup C$

the second one being a consequence of the first. Negligibility expresses the non-compensatory nature of possibility measures, according to which the union of unlikely singletons cannot override a very plausible one.

Necessity relations obey counterparts of (Q) and (N):

Dual qualitativeness (Q^d): If $A \succ_N B \cap C$ and $B \succ_N A \cap C$ then $A \cap B \succ_N C$
Adjunction (A): If $A \succ_N C$ and $B \succ_N C$ then $A \cap B \succ_N C$

These properties make it clear that the family of sets $\{A : A \succ_N C\}$ is a filter (closed under inclusion and intersection) or in terms of propositions, deductively closed. This closure property for confidence measures is also characteristic of necessity relations for complete and transitive confidence relations [15].

Preadditive complete and transitive confidence relations $\succeq_P$, called comparative probabilities, behave very differently. Given a probability measure on S, the relation $A \succeq B$ if and only if $P(A) \geq P(B)$ for some probability measure on a finite set is indeed preadditive, complete and transitive. However the converse is false, namely it has been known since the 1950's [26] that there are comparative probability relations that are not representable by a probability measure; see also [33]. Nevertheless comparative probability relations are self dual, in the sense of the following property:

Self-duality (D) $A \succeq_P B$ iff $\overline{B} \succeq_P \overline{A}$

However the fact that comparative probability relations are more general than confidence relations induced by probabilities highlights the fact that, contrary to comparative possibility and necessity relations, they cannot be defined by a complete preorder on S: the restriction of $\succeq_P$ on singletons is not enough to reconstruct it. In fact comparative probabilities can be represented by special kinds of belief functions inducing a self-dual order [38].

Interestingly, there are comparative probability relations that satisfy the qualitativeness properties. It is proved in [3] that they correspond to so-called big-stepped probabilities on S: there is a probability distribution p such that $p(s_1) > p(s_2) > \cdots > p(s_{n-1}) > p(s_n)$, with $\forall i = 1, \ldots, n-1, p(s_i) > \sum_{j=i+1}^{n} p(s_j)$, and then $A \succeq_P B$ if and only if $P(A) \geq P(B)$. The probabilities of singletons form a super-increasing sequence. Moreover if we consider the possibility ordering $s_1 >_\pi s_2 >_\pi \cdots >_\pi s_{n-1} >_\pi s_n$, then, for non-elementary events A, B we have that $A \succ_\Pi B$ implies $A \succ_P B$. In other words, the comparative probability relation induced by a big-stepped probability refines the possibility relation (see also [12]).

In this paper, we generalize possibility relations and necessity relations to partial orders on S, and consider their preadditive refinements.

2.2 Partial qualitative confidence relations

In this section we consider partial confidence relations satisfying property (Q). The four properties (CI), (O), (QT) and (Q) are not independent [25, 5].

Proposition 1. *If a relation on $\wp(S)$ satisfies (Q) and (O), this relation and its dual are transitive.*

Proof of Proposition 1:
We use a relation denoted by $\triangleright$ that can stand for $\succeq$ or its strict part. Suppose $A \triangleright B$ and $B \triangleright C$. Then, from (O), $A \cup C \triangleright B$ and $A \cup B \triangleright C$, and from (Q): $A \triangleright B \cup C$, then by (O), $A \triangleright C$. A similar proof holds for the dual relation. □

Partial confidence relations satisfying property (Q) generalize comparative possibilities. However, in the following we consider asymmetric relations of this kind, to which (CI) does not apply:

Definition 2. *A* qualitative plausibility relation *is an asymmetric relation $\succ_{pl}$ on $\wp(S)$ that satisfies (Q) and (O).*

Due to Proposition 1, a qualitative plausibility relation is indeed a strict partial order on $\wp(S)$ since it is transitive. Moreover,

Proposition 2. *A qualitative plausibility relation satisfies (N), and (SU) in contrapositive form: If $A \cup C \succ_{pl} B \cup C$ then $A \succ_{pl} B$.*

Proof of Proposition 2:
(N) is an obvious consequence of (Q) and (O). For (SU), suppose $A \cup C \succ_{pl} B \cup C$. By (O), we infer that $A \cup (B \cup C) \succ_{pl} C$. Applying (Q) yields $A \succ_{pl} B \cup C$, which by (O), results in $A \succ_{pl} B$ [7]. □

Another useful property related to (SU) is:

Proposition 3. *A qualitative plausibility relation is such that: If $A \succ_{pl} B$ and $C \succ_{pl} D$ then $A \cup C \succ_{pl} B \cup D$.*

Proof of Proposition 3:
Due to (O), $A \succ_{pl} B$ and $C \succ_{pl} D$ imply $A \cup C \cup D \succ_{pl} B$ and $A \cup C \cup B \succ_{pl} D$, and then by (Q), $A \cup C \succ_{pl} B \cup D$ follows. □

Now we introduce another partial order on a set of events, called a qualitative certainty relation:

Definition 3. *A* qualitative certainty relation, *denoted by $\succ_{cr}$, is an asymmetric relation on $\wp(S)$ that satisfies Q^d and O.*

It is clear that $\succ_{cr}$ is a qualitative certainty relation if and only if its dual relation is a qualitative plausibility relation. In particular, from the above results, it easily follows that a qualitative certainty relation is transitive, satisfies adjunction, and (SI) under the form: If $A \cap C \succ_{cr} B \cap C$ then $A \succ_{cr} B$. Moreover, if $A \succ_{cr} B$ and $C \succ_{cr} D$ then $A \cap C \succ_{cr} B \cap D$.

Contrary to the terminology used in [23], the use of *plausibility* vs. *certainty* to name confidence relations satisfying (Q) vs. its dual property (Q^d) makes the point that such relations are dual to each other, and reflect the dual pairs (possibility, necessity), (plausibility, belief) in other uncertainty theories, where the second concept in each pair is more committing than the first one.

Qualitative plausibility and certainty relations are instrumental for defining a semantics for non-monotonic reasoning (as explained in [22, 15]). Namely consider the following properties for a partial order on $\wp(S)$, inspired from [27]:

- **Conditional Closure by Implication (CCI)** If $A \subseteq B$ and $A \cap C \succ \overline{A} \cap C$ then $B \cap C \succ \overline{B} \cap C$

- **Conditional Closure by Conjunction (CCC)** If $C \cap A \succ C \cap \overline{A}$ and $C \cap B \succ C \cap \overline{B}$ then $C \cap (A \cap B) \succ C \cap \overline{(A \cap B)}$

- **Left Disjunction (OR)** If $A \cap C \succ A \cap \overline{C}$ and $B \cap C \succ B \cap \overline{C}$ then $(A \cup B) \cap C \succ (A \cup B) \cap \overline{C}$

- **Cut (CUT)** If $A \cap B \succ A \cap \overline{B}$ and $A \cap B \cap C \succ A \cap B \cap \overline{C}$ then $A \cap C \succ A \cap \overline{C}$

- **Cautious Monotony (CM)** If $A \cap B \succ A \cap \overline{B}$ and $A \cap C \succ A \cap \overline{C}$ then $A \cap B \cap C \succ A \cap B \cap \overline{C}$

These properties are intuitive when $A \succ \overline{A}$ is interpreted as "A is an accepted belief", and $A \cap C \succ \overline{A} \cap C$ as "A is an accepted belief in the context C", hence the name "acceptance relations" for qualitative plausibility relations in [15]. In that work, it has been proved that:

Proposition 4.

- *(O) implies (CCI).*

- *If a relation between subsets of S satisfies (Q) and (O), then it satisfies (CCI), (CCC), (OR), (CUT), (CM).*

- *For any relation that satisfies (O), (CCC) is equivalent to (Q).*

See also [6] for the two first results.

It is clear that qualitative plausibility relations satisfy all these properties and are ideally fit for non-monotonic reasoning with conditional assertions of the form $A \mathrel{|\!\sim} B$, which stand for $A \cap B \succ_{pl} A \cap \overline{B}$ [27]. Note that properties (CCI), (CCC), (OR), (CUT), (CM) only involve the comparison of disjoint subsets. It is proved in [15], and follows from Proposition 4 that if the restriction of a confidence relation to disjoint subsets satisfies (CCI), (CCC), (OR), (CUT), (CM) then it is the restriction of a qualitative plausibility relation.

One way to construct a qualitative plausibility relation is to proceed as suggested by Halpern [25]. Let $(S, \triangleright)$ be a partially ordered set, where $\triangleright$ is an asymmetric and transitive relation. Various possible definitions for extending the comparative possibility to qualitative plausibility relations have been reviewed in [6] and arguments have been given for selecting one of them. Here, like in our previous paper [36] we consider the extensions (1) and (2) of the strict part of $\geq_\pi$ to build a partial order between subsets. It turns out they are no longer equivalent, and the one possessing the greatest number of properties is:

Definition 4 (Weak optimistic strict dominance). *Let $\triangleright$ be an asymmetric and transitive relation on S. Then $A \succ_{wos}^{\triangleright} B$ iff $A \neq \emptyset$ and $\forall b \in B, \exists a \in A, a \triangleright b$.*

It is clear that if $\triangleright$ is the strict part of a complete preorder on S encoded by a possibility distribution π, $A \succ_{wos}^{>\pi} B$ if and only if $\Pi(A) > \Pi(B)$. In the partially ordered setting, the following properties have been established [25, 6, 36]:

Proposition 5. *The weak optimistic strict dominance $\succ_{wos}^{\triangleright}$ is a strict partial order that satisfies Qualitativeness (Q) and Orderliness (O).*

Unfortunately, contrary to the totally ordered case, not all qualitative plausibility relations can be generated from a partial order on S. This is because knowing only the restriction to the singletons of S of a qualitative plausibility relation $\succ_{pl}$ on $\wp(S)$ is insufficient to reconstruct $\succ_{pl}$. Namely, let a partial order on S be defined by $s_1 \triangleright_{pos} s_2$ if and only if $\{s_1\} \succ_{pl} \{s_2\}$, where $\succ_{pl}$ satisfies (Q) and (O). Consider the relation $\succ_{wos}^{\triangleright_{pos}}$ induced by $\triangleright_{pos}$ via Definition 4. Then $A \succ_{wos}^{\triangleright_{pos}} B$ implies $A \succ_{pl} B$, but generally the converse does not hold [6].

Example 1 (due to Halpern). *Let $S = \{a, b, c\}, A = \{a\}, B = \{b\}, C = \{c\}$. Suppose relation $\succ$ is the smallest asymmetric partial order relation including constraints $B \cup C \succ A, A \succ \emptyset, B \succ \emptyset, C \succ \emptyset$, and that is closed for (O) and (T). It obviously satisfies (Q). It is a qualitative plausibility relation. Define the partial order on S as $s_1 \triangleright_{pos} s_2$ if and only if $\{s_1\} \succ \{s_2\}$. Then elements a, b, c are not comparable. So we do not have $\{b, c\} \succ_{wos}^{\triangleright_{pos}} \{a\}$ and we cannot retrieve $B \cup C \succ A$.*

Remark A result due to Halpern [25] says that qualitative plausibility relations on $\wp(S)$ can be generated from a partial order on a set larger than S, which stands as a refinement of it. Namely, for any qualitative plausibility relation $\succ_{pl}$ on S, there is a set Ω, a surjective map $f : \Omega \to S$, and a partial order $\triangleright$ on Ω such that, if A, B are subsets of S, $A \succ_{pl} B$ if and only if $f^{-1}(A) \succ_{wos}^{\triangleright} f^{-1}(B)$. This is in fact the semantics adopted by Lehmann and colleagues [27] for non-monotonic relations from conditional assertions.

Another way to generate qualitative plausibility relations is to start from a family $\mathcal{L}$ of linear orders $>_\sigma$ on S defined by permutations σ of elements $(s_{\sigma(1)} >_\sigma s_{\sigma(2)} >_\sigma \cdots >_\sigma s_{\sigma(n)})$ and let the relation $\succ_\mathcal{L}$ on $\wp(S)$ be defined as follows:

$$A \succ_\mathcal{L} B \iff \forall >_\sigma \in \mathcal{L}, A \succ_\Pi^\sigma B$$

where $\succ_\Pi^\sigma$ is the strict part of the comparative possibility relation induced by $>_\sigma$ on S [3]. It is easy to check that the relation $\succ_\mathcal{L}$ is a qualitative plausibility relation, i.e., it satisfies the properties (Q) and (O). An interesting question addressed below is whether any qualitative plausibility relation can be generated in this way. To this end, we introduce two more properties of relations between sets:

Non-Dogmaticism (NoD) $\forall A \neq \emptyset, A \succ \emptyset$

Semi-Cancellativity (SC) $A \succ B$ if and only if $A \setminus B \succ B$

We can establish the following proposition:

Proposition 6. *A qualitative plausibility relation is semi-cancellative.*

Proof of Proposition 6:
It is clear that by (O), $A \setminus B \succ_{pl} B$ implies $A \succ_{pl} B$. The less obvious part is the converse: suppose $A \succ_{pl} B$. It can be written as

- $(A \setminus B) \cup (A \cap B) \succ_{pl} B$
- and also as $(A \setminus B) \cup B \succ_{pl} B$ which implies $(A \setminus B) \cup B \succ_{pl} A \cap B$.

Now applying (Q) yields $A \setminus B \succ_{pl} B \cup (A \cap B) = B$. □

It is proved in [15] that for any non-dogmatic, semi-cancellative qualitative plausibility relation $\succ_{pl}$, there exists a family $\mathcal{L}$ of linear orders on S, such that $\succ_{pl}$ coincides with the relation $\succ_\mathcal{L}$ on disjoint subsets.

Using this result, we get the representation theorem for qualitative plausibility relations as follows:

Corollary 1. *A non-dogmatic relation $\succ$ between sets is a qualitative plausibility relation if and only if there is a family $\mathcal{L}$ of linear orders $>_\sigma$ on S, such that $A \succ B$ if and only if $A \succ_\mathcal{L} B$.*

Proof of Corollary 1:
Let $\succ_{pl}$ be a non-dogmatic qualitative plausibility relation. From [15], there exists a family $\mathcal{L}$ of linear orders $>_\sigma$ on S, such that $\succ_{pl}$ coincides with the relation $\succ_\mathcal{L}$ on disjoint subsets. $A \succ_{pl} B$

if and only if $A \setminus B \succ_{pl} B$ (by semi-cancellativity). So $A \succ_{pl} B$ if and only if $A \setminus B \succ_{\mathcal{L}} B$ if and only if $\forall >_\sigma \in \mathcal{L}, A \setminus B \succ_\Pi^\sigma B$, if and only if $\forall >_\sigma \in \mathcal{L}, A \succ_\Pi^\sigma B$ if and only if $A \succ_{\mathcal{L}} B$. For the converse it has been already said that the relation $\succ_{\mathcal{L}}$ built from a family of linear orders is a qualitative plausibility relation. □

This is the answer to the question of whether any qualitative plausibility relation can be constructed from a family of possibility orderings.

2.3 Preadditive substitutes of confidence relations

The property of preadditivity considers that the common part of two sets should play no role in their comparison. This is the idea behind Savage sure thing principle [35], which applies to the comparison of more general functions than characteristic functions of sets. One may say that preadditivity is precisely an instance of this principle. Preadditivity is a sufficient condition to make a relation between subsets self-dual:

Proposition 7. *For any relation $\succ$ on $\wp(S)$, (P) implies (D).*

Proof of Proposition 7:
Let $A \succ B$. $A = (A \setminus B) \cup (A \cap B)$ and similarly $B = (B \setminus A) \cup (A \cap B)$. Applying (P) produces $(A \setminus B) \succ (B \setminus A)$. Applying (P) again yields $(A \setminus B) \cup \overline{(A \cup B)} \succ (B \setminus A) \cup \overline{(A \cup B)}$. That is $\overline{B} \succ \overline{A}$. □

As a direct consequence, we have an equivalent form of (P), which is to (P) what (SI) is to (SU):

$$(P) \Leftrightarrow \text{If } A \cup (B \cap C) = S \text{ then } (B \succ C \text{ iff } A \cap B \succ A \cap C)$$

Moreover, the two following properties are direct consequences of (P):

- $B \succ C$ iff $B \setminus C \succ C \setminus B$ (a stronger property than semi-cancellativity)

- $B \succ C$ iff $B \cup \overline{C} \succ C \cup \overline{B}$

A preadditive approach for comparing two sets A and B then consists in eliminating the common part and then comparing $A \setminus B$ and $B \setminus A$. This is not a new idea (see [24], [25]). Given a partial order $\succ$ on $\wp(S)$ one can define a preadditive ordering $\succ^+$, called *preadditive substitute* of $\succ$ as follows:

$$A \succ^+ B \text{ if and only if } A \setminus B \succ B \setminus A$$

Clearly $\succ^+$ and $\succ$ coincide on pairs of disjoint subsets, and it is obvious that $\succ^+$ is preadditive, which implies it is self-dual, due to Proposition 7.

Consider a confidence relation $\succeq$ in the sense of Definition 1 and its preadditive substitute $\succeq^+$. It is obvious that, as soon as the strict part of this relation is non-dogmatic (which means that all elements in S are in some sense useful or possible), the latter satisfies a strong form of compatibility with inclusion:

Strict Compatibility with Inclusion (SCI) If $B \subset A$ then $A \succ B$

Proposition 8. *If a relation $\succ$ between subsets satisfies Preadditivity, then SCI is equivalent to its weak form: If $A \neq \emptyset$ then $A \succ \emptyset$ (NoD)*

Proof of Proposition 8:
Assume that $\succ$ satisfies (P) and (NoD). Let $B \subset A$. We have $B \setminus A = \emptyset$ and $A \setminus B \neq \emptyset$. By (NoD) we obtain $(A \setminus B) \succ (B \setminus A)$. By (P), we add $A \cap B = B$ to each side and we obtain $((A \setminus B) \cup B) = A \succ ((B \setminus A) \cup B) = B$. □

The following relaxed versions of properties (Q) and (N) are appropriate for preadditive relations.

- **Qualitativeness for disjoint sets (QD)** If $A \cup C \succ B$ and $A \cup B \succ C$ then $A \succ B \cup C$, provided that $A \cap B = A \cap C = B \cap C = \emptyset$

- **Negligibility for disjoint sets (ND)** If $A \succ B$ and $A \succ C$ then $A \succ B \cup C$, provided that $A \cap B = A \cap C = \emptyset$

It is easy to verify:

Proposition 9. *The properties (Q) and (QD) are equivalent when $\succ$ is applied to disjoint sets.*

Proof of Proposition 9:
Obviously, (Q) implies (QD).
Conversely, let us assume that $\succ$ satisfies (QD) and consider that $A \cup C \succ B$ and $A \cup B \succ C$, with $(A \cup C) \cap B = (A \cup B) \cap C = \emptyset$. As $(A \cup C) \cap B = \emptyset$, we have that $A \cap B = C \cap B = \emptyset$. Similarly, we have $A \cap C = \emptyset$. So (QD) can be applied, producing $A \succ B \cup C$. □

Proposition 10. *For any relation $\succ$ on $\wp(S)$, if $\succ$ satisfies:*

- *transitivity (T) and (SCI), then it satisfies (O);*

- *(QD) and (O), then it satisfies (ND);*

- (QD) and (O), then it satisfies (CCI), (CCC), (OR), (CUT), (CM);
- (CCC), then it satisfies (QD).

Proof of Proposition 10:
T, SCI ⇒ O: Assume that $A \succ B$, $A \subseteq A'$, and $B' \subseteq B$. We have to prove that $A' \succ B'$.
If $A = A'$ we have $A' \succ B$. If $A \subset A'$ we obtain $A' \succ A$ by (SCI) and then $A' \succ B$ by transitivity (T).
Now, if $B = B'$ we obtain $A' \succ B'$. Otherwise $B' \subset B$, so $B \succ B'$ by (SCI) and by transitivity we obtain $A' \succ B'$.
O, QD ⇒ ND: Assume that $A \cap B = A \cap C = \emptyset$, $A \succ B$ and $A \succ C$. We have to prove that $A \succ (B \cup C)$. From $A \succ B$ and (O): $(A \cup C) \succ (B \setminus C)$ (1). From $A \succ C$ and (O), we obtain $A \cup (B \setminus C) \succ C$ (2).
Due to the assumptions, we have $A \cap (B \setminus C) = A \cap C = \emptyset$ and obviously $C \cap (B \setminus C) = \emptyset$. Applying (QD) from (1) and (2) yields $A \succ (C \cup (B \setminus C))$ that is $A \succ (B \cup C)$.
O, QD ⇒ CCI, CCC, OR, CUT, CM: As $A \cap C$ and $\overline{A} \cap C$ (resp. $B \cap C$ and $\overline{B} \cap C$) are disjoint sets, the proof of Proposition 4 can be used.
(CCC) ⇒ (QD) This is Theorem 1 in [15]. □

As (P) implies Self-duality, it follows that the property (QD) possesses a dual property (QD^d) equivalent to the former for preadditive relations:

QD^d: If $A \cup B = A \cup C = B \cup C = S$, then if $C \succ A \cap B$ and $B \succ A \cap C$, then
$$B \cap C \succ A$$

So, for preadditive substitutes, we can use the dual property (QD^d), in place of (QD).

An important question is whether a strict confidence relation $\succ$ is refined or not by its preadditive substitute. We can prove this property for confidence relations that obey the following weak form of both preadditivity and stability for disjunction (first proposed in [11] for weak – transitive and complete – orders):

Stability for Disjoint Union (SDU) If $A \cap (B \cup C) = \emptyset$ then $A \cup B \succ A \cup C$ implies
$$B \succ C$$

Proposition 11. *If an asymmetric relation $\succ$ satisfies (SDU), then its preadditive substitute $\succ^+$ is a self-dual refinement of $\succ$ and of its dual.*

Proof of Proposition 11:
The result is obvious from $\succ$ to $\succ^+$ since by (SDU), if $A \succ B$ then $A \setminus B \succ B \setminus A$ which is $A \succ^+ B$. For the dual relation $\succ^d$, $A \succ^d B$ means $\overline{B} \succ \overline{A}$ which also reads $(A \setminus B) \cup (\overline{A} \cap \overline{B}) \succ (B \setminus A) \cup (\overline{A} \cap \overline{B})$, which by (SDU) implies $A \succ^+ B$. □

2.4 From qualitative plausibility to qualitative likelihood and back

Consider now qualitative plausibility relations $\succ_{pl}$ and their preadditive substitutes $\succ_{pl}^+$. It is obvious that $\succ_{pl}^+$ satisfies (O), (QD) and (P). Moreover, due to Propositions 7 and 10, the preadditive relation $\succ_{pl}^+$ also satisfies the properties of Self-duality (D), Negligibility for disjoint sets (ND), and also Conditional Closure by Implication (CCI), Conditional Closure by Conjunction (CCC), Left Disjunction (OR), (CUT), (CM). The use of $\succ_{pl}$ or $\succ_{pl}^+$ for non-monotonic inference is immaterial as it only involves disjoint subsets.

We can prove that the preadditive substitute of a qualitative plausibility relation is transitive.

Proposition 12. *If $A \succ_{pl}^+ B$ and $B \succ_{pl}^+ C$, then $A \succ_{pl}^+ C$.*

Proof of Proposition 12:
We can write the two assumptions as (we omit the intersection symbol for simplicity): $A\bar{B}C \cup A\bar{B}\bar{C} \succ_{pl} \bar{A}BC \cup \bar{A}B\bar{C}$ and $AB\bar{C} \cup \bar{A}B\bar{C} \succ_{pl} \bar{A}BC \cup \bar{A}\bar{B}C$. We must prove that $AB\bar{C} \cup A\bar{B}\bar{C} \succ_{pl} \bar{A}BC \cup \bar{A}\bar{B}C$. Taking the union on both sides it yields, using Proposition 3:

$$A\bar{B}C \cup A\bar{B}\bar{C} \cup AB\bar{C} \cup \bar{A}B\bar{C} \succ_{pl} \bar{A}BC \cup \bar{A}B\bar{C} \cup \bar{A}BC \cup \bar{A}\bar{B}C$$

Due to property (SU) contraposed (Prop. 2) we can cancel $A\bar{B}C$ and $\bar{A}B\bar{C}$ which yields $AB\bar{C} \cup A\bar{B}\bar{C} \succ_{pl} \bar{A}BC \cup \bar{A}\bar{B}C$. □

Remark: Due to the representation result in Corollary 1, there is an alternative proof that goes as follows: there exists a family $\mathcal{L}$ of linear orders $>_\pi$ on S that generates $\succ_{pl}$ in the sense that $A \succ_{pl} B$ if and only if $A \succ_\Pi B, \forall >_\pi \in \mathcal{L}$. Then suppose $A \succ_{pl}^+ B$, which means $A \setminus B \succ_{pl} B \setminus A$, which means $A \setminus B \succ_\Pi B \setminus A, \forall >_\pi \in \mathcal{L}$. Likewise with $B \succ_{pl}^+ C$. Using transitivity of $\succ_\Pi^+$ (claimed in [14]), we conclude that $A \setminus C \succ_\Pi C \setminus A, \forall >_\pi \in \mathcal{L}$, which is $A \succ_{pl}^+ B$. However the use of linear orders on S presupposes a non-dogmatic qualitative plausibility relation.

Since a qualitative plausibility relation $\succ_{pl}$ satisfies a strong form of axiom (SDU) (without the condition $A \cap (B \cup C) = \varnothing$), we get the following result, which is a direct consequence of Proposition 11:

Corollary 2. *The preadditive relation $\succ_{pl}^+$ is a self-dual refinement of $\succ_{pl}$ and of its dual:*

- *If $A \succ_{pl} B$ then $A \succ_{pl}^+ B$.*
- *If $\overline{B} \succ_{pl} \overline{A}$ then $A \succ_{pl}^+ B$.*

This fact has already been known for a long time for comparative possibility and necessity relations [14]. But it is not valid for any kind of confidence relation. For instance it is easy to find capacities for which $f(A) > f(B)$ but $f(A \cup C) < f(B \cup C)$, for disjoint A, B, C. So using the order $\succ_f$ induced by f, one would have $B \cup C \succ_f A \cup C$ but $A \cup C \succ_f^+ B \cup C$.

These results can be applied to special cases of qualitative plausibility relations $\succ_{pl}$:

- Comparative possibility relations $\succ_\Pi$

- Weak optimistic strict dominance relations $\succ_{wos}^{\triangleright}$ (renamed as $\succ_{wos}$ for short in the following)

In particular, we can consider the preadditive substitute of a comparative possibility relation. It is a special case of the discrimax relation for comparing vectors of values in a totally ordered scale [13]. It is defined equivalently as follows in terms of a possibility distribution: $A \succ_\Pi^+ B$ if and only if $\max_{s \in A \setminus B} \pi(s) > \max_{s \in B \setminus A} \pi(s)$ [14]. It is a transitive refinement of the comparative possibility relation (as pointed out, in [13, 14], but not proved for transitivity).

The preadditive substitute of a weak optimistic strict dominance relation is as follows:

Definition 5 (Weak preadditive strict dominance). $A \succ_{wos}^+ B$ if and only if $A \neq B$ and $A \setminus B \succ_{wos} B \setminus A$.

This relation has been thoroughly studied in [6][2]. It coincides with $\succ_{wos}$ on disjoint sets. The above results can also be applied to the weak preadditive strict dominance.

Proposition 13. *The weak preadditive strict dominance $\succ_{wos}^+$ is a strict partial order that satisfies Preadditivity (P), Strict Compatibility with Inclusion (SCI) and Qualitativeness for disjoint sets (QD).*

Proof of Proposition 13:
(T), (P) and (QD) hold due to the above results about $\succ_{pl}^+$. Transitivity has already been proved in [8] (see also [7], Proposition 30, p. 35). (SCI) follows from the fact that $C \succ_{wos}^+ \emptyset$ when $C \neq \emptyset$. □

As a consequence of Corollary 2, the weak optimistic dominance is also refined by its preadditive substitute.

Corollary 3. $\succ_{wos}^+$ *refines* $\succ_{wos}$ *and its dual variant:*

- *If $A \succ_{wos} B$ then $A \succ_{wos}^+ B$.*

- *If $\overline{B} \succ_{wos} \overline{A}$ then $A \succ_{wos}^+ B$.*

---
[2] A loose preadditive dominance has also been studied in [7].

These results lead us to define a qualitative likelihood relation as follows:

Definition 6. *A qualitative likelihood relation is an asymmetric relation $\succ^+$ on $\wp(S)$ that satisfies (O), (P) and (QD).*

Due to the above results, the preadditive substitute of a qualitative plausibility relation is a qualitative likelihood relation. More importantly, we can get a representation theorem for qualitative likelihood relations as follows:

Proposition 14. *Any qualitative likelihood relation is the preadditive substitute of a qualitative plausibility relation.*

Proof of Proposition 14:
Let $\succ$ be a qualitative likelihood relation. Let us define $\rhd$ as $A \rhd B$ whenever $A \setminus B \succ B$. We have to prove that $\rhd$ satisfies (Q) and (O) and that the preadditive substitute of $\rhd$ is $\succ$.

- First we show that $\rhd$ satisfies (Q). That is: if $A \cup C \rhd B$ and $B \cup C \rhd A$, then $C \rhd A \cup B$.
  Due to the definition of $\rhd$, we must prove that if $(A \cup C) \setminus B \succ B$ and $(B \cup C) \setminus A \succ A$, then $C \setminus (A \cup B) \succ A \cup B$. Let C' denote $C \setminus (A \cup B)$ and AB denote $A \cap B$.
  The hypothesis can be written as $(A \setminus B) \cup C' \succ (B \setminus A) \cup AB$ (1) and $(B \setminus A) \cup C' \succ (A \setminus B) \cup AB$ (2). The conclusion can be written as $C' \succ AB \cup A\Delta B$.
  Applying (O) to (1) and (2) produces $(A \setminus B) \cup C' \succ (B \setminus A)$ and $(B \setminus A) \cup C' \succ (A \setminus B)$. Now using (QD) we obtain $C' \succ (A \setminus B) \cup (B \setminus A)$ or equivalently $C' \succ A\Delta B$ (3).
  Applying (O) to (3) produces $AB \cup C' \succ A\Delta B$. Using (O) once again from (1) yields $(A\Delta B) \cup C' \succ AB$. From (QD) we obtain $C' \succ (A\Delta B) \cup AB$ which is exactly the expected conclusion.

- $\rhd$ satisfies (O). Assume that $A \rhd B$, $A \subseteq A'$ and $B' \subseteq B$. Due to the definition of $\rhd$, we have $A \setminus B \succ B$. Obviously, $A \setminus B \subseteq A' \setminus B'$. As $\succ$ satisfies (O), we conclude that $A' \setminus B' \succ B'$ which is exactly $A' \rhd B'$.

- It remains to prove that the preadditive substitute of $\rhd$, say $\rhd^+$, is $\succ$. By definition, $A \rhd^+ B$ iff $A \setminus B \rhd B \setminus A$ iff $A \setminus B \succ B \setminus A$ since $A \setminus B$ and $B \setminus A$ are disjoint. As $\succ$ satisfies (P), $A \setminus B \succ B \setminus A$ is equivalent to $A \succ B$. So we have proved that $A \rhd^+ B$ iff $A \succ B$.

□

As a corollary of Propositions 12 and 14, we conclude that any *qualitative likelihood relation is transitive*, which was not obvious from its definition. In fact what this result shows is that the application $\rho : \succ_{pl} \mapsto \succ^+_{pl}$ that assigns to each qualitative plausibility relation its preadditive refinement is a bijection between the set of qualitative plausibility relations $\succ_{pl}$ and the set of qualitative likelihood relations $\succ^+$, namely:

- $A \succ^+ B$ such that $\succ^+ = \rho(\succ_{pl})$ is defined as $A \setminus B \succ_{pl} B \setminus A$.

- $A \succ_{pl} B$ such that $\succ_{pl} = \mu(\succ^+)$ is defined as $A \setminus B \succ^+ B$.

Then, relation $\rho(\succ_{pl})$ is a qualitative likelihood relation, and relation $\mu(\succ^+)$ is a qualitative plausibility relation. Moreover: $\mu(\rho(\succ_{pl})) = \succ_{pl}$ and $\rho(\mu(\succ^+)) = \succ^+$.

3 Relative confidence and certainty logics

In [36], a logic for reasoning about partially ordered bases has been proposed, with inference rules inspired from the properties of a qualitative certainty relation.

In the following, we define a logical language capable of expressing relative confidence between logical propositions, and a semantics based on confidence relations between sets of intepretations. An example of such a logic is the one in [36]. After recalling this logic, we consider a logic for qualitative likelihood, for which the preadditivity axiom holds. The results in the previous section indicate that the relative certainty logic and qualitative likelihood logic are closely related due to the bijection between the two notions. Especially they will coincide for pairs of formulas whose disjunction is a tautology.

3.1 A logical framework for confidence relations

We consider a propositional language $\mathcal{L}$ where formulas are denoted by ϕ, ψ etc., and Ω is the set of its interpretations. $[\phi]$ denotes the set of models of ϕ, a subset of Ω. We denote by $\models$ the classical semantic inference. We also denote by $\vdash_X$ the syntactic inference in the proof system X.

Let $\mathcal{K} \subseteq \mathcal{L}$ be a finite set of formulas equipped with a relation $>$. The idea is that this relation should represent a fragment of a strict partial ordering. We call $(\mathcal{K}, >)$ a partially ordered belief base (po-base, for short) where $\phi > \psi$ is supposed to express that ϕ is more prone to being true than ψ, for an agent. The standard language $\mathcal{L}$ is encapsulated inside a language equipped with a binary connective $>$ (interpreted as a partial order relation). Formally, an atom $\Phi \in \mathcal{L}_>$ is of the form $\phi > \psi$ where ϕ and ψ are formulas of $\mathcal{L}$. A formula of $\mathcal{L}_>$ is either an atom Φ of $\mathcal{L}_>$, or a conjunction of formulas, that is, $\Psi \wedge \Phi \in \mathcal{L}_>$ if $\Psi, \Phi \in \mathcal{L}_>$. We also have the formulas $\bot$ and $\top$ in $\mathcal{L}_>$. In contrast with Halpern [25], we exclude negations and disjunctions of atomic formulas just like in basic possibilistic logic, where we do not use negations nor disjunctions of weighted formulas.

A relative confidence base $\mathcal{B}$ is a finite subset of $\mathcal{L}_>$. We associate to a po-base $(\mathcal{K}, >)$ the set of formulas of the form $\phi > \psi$ and forming a base $\mathcal{B}_{(\mathcal{K},>)} \subset \mathcal{L}_>$. In the following, we shall often write $(\mathcal{K}, >)$ instead of $\mathcal{B}_{(\mathcal{K},>)}$ for simplicity.

We consider a semantics defined by a strict confidence relation between sets of interpretations. The idea is to interpret the formula $\phi > \psi$ on 2^Ω by $[\phi] \succ [\psi]$ for a strict confidence relation $\succ$ (Definition 1). A relative confidence model $\mathcal{M}$ is a structure $(2^\Omega, \succ)$ where $\succ$ is a strict confidence relation on 2^Ω (that is a strict partial order on 2^Ω satisfying the properties O and T).

We define the satisfiability of a formula $\phi > \psi \in \mathcal{L}_>$ in $\mathcal{M}$ as $\mathcal{M} \models \phi > \psi$ iff $[\phi] \succ [\psi]$. The satisfiability of the set of formulas $\mathcal{B}_{(\mathcal{K},>)}$ is defined by $\mathcal{M} \models \mathcal{B}_{(\mathcal{K},>)}$ iff $\mathcal{M} \models \phi_i > \psi_i, \forall \phi_i > \psi_i \in \mathcal{B}_{(\mathcal{K},>)}$. Note that there is not always a relative confidence model of a po-base $(\mathcal{K}, >)$. For instance, if $\phi > \psi \in \mathcal{B}_{(\mathcal{K},>)}$ such that $\phi \models \psi$, it is impossible to find a confidence relation $\succ$ such that $[\phi] \succ [\psi]$ since $\succ$ should satisfy property O. This comes down to saying that no model of this formula in $\mathcal{L}_>$ exists for the semantics of relative confidence.

We say that $(\mathcal{K}, >)$ is *inconsistent with respect to the relative confidence semantics*, in short *rc-inconsistent*, iff there is no relative confidence model for $\mathcal{B}_{(\mathcal{K},>)}$.

A logic for relative confidence, denoted by CO, can be defined as follows: It directly interprets the atoms $\phi > \psi$ in $\mathcal{L}_>$ by means of the strict confidence relation $\succ$ having properties (O) and (T) for comparing the sets of models $[\phi]$ and $[\psi]$. The idea behind the proof system is to use the characteristic properties of the confidence relation $\succ$, expressed in terms of inference rules that define the syntactic entailment $\vdash_{CO}$. We need one axiom and three inference rules:

Axiom
ax_{NT}: $\top > \bot$

Inference rules

RI_O : If $\phi \models \phi'$ and $\psi' \models \psi$ then $\phi > \psi \vdash \phi' > \psi'$ (O)

RI_T : $\{\phi > \psi, \psi > \chi\} \vdash \phi > \chi$ (T)

RI_{AS} : $\{\phi > \psi, \psi > \phi\} \vdash \bot$ (AS)

The axiom says that the order relation is not trivial [3]. Rules RI_O and RI_T correspond to the properties of Orderliness and Transitivity. Rule RI_{AS} expresses the asymmetry of the relation $>$. The proof system of the logic of relative confidence is composed of the axiom ax_{NT} and the three inference rules $RI_O - RI_{AS}$.

Remark 1. *The order relation $>$ does not contradict classical inference. Indeed, if we have $\psi \models \phi$ and $\psi > \phi \in \mathcal{B}_{(\mathcal{K},>)}$, we prove that $\phi > \phi$ by RI_O and the contradiction by RI_{AS}.*

The associated semantic consequence $\models_{CO}$ can then be defined in the usual way:

$$(\mathcal{K}, >) \models_{CO} \phi > \psi \text{ iff } \forall \mathcal{M}, \text{ if } \mathcal{M} \models \mathcal{B}_{(\mathcal{K},>)} \text{ then } \mathcal{M} \models \phi > \psi. \tag{5}$$

The proof system of the logic of relative confidence is sound and complete for the relative confidence semantics:

---

[3] This axiom could be replaced by $\phi \vee \neg \phi > \psi \wedge \neg \psi$, in the presence of the inference rule RI_T.

Proposition 15. *Let $(\mathcal{K}, >)$ be a partially ordered base and $\phi, \psi \in \mathcal{L}$.*

- **Soundness:**
  If $(\mathcal{K}, >) \vdash_{CO} \phi > \psi$ then $(\mathcal{K}, >) \models_{CO} \phi > \psi$

- **Completeness:**
  If $(\mathcal{K}, >)$ is rc-consistent and $(\mathcal{K}, >) \models_{CO} \phi > \psi$ then $(\mathcal{K}, >) \vdash_{CO} \phi > \psi$
  If $(\mathcal{K}, >)$ is rc-inconsistent then $(\mathcal{K}, >) \vdash_{CO} \bot$

Proof of Proposition 15:

Let $\mathcal{B}_{(\mathcal{K},>)} = \{(\phi_i > \psi_i), i = 1 \cdots n\}$.

- **Soundness:**
  Let $\succ$ be a strict partial order on 2^Ω satisfying O. We must show that if $\forall i = 1 \cdots n, [\phi_i] \succ [\psi_i]$ then $[\phi] \succ [\psi]$. We assume that $\phi > \psi$ was obtained from $(\phi_i > \psi_i)$ by inference rules RI_O, RI_T, RI_{AS} and the axiom. So we just have to show that each of the rules is sound and that the axiom ax_{NT} is valid.

  ax_{NT}: It holds because $S \succ \emptyset$ for a confidence relation.

  RI_O: It holds because $\succ$ satisfies (O)

  RI_T: It holds because $\succ$ is transitive.

  RI_{AS}: The presence of both $\phi > \psi$ and $\psi > \phi$ leads to a semantic contradiction because the relation $\succ$ being asymmetric, we can not have both $[\psi] \succ [\phi]$ and $[\phi] \succ [\psi]$.

- **Completeness:**
  We assume that $(\mathcal{K}, >)$ is rc-consistent. We suppose that for each strict partial order $\succ$ on 2^Ω satisfying O, if $\forall i = 1 \cdots n, \phi_i \succ \psi_i$ then $[\phi] \succ [\psi]$. We must show that $(\mathcal{K}, >) \vdash_{CO} \phi > \psi$. If $\phi > \psi$ appears in $\mathcal{B}_{(\mathcal{K},>)}$, it is proven.
  Otherwise, consider the strict partial order $\succ$ defined on 2^Ω as the smallest order containing pairs $[\phi_i] \succ [\psi_i]$ and closed for the properties O, T.
  This relation exists because $(\mathcal{K}, >)$ is rc-consistent. According to the hypothesis, we have $[\phi] \succ [\psi]$. And, by definition of $\succ$, the pair $([\phi], [\psi])$ is obtained by successive applications of the properties O, T. This amounts to getting $\phi > \psi$ by successive applications of inference rules RI_O, RI_T.
  It remains to prove that if $(\mathcal{K}, >)$ is rc-inconsistent, then $(\mathcal{K}, >) \vdash_{CO} \bot$.
  Note that, as $\mathcal{L}_>$ contains only atomic comparison constraints and their conjunctions, the only form of syntactic inconsistency is the presence of both $\phi > \psi$ and $\psi > \phi$ derived from $(\mathcal{K}, >)$. This is the only way to get $(\mathcal{K}, >) \vdash_{CO} \bot$. In this case, we know that $\mathcal{B}_{(\mathcal{K},>)}$ does not have a model of relative confidence. So if $(\mathcal{K}, >) \vdash_{CO} \bot$ does not hold, then the relation $>$ obtained on $\mathcal{L}_>$ by the syntactic closure is asymmetric and transitive, and so is the relation $\succ$ on 2^Ω defined by $[\phi] \succ [\psi]$ if and only if $(\mathcal{K}, >) \vdash_{CO} \phi > \psi$. In addition, $\succ$ will be the smallest relation containing the pairs $([\phi_i], [\psi_i])$ with $\phi_i > \psi_i$ in $(\mathcal{K}, >)$, and closed for the properties O, T. It is a model of $\mathcal{B}_{(\mathcal{K},>)}$, which is rc-consistent.

Example 2. $\mathcal{K}_1 = \{\phi \wedge \psi, \phi \wedge \neg\psi, \neg\phi\}$ with $\phi \wedge \psi > \phi \wedge \neg\psi > \neg\phi$.
Then we let $\mathcal{B}_{(\mathcal{K}_1,>)} = \{\phi \wedge \psi > \phi \wedge \neg\psi, \phi \wedge \neg\psi > \neg\phi\}$. With the proof system of CO, by RI_T, we deduce $\phi \wedge \psi > \neg\phi$ and by RI_O, $\psi > \neg\phi$.

Next is a case where inconsistency can be detected.

Example 3. $\mathcal{K}_2 = \{\phi, \phi \wedge \psi\}$ with $\mathcal{B}_{(\mathcal{K}_2,>)} = \{\phi \wedge \psi > \phi\}$. With the proof system of CO, we obtain a contradiction by RI_O (we have $(\mathcal{K}_2, >) \vdash_{CO} \phi \wedge \psi > \phi \wedge \psi$) and RI_{AS}.

3.2 Axioms and inference rules for relative certainty logic

The logic for relative certainty described in [36], here denoted by $\mathcal{C}$, directly interprets the atoms $\phi > \psi$ in $\mathcal{L}_>$ by means of a qualitative certainty relation $\succ_{cr}$ having properties (O) and (Q^d) for comparing the sets of models $[\phi]$ and $[\psi]$. A relative certainty model is a structure $(2^\Omega, \succ_{cr})$ where $\succ_{cr}$ is a qualitative certainty relation on 2^Ω.

The idea behind the proof system is again to use the characteristic properties of the relation $\succ_{cr}$, expressed in terms of inference rules. We need again one axiom and three inference rules in the language $\mathcal{L}_>$: the same axiom as for the relative confidence logic above, and we can add the following inference rule to the inference rules RI_O and RI_{AS} of the confidence relation logic:

$$RI_{Q^d} : \{\chi > \phi \wedge \psi, \psi > \phi \wedge \chi\} \vdash \psi \wedge \chi > \phi \tag{Q^d}$$

This rule corresponds to the properties of dual Qualitativeness. So the relative certainty logic proof system is made of axiom ax_{NT}, and rules RI_O, RI_{AS} and RI_{Q^d}.

The inference rule RI_T can be derived in this system (see also Proposition 1), as well as the following inference rules, some of which are established in [36]:

$$RI_A : \{\psi > \phi, \chi > \phi\} \vdash \psi \wedge \chi > \phi \tag{A}$$

$$RI_{OR^d} : \{\phi \to \chi > \phi \to \neg\chi, \psi \to \chi > \psi \to \neg\chi\} \vdash (\phi \vee \psi) \to \chi > (\phi \vee \psi) \to \neg\chi \tag{OR^d}$$

$$RI_{CCC^d} : \{\chi \to \phi > \chi \to \neg\phi, \chi \to \psi > \chi \to \neg\psi\} \vdash \chi \to (\phi \wedge \psi) > \chi \to \neg(\phi \wedge \psi) \tag{CCC^d}$$

$$RI_{CUT^d} : \{\phi \to \psi > \phi \to \neg\psi, (\phi \wedge \psi) \to \chi > (\phi \wedge \psi) \to \neg\chi\} \vdash \phi \to \chi > \phi \to \neg\chi \tag{CUT^d}$$

$RI_{CM^d} : \{\phi \to \psi > \phi \to \neg\psi, \phi \to \chi > \phi \to \neg\chi\} \vdash (\phi \wedge \psi) \to \chi > (\phi \wedge \psi) \to \neg\chi$
(CM^d)

$RI_{Nec} : \phi > \bot \vdash \phi > \neg\phi$

$RI_{SC^d} : \phi > \psi \vdash \phi > \neg\phi \vee \psi$ (semi-cancellativity).

The first derived rule expresses adjunction and ensures that formulae that are more certain than another one will form a deductively closed set. The next four rules are key inference properties in non-monotonic logic of the KLM type [27]. Rule RI_{Nec} results from applying RI_{Q^d} to $\phi > \phi \wedge \neg\phi$, and reminds of the property $\min(N(A), N(\overline{A})) = 0$ of necessity measures N in possibility theory. The last rule can be proved by implementing the proof of Proposition 6 in $\mathcal{C}$.

Example 4. Let $\mathcal{K}_3 = \{\phi, \neg\phi, \psi, \neg\psi\}$ with $\mathcal{B}_{(\mathcal{K}_3,>)} = \{\phi > \neg\phi, \psi > \neg\psi\}$. Using RI_{CCC^d} by considering $\phi > \neg\phi$ as $\top \to \phi > \top \to \neg\phi$ and $\psi > \neg\psi$ as $\top \to \psi > \top \to \neg\psi$, we have $\phi \wedge \psi > \neg\phi \vee \neg\psi$. Then by RI_O we obtain $\psi > \neg\phi$. And similarly we obtain $\phi > \neg\psi$.

The proof system of the relative certainty logic $\mathcal{C}$ has been proved sound and complete [36] for the semantics of relative certainty. Namely, define $(\mathcal{K}, >) \vDash_\mathcal{C} \phi > \psi$ to mean: for each qualitative certainty relation $\succ_{cr}$, if $[\phi_i] \succ_{cr} [\psi_i], \forall i$ s.t. $\phi_i > \psi_i \in \mathcal{B}_{(\mathcal{K},>)}$, then $[\phi] \succ_{cr} [\psi]$. Moreover $(\mathcal{K}, >)$ is said to be rcr-consistent if it has a relative certainty model $(2^\Omega, \succ_{cr})$. Then we have proved in [36]:

Proposition 16. Let $(\mathcal{K}, >)$ be a partially ordered base and $\phi, \psi \in \mathcal{L}$.

- **Soundness:**
  If $(\mathcal{K}, >) \vdash_\mathcal{C} \phi > \psi$ then $(\mathcal{K}, >) \vDash_\mathcal{C} \phi > \psi$.

- **Completeness:**
  If $(\mathcal{K}, >)$ is rcr-consistent and $(\mathcal{K}, >) \vDash_\mathcal{C} \phi > \psi$ then $(\mathcal{K}, >) \vdash_\mathcal{C} \phi > \psi$.
  If $(\mathcal{K}, >)$ is rcr-inconsistent then $(\mathcal{K}, >) \vdash_\mathcal{C} \bot$.

4 Qualitative likelihood logic

In this section, we will present the preadditive version of the relative certainty logic. As done for relative certainty, we propose an inference system for qualitative likelihood relations, which is preadditive, with a semantics defined by a relation between sets of interpretations. As before, we interpret a partially ordered base as a fragment of a qualitative likelihood ordering. We propose a logic system for reasoning with comparative statements interpreted by such a relation. We keep the syntax as defined in the previous section.

4.1 Semantics of qualitative likelihood

Let us interpret the formula $\phi > \psi$ on 2^Ω by $[\phi] \succ^+ [\psi]$ for a qualitative likelihood relation $\succ^+$ in the sense of Definition 6. We assume it is non-dogmatic. A qualitative likelihood model $\mathcal{M}^+$ is a structure $(2^\Omega, \succ^+)$ where $\succ^+$ is a non-dogmatic qualitative likelihood relation on 2^Ω. We define the satisfiability of a formula $\phi > \psi \in \mathcal{L}_>$ in $\mathcal{M}^+$ as $\mathcal{M}^+ \vDash \phi > \psi$ iff $[\phi] \succ^+ [\psi]$. The satisfiability of the set of formulas $\mathcal{B}_{(\mathcal{K},>)}$ is defined by $\mathcal{M}^+ \vDash \mathcal{B}_{(\mathcal{K},>)}$ iff $\mathcal{M}^+ \vDash (\phi_i > \psi_i), \forall \phi_i > \psi_i \in \mathcal{B}_{(\mathcal{K},>)}$. The associated semantic consequence $\vDash_+$ can then be defined in the usual way:

Definition 7. $(\mathcal{K}, >) \vDash_+ \phi > \psi$ iff $\forall \mathcal{M}^+$, if $\mathcal{M}^+ \vDash \mathcal{B}_{(\mathcal{K},>)}$ then $\mathcal{M}^+ \vDash \phi > \psi$.

In other words, $(\mathcal{K}, >) \vDash_+ \phi > \psi$ iff for every strict partial order $\succ^+$ on 2^Ω verifying O, P, QD, NoD, if $\forall i = 1 \cdots n, [\phi_i] \succ^+ [\psi_i]$ then $[\phi] \succ^+ [\psi]$.

We say that $(\mathcal{K}, >)$ is inconsistent with respect to the qualitative likelihood semantics, in short *ql-inconsistent*, iff there is no qualitative likelihood model for $\mathcal{B}_{(\mathcal{K},>)}$.

4.2 Proof system

The logic for qualitative likelihood directly interprets the atoms $\phi > \psi$ in $\mathcal{L}_>$ by means of a qualitative likelihood relation $\succ^+$ for comparing the sets of models $[\phi]$ and $[\psi]$. The idea behind the proof system is again to use the characteristic properties of the relation $\succ^+$, expressed in terms of formulas, as inference rules. Indeed, we need one axiom and four inference rules in the language $\mathcal{L}_>$, owing to Proposition 8, that indicates that SCI is a derived property in this setting.

Axiom
ax_{NoD}: If $\phi \nvDash \bot$ then $\phi > \bot$ \hfill (NoD)

Inference rules: RI_O, RI_{AS} and

RI_{QD^d} : If $\vDash \phi \vee \psi, \vDash \phi \vee \chi$ and $\vDash \psi \vee \chi$, then $\{\chi > \phi \wedge \psi, \psi > \phi \wedge \chi\} \vdash \psi \wedge \chi > \phi$ \hfill (QD^d)

RIP_1 : If $\neg \chi \vDash \phi \wedge \psi$ then $\phi > \psi \vdash \phi \wedge \chi > \psi \wedge \chi$ \hfill ($\Rightarrow$ P)

RIP_2 : If $\neg \chi \vDash \phi \wedge \psi$ then $\phi \wedge \chi > \psi \wedge \chi \vdash \phi > \psi$ \hfill ($\Leftarrow$ P)

We denote by QL this logic and by $\vdash_+$ the associated syntactic inference.
Note that the axiom ax_{NoD} encodes property (NoD). Besides, (RI_{QD^d}) could be replaced

by (RI_{QD}) of the form

RI_{QD}: If $\phi\wedge\psi \models \bot$ and $\phi\wedge\chi \models \bot$ and $\psi\wedge\chi \models \bot$, then $\{\phi\vee\psi > \chi, \phi\vee\chi > \psi\} \vdash \phi > \psi\vee\chi$

since, in the presence of (P), the properties (QD) and (QD^d) are equivalent.

Due to Propositions 8 and 10, it can be proved that other rules can be derived from the rules of the proof system of QL. Some of these derived rules are theorems of the proof system of C: RI_T, RI_{OR^d}, RI_{CCC^d}, RI_{CUT^d}, RI_{CM^d}.

Other derived rules are new:

RI_{ND^d} : If $\models \phi \vee \psi$ and $\models \phi \vee \chi$, then $\{\psi > \phi, \chi > \phi\} \vdash \psi \wedge \chi > \phi$ \hfill (ND^d)

RI_D : $\{\phi > \psi\} \vdash \neg\psi > \neg\phi$ \hfill (Self-duality D)

RI_{SCI} : If $\psi \models \phi$ and not $\phi \models \psi$ then $\phi > \psi$ \hfill (SCI)

RI_{SD} : If $\phi \wedge \psi = \chi \wedge \xi = \bot$, $\{\phi > \chi, \psi > \xi\} \vdash \phi \vee \psi > \chi \vee \xi$ \hfill (SD)

RI_{STP} : $\phi > \psi$ iff $\phi \wedge \neg\psi > \psi \wedge \neg\phi$ \hfill (direct consequence of P)

The last rule is a consequence of RIP_1 and RIP_2 (taking $\neg\chi = \phi \wedge \psi$), that expresses the sure thing principle for events.

Example 5. Let $\mathcal{K}_4 = \{\neg\phi\vee\neg\psi, \neg\phi, \phi\wedge\psi, \phi\}$ with $\neg\phi\vee\neg\psi > \phi\wedge\psi > \neg\phi$ and $\phi > \neg\phi$. So $\mathcal{B}_{(\mathcal{K}_4,>)} = \{\neg\phi \vee \neg\psi > \phi \wedge \psi, \phi \wedge \psi > \neg\phi, \phi > \neg\phi\}$. Using RI_D (Self-duality) we obtain $\phi > \neg\phi \vee \neg\psi$ and so we get the chain $\phi > \neg\phi \vee \neg\psi > \phi \wedge \psi > \neg\phi$.

Of interest is to prove rule RI_{SD} and RI_T. The derivation of RI_{SD} follows from the following lemma.

Lemma 1. *If A, B, C, D satisfy $A \cap B = A \cap C = B \cap D = C \cap D = \emptyset$ then, for any qualitative likelihood relation $\succ^+$, it holds that whenever $A \succ^+ B$ and $C \succ^+ D$ then $A \cup C \succ^+ B \cup D$.*

Proof of Lemma 1:
Let $\succ^+$ be a qualitative likelihood relation. Let $\succ_{pl}$ denote the plausibility relation which is refined by $\succ^+$, as defined in Proposition 14. We have $\succ^+ = \rho(\succ_{pl})$ and $\succ_{pl} = \mu(\succ^+)$.
So, $A \succ^+ B$ and $C \succ^+ D$ can be written as $A \setminus B \succ_{pl} B \setminus A$ and $C \setminus D \succ_{pl} D \setminus C$. Moreover as $A \cap B = C \cap D = \emptyset$, we obtain $A \succ_{pl} B$ and $C \succ_{pl} D$.
From Proposition 3, it follows that $(A \cup C) \succ_{pl} (B \cup D)$. Then from Proposition 2, we obtain $(A \setminus D) \cup (C \setminus B) \succ_{pl} (B \setminus C) \cup (D \setminus A)$ (deleting $(A \cap D) \cup (B \cap C)$ on both sides). As $\succ_{pl} = \mu(\succ^+)$ and the sets are disjoint, we also have $(A \setminus D) \cup (C \setminus B) \succ^+ (B \setminus C) \cup (D \setminus A)$. Then

applying (P) we get $A \cup C \succ^+ B \cup D$ (adding $(A \cap D) \cup (B \cap C)$ to both sides). □

Then the derivation of RI_T goes as follows:

- $\{\phi > \psi, \psi > \chi\} \vdash_{QL} \{\phi \wedge \neg \psi > \psi \wedge \neg \phi, \psi \wedge \neg \chi > \chi \wedge \neg \psi\}$ (using RI_{STP})

- $\{\phi \wedge \neg \psi > \psi \wedge \neg \phi, \psi \wedge \neg \chi > \chi \wedge \neg \psi\} \vdash_{QL} (\phi \wedge \neg \psi) \vee (\psi \wedge \neg \chi) > (\psi \wedge \neg \phi) \vee (\chi \wedge \neg \psi)$ (using RI_{SD})

- $(\phi \wedge \neg \psi) \vee (\psi \wedge \neg \chi) > (\psi \wedge \neg \phi) \vee (\chi \wedge \neg \psi) \vdash_{QL} \phi \wedge \neg \chi > \neg \phi \wedge \chi$ (using RI_{STP})

- $\phi \wedge \neg \chi > \neg \phi \wedge \chi \vdash_{QL} \phi > \chi$ (using RI_{STP}).

The proof system of QL is sound and complete for the semantics of qualitative likelihood.

Proposition 17. *Let $(\mathcal{K}, >)$ be a partially ordered base and $\phi, \psi \in \mathcal{L}$.*

- **Soundness:**
  If $(\mathcal{K}, >) \vdash_+ \phi > \psi$ then $(\mathcal{K}, >) \models_+ \phi > \psi$

- **Completeness:**
  If $(\mathcal{K}, >)$ is ql-consistent and $(\mathcal{K}, >) \models_+ \phi > \psi$ then $(\mathcal{K}, >) \vdash_+ \phi > \psi$
  If $(\mathcal{K}, >)$ is ql-inconsistent then $(\mathcal{K}, >) \vdash_+ \bot$

Proof of Proposition 17:

The proof follows the same pattern as for the soundness and completeness of the relative confidence proof system (proof of Proposition 15).

- **Soundness:**
  Let $\succ^+$ be a strict partial order on 2^Ω satisfying O, P, QD^d and NoD. We must show that if $\forall i = 1 \cdots n, [\phi_i] \succ^+ [\psi_i]$ then $[\phi] \succ^+ [\psi]$. We do it for axioms and rules not previously encountered.

  ax_{NoD} We must show that $\forall \mathcal{M}^+$, if $\phi \not\models \bot$ then $\mathcal{M}^+ \models \phi > \bot$. Or equivalently, for any strict relation $\succ^+$ on 2^Ω that satisfies the properties O, P, QD^d and NoD, if $\phi \not\models \bot$ then $[\phi] \succ^+ [\bot]$. It follows from Proposition 8 since $[\phi] \neq \emptyset$ when $\phi \not\models \bot$.

  RIP_1 We must show that if $[\phi] \succ^+ [\psi]$ and $\neg \chi \models \phi \wedge \psi$ then $[\phi \wedge \chi] \succ^+ [\psi \wedge \chi]$. This is true since the relation $\succ^+$ is preadditive.

  RIP_2 We must show that if $[\phi \wedge \chi] \succ^+ [\psi \wedge \chi]$ and $\neg \chi \models \phi \wedge \psi$ then $[\phi] \succ^+ [\psi]$. This is true since the relation $\succ^+$ is preadditive.

- **Completeness:**
  The proof is exactly the same as for the relative confidence logic CO, (Proposition 15) replacing (O) by (O), (P), (QD^d) and (NoD). Note that the only possible form of syntactic inconsistency that can be detected in $(\mathcal{K}, >)$ is again when $(\mathcal{K}, >) \vdash_+ \phi > \psi$ and $(\mathcal{K}, >) \vdash_+ \psi > \phi$.

  □

5 Comparison between proof systems of QL and $\mathcal{C}$

Based on results from the previous sections, it is interesting to compare relative certainty and qualitative likelihood logics $\mathcal{C}$ and QL in terms of strength of their proof systems.

5.1 Is one system more productive than the other?

Recall that for any relative certainty relation, there is a qualitative likelihood relation that refines it. Due to Proposition 2 and Proposition 14, we have: if $\mathcal{M}$ is a relative certainty model and $\mathcal{M}^+$ its associated qualitative likelihood model, $\mathcal{M} \vDash \phi > \psi$ implies $\mathcal{M}^+ \vDash \phi > \psi$ and so $\mathcal{M} \vDash \mathcal{B}_{(\mathcal{K},>)}$ implies $\mathcal{M}^+ \vDash \mathcal{B}_{(\mathcal{K},>)}$. However it does not imply that, applied to a set of constraints in the form of a partially ordered set of formulas, the system QL will produce more comparative statements than $\mathcal{C}$.

Indeed, the following points must be noticed:

- Inference rules RI_O and RI_{AS} belong to both systems.

- Axiom ax_{NoD} is stronger that axiom ax_{NT}.[4]

- The proof system of QL adds two preadditivity rules that are not part of $\mathcal{C}$.

- $\mathcal{C}$ uses rule RI_{Q^d}, but the qualitativeness rule RI_{QD^d} used in QL is weaker than RI_{Q^d} as it only applies to relative confidence statements when the disjunction of the two compared formulas forms a tautology.

Note that due to semi-cancellativity, $\phi > \psi$ in QL is equivalent to $\phi > \neg \phi \vee \psi$ in $\mathcal{C}$, and the latter statement obeys the condition that the disjunction of the two formulas forms a tautology, which enables the use of RI_{QD^d}. But the form of the obtained statements does not allow to apply it directly. So it seems that the two logics are not comparable. Moreover, if a partially ordered base is inconsistent for relative certainty semantics, it may be consistent for qualitative likelihood semantics. The following example illustrates this point.

---
[4]but non-dogmaticism is not compulsory: one can specialize system $\mathcal{C}$ adding it in the form $\top > \phi$ if $\top \not\vdash \phi$, or weaken system QL by using ax_{NT} in place of ax_{NoD}.

Example 6. $\mathcal{K}_4 = \{\neg\phi \vee \neg\psi, \neg\phi, \phi \wedge \psi, \phi\}$ with $\neg\phi \vee \neg\psi > \phi \wedge \psi > \neg\phi$ and $\phi > \neg\phi$, *(the same case as in Example 5).*

With system $\mathcal{C}$, we obtain a contradiction: By RI_T we obtain $\neg\phi \vee \neg\psi > \neg\phi$. Then by RI_A we obtain $(\neg\phi \vee \neg\psi) \wedge (\phi \wedge \psi) > \neg\phi$. Applying RI_O produces $\bot > \neg\phi$ and applying RI_O again produces $\neg\phi > \neg\phi$, which by RI_{AS} yields a contradiction.

With system QL, we obtain $\phi > \neg\phi \vee \neg\psi > \phi \wedge \psi > \neg\phi$ (see Example 5).

Note that the reason why we get a contradiction in $\mathcal{C}$ is because we have $\neg\phi \vee \neg\psi > \phi \wedge \psi > \bot$, of the form $\neg\varphi > \varphi > \bot$ which is forbidden in $\mathcal{C}$ due to the inference $\{\varphi > \bot, \neg\varphi > \bot\} \vdash_\mathcal{C} \bot$ valid in $\mathcal{C}$ (just use RI_A).

However, nothing prevents $\neg\varphi > \varphi > \bot$ in QL (e.g., $\varphi > \bot$ is axiom ax_{NoD}). Note that we cannot apply rule RI_A to $\{\neg\varphi > \bot, \varphi > \bot\}$ in QL.

In this example it can be shown that the resulting total order in the case of the system QL is the refinement of a relative certainty ordering that differs from the set of constraints given in the original $(\mathcal{K}_4, >)$. Namely, consider a big-stepped probability that represents the linear order $\phi > \neg\phi \vee \neg\psi > \phi \wedge \psi > \neg\phi$, letting

$$p_1 = P([\phi \wedge \neg\psi]) \propto 8; p_2 = P([\phi \wedge \psi]) \propto 4; p_3 = P([\neg\phi \wedge \psi]) \propto 2; p_4 = P([\neg\phi \wedge \neg\psi]) \propto 1.$$

Then, the reader can check that $P([\phi]) > P([\neg\phi \vee \neg\psi]) > P([\phi \wedge \psi]) > P([\neg\phi])$. It ensures ql-consistency of the linear order. The big-stepped probability assignment viewed as a possibility ordering corresponds to the strict constraints (using the max instead of the sum)

$$\Pi([\neg\phi \vee \neg\psi]) > \Pi([\phi \wedge \psi]) > \Pi([\neg\phi]) \text{ and } \Pi([\phi]) > \Pi([\phi \wedge \psi]).$$

Indeed, $\Pi([\phi]) = \Pi([\neg\phi \vee \neg\psi])$. The corresponding plausibility ordering, expressed in terms of a partial certainty relation, leads to the new set of constraints obtained by duality from the possibility constraints:

$$\mathbb{C}: \phi \gg \neg\phi \vee \neg\psi \gg \phi \wedge \psi \text{ and } \neg\phi \vee \neg\psi \gg \neg\phi.$$

This new partially ordered base is no longer $\mathcal{C}$-inconsistent and is refined by means of the QL logic. In $\mathcal{C}$, we cannot prove that $\mathbb{C}$ implies $\phi \wedge \psi \gg \neg\phi$, while this is obtained in QL logic using self-duality rule RI_D applied to $\phi \gg \neg\phi \vee \neg\psi$.

However, even rcr-consistent bases do not necessarily produce less inferences using system $\mathcal{C}$ than using system QL, as shown now.

Example 7. $\mathcal{K}_5 = \{\phi, \phi \wedge \psi\}$ *with the constraint $\phi > \phi \wedge \psi$.*
With system $\mathcal{C}$, we obtain $\phi > \psi$, using RI_{Q^d} but we do not have that $\psi > \phi \wedge \psi$.
With system QL, the partially ordered base $(\mathcal{K}_5, >)$ gives no information. Indeed from RI_{SCI} and ax_{NoD}, $\phi > \phi \wedge \psi$ and $\psi > \phi \wedge \psi$ are theorems of QL. But we cannot infer $\phi > \psi$ in QL.

5.2 Using system $\mathcal{C}$ to compute inference in QL

As seen above, we do not have that $(\mathcal{K}, >) \models_\mathcal{C} \phi > \psi$ implies $(\mathcal{K}, >) \models_+ \phi > \psi$. Indeed, while the proof system of QL contains inference rules that are not in the proof system of $\mathcal{C}$ (the preadditivity property which is translated into the inference rules RIP_1 and RIP_2), it contains one less powerful inference rule (qualitativeness for disjoint sets, which is translated into RI_{QD^d}) than RI_{Q^d} for system $\mathcal{C}$. Examples above have shown that if applied to a bunch of comparative confidence statements, one system is, strictly speaking, not more powerful than the other. Nevertheless we can try to use $\mathcal{C}$ to compute inferences in QL, provided that we modify the original base in a suitable way.

The idea is to exploit Proposition 14 that says that qualitative likelihood orderings are in bijection with qualitative plausibility ones. Due to Proposition 9, the properties (Q) and (QD) are equivalent when considering disjoint sets. By duality, it follows that the properties (Q^d) and (QD^d) are equivalent on pairs (A, B) such that $\overline{A} \cap \overline{B} = \emptyset$. As a consequence, the rules RI_{Q^d} and RI_{QD^d} are equivalent for bases consisting of $\phi > \psi$ such that $\overline{[\phi]} \cap \overline{[\psi]} = \emptyset$, or equivalently such that $\models \phi \vee \psi$ (two such formulas are said to be subcontraries). So, the first step is to transform a partially ordered base understood as a fragment of a qualitative likelihood ordering, into a partially ordered base with comparative propositions involving only subcontraries.

More precisely, applying the transformation in Proposition 14, if $>$ is interpreted as a qualitative likelihood ordering, we consider $>_{pl}$ the qualitative plausibility ordering that is refined by $>$, and denote by $>_{cr}$ the certainty ordering dual of $>_{pl}$. We have $> = \rho(>_{pl})$ and $>_{pl} = \mu(>)$. So the formula $\phi > \psi$ stands for $\phi \wedge \neg\psi >_{pl} \psi \wedge \neg\phi$ and can be written as $\phi \vee \neg\psi >_{cr} \psi \vee \neg\phi$.

Once we obtain such a relative certainty base, the inference rules of $\mathcal{C}$ can be applied. Then, applying the converse transformation in Proposition 14, we obtain formulas belonging to the QL-closure $QL(\mathcal{B}_{(\mathcal{K},>)})$. More precisely, if the formula $\phi' >_{cr} \psi'$ is produced using $\mathcal{C}$, as $>_{cr}$ is the certainty ordering dual of $>_{pl}$ and $>_{pl} = \mu(>)$, $\phi' >_{cr} \psi'$ stands for $\phi' \wedge \neg\psi' > \neg\phi'$.

The strategy is summarized as follows. Starting from a partially ordered QL-base $(\mathcal{K}, >)$:

1. Turn $(\mathcal{K}, >)$ into a new partially ordered $\mathcal{C}$-base $\mu(\mathcal{K}, >) = (\mathcal{K}', \gg)$, replacing each $\phi > \psi$ by $\phi \vee \neg\psi \gg \psi \vee \neg\phi$.

2. Apply the rules of $(\mathcal{C})$ to the base $\mathcal{B}_{(\mathcal{K}', \gg)}$, thus obtaining the closure $\mathcal{C}(\mathcal{B}_{(\mathcal{K}', \gg)})$.

3. Turn $\mathcal{C}(\mathcal{B}_{(\mathcal{K}', \gg)})$ into a new base $\rho(\mathcal{C}(\mathcal{B}_{(\mathcal{K}', \gg)}))$ by replacing each $\phi' \gg \psi'$ by $\phi' \wedge \neg\psi' > \neg\phi'$. (Note that this is equivalent to applying rule RI_{SC^d} (semi-cancellativity), a rule of system $\mathcal{C}$).

Obviously, following the above strategy we obtain only a subset of $QL(\mathcal{B}_{(\mathcal{K},>)})$.

Example 8. *Let $\mathcal{K} = \{\phi, \psi\}$ with $\phi > \psi$.
As $> = \rho(>_{pl})$ the formula $\phi > \psi$ stands for $\phi \wedge \neg\psi >_{pl} \psi \wedge \neg\phi$. Conversely, as $>_{pl} = \mu(>)$, we obtain the formula $\phi \wedge \neg\psi > \psi \wedge \neg\phi$. So we do not recover the initial formula $\phi > \psi$. The rule RI_{STP} must be used for that purpose.*

Example 9. *Let $\mathcal{K} = \{\phi, \neg\phi \vee \psi, \neg\psi\}$ with $\mathcal{B}_{(\mathcal{K},>)} = \{\phi > \neg\psi, \neg\phi \vee \psi > \neg\psi\}$ interpreting $>$ as qualitative likelihood.
First, we transform the QL-base into a $\mathcal{C}$-base:*

- $\phi > \neg\psi$ will be turned into $\phi \vee \psi \gg \neg\psi \vee \neg\phi$
- $\neg\phi \vee \psi > \neg\psi$ into $\neg\phi \vee \psi \gg \neg\psi$, which remains unchanged

*Then we use $\mathcal{C}$. By RI_O on $\phi \vee \psi \gg \neg\psi \vee \neg\phi$ we obtain $\phi \vee \psi \gg \neg\psi$. Then by RI_A and RI_O again we obtain $\psi \gg \neg\psi$. Finally, applying RI_{SC^d} produces no other formula. As $\phi \vee \psi, \neg\psi$ are subcontraries, and so are $\psi, \neg\psi$, we do get $\phi \vee \psi > \neg\psi$ and $\psi > \neg\psi$.
By QL we directly compute the partial preadditive deductive closure.*

- *By RI_O we obtain $\phi \vee \psi > \neg\psi$*
- *By RI_{ND^d} we obtain $\psi > \neg\psi$.*

To conclude, applying the proof system of QL to a comparative base does not give the same results as applying system $\mathcal{C}$ first and then QL (see Example 7). However, by changing a QL-base into a $\mathcal{C}$-base, applying the transformation in Proposition 14 enables us to derive QL consequences using inference rules of system $\mathcal{C}$. It is yet to be proved whether adding axiom ax_{NoD} and using preadditivity rules (or just the sure thing principle rule RI_{STP}) to a QL base, completed by its consequences obtained applying system $\mathcal{C}$ to the transformed original base, will generate the whole QL closure of the latter.

That it can be conjectured relies on the following reasoning. If we consider a qualitative likelihood relation $\succ^+$ and its associated plausibility relation $\succ_{pl} = \mu(\succ^+)$, these relations coincide on pairs of disjoint sets. Consider a relation $\succ$ relating only A, B such that $A \cap B = \emptyset$; it is clear that

- $\succ^+$ can be obtained from $\succ$ using $C \succ^+ D$ if and only if $C \setminus D \succ D \setminus C$, for C, D not disjoint.

- $\succ_{pl}$ can be obtained from $\succ$ using $C \succ_{pl} D$ if and only if $C \setminus D \succ D$.

So if a QL base $(K, >)$ is changed into a $\mathcal{C}$-base using the transformation μ (and taking the certainty relation dual of $\mu(>)$), we can extract from the $\mathcal{C}$-closure of the transformed base all statements $\phi > \psi$ where ϕ, ψ are subcontraries. Call this set of comparative statements $SC(K, >)$. All statements in $SC(K, >)$ are in the QL-closure of $(K, >)$ and we can argue that the C-closure of the transformed base contains all QL-consequences of $(K, >)$ involving subcontraries. So if we apply the sure thing principle rule RI_{STP} to $SC(K, >)$, we can hope to recover the QL-closure $(K, >)$.

5.3 Case of a flat base

One interesting issue is whether classical propositional logic is a special case of the logics of relative certainty and of qualitative likelihood. To see it, we can encode a propositional knowledge base in the syntax of these logics, and show that the standard closure of the original propositional knowledge base can be recovered respectively from the C-closure, and the QL-closure, of the set of comparative statements obtained by such encodings.

Consider a flat propositional base of the form $\mathcal{K} = \{\phi_1, \cdots, \phi_n\}$, where each formula ϕ_i expresses a piece of information given by an agent. We thus suppose that each formula is certain. In consequence a natural encoding of $\mathcal{K}$ in terms of comparative statements consists in translating each formula ϕ_i into $\phi_i > \neg \phi_i$. Let $\mathcal{B}_\mathcal{K} = \{\phi_1 > \neg \phi_1, \cdots, \phi_n > \neg \phi_n\}$.

We try to show that introducing the comparative statement $\phi_i > \neg \phi_i$ for each formula ϕ_i of the flat base $\mathcal{K}$, we can recover a classical consequence ψ of $\mathcal{K}$ as the consequence $\psi > \neg \psi$ of $\mathcal{B}_\mathcal{K}$. We will successively study the deductive closures of $\mathcal{B}_\mathcal{K}$ in the sense of relative certainty and qualitative likelihood logics.

Example 10. *Let $\mathcal{K} = \{\phi, \neg\phi \vee \psi\}$ be a classical base. So, $\mathcal{B}_\mathcal{K} = \{\phi > \neg\phi, \neg\phi \vee \psi > \phi \wedge \neg\psi\}$.*
By modus ponens on $\mathcal{K}$, ψ can be derived. So, we would like to obtain $\psi > \neg\psi$ from $\mathcal{B}_\mathcal{K}$.

- *We compute the C-closure: by RI_{CCC^d} on $\phi > \neg\phi$ and $\neg\phi \vee \psi > \phi \wedge \neg\psi$ we obtain $\phi \wedge \psi > \neg\psi \vee \neg\phi$. Then by RI_O we obtain $\psi > \neg\psi$. The C-closure also contains:*
  - *$\phi \wedge \psi > \neg\psi$, $\phi \wedge \psi > \neg\phi$, $\psi > \neg\psi$, $\phi > \neg\psi$ and $\psi > \neg\phi$.*
  - *$\phi > \neg\phi \vee \neg\psi$ and $\psi > \neg\phi \vee \neg\psi$.*

- *We compute the QL-closure. Each formula is of the form $\phi_i > \neg\phi_i$, so $\mathcal{B}_\mathcal{K}$ contains only pairs of disjoint formulas, that are also subcontraries. Inference rule RI_{CCC^d} can still be applied and so the same conclusion $\psi > \neg\psi$ can be inferred. Other comparative formulas can be inferred such as*
  - *By axiom ax_{NoD}, we obtain $\phi \wedge \neg\psi > \bot$ and $\psi \wedge \neg\phi > \bot$ if ϕ and ψ are not equivalent. So we have $\phi \wedge \neg(\phi \wedge \psi) > \neg\phi \wedge (\phi \wedge \psi)$.*
  - *By RI_{STP}, we obtain $\phi > \phi \wedge \psi$ and similarly $\psi > \phi \wedge \psi$.*

With both systems, we obtain $\psi > \neg\psi$. Moreover the QL-closure contains $\phi > \phi \wedge \psi$ and $\psi > \phi \wedge \psi$ (when ϕ and ψ are not equivalent).

What the above example suggests holds more generally:

Proposition 18. *If $\{\phi_1, \ldots \phi_n\} \vdash \phi$ then $\{\phi_1 > \neg\phi_1, \ldots \phi_n > \neg\phi_n\} \vdash_X \phi > \neg\phi$ for $X \in \{\mathcal{C}, QL\}$.*

Proof of Proposition 18:
In both systems $\mathcal{C}$ and QL, we can apply inference rule RI_{CCC^d} to $\{\phi_1 > \neg\phi_1, \ldots \phi_n > \neg\phi_n\}$ and get the consequence $\phi_1 \wedge \cdots \wedge \phi_n > \neg\phi_1 \vee \cdots \vee \neg\phi_n$. And indeed, $\{\phi_1, \ldots \phi_n\} \vdash \phi_1 \wedge \cdots \wedge \phi_n$.

Now it is well-known that $\{\phi_1, \ldots \phi_n\} \vdash \phi$ if and only if $\phi_1 \wedge \cdots \wedge \phi_n \vdash \phi$. In this case $\{\phi_1 > \neg\phi_1, \ldots \phi_n > \neg\phi_n\} \vdash_X \phi > \neg\phi$ also holds using RI_O, valid for $X \in \{\mathcal{C}, QL\}$. □

For the converse proposition, the situation is different between $\mathcal{C}$ and QL. Note that

Lemma 2. *In $\mathcal{C}$, $\phi_i > \neg\phi_i$ is equivalent to $\phi_i > \bot$.*

Proof of Lemma 2:
Rule RI_{Nec} expresses that $\phi_i > \bot$ implies $\phi_i > \neg\phi_i$, and for the converse, apply RI_O. □

So we can prove:

Proposition 19. *If $\{\phi_1 > \neg\phi_1, \ldots \phi_n > \neg\phi_n\} \vdash_\mathcal{C} \phi > \neg\phi$ then $\{\phi_1, \ldots \phi_n\} \vdash \phi$.*

Proof of Proposition 19:
In $\mathcal{C}$, the knowledge base $\{\phi_1 > \neg\phi_1, \ldots \phi_n > \neg\phi_n\}$ is equivalent to $\{\phi_1 > \bot, \ldots \phi_n > \bot\}$. Only rules RI_A and RI_O can be used to the latter base, which ensures that $\{\phi_1 > \bot, \ldots \phi_n > \bot\} \vdash_\mathcal{C} \phi > \bot$ only when $\{\phi_1, \ldots \phi_n\} \vdash \phi$, so that $\{\phi_1 > \neg\phi_1, \ldots \phi_n > \neg\phi_n\} \vdash_\mathcal{C} \phi > \neg\phi$ implies $\{\phi_1, \ldots \phi_n\} \vdash \phi$. □

In QL, the base $\{\phi_1 > \bot, \ldots \phi_n > \bot\}$ brings no information as it follows from non-dogmaticism axiom ax_{NoD}, so it is not equivalent to $\{\phi_1 > \neg\phi_1, \ldots \phi_n > \neg\phi_n\}$. Moreover, we cannot apply the QL rule RI_{Q^d} to the knowledge base $\{\phi_1 > \neg\phi_1, \ldots \phi_n > \neg\phi_n\}$. We can only apply inference rules RI_{CCC^d} and RI_O. But then what we get is again the $\mathcal{C}$-closure. The inference rules we can use on top are RIP_1 and RIP_2, or better the sure thing principle R_{STP}. However they would only deduce statements of the form $\phi \vee \psi > \phi$ whenever $\psi \not\vdash \phi$ from axiom ax_{NoD}. But note that we cannot apply R_{STP} to statements of the form $\phi_i > \neg\phi_i$. So inference from such statements in QL is again equivalent to inference in classical logic.

6 Conclusion

In their early survey on qualitative approaches to probabilistic reasoning, Walley and Fine [37] pointed out in 1979 that

> there is a uniform disregard for the formal analysis of probability concepts that cannot be reduced in some fashion to numerical probability.

Due to the assumption that probability is intrinsically numerical, most logical approaches to reasoning with absolute or comparative probability statements in a symbolic framework still reject the adjunction principle according to which the conjunction of two beliefs is still a belief (see for instance the logic of risky knowledge [30], or yet Burgess comparative probability logic [4]). In this paper we have tried to reconcile two uncertain reasoning traditions in a symbolic framework, namely the non-monotonic reasoning approach of the Kraus, Lehman and Magidor style [27] as captured in the possibility theory setting, and the probabilistic reasoning approach as captured via the sure thing principle. There is a clash of intuitions between the two frameworks as the first one respects deductive closure for beliefs, while the latter often rejects it, for instance on the basis of the lottery paradox, originally introduced by Kyburg [28]. In this example, a conjunction of strong beliefs may turn out to be inconsistent. As explained in [15], the lottery paradox is less convincing in situations where some possible worlds are much more frequent than other ones, and probabilities tend to be big-stepped on a suitable partition, which brings probability orderings much closer to possibilistic orderings. However, if the considered probability ordering is total, a certain trivialization results from adopting the adjunction principle, as it enforces a linear order of possible worlds ([15] again).

In this paper, we restrict to partial orders expressing relative likelihood, giving up the reference to numerical probabilities, thus avoiding this trivialization. We show that strict partial comparative plausibility and qualitative likelihood relations coincide on pairs of disjoint sets and are in bijection with one another, and we provide a logic for relative likelihood that is both adjunctive and respects the sure thing principle.

A possible extension of this work would be to consider similar notions dropping the asymmetry property, so as to capture equal likelihoods between propositions as distinct from incomparability due to incompleteness, as studied in [12]. However it is clear that such a logic should then allow for negation and disjunction of comparative statements, in order to express relations between strict and weak preference, which would make the language more complex. Another line of further research would be to extend QL to comparative conditional statements.

References

[1] E.W. Adams and H.P. Levine. On the uncertainties transmitted from premises to conclusions in deductive inferences. *Synthese*, 30:429–460, 1975.

[2] S. Benferhat, D. Dubois, and H. Prade. Nonmonotonic reasoning, conditional objects and possibility theory. *Artif. Intell.*, 92(1-2):259–276, 1997.

[3] S. Benferhat, D. Dubois, and H. Prade. Possibilistic and standard probabilistic semantics of conditional knowledge bases. *Journal of Logic and Computation*, 9(6):873–895, 1999.

[4] J. P. Burgess. Axiomatizing the logic of comparative probability. *Notre Dame Journal of Formal Logic*, 51(1):119–126, 2010.

[5] C. Cayrol, D. Dubois, and F. Touazi. Fermeture déductive d'une base partiellement ordonnée. Research report RR–2014-08–FR, IRIT, Université Paul Sabatier, Toulouse, November 2014.

[6] C. Cayrol, D. Dubois, and F. Touazi. On the semantics of partially ordered bases. In C. Beierle and C. Meghini, editors, *Foundations of Information and Knowledge Systems*, volume 8367 of *Lecture Notes in Computer Science*, pages 136–153. Springer, 2014.

[7] C. Cayrol, D. Dubois, and F. Touazi. Ordres Partiels entre Sous-Ensembles d'un Ensemble Partiellement Ordonné. Research report RR–2014-02–FR, IRIT, Université Paul Sabatier, Toulouse, February 2014.

[8] C. Cayrol, V. Royer, and C. Saurel. Management of preferences in assumption based reasoning. In *Information Processing and the Management of Uncertainty in Knowledge based Systems (IPMU'92)*, volume 682 of *Lecture Notes in Computer Science*, pages 13–22. Springer, 1993.

[9] G. Coletti and R. Scozzafava. *Probabilistic Logic in a Coherent Setting*. Kluwer Academic Pub, 2002.

[10] B. de Finetti. La prévision : ses lois logiques, ses sources subjectives. *Annales Institut Poincaré*, 7:1–68, 1937.

[11] D. Dubois. Belief structures, possibility theory and decomposable confidence measures on finite sets. *Computers and Artificial Intelligence (Bratislava)*, 5:403–416, 1986.

[12] D. Dubois and H. Fargier. A unified framework for order-of-magnitude confidence relations. In *Proceedings of the 20th Conference in Uncertainty in Artificial Intelligence*, pages 138–145. AUAI Press, 2004.

[13] D. Dubois, H. Fargier, and H. Prade. Refinements of the maximin approach to decision-making in fuzzy environment. *Fuzzy Sets and Systems*, 81:103–122, 1996.

[14] D. Dubois, H. Fargier, and H. Prade. Possibilistic likelihood relations. In *Proceedings of 7th International Conference on Information Processing and Management of Uncertainty in Knowledge-based Systems (IPMU'98)*, pages 1196–1202, Paris, 1998. Editions EDK.

[15] D. Dubois, H. Fargier, and H. Prade. Ordinal and probabilistic representations of acceptance. *J. Artif. Intell. Res. (JAIR)*, 22:23–56, 2004.

[16] D. Dubois and H. Prade. *Possibility Theory: An Approach to Computerized Processing of Uncertainty*. Plenum Press, New York, 1988.

[17] D. Dubois and H. Prade. Epistemic entrenchment and possibilistic logic. *Artificial Intelligence*, 50(2):223–239, 1991.

[18] D. Dubois and H. Prade. Formal representations of uncertainty. In D. Bouyssou, D. Dubois, M. Pirlot, and H. Prade, editors, *Decision-making - Concepts and Methods*, chapter 3, pages 85–156. ISTE & Wiley, London, 2009.

[19] D. Dubois and H. Prade. Possibilistic logic - an overview. In D. Gabbay, J. Siekmann, and J. Woods, editors, *Computational logic*, volume 9 of *Handbook of the History of Logic*, pages 283–342. elsevier, 2014.

[20] T. Fine. *Theories of Probability*. Academic Press, New York, 1983.

[21] P. C. Fishburn. The axioms of subjective probability. *Statistical Science*, 1(3):335–358, 1986.

[22] N. Friedman and J. Halpern. Plausibility measures and default reasoning. In *Proc of the 13th National Conf. on Artificial Intelligence*, pages 1297–1304, Portland, OR, 1996.

[23] N. Friedman and J. Y. Halpern. Plausibility measures: A user's guide. In *Proc of the Eleventh Annual Conference on Uncertainty in Artificial Intelligence, Montreal, Quebec, August 18-20*, pages 175–184, 1995.

[24] H. Geffner. *Default reasoning: Causal and Conditional Theories*. MIT Press, 1992.

[25] J. Y. Halpern. Defining relative likelihood in partially-ordered preferential structures. *Journal of Artificial intelligence Research*, 7:1–24, 1997.

[26] C.H. Kraft, J.W. Pratt, and A. Seidenberg. Intuitive probability on finite sets. *Ann. Math. Stat.*, 30:408–419, 1959.

[27] S. Kraus, D. Lehmann, and M. Magidor. Nonmonotonic reasoning, preferential models and cumulative logics. *Artificial Intelligence*, 44:167–207, 1990.

[28] H. E. Kyburg Jr. Probabilistic acceptance. In *UAI '97: Proceedings of the Thirteenth Conference on Uncertainty in Artificial Intelligence, Brown University, Providence, Rhode Island, USA, August 1-3, 1997*, pages 326–333, 1997.

[29] H. E. Kyburg, Jr and H. E. Smokler, editors. *Studies in Subjective Probability*. Wiley, New York, 1964. Second edition (with new material) 1980.

[30] H. E. Kyburg Jr. and C-M. Teng. The logic of risky knowledge, reprised. *Int. J. Approx. Reasoning*, 53(3):274–285, 2012.

[31] D. Lewis. *Counterfactuals. Basil Blackwell*, 1973.

[32] D. Lewis. Counterfactuals and comparative possibility. *Journal of Philosophical Logic*, 2(4):418–446, 1973.

[33] R.D. Luce, D.H. Krantz, P. Suppes, and A. Tversky. *Foundations of measurement*. Academic Press, New York, 1990.

[34] N. J. Nilsson. Probabilistic logic. *Artificial Intelligence*, 28(1):71 – 87, 1986.

[35] L. Savage. *The foundations of statistics*. Dover, New-York, 1972.

[36] F. Touazi, C. Cayrol, and D. Dubois. Possibilistic reasoning with partially ordered beliefs. *J. Applied Logic*, 13(4):770–798, 2015.

[37] P. Walley and T. Fine. Varieties of modal (classificatory) and comparative probabilities. *Synthese*, 41:321–374, 1979.

[38] S. K. M. Wong, P. Bollmann Y. Y. Yao, and H. C. Burger. Axiomatization of qualitative belief structure. *IEEE transactions on SMC*, 21(34):726–734, 1991.

[39] L.A. Zadeh. Fuzzy sets as a basis for a theory of possibility. *Fuzzy Sets and Systems*, 1:3–28, 1978.

Received 9 January 2017

Characterization of a New Subquasivariety of Residuated Lattice

Saeed Rasouli
*Persian Gulf University,
Bushehr, 75169, Iran.*
srasouli@pgu.ac.ir

Zeinab Zarin and Abass Hasankhan
Shahid Bahonar University, Kerman, Iran.
zeinabzarin@yahoo.com, abhasan@uk.ac.ir

Abstract

The paper is devoted to study the notions of right and left stabilizers in residuated lattices relative to a filter. We establish a connection between right and left stabilizers in residuated lattices relative to a filter and (contravariant) Galois connection. We define a new class of residuated lattices, called $\mathcal{RS} - \mathcal{RL}$ and we show this class is a subquasivariety of the residuated lattices variety.

1 Introduction

It is well known that certain information processing, especially inferences based on certain information, is based on the classical logic. Naturally, it is necessary to establish some rational logical systems as the logical foundation for uncertain information processing. For this reason, various kinds of non-classical logical systems have been extensively proposed and researched. In fact, non-classical logic has become a formal and useful tool for computer science to deal with uncertain information and fuzzy information. On the other hand, various logical algebras have been proposed as the semantical systems of non-classical logical systems, for example, residuated lattices, divisible residuated lattices, MTL algebras, Girard monoids, BL algebras, Gödel algebras, etc. Among these algebras, residuated lattices are very basic and important algebraic structures because the other logical algebras are all particular cases of residuated lattices.

In Gentzen-style systems, a structural rule is an inference rule that does not refer to any logical connective. Substructural logics were introduced as logics which, when formulated as Gentzen-style systems, lack some of the three basic structural rules as follows:

Weakening rule:
$$\frac{\Gamma, \Delta \Rightarrow \phi}{\Gamma, \alpha, \Delta \Rightarrow \phi}.$$

Contraction rule:
$$\frac{\Gamma, \alpha, \alpha, \Delta \Rightarrow \phi}{\Gamma, \alpha, \Delta \Rightarrow \phi}.$$

Exchange rule:
$$\frac{\Gamma, \alpha, \beta, \Delta \Rightarrow \phi}{\Gamma, \beta, \alpha, \Delta \Rightarrow \phi}.$$

Commutative residuated lattices are the algebraic counterpart of logics without contraction rule. The concept of commutative residuated lattice firstly introduced by W. Krull in [31] who discussed decomposition into isolated component ideals. After him, they were investigated by M. Ward and R. P. Dilworth in a series of important papers [13, 14, 37, 38, 39, 40, 41], as the main tool in the abstract study of ideal lattices in ring theory. These lattices have been known under many names: BCK latices in [23], full BCK algebras in [31], FL_{ew} algebras in [33], and integral, residuated, commutative ℓ-monoids in [5].

Apart from their logical interest, residuated lattices have interesting algebraic properties. The properties of residuated lattices were presented in [18, 30, 34]. For a survey of residuated lattices we refer to [29].

The deductive system theory of the logical algebras plays an important role in studying these algebras and the completeness of the corresponding non-classical logics. From a logical point of view, various deductive systems correspond to various sets of provable formulas. Since deductive systems correspond to subsets closed with respect to Modus Ponens so they are sometimes called (implicative) filters.

Di Nola, Georgescu and Iorgulescu in [15] introduced the notion of left stabilizers in pseudo-BL algebras. After that Haveshki and Mohamadhasani in [25] generalized the notion of stabilizers to the stabilizers with respect to a subset and introduced the notion of left stabilizer with respect to a subset in BL-algebras. Borzooei and Paad in [4] introduced some new types of stabilizers in BL-algebras. Borumand and Mohtashamnia in [3] introduced the notion of right and left stabilizer in (commutative) residuated lattices. Haveshki in [22] improved some results in [3]. Ahadpanah and Torkzadeh in [2] introduced the normal residuated lattices and studied them.

Motamed and Torkzadeh in [32] introduced the notion of right stabilizers in BL-algebras and define a class of BL-algebras, called RS-BL-algebra. In this paper we study the notions of right and left stabilizers in residuated lattices relative to a filter and we establish a connection between them and (contravariant) Galois connection. Also, we introduce a new quasi subvariety of the variety $\mathcal{RL}$.

This paper is organized in five sections. In Section 2, we recall some definitions and facts about residuated lattices and Galois connection that we use in the sequel. In Section 3, we introduce the notion of left and right stabilizer of a nonempty subset relative to a filter of a residuated lattice and study the relationship between them. In Section 4, we establish a connection between Galois connection and stabilizers in a residuated lattices. In Section 5, we introduce the notion of right stabilizer residuated lattices relative to a filter and we show that the class of right stabilizer residuated lattices is a quasivariety.

2 A brief excursion into residuated lattices and Galois connections

In this section we recall some definitions, properties and results relative to residuated lattices and Galois connection which will be used in the following sections of this paper.

2.1 residuated Lattices

Definition 2.1. *[37] A residuated lattice is an algebraic structure* $\mathfrak{A} = (A; \vee, \wedge, \odot, \rightarrow, 0, 1)$ *of type* $(2, 2, 2, 2, 0, 0)$ *satisfying the following conditions:*

RL_1 $(A; \vee, \wedge, 0, 1)$ *is a bounded lattice.*

RL_2 $(A, \odot, 1)$ *is a commutative monoid.*

RL_3 $x \odot y \leq z$ *if and only if* $x \leq y \rightarrow z$.

The operation $\rightarrow$ is referred to as the residual of $\odot$. A residuated lattice $\mathfrak{A}$ is nontrivial if and only if $0 \neq 1$. We denote by $\mathcal{RL}$ the class of residuated lattices. In a residuated lattice $\mathfrak{A}$, for any $a \in A$, we put $\neg a := a \rightarrow 0$ and $x^0 = 1$ and for any natural number n, we define $x^n = x^{n-1} \odot x$.

A residuated lattice $\mathfrak{A}$ is called an *MTL algebra* [10] if it satisfies the pre-linearity condition (denoted by prel):

$(prel)$ $(x \rightarrow y) \vee (y \rightarrow x) = 1$.

It is easy to see that each linearly-ordered residuated lattice is an MTL algebra. We denote by $\mathcal{MTL}$ the class of MTL algebras. Obviously, the class $\mathcal{MTL}$ of MTL algebras is equational, hence it forms a subvariety of the variety $\mathcal{RL}$.

A residuated lattice $\mathfrak{A}$ is called a *divisible residuated lattice* [24] if it satisfies the divisibility condition (denoted by div):

(div) $x \odot (x \to y) = x \wedge y$.

We denote by $\mathcal{DRL}$ the class of divisible residuated lattice. Obviously, the class $\mathcal{DRL}$ of divisible residuated lattice is equational, hence it forms a subvariety of the variety $\mathcal{RL}$. A residuated lattice $\mathfrak{A}$ in which $x \odot y = x \wedge y$ (or equivalently, $x^2 = x$) for all $x, y \in A$ is called a *Heyting algebra* or *pseudo-Boolean algebra* [36]. A Heyting algebra is a particular case of divisible residuated lattice.

A residuated lattice $\mathfrak{A}$ is called a *BL algebra* [24] if it satisfies both $(prel)$ and (div). Denote by $\mathcal{BL}$ the class of BL algebras. A residuated lattice is called proper if it is not a MTL algebra, a divisible residuated lattice or a BL algebra, i.e. if $(prel)$ and (div) do not hold. A MTL algebra is called proper if it is not a BL algebra, i.e. if (div) does not hold. A divisible residuated lattice is called proper if it is not a BL algebra, i.e. if $(prel)$ does not hold.

A BL-algebra $\mathfrak{A}$ is called an MV-algebra [27] if it is an involutive (or regular) i.e. $\neg\neg x = x$. Denote by $\mathcal{MV}$ the class of MV algebras. According to [42] a residuated lattice $\mathfrak{A}$ is an MV-algebra if and only if it satisfies the following assertions:

mv $(x \to y) \to y = (y \to x) \to x$.

A BL algebra is called proper if it is not an MV algebra, i.e. if mv does not hold.

Note that $\mathcal{MTL}, \mathcal{DRL}, \mathcal{BL}$ and $\mathcal{MV}$ are all subvarieties of $\mathcal{RL}$, connected as Figure 3.

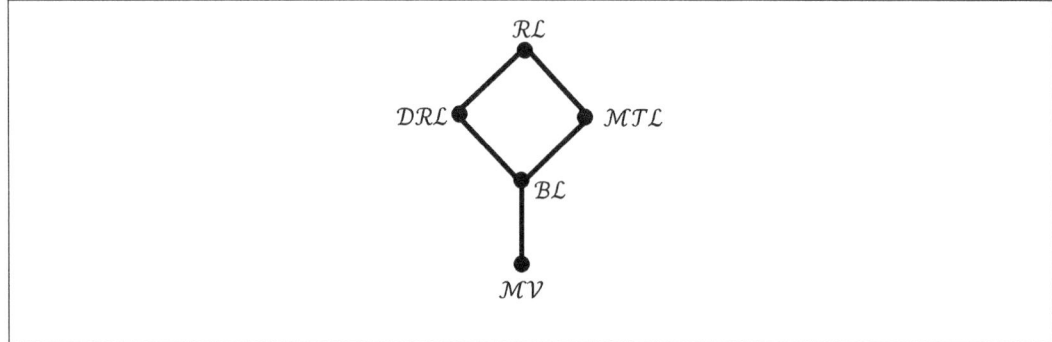

Figure 1: Inclusions between some subvarieties of $\mathcal{RL}$

Proposition 2.2. *[30] Let $\mathfrak{A}$ be a residuated lattice. Then the following conditions are satisfied for any $x, y, z \in A$:*

r_1 $x \leq y$ if and only if $x \to y = 1$.

r_2 $x \to x = 0 \to x = x \to 1 = 1$ and $1 \to x = x$.

r_3 $x \to (y \to z) = (x \odot y) \to z = y \to (x \to z)$.

r_4 $x \odot y \leq x \odot (x \to y) \leq x \wedge y$. In particular, $x \leq y \to x$ and $x \leq (x \to y) \to y$.

r_5 $x \leq y$ implies $x \odot z \leq y \odot z$.

r_6 $x \leq y$ implies $z \to x \leq z \to y$ and $y \to z \leq x \to z$.

r_7 $x \to y \leq (y \to z) \to (x \to z)$.

r_8 $x \to y \leq (z \to x) \to (z \to y)$.

r_9 $x \to (y \wedge z) = (x \to y) \wedge (x \to z)$. In particular, $x \to y = x \to (x \wedge y)$.

r_{10} $((x \to y) \to y) \to y = x \to y$.

In the following, we give some examples of residuated lattice.

Example 2.3. *Let $A_7 = \{0, a, b, c, d, e, 1\}$ be a lattice whose Hasse diagram is below (see Figure 2). Define $\odot$ and $\to$ on A_7 as follows:*

$\odot$	0	a	b	c	d	e	1
0	0	0	0	0	0	0	0
a	0	a	a	a	a	a	a
b	0	a	a	a	a	a	b
c	0	a	a	c	c	c	c
d	0	a	a	c	c	c	d
e	0	a	a	c	c	e	e
1	0	a	b	c	d	e	1

$\to$	0	a	b	c	d	e	1
0	1	1	1	1	1	1	1
a	0	1	1	1	1	1	1
b	0	e	1	e	1	1	1
c	0	b	b	1	1	1	1
d	0	b	b	e	1	1	1
e	0	a	b	c	d	1	1
1	0	a	b	c	d	e	1

Routine calculation shows that $\mathfrak{A}_7 = (A_7; \vee, \wedge, \odot, \to, 0, 1)$ is a proper residuated lattice, because the property (prel) does not hold: $(b \to c) \vee (c \to b) = e \vee b = e \neq 1$ and the property (div) also does not hold: $d \odot (d \to b) = d \odot b = a \neq d \wedge b$.

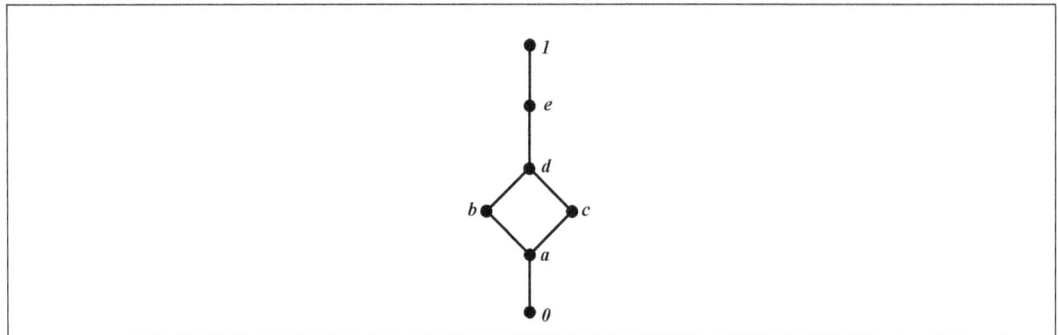

Figure 2: The Hasse diagram of $\mathfrak{A}_7$.

Example 2.4. *Let $A_5 = \{0, a, b, c, 1\}$ be a lattice whose Hasse diagram is below (see Figure 3). Define $\odot$ and $\rightarrow$ on A_5 as follows:*

$\odot$	0	a	b	c	1
0	0	0	0	0	0
a	0	a	0	a	a
b	0	0	0	0	b
c	0	a	0	a	c
1	0	a	b	c	1

$\rightarrow$	0	a	b	c	1
0	1	1	1	1	1
a	b	1	b	1	1
b	c	c	1	1	1
c	b	c	b	1	1
1	0	a	b	c	1

Routine calculation shows that $\mathfrak{A}_5 = (A_5; \vee, \wedge, \odot, \rightarrow, 0, 1)$ is a proper residuated

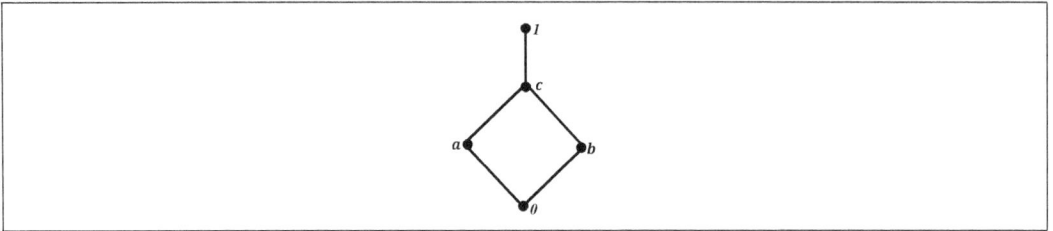

Figure 3: The Hasse diagram of $\mathfrak{A}_5$.

lattice, because the property (prel) does not hold: $(a \rightarrow b) \vee (b \rightarrow a) = b \vee c = c \neq 1$ and the property (div) also does not hold: $c \odot (c \rightarrow b) = c \odot b = 0 \neq b = c \wedge b$.

Let $\mathfrak{A}$ be a residuated lattice and F be a subset of A. For convenience, we enumerate some conditions which will be used in this paper.

$c_\emptyset$ $F \neq \emptyset$.

c_1 $1 \in F$.

$c_\odot$ $x, y \in F \Rightarrow x \odot y \in F.$

$c_\leq$ $x \leq y, x \in F \Rightarrow y \in F.$

$c_\vee$ $x \in F$ and $y \in A \Rightarrow x \vee y \in F.$

c_l $x, x \to_l y \in F \Rightarrow y \in F.$

c_r $x, x \to_r y \in F \Rightarrow y \in F.$

Definition 2.5. Let $\mathfrak{A}$ be a residuated lattice and F be a subset of A.

- F is called an ordered-filter of $\mathfrak{A}$ if it satisfies $c_\emptyset$ and $c_\leq$.
- F is called a filter of $\mathfrak{A}$ if it satisfies $c_\emptyset$, $c_\odot$ and $c_\leq$.
- F is called a 1-ideal of $\mathfrak{A}$ if it satisfies $c_\emptyset$, $c_\odot$ and $c_\vee$.
- F is called a left deductive system of $\mathfrak{A}$ if it satisfies c_1 and c_l.
- F is called a right deductive system of $\mathfrak{A}$ if it satisfies c_1 and c_r.

Proposition 2.6. Let $\mathfrak{A}$ be a residuated lattice and F be a subset of A containing 1. Then the following assertions are equivalent for any $x, y, z \in A$:

F_1 F is a filter.

F_2 F is a 1-ideal.

F_3 F is a left deductive system.

F_4 F is a right deductive system

F_5 $x \to_l y, y \to_l z \in F \Rightarrow x \to_l z \in F.$

F_6 $x \to_r y, y \to_r z \in F \Rightarrow x \to_r z \in F.$

F_7 $x \to_l y, x \odot z \in F \Rightarrow y \odot z \in F.$

F_8 $x \to_r y, z \odot x \in F \Rightarrow z \odot y \in F.$

F_9 $x \to_l y, \neg_l y \in F \Rightarrow \neg_l x \in F.$

F_{10} $x \to_r y, \neg_r y \in F \Rightarrow \neg_r x \in F.$

F_{11} $x, y \in F$ and $x \leq y \to_l z \Rightarrow z \in F.$

F_{12} $x, y \in F$ and $x \leq y \to_r z \Rightarrow z \in F.$

Proof. It is straightforward by Proposition 2.2. □

The set of ordered-filters and filters of a residuated lattice $\mathfrak{A}$ will be denoted by $OF(\mathfrak{A})$ and $F(\mathfrak{A})$, respectively. It is clear that $F(\mathfrak{A}) \subseteq OF(\mathfrak{A})$. Trivial examples of filters are $\mathbf{1} = \{1\}$ and A. A filter F of $\mathfrak{A}$ is proper if $F \neq A$. Clearly, F is a proper filter if and only if $0 \notin F$.

Example 2.7. *Consider the proper residuated lattice $\mathfrak{A}_7$ from Example 2.3. Then $F(\mathfrak{A}_7) = \{F_1 = 1, F_2 = \{e, 1\}, F_3 = \{c, d, e, 1\}, F_4 = \{a, b, c, d, e, 1\}, F_5 = A_7\}$.*

Example 2.8. *Consider the proper residuated lattice $\mathfrak{A}_5$ from Example 2.4. Then $F(\mathfrak{A}_5) = \{F_1 = 1, F_2 = \{a, c, 1\}, F_3 = A_5\}$.*

It is obvious that $(A; F(\mathfrak{A}))$ is an algebraic closed set system. The closure operator associated with the closed set system $(A; F(\mathfrak{A}))$ is denoted by $Fi^{\mathfrak{A}} : \mathcal{P}(A) \longrightarrow \mathcal{P}(A)$. Thus for any subset X of A, $Fi^{\mathfrak{A}}(X) = \cap \{F \in F(\mathfrak{A}) | X \subseteq F\}$ is the smallest filter of $\mathfrak{A}$ containing X. $Fi^{\mathfrak{A}}(X)$ is called the filter generated by X. For each $x \in A$, the filter generated by $\{x\}$ is denoted by $Fi^{\mathfrak{A}}(x)$ and it is called the principle filter of $\mathfrak{A}$. When there is no ambiguity we will drop the superscript $\mathfrak{A}$.

If $\mathcal{F} = \{F_i\}_{i \in I}$ is a family of all filters of $\mathfrak{A}$, we define $\overline{\wedge} \mathcal{F} = \cap \mathcal{F}$ and $\underline{\vee} \mathcal{F} = Fi(\cup \mathcal{F})$. According to [11], $(F(\mathfrak{A}), \overline{\wedge}, \underline{\vee}, 1, A)$ is a bounded complete distributive lattice.

Proposition 2.9. *[11] Let $\mathfrak{A}$ be a residuated lattice and X be a subset of A. Then we have*

$$Fi(X) = \{a \in A | x_1 \odot \cdots \odot x_n \leq a, \text{ for some integer } n, \ x_1, \cdots, x_n \in X\}.$$

Proposition 2.10. *[11] Let $\mathfrak{A}$ be a residuated lattice, F and G be two filters of $\mathfrak{A}$ and $x, y \in A$. The following assertions hold:*

(1) $F \underline{\vee} G = \{a \in A | f \odot g \leq a, \text{ for some, } f \in F, g \in G\}$.

(2) $Fi(x \vee y) = Fi(x) \cap Fi(y)$.

(3) $x \leq y$ implies $Fi(y) \subseteq Fi(x)$.

(4) $Fi(x) \underline{\vee} Fi(y) = Fi(x \wedge y) = Fi(x \odot y)$.

Let $\mathfrak{A}$ be a residuated lattice. We put $d(a, b) = (a \to b) \odot (b \to a)$. With any filter of a residuated lattice $\mathfrak{A}$ we associate two binary relations $\equiv_F$ on A by defining

$$(a, b) \in \equiv_F \text{ if and only if } d(a, b) \in F.$$

It is easy to check that the binary relations $\equiv_F$ is a equivalence relations on A. $\equiv_F$ are called the equivalence relation induced by F. In the following, for any $a \in A$ the equivalence classes $a/\equiv_F$ and is denoted by $[a]_F$.

Definition 2.11. *[9] Let $\mathfrak{A}$ be a residuated lattice. A filter F of $\mathfrak{A}$ is called an MV filter (filter of type MV) if $\mathfrak{A}/F \in \mathcal{MV}$.*

Proposition 2.12. *[9] Let $\mathfrak{A}$ be a residuated lattice and F be a subset of $\mathfrak{A}$. The following assertions are equivalent:*

1. F is an MV filter.
2. F is a filter and $x \to y \in F$ implies $((y \to x) \to x) \to y \in F$, for any $x, y \in A$.
3. $1 \in F$ and $z, z \to (x \to y) \in F$ implies $((y \to x) \to x) \to y \in F$, for any $x, y, z \in A$.
4. F is a filter and $((x \to y) \to y) \to (x \vee y) \in F$, for any $x, y \in A$.
5. F is a filter and $((y \to x) \to x) \to ((x \to y) \to y) \in F$, for any $x, y \in A$.

Let $\mathfrak{A}$ be a residuated lattice. The set of all complemented elements in the lattice reduct of $\mathfrak{A}$ is denoted by $B(\mathfrak{A})$ and it is called the Boolean center of $\mathfrak{A}$. Complements are generally not unique unless the lattice is distributive. In residuated lattices however, although the underlying lattices need not be distributive, according to [11], the complements are unique.

Proposition 2.13. *[11] Let $\mathfrak{A}$ be a residuated lattice, $e \in B(\mathfrak{A})$ and $a \in A$. The following assertions hold:*

1. $e^c = \neg e$ and $\neg\neg e = e$ and $e^2 = e$.
2. $\neg e \to e = e$
3. $e \odot a = e \wedge a$.
4. $(e \to a) \to e = e$.
5. $e \vee \neg e = 1$.
6. $e \wedge \neg e = 0$.

Proposition 2.14. *[12] Let $\mathfrak{A}$ be a residuated lattice and $e \in B(\mathfrak{A})$. Then $Fi(e)$, is a normal filter of $\mathfrak{A}$ and we have*

$$Fi(e) = \{a \in A | e \leq a\}.$$

Let $\mathfrak{A}$ be a residuated lattice, F be a filter of A and $X \subseteq A$. In the following the filter generated by $F \cup X$, i. e. $F \veebar Fi(X)$, will be denoted by F_X.

Proposition 2.15. *Let $\mathfrak{A}$ be a residuated lattice, F be a filter of A and $e \in B(\mathfrak{A})$. Then we have*

$$F_e = \{a \in A | f \odot e \leq a \text{ for some } f \in F\}.$$

Let $\mathfrak{A}$ and $\mathfrak{B}$ be two residuated lattices. A mapping $h : A \longrightarrow B$ is called a homomorphism, in symbols $h : \mathfrak{A} \longrightarrow \mathfrak{B}$, if it preserves the fundamental operations. If $h : \mathfrak{A} \longrightarrow \mathfrak{B}$ is a homomorphism we put $coker(h) = h^{\leftarrow}(1)$. It is easy to check that $coker(h)$ is a normal filter of $\mathfrak{A}$. Also, it is obvious that h is a monomorphism if and only if $coker(h) = \{1\}$.

Proposition 2.16. *Let $h : \mathfrak{A} \longrightarrow \mathfrak{B}$ be a homomorphism.*

(1) *If h is surjective and $F \in F(\mathfrak{A})(F \in F_n(\mathfrak{A}))$ such that $coker(h) \subseteq F$ then $h(F) \in F(\mathfrak{B})(h(F) \in F_n(\mathfrak{B}))$.*

(2) *If $F \in F(\mathfrak{B})(F \in F_n(\mathfrak{B}))$ then $h^{\leftarrow}(F) \in F(\mathfrak{A})(h^{\leftarrow}(F) \in F_n(\mathfrak{A}))$ and $coker(h) \subseteq h^{\leftarrow}(F)$.*

Proof. It is straightforward. □

Let $\mathfrak{A}$ be a residuated lattice and F be a filter of $\mathfrak{A}$ and X be a subset of A. The generalized co-annihilator of X (relative to F) is denoted by $(F : X)$ and defined as follow:
$$(F : X) = \{a \in A | x \vee a \in F, \forall x \in X\}.$$
In the following proposition, we collect the properties of generalized co-annihilators:

Proposition 2.17. *[35] Let $\mathfrak{A}$ be a residuated lattice, F, G be filters of $\mathfrak{A}$ and X, Y be subsets of A. Then the following conditions satisfy:*

(1) $(F : X)$ *is a filter of $\mathfrak{A}$.*

(2) $F \subseteq (F : X)$.

(3) $(F : X) = A$ *if and only if $X \subseteq F$.*

(4) $X \subseteq (F : (F : X))$.

Definition 2.18. *[9, Definition 7.] Let $\mathcal{V}$ be a subvariety of the variety $\mathcal{RL}$ of residuated lattices and $\mathfrak{A} \in \mathcal{V}$. A filter F of $\mathfrak{A}$ will be called a $\mathcal{V}$-filter (or filter of type $\mathcal{V}$) if $\mathfrak{A}/F \in \mathcal{V}$. We denote by $F_\mathcal{V}(\mathfrak{A})$ the set of all $\mathcal{V}$-filters of $\mathfrak{A}$.*

2.2 Galois connection

This section is devoted to recall some definitions, properties and results relative to Galois connection.

Definition 2.19. *Let $\mathscr{A} = (A; \leq)$ and $\mathscr{B} = (B; \preccurlyeq)$ be posets and $f : A \longrightarrow B$ be a map between posets.*

1. f is monotone if $a_1 \leq a_2$ implies $f(a_1) \preccurlyeq f(a_2)$, for all $a_1, a_2 \in A$.

2. f is antitone if $a_1 \leq a_2$ implies $f(a_2) \preccurlyeq f(a_1)$, for all $a_1, a_2 \in A$.

In particular case which $\mathscr{A} = \mathscr{B}$,

1. f is inflationary (also called extensive) if $a \leq f(a)$ for all $a \in A$.

2. f is idempotent if $f^2 = f$.

3. f is a closure operator on $\mathscr{A}$ if it is inflationary, monotone and idempotent. A fixpoint of the closure operator f, i.e. an element a of A that satisfies $f(a) = a$, is called a closed element of f. The set of closed elements of the closure operator f will be denoted by $\mathscr{C}_f$.

Definition 2.20. Let $\mathscr{A} = (A; \leq)$ and $\mathscr{B} = (B; \preccurlyeq)$ be posets. Suppose that $f : A \longrightarrow B$ and $g : B \longrightarrow A$ are functions such that for all $a \in A$ and $b \in B$ we have

$$a \leq g(b) \text{ if and only if } b \preccurlyeq f(a).$$

Then the pair (f, g) is called a (contravariant or antitone) Galois connection between $\mathscr{A}$ and $\mathscr{B}$.

Proposition 2.21. Let $\mathscr{A}$ and $\mathscr{B}$ be posets and $f : A \longrightarrow B$ and $g : B \longrightarrow A$ be two functions. Then the pair (f, g) forms a Galois connection between $\mathscr{A}$ and $\mathscr{B}$ if and only if the following assertions hold:

(1) gf and fg are inflationary functions.

(2) f and g are antitone functions.

Proof. Let (f, g) forms a Galois connection between $\mathscr{A}$ and $\mathscr{B}$. Consider $a \in A$. We have $f(a) \preccurlyeq f(a)$ and it implies that $a \leq g(f(a))$. So gf is an inflationary function. Analogously, we can show that fg is an inflationary function. If we have $a_1 \leq a_2$ then we have $a_1 \leq g(f(a_2))$ and it states that $f(a_2) \preccurlyeq f(a_1)$. In a similar way, we can obtain that g is an antitone function.

Now, let (1) and (2) holds. Assume that $a \leq g(b)$ for some $a \in A$ and $b \in B$. So we have $b \preccurlyeq f(g(b)) \preccurlyeq f(a)$. Analogously, we can show that $b \preccurlyeq f(a)$ implies $a \leq g(b)$. Therefore, (f, g) forms a Galois connection between $\mathscr{A}$ and $\mathscr{B}$. □

Proposition 2.22. Let $\mathscr{A}$ and $\mathscr{B}$ be posets and (f, g) forms a Galois connection between $\mathscr{A}$ and $\mathscr{B}$. Then the following assertions hold:

(1) $fgf = f$ and $gfg = g$.

(2) If $\vee X$ exists for some $X \subseteq A$ then $\wedge f(X)$ exists and $\wedge f(X) = f(\vee X)$.

(3) If $\vee Y$ exists for some $Y \subseteq B$ then $\wedge g(Y)$ exists and $\wedge g(Y) = g(\vee Y)$.

(4) $f(a) = \max\{b \in B | a \leq g(b)\}$, $g(b) = \max\{a \in A | b \preccurlyeq f(a)\}$

(5) gf is a closure operator on $\mathscr{A}$ and $\mathscr{C}_{gf} = g(B)$.

(6) fg is a closure operator on $\mathscr{B}$ and $\mathscr{C}_{fg} = f(A)$.

Proof. 1. Let $a \in A$. By Proposition 2.21(1) we have $a \leq g(f(a))$ and $f(a) \preccurlyeq f(g(f(a)))$ and by 2.21(2) we get that $f(g(f(a))) \preccurlyeq f(a)$. It shows that $f = fgf$. Analogously, we can show that $g = gfg$.

2. Let $x \in X$. Then $x \leq \vee X$ and it implies that $f(\vee X) \preccurlyeq f(x)$ and this means that $f(\vee X)$ is a lower bound of the set $f(X)$. Assume that $b \preccurlyeq f(x)$ for any $x \in X$. So we obtain that $x \leq g(b)$ for any $x \in X$. So we have $\vee X \leq g(b)$ and this states that $b \leq f(\vee X)$. Therefore, $f(\vee X) = \wedge f(X)$.

3. Let $a \in A$. By Proposition 2.21(1) we obtain that $f(a) \in \{b \in B | a \leq g(b)\}$. Assume that $b \in \{b \in B | a \leq g(b)\}$. Then $a \leq g(b)$ and it implies that $b \leq f(a)$. Analogously, we can show that $g(b) = \max\{a \in A | b \preccurlyeq f(a)\}$.

4. By Proposition 2.21(1), gf is inflationary and by Proposition 2.21(1), gf is isotone. Also, by (1) we can conclude that gf is idempotent. It states that gf is a closure operator on $\mathscr{A}$.

Let $b \in B$. By (1) we have $g(B) \subseteq \mathscr{C}_{gf}$. Also, for each $a \in A$, $a \in \mathscr{C}_{gf}$ implies $a = g(f(a)) \in g(B)$ and this shows that $\mathscr{C}_{gf} \subseteq g(B)$. Hence, we have $\mathscr{C}_{gf} = g(B)$. Analogously, we can show that fg is a closure operator on $\mathscr{B}$ and $\mathscr{C}_{fg} = f(A)$. $\square$

Theorem 2.23. *[6] Let A be set and $f : \mathcal{P}(A) \longrightarrow \mathcal{P}(A)$ be a closure operator. Then the set of closed elements of f, $\mathscr{C}_f$, is a complete lattice with respect to the following operations:*

$$\wedge^f : \mathscr{C}_f \times \mathscr{C}_f \longrightarrow \mathscr{C}_f \qquad \vee^f : \mathscr{C}_f \times \mathscr{C}_f \longrightarrow \mathscr{C}_f$$
$$(X, Y) \longmapsto X \cap Y, \qquad\qquad (X, Y) \longmapsto f(X \cup Y).$$

Corollary 2.24. *Let A be a set and (f, g) forms a Galois connection between $\mathcal{P}(A)$ and $\mathcal{P}(B)$. Then the following assertions hold.*

(1) $\mathscr{L}_g = (g(\mathcal{P}(B)); \wedge^g, \vee^g, 0 = g(B), 1 = g(\emptyset))$ is a complete lattice where $\wedge^g_{i \in I} g(Y_i) = g(\cup_{i \in I} Y_i)$ and $\vee^g_{i \in I} g(Y_i) = g(\cap_{i \in I} fg(Y_i))$ for any family $\{Y_i\}_{i \in I} \in \mathcal{P}(B)$.

(2) $\mathscr{L}_f = (f(\mathcal{P}(A)); \wedge^f, \vee^f, 0 = f(A), 1 = f(\emptyset))$ is a complete lattice where $\wedge^f_{i \in I} f(X_i) = f(\cup_{i \in I} X_i)$ and $\vee^f_{i \in I} f(X_i) = f(\cap_{i \in I} gf(X_i))$ for any family $\{X_i\}_{i \in I} \in \mathcal{P}(B)$.

Proof. By Proposition 2.22(5), gf is a closure operator on $\mathcal{P}(A)$ and $\mathscr{C}_{gf} = g(\mathcal{P}(B))$. So by Theorem 2.23, $(g(\mathcal{P}(B)); \wedge^{gf}, \vee^{gf})$ is a complete lattice where $\wedge^{gf}_{i \in I} g(Y_i) = \cap_{i \in I} g(Y_i)$ and $\vee^{gf}_{i \in I} g(Y_i) = gf(\cup_{i \in I} g(Y_i))$ for any family $\{Y_i\}_{i \in I} \in \mathcal{P}(B)$. Now, let $\{Y_i\}_{i \in I}$ be a family of subset of the set B. By Proposition 2.22(3) we have $\cap_{i \in I} g(Y_i) = g(\cup_{i \in I} Y_i)$ and this shows that $\wedge^{gf}_{i \in I} g(Y_i) = \wedge^g_{i \in I} g(Y_i)$. Also, we have $gf(\cup_{i \in I} g(Y_i)) = g(\cap_{i \in I} fg(Y_i))$ and it implies that $\vee^{gf}_{i \in I} g(Y_i) = \vee^g_{i \in I} g(Y_i)$. Since g is an antitone function so we have $g(B) \subseteq g(Y) \subseteq g(\emptyset)$ for any $Y \subseteq B$. Therefore, $(g(\mathcal{P}(B)); \wedge, \vee, 0 = g(B), 1 = g(\emptyset))$ is a complete lattice. Analogously, we can show that (2) holds. □

3 Stabilizer in residuated lattice

In this section we introduce and investigate the notion of stabilizer relative to a filter in residuated lattices.

Definition 3.1. *[3] Let $\mathfrak{A}$ be a residuated lattice, F be a filter of $\mathfrak{A}$ and X be a subset of A. The left stabilizer and the right stabilizer of X relative to F is denoted by $(F : X)_l$ and $(F : X)_r$, respectively and defined as follows.*

1. $(F : X)_l = \{a \in A | (a \to x) \to x \in F, \forall x \in X\}$.

2. $(F : X)_r = \{a \in A | (x \to a) \to a \in F, \forall x \in X\}$.

Also, $(F : X)_s = (F : X)_l \cap (F : X)_r$ is called the stabilizer of X relative to F. Let $\square \in \{l, r, s\}$. If $X = \{x\}$ then $(F : \{x\})_\square$ is denoted by $(F : x)_\square$. Also, $(1, X)_\square$ is called the stabilizer of X and it is denoted by $(X)_\square$.

Example 3.2. *Consider the proper residuated lattice $\mathfrak{A}_7$ from Example 2.3 and its filters from Example 2.7. In Table 1 we calculate the right and left stabilizers relative to all filters of $\mathfrak{A}_7$.*
Also, we have $(F_4 : 0)_l = (F_4 : 0)_r = F_4$, $(F_5 : 0)_l = (F_5 : 0)_r = F_5$ and for any element $0 \neq x \in A$ we have $(F_i : x)_l = (F_i : x)_r = F_5$, for $i = 4, 5$.

		0	a	b	c	d	e	1
F_1	l	F_4	F_2	F_3	F_2	F_2	F_1	F_5
	r	F_1	$\{0,1\}$	$\{0,1\}$	$\{0,1\}$	$\{0,1\}$	$\{0,1\}$	$\{0,1\}$
F_2	l	F_4	F_3	F_3	F_2	F_2	F_5	F_5
	r	F_2	$\{0,e,1\}$	$\{0,e,1\}$	$\{0,a,b,e,1\}$	$\{0,a,b,e,1\}$	F_5	F_5
F_3	l	F_4	F_3	F_3	F_5	F_5	F_5	F_5
	r	F_3	$\{0,c,d,e,1\}$	F_5	F_5	F_5	F_5	F_5

Table 1: Table of stabilizers of the residuated lattice $\mathfrak{A}_7$.

Example 3.3. *Consider the proper residuated lattice $\mathfrak{A}_5$ from Example 2.4 and its filters from Example 2.8. Then we have $(F_2:0)_r = F_2$, $(F_2:a)_r = F_3$, $(F_2:b)_r = F_2$, $(F_2:c)_r = F_3$ and $(F_2:1)_r = F_3$.*

In the following proposition, we collect some properties of stabilizers:

Proposition 3.4. *Let $\mathfrak{A}$ be a residuated lattice. Then the following assertions hold for any family $\{X\} \cup \{Y\} \cup \{X_i\}_{i \in I} \in \mathcal{P}(A)$, $\{F\} \cup \{G\} \cup \{F_i\}_{i \in I} \in Fi(\mathfrak{A})$ and $\square \in \{l, r, s\}$:*

(1) $X_l = \{a \in A | a \to x = x, \forall x \in X\}$ and $X_r = \{a \in A | x \to a = a, \forall x \in X\}$.

(2) $(F:X) \subseteq (F:X)_s$. *In particular, $F \subseteq (F:X)_s$.*

(3) $X \subseteq Y$ *implies* $(F:Y)_\square \subseteq (F:X)_\square$.

(4) $(F:Fi(X))_\square \subseteq (F:X)_\square$.

(5) $F \subseteq G$ *implies* $(F:X)_\square \subseteq (G:X)_\square$.

(6) $X \cap (F:X)_\square \subseteq F$. *In particular, if X contains F then $X \cap (F:X)_\square = F$.*

(7) $(F:X)_\square = A$ *if and only if* $X \subseteq F$. *Consequently, $(F:\emptyset)_s = (F:1)_s = (F:F)_s = A$.*

(8) $X \subseteq (F:(F:X)_l)_r$, $(F:(F:X)_r)_l$, $(F:(F:X)_s)_s$.

(9) $\cap_{i \in I}(F_i:X)_\square = (\cap_{i \in I} F_i : X)_\square$.

(10) $(F:0)_r = (F:A)_\square = F$.

(11) $(F:0)_l = \{a \in A | \neg\neg a \in F\}$. *In particular, $(0)_l = D_s(\mathfrak{A})$.*

Proof. 1. By r_1 we have $X_l = \{a \in A | (a \to x) \to x = 1, \forall x \in X\} = \{a \in A | a \to x \leq x, \forall x \in X\}$. On the other hand, by r_4 we have $x \leq a \to x$, for any $a, x \in A$. It implies that $X_l = \{a \in A | a \to x = x, \forall x \in X\}$. It shows that $X_l = \{a \in A | a \to x = x, \forall x \in X\}$. Analogously, we can show that $X_r = \{a \in A | x \to a = a, \forall x \in X\}$.

2. Let $a \in (F : X)$. Then for any $x \in X$ we have $a \vee x \in F$. By r_4 we have $a \vee x \leq ((a \to x) \to x) \wedge ((x \to a) \to a)$. Since F is a filter so we have $((a \to x) \to x) \wedge ((x \to a) \to a) \in F$. It shows that $a \in (F : X)_l \cap (F : X)_r = (F : X)_s$. By Proposition 2.17(2) we can conclude that $F \subseteq (F : X)_s$.

3. Let $X \subseteq Y$ and $a \in (F : Y)_l$. Then for any $x \in X$, since $X \subseteq Y$, we have $(a \to x) \to x \in F$ and it shows that $a \in (F : X)_l$. So $(F : Y)_l \subseteq (F : X)_l$. Analogously, we can obtain the other cases.

4. It is an immediate consequence of (3).

5. Let $F \subseteq G$ and $a \in (F : X)_l$. For any $x \in X$ we have $(a \to x) \to x \in F$. It implies that $(a \to x) \to x \in G$ and it shows that $a \in (G : X)_l$. Analogously, we can show that the other cases.

6. Let $a \in X \cap (F : X)_l$. Then $(a \to x) \to x \in F$ for any $x \in X$. Let $x = a$. By r_2 we have $a \in F$ and it implies that $X \cap (F : X)_l \subseteq F$. Similarly, we can show the other cases. In particular, if X contains F by (2) we conclude that $X \cap (F : X)_\square = F$.

7. Let $(F : X)_l = A$ and $x \in X$. We have $x = 1 \to x = (x \to x) \to x \in F$ and it shows that $X \subseteq F$. Conversely, if $X \subseteq F$, then by Proposition 2.17(3) we have $(F : X) = A$ and (2) implies that $(F : X)_l = A$. Analogously, we can obtain the other cases.

8. Let $x \in X$. Then for any $a \in (F : X)_l$ we have $(a \to x) \to x \in F$ and it implies that $x \in (F : (F : X)_l)_r$. Analogously, we can show that $X \subseteq (F : (F : X)_r)_l$, $(F : (F : X)_s)_s$.

9. By (5), for each $i \in I$ we have $(\cap_{i \in I} F_i : X)_\square \subseteq (F_i : X)_\square$ and it shows that $(\cap_{i \in I} F_i : X)_\square \subseteq \cap_{i \in I}(F_i : X)_\square$. Conversely, let $a \in \cap_{i \in I}(F_i : X)_l$. Thus, for any $i \in I$ and $x \in X$ we have $(a \to x) \to x \in F_i$ and it implies that we have $(a \to x) \to x \in \cap_{i \in I} F_i$ for any $x \in X$. Hence we obtain that $a \in (\cap_{i \in I} F_i : X)_l$. Analogously, we can obtain the other cases.

10. By (2) we know that $F \subseteq (F:0)_r, (F:A)_\square$. Let $a \in (F:0)_r$. Thus we have $a = 1 \to a = (0 \to a) \to a \in F$. It means that $(F:0)_r = F$. If $a \in (F:A)_l$, then for each $x \in A$ we have $(a \to x) \to x \in F$. Consider $x = a$. So we have $a = 1 \to a = (a \to a) \to a \in F$. It shows that $(F:A)_l = F$. Analogously, we can obtain the other cases.

11. It is straightforward.

$\square$

Proposition 3.5. *[25, Theorem 3] Let $\mathfrak{A}$ be a residuated lattice and F be filters of $\mathfrak{A}$. Then $(F:X)_l$ is a filter of $\mathfrak{A}$ for any $X \subseteq A$.*

Proposition 3.6. *Let $\mathfrak{A}$ be a residuated lattice, F be a filter of $\mathfrak{A}$ and $x \in A$. Then $(F:x/F)_\square = (F:x)_\square$.*

Proof. By Proposition 3.4(3), it is obvious that $(F:x/F)_\square \subseteq (F:x)_\square$. Now, let $a \in (F:X)_l$ and $y \in x/F$. Therefore, $d(x,y) \in F$ and this means $d((a \to x) \to x, (a \to y) \to y) \in F$. On the other hand, we have $(a \to x) \to x \in F$ and this implies that $(a \to y) \to y \in F$. Thus $a \in (F:x/F)_l$ and this shows that the equality holds. Analogously, we can show that $(F:x)_r \subseteq (F:x/F)_r$ and $(F:x)_s \subseteq (F:x/F)_s$. $\square$

Proposition 3.7. *Let $h: \mathfrak{A} \longrightarrow \mathfrak{B}$ be a surjective homomorphism and $\square \in \{l, r, s\}$.*

1. *If F is a filter of $\mathfrak{A}$ containing $coker(h)$ and $X \subseteq A$ then $h((F:X)_\square) = (h(F):h(X))_\square$.*

2. *If F is a filter of $\mathfrak{B}$ and $Y \subseteq B$ then $h^\leftarrow((F:Y)_\square) = (h^\leftarrow(F):h^\leftarrow(Y))_\square$.*

Proof. 1. Let F be a filter of $\mathfrak{A}$ and $X \subseteq A$. By Proposition 2.16(1), $h(F)$ is a filter of $\mathfrak{B}$. If $X = \emptyset$ then by Proposition 3.4(7) we have $(F:X) = A$ and $(h(F):h(X)) = B$. Since h is surjective so the equality holds. So let X be a nonempty subset of A. Assume that $b \in (h(F):h(X))_l$. So for each $y \in h(X)$ we have $(b \to y) \to y \in h(F)$. Hence, there are $x \in X$, $a \in A$ and $f \in F$ such that $h(x) = y$, $h(a) = b$ and $(b \to y) \to y = h(f)$. It means that $(h(a) \to h(x)) \to h(x) = h(f)$ and it implies that $f \to ((a \to x) \to x) \in coker(h) \subseteq F$. Since F is a filter so we can conclude that $(a \to x) \to x \in F$. Thus $a \in (F:X)_l$ and it states that $b \in h((F:X)_l)$.

Now, let $b \in h((F:X)_l)$ and $y \in h(X)$. So there are $a \in (F:X)_l$ and $x \in X$ such that $h(a) = b$ and $h(x) = y$ and it results that $(a \to x) \to x \in F$. Therefore, $(h(a) \to h(x)) \to h(x) \in h(F)$ and it implies that $b \in (h(F):h(X))_l$. Analogously, we can show that $h((F:X)_r) = (h(F):h(X))_r$ and $h((F:X)_s) = (h(F):h(X))_s$.

2. Let F be a filter of $\mathfrak{B}$ and $Y \subseteq A$. By Proposition 2.16(2), $h(F)$ is a filter of $\mathfrak{A}$. If $Y = \emptyset$ then we have $(F : Y)_l = B$ and $(h^{\leftarrow}(F) : h^{\leftarrow}(Y))_l = A$. Since h is surjective so the equality holds. Suppose that $a \in (h^{\leftarrow}(F) : h^{\leftarrow}(Y))_l$. Consider $y \in Y$. So there is $x \in A$ such that $h(x) = y$. We have $(h(a) \to y) \to y = (h(a) \to h(x)) \to h(x) = h((a \to x) \to x)$. On the other hand, we have $x \in h^{\leftarrow}(Y)$ and it implies that $(a \to x) \to x \in h^{\leftarrow}(F)$. Therefore, $(h(a) \to y) \to y \in F$ for each $y \in Y$ and it states that $h(a) \in (F : Y)_l$. It shows that $a \in h^{\leftarrow}((F : Y)_l)$.

Conversely, assume that $a \in h^{\leftarrow}((F : Y)_l)$ and $x \in h^{\leftarrow}(Y)$. Hence $h(a) \in (F : Y)_l$ and $h(x) \in Y$. It implies that $h((a \to x) \to x) = (h(a) \to y) \to y \in F$. Therefore, $(a \to x) \to x \in h^{\leftarrow}(F)$ and it concludes that $a \in (h^{\leftarrow}(F) : h^{\leftarrow}(Y))_l$. Analogously, we can show that $h^{\leftarrow}((F : Y)_r) = (h^{\leftarrow}(F) : h^{\leftarrow}(Y))_r$ and $h^{\leftarrow}((F : Y)_s) = (h^{\leftarrow}(F) : h^{\leftarrow}(Y))_s$.

□

Let $\mathfrak{A}$ be a residuated lattice and F be a normal filter of $\mathfrak{A}$. The mapping $\pi_F^{\mathfrak{A}} : \mathfrak{A} \longrightarrow \mathfrak{A}/F$ defined by $\pi_F^{\mathfrak{A}}(a) = a/F$ is called the natural homomorphism. It is obvious that the natural homomorphism $\pi_F^{\mathfrak{A}}$ is surjective and $coker(\pi_F^{\mathfrak{A}}) = F$. Therefore, by Proposition 2.16 we have

$$F(\mathfrak{A}/F) = \{H/F | F \subseteq H \in F(\mathfrak{A})\}.$$

Lemma 3.8. *Let $\mathfrak{A}$ be a residuated lattice and F be a normal filter of $\mathfrak{A}$. Then for any filter G of $\mathfrak{A}$ containing F and for any subset X of A $(G : X)_l/F$ is a filter of $\mathfrak{A}/F$.*

Proof. Let G be a filter of $\mathfrak{A}$ contains F and X be a subset of A. By Proposition 3.4(2) we have $F \subseteq (G : X)_l$ and by Proposition 3.5 we have $(G : X)_l \in F(\mathfrak{A})$. So $(G : X)_l/F$ is a filter of $\mathfrak{A}/F$. □

Corollary 3.9. *Let $\mathfrak{A}$ be a residuated lattice, F be a normal filter of $\mathfrak{A}$, G be a filter of $\mathfrak{A}$ containing F and X be a subset of A containing F. Then we have*

$$(G/F : X/F)_l = (G : X)_l/F.$$

Proof. Consider the natural epimorphism π_F in Proposition 3.7. Then we have $\pi_F^{\leftarrow}(G/F : X/F)_l = (G : \pi_F^{\leftarrow}(\pi_F(X)))_l$. By Proposition 3.4((3)), we have $(G : \pi_F^{\leftarrow}(\pi_F(X)))_l \subseteq (G : X)_l$ so $(G/F : X/F)_l \subseteq (G : X)_l/F$.

Now, assume that $a/F \in (G : X)_l/F$. By Lemma 3.8, $(G : X)_l/F$ is a filter of $\mathfrak{A}/F$ and it implies that $a \in (G : X)_l$. Consider $y/F \in X/F$. So there is $x \in X$ such that $y/F = x/F$ and it implies that $(a/F \to y/F) \to y/F = (a/F \to x/F) \to$

$x/F = ((a \to x) \to x)/F \in G/F$. Hence, $a/F \in (G/F : X/F)_l$ and it shows that $(G : X)_l/F \subseteq (G/F : X/F)_l$. □

4 Galois connection of stabilizers in residuated lattice

Let $\mathfrak{A}$ be a residuated lattice, F be a filter of $\mathfrak{A}$ and $\square \in \{l, r, s\}$. We define the following function.
$$\begin{aligned} F_\square : \mathcal{P}(A) &\longrightarrow \mathcal{P}(A) \\ X &\longmapsto (F : X)_\square. \end{aligned}$$

Proposition 4.1. *Let $\mathfrak{A}$ be a residuated lattice and F be a filter of $\mathfrak{A}$. Then the following pairs (F_l, F_r) and (F_s, F_s) are Galois connections on $\mathcal{P}(A)$.*

Proof. By Proposition 3.4(3), functions F_l and F_r are antitone and by 3.4(8), $F_l F_r$ and $F_r F_l$ are inflationary functions. So by Proposition 2.21 we obtain that (F_l, F_r) is a Galois connection on $\mathcal{P}(A)$. Analogously, we can show that (F_s, F_s) is Galois connections on $\mathcal{P}(A)$. □

Corollary 4.2. *Let $\mathfrak{A}$ be a residuated lattice and F be a filter of $\mathfrak{A}$. Then for any $X, Y \subseteq A$ the following assertions hold:*

(1) $X \subseteq (F : Y)_l$ *if and only if* $Y \subseteq (F : X)_r$.

(2) $X \subseteq (F : Y)_s$ *if and only if* $Y \subseteq (F : X)_s$.

Proof. It follows by Proposition 4.1 and Definition 2.20. □

Corollary 4.3. *Let $\mathfrak{A}$ be a residuated lattice and F be a filter of $\mathfrak{A}$. Then the following assertions hold for any $X \subseteq A$:*

(1) $(F : X)_{l(r)} = (F : (F : (F : X)_{l(r)})_{r(l)})_{l(r)}$.

(2) $(F : X)_s = (F : (F : (F : X)_s)_s)_s$.

Proof. It follows by Proposition 4.1 and Proposition 2.22(1). □

Corollary 4.4. *Let $\mathfrak{A}$ be a residuated lattice and F, G be filters of $\mathfrak{A}$. Then the following assertions hold for any family $\{X\} \cup \{X_i\}_{i \in I} \in \mathcal{P}$ and $\square \in \{l, r, s\}$:*

(1) $(F : \cup_{i \in I} X_i)_\square = \cap_{i \in I}(F : X_i)_\square$.

(2) $(F : X)_\square = \cap_{x \in X}(F : x)_\square$.

(3) $(F : X)_r = (F : Fi(X))_r$.

(4) $(F:X)_r \cap Fi(X) \subseteq F$.

(5) $(F:X)_\Box = (F:X-F)_\Box$.

(6) $(F:F_X)_r = (F:X)_r$.

Proof. 1. It is straightforward by Proposition 4.1 and Proposition 2.22((2) and (3)).

2. By taking $X = \cup_{x \in X}\{x\}$ it follows by (1).

3. By Proposition 3.4(4) we have $(F:Fi(X))_r \subseteq (F:X)_r$. Assume that $a \in (F:X)_r$. By Proposition 4.2(1) we obtain that $X \subseteq (F:a)_l$ and since $(F:a)_l$ is a filter it states that $Fi(X) \subseteq (F:a)_l$. Thus we have $a \in (F:Fi(X))_r$.

4. It follows by (3) and Proposition 3.4(6).

5. By (1) we have $(F:X)_\Box = (F:(X-F)\cap(X\cap F))_\Box = (F:X-F)_\Box \cap (F:X\cap F)_\Box$ and by Proposition 3.4(7) we have $(F:X\cap F)_\Box = A$. It states that $(F:X)_\Box = (F:X-F)_\Box$.

6.

$\Box$

Proposition 4.5. *Let $\mathfrak{A}$ be a residuated lattice, F be a filter of $\mathfrak{A}$ and $x,y \in A$. Then the following assertions hold:*

(1) $x \leq y$ implies $(F:x)_r \subseteq (F:y)_r$.

(2) $(F: x \odot y)_r = (F: x \wedge y)_r = (F:x)_r \cap (F:y)_r = (F:\{x,y\})_r$.

Proof. 1. Let $x \leq y$ and $a \in (F:x)_r$. By Proposition 4.2(1) we obtain that $x \subseteq (F:a)_l$. Since $(F:a)_l$ is a filter so $y \in (F:a)_l$ and it implies that $a \in (F:y)_l$.

2. By (1) follows that $(F: x \odot y)_r \subseteq (F: x \wedge y)_r \subseteq (F:x)_r \cap (F:y)_r$ and by Proposition 4.4(2) we have $(F:x)_r \cap (F:y)_r = (F:\{x,y\})_r$. If $a \in (F:\{x,y\})_r$ then $\{x,y\} \subseteq (F:a)_l$ and it implies that $x \odot y \in (F:a)_l$. It states that $a \in (F: x \odot y)_r$.

$\Box$

Corollary 4.6. *Let $\mathfrak{A}$ be a residuated lattice and F be a filter of $\mathfrak{A}$. Then the following assertions hold for any $X \subseteq A$:*

(1) $(F:X)_{l(r)} = \cup\{Y \in \mathcal{P}(A) | X \subseteq (F:Y)_{r(l)}\}$.

(2) $(F:X)_s = \cup\{Y \in \mathcal{P}(A) | X \subseteq (F:Y)_s\}$.

Proof. Let $X \subseteq A$. By Proposition 4.1 and Proposition 2.22(4) we have $(F:X)_{l(r)} = \max\{Y \in \mathcal{P}(A) | X \subseteq (F:Y)_{r(l)}\}$. Let $\Gamma = \{Y \in \mathcal{P}(A) | X \subseteq (F:Y)_{r(l)}\}$. We have $\max \Gamma \subseteq \cup \Gamma$. By considering $Y \in \Gamma$ we obtain that $X \subseteq (F:Y)_{r(l)}$ and it implies $Y \subseteq (F:X)_{l(r)}$ by Proposition 4.2. Therefore, $\cup \Gamma \subseteq (F:X)_{l(r)}$ and by Proposition 4.2 we obtain that $X \subseteq (F:\cup\Gamma)_{r(l)}$. So $\cup \Gamma \in \Gamma$ and it means that $\cup \Gamma = \max \Gamma$. Similarly, (2) holds. □

Corollary 4.7. *Let $\mathfrak{A}$ be a residuated lattice and F be a filter of $\mathfrak{A}$. Then the following assertions hold for any $\square \in \{l, r\}$:*

(1) $F_{l(r)}F_{r(l)}$ *is a closure operator on $\mathcal{P}(A)$ and $\mathscr{C}_{F_{l(r)}F_{r(l)}} = \{(F:X)_{l(r)} | X \subseteq A\}$.*

(2) $F_s F_s$ *is a closure operator on $\mathcal{P}(A)$ and $\mathscr{C}_{F_s F_s} = \{(F:X)_s | X \subseteq A\}$.*

Proof. By Proposition 4.1 and Proposition 2.22((5) and (6)) we obtain that $F_{l(r)}F_{r(l)}$ is a closure operator on $\mathcal{P}(A)$ and $\mathscr{C}_{F_{l(r)}F_{r(l)}} = \{F_{l(r)}(X) | X \subseteq A\} = \{(F:X)_{l(r)} | X \subseteq A\}$. Analogously, (2) holds. □

Corollary 4.8. *Let $\mathfrak{A}$ be a residuated lattice and F be a filter of $\mathfrak{A}$. Then the following assertions hold:*

(1) $\mathscr{L}_{F_{l(r)}} = (F_{l(r)}(\mathcal{P}(A)); \wedge^{F_{l(r)}}, \vee^{F_{l(r)}}, F, A)$ *is a complete lattice where the operations $\wedge^{F_{l(r)}}$ and $\vee^{F_{l(r)}}$ are defined as follows:*

$$\wedge_{i \in I}^{F_{l(r)}} (F:X_i)_{l(r)} = (F : \cup_{i \in I} X_i)_{l(r)},$$

and

$$\vee_{i \in I}^{F_{l(r)}} (F:X_i)_{l(r)} = (F : \cap_{i \in I}(F:(F:X_i)_{l(r)})_{r(l)})_{l(r)}.$$

(2) $\mathscr{L}_{F_s} = (F_s(\mathcal{P}(A)); \wedge^{F_s}, \vee^{F_s}, F, A)$ *is a complete lattice where the operations $\wedge^{F_s}$ and $\vee^{F_s}$ are defined as follows:*

$$\wedge_{i \in I}^{F_s} (F:X_i)_s = (F : \cup_{i \in I} X_i)_s,$$

and

$$\vee_{i \in I}^{F_s} (F:X_i)_s = (F : \cap_{i \in I}(F:(F:X_i)_s)_s)_s.$$

Proof. By Proposition 4.1, (F_l, F_r) is a Galois connection and by Proposition 4.7(1), $F_{l(r)}F_{r(l)}$ is a closure operator on $\mathcal{P}(A)$ and $\mathscr{C}_{F_{l(r)}F_{r(l)}} = F_{l(r)}(\mathcal{P}(A))$. So by Proposition 2.24, $\mathscr{L}_{F_{l(r)}} = (F_{l(r)}(\mathcal{P}(A)); \wedge^{F_{l(r)}}, \vee^{F_{l(r)}}, F_{l(r)}^{\square}(A), F_{l(r)}^{\square}(\emptyset))$ is a complete lattice. Also, by Proposition 3.4(7) we have $F_{l(r)}(\emptyset) = (F:\emptyset)_{l(r)} = A$ and by Proposition 3.4(10) we have $F_{l(r)}(A) = (F:A)_{l(r)} = F$. Analogously, we can show that (2) holds. $\square$

Proposition 4.9. *Let $\mathfrak{A}$ be a residuated lattice and F be a filter of $\mathfrak{A}$. Then the following assertion holds:*

$$\mathscr{C}_{F_r F_l} = \{(F:G)_r | G \in F(\mathfrak{A})[F,A]\}.$$

Proof. It is obvious that $\{(F:G)_r | G \in F(\mathfrak{A})[F,A]\} \subseteq \mathscr{C}_{F_r F_l}$. Now, let $H = (F:X)_r$ for some $X \subseteq A$. By Proposition 4.3(1) we have $(F:(F:H)_l)_r = H$ and by Proposition 3.5 we have $(F:H)_l \in F(\mathfrak{A})[F,A]$. It shows that $\mathscr{C}_{F_r F_l} \subseteq \{(F:G)_r | G \in F(\mathfrak{A})[F,A]\}$. $\square$

Proposition 4.10. *Let $\mathfrak{A}$ be a residuated lattice and F be a filter of $\mathfrak{A}$. Also let F_1 and F_2 be two ordered filters of $\mathfrak{A}$ such that $F \subseteq F_1 \cap F_2$. Then the following assertions are equivalent:*

(1) $F_1 \cap F_2 = F$.

(2) $F_1 \subseteq (F:F_2)$.

(3) $F_1 \subseteq (F:F_2)_s$.

(4) $F_1 \subseteq (F:F_2)_l$.

(5) $F_1 \subseteq (F:F_2)_r$.

Proof. Let $a_1 \in F_1$ and $a_2 \in F_2$. We have $a_1, a_2 \leq a_1 \vee a_2$ and it states that $a_1 \in (F:F_2)$. Thus we have $F_1 \subseteq (F:F_2)$. Therefore (1) implies (2). By Proposition 3.4(2), (2) implies (3), (4) and (5). Now, let $a \in F_1 \cap F_2$. So we have $a = 1 \rightarrow a = (a \rightarrow a) \rightarrow a$ and it shows that (4), (5) and consequently (2) and (3) implies (1). $\square$

Corollary 4.11. *Let $\mathfrak{A}$ be a residuated lattice and F be a filter of $\mathfrak{A}$. Also let G be an ordered-filter of $\mathfrak{A}$ containing F. Then the following assertions hold:*

(1) $(F:G) = (F:G)_s = (F:G)_l \subseteq (F:G)_r$.

(2) $(F:G)_s$ *is a filter of* $\mathfrak{A}$.

(3) $G \subseteq (F : (F : G)_l)_l \cap (F : (F : G)_s)_l \cap (F : (F : G)_s)_r$.

Proof. 1. Let G be an ordered-filter of $\mathfrak{A}$ containing F. By Proposition 3.5 we know that $(F : G)_l$ is a filter of $\mathfrak{A}$. Also, by hypothesis and Proposition 3.4(6) we have $(F : G)_l \cap G = F$. So by Proposition 4.10 we obtain that $(F : G)_l \subseteq (F : G)$. It shows that $(F : G) = (F : G)_s = (F : G)_l \subseteq (F : G)_r$.

2. It follows by (1).

3. It follows by (1) and Proposition 2.17(4). $\square$

Proposition 4.12. *Let $\mathfrak{A}$ be a residuated lattice and F be a filter of $\mathfrak{A}$. Then the meet-semilattice $\mathscr{L}_{F_l}(\mathfrak{A}) = (F_l(\mathcal{P}(A)); \wedge^{F_l}, (F : -), F)$ is pseudocomplemented.*

Proof. Let $(F : X)_l \in F_l(\mathcal{P}(A))$. By Proposition 3.4((2) and (6)) we have $(F : X)_l \cap (F : (F : X)_l)_l = F$. Also, Proposition 3.5 states that $(F : X)_l$ is a filter of $\mathfrak{A}$. So by Corollary 4.11(2) we obtain that $(F : (F : X)_l)_l = (F : (F : X)_l)$. It shows that $(F : X)_l \cap (F : (F : X)_l) = F$. Now, let $(F : X)_l \cap (F : Y)_l = F$. Since, $(F : X)_l$ and $(F : Y)_l$ are filters of $\mathfrak{A}$ containing F so by Proposition 4.10 we obtain that $(F : Y)_l \subseteq (F : (F : X)_l)$. It shows that the meet-semilattice $\mathscr{L}_{F_l}(\mathfrak{A})$ is pseudocomplemented $\square$

According to [20], if $\mathfrak{A} = (A; \wedge, ^*, 0)$ is a pseudocomplemented meet-semilattice and $S(A) = \{a^* | a \in A\}$ then $S(\mathfrak{A}) = (S(A); \wedge, \vee, 0, 1 = 0^*)$ is a Boolean lattice where for any $x, y \in S(\mathfrak{A})$, the join in $S(\mathfrak{A})$ is described by $x \vee y := (x^* \wedge y^*)^*$.

Corollary 4.13. *Let $\mathfrak{A}$ be a residuated lattice and F be a filter of $\mathfrak{A}$. Then $S(\mathscr{L}_{F_l}(\mathfrak{A}))$ is a Boolean lattice.*

5 Right Stabilizer residuated lattice

Definition 5.1. *Let $\mathfrak{A}$ be a residuated lattice and F be a filter of $\mathfrak{A}$. $\mathfrak{A}$ is called a right stabilizer residuated lattice relative to F (or RS_F-residuated lattice) if $(F : a)_r$ is a filter of $\mathfrak{A}$ for any $a \in A$. $\mathfrak{A}$ is called a right stabilizer residuated lattice (or RS-residuated lattice) if it is a RS_F-residuated lattice for any filter F of $\mathfrak{A}$. The class of right stabilizer residuated lattices will be denoted by $\mathcal{RS} - \mathcal{RL}$.*

Example 5.2. *Consider Example 3.2. Then $\mathfrak{A}_7$ is a right stabilizer residuated lattice relative to F_4.*

Example 5.3. *Consider Example 3.3. Then $\mathfrak{A}_5$ is a right stabilizer residuated lattice relative to F_2.*

Proposition 5.4. *Let $\mathfrak{A}$ be a RS_F-residuated lattice for some filter F of $\mathfrak{A}$. Then the following assertions hold for any $X \subseteq A$, $x, y \in A$ and $\square \in \{l, r, s\}$:*

(1) $(F : X)_\square = (F : Fi(X))_\square$.

(2) $(F : X)_\square \cap Fi(X) \subseteq F$.

(3) $x \le y$ implies $(F : x)_\square \subseteq (F : y)_\square$.

(4) $(F : x \odot y)_\square = (F : x \wedge y)_\square = (F : x)_\square \cap (F : y)_\square = (F : \{x,y\})_\square$.

Proof. The proof is similar to the proof of Corollary 4.4 and Proposition 4.5. □

Theorem 5.5. *Any subalgebra of a RS residuated lattice is a RS residuated lattice.*

Proof. Let $\mathfrak{A}$ be a RS residuated lattice and $\mathfrak{B}$ be a subalgebra of $\mathfrak{A}$. Assume that F is a filter of $\mathfrak{B}$. By Proposition 2.9 we have $Fi^{\mathfrak{A}}(F) \cap B = F$. Consider $b \in B$. By Proposition 3.4(2) we have $1 \in (F : b)_r$. Now let $x, x \to y \in (F : b)_r$ for some $x, y \in B$. Thus we have $(b \to x) \to x \in F$ and $(b \to (x \to y)) \to (x \to y) \in F$. So we have $(b \to x) \to x \in Fi^{\mathfrak{A}}(F)$ and $(b \to (x \to y)) \to (x \to y) \in Fi^{\mathfrak{A}}(F)$. It implies that $x \in (Fi^{\mathfrak{A}}(F) : b)_r$ and $x \to y \in (Fi^{\mathfrak{A}}(F) : b)_r$. $\mathfrak{A} \in \mathcal{RS} - \mathcal{RL}$ states that $y \in (Fi^{\mathfrak{A}}(F) : b)_r$. Now we have $(b \to y) \to y \in Fi^{\mathfrak{A}}(F)$. Since $\mathfrak{B}$ is a subalgebra of $\mathfrak{A}$ so $(b \to y) \to y \in B$. It shows that $(b \to y) \to y \in Fi^{\mathfrak{A}}(F) \cap B - F$ and it means $y \in (F : b)_r$. Hence $\mathfrak{B}$ is a right stabilizer residuated lattice. □

Theorem 5.6. *Any homomorphic image of a RS residuated lattice is a RS residuated lattice.*

Proof. Let $h : \mathfrak{A} \longrightarrow \mathfrak{B}$ be an epimorphism of residuated lattices and $\mathfrak{A}$ is a RS residuated lattice. Let F be a filter of $\mathfrak{B}$ and $b \in B$. By Proposition 3.4(2) we have $1 \in (F : b)_r$. Now let $y_1, y_1 \to y_2 \in (F : b)_r$ for some $y_1, y_2 \in B$. Thus we have $(b \to y_1) \to y_1 \in F$ and $(b \to (y_1 \to y_2)) \to (y_1 \to y_2) \in F$. Since h is surjective so there are $a, x_1, x_2 \in A$ such that $h(a) = b$, $h(x_1) = y_1$ and $h(x_2) = y_2$. It implies that $(a \to x_1) \to x_1 \in h^{\leftarrow}(F)$ and $(a \to (x_1 \to x_2)) \to (x_1 \to x_2) \in h^{\leftarrow}(F)$. So we have $x_1, x_1 \to x_2 \in (h^{\leftarrow}(F) : a)_r$ and it states that $x_2 \in (h^{\leftarrow}(F) : a)_r$. Therefore, $(a \to x_2) \to x_2 \in h^{\leftarrow}(F)$ and it shows that $(b \to y_2) \to y_2 \in F$. Thus $y_2 \in (F : b)_r$ and it shows that $(F : b)_r$ is a filter of $\mathfrak{B}$. □

Proposition 5.7. *Let $\mathfrak{A}$ be a residuated lattice and F be a filter of $\mathfrak{A}$. Then $\mathfrak{A}$ is a RS_F-residuated lattice if and only if for any filter $G \in Fi(\mathfrak{A})[F, A]$ we have $(F : G)_r$ is a filter of $\mathfrak{A}$.*

Proof. It is an immediate consequence of Proposition 4.9. □

Lemma 5.8. *Let $\mathfrak{A}$ be a RS_F-residuated lattice for some filter F of $\mathfrak{A}$. If G is an ordered-filter of $\mathfrak{A}$ containing F, then the following assertion holds:*

$$(F:G) = (F:G)_s = (F:G)_l = (F:G)_r.$$

Proof. The proof is similar to the proof of Corollary 4.11(1). □

Proposition 5.9. *Let $\mathfrak{A}$ be a RS_F-residuated lattice for some filter F of $\mathfrak{A}$. Then the following assertion holds:*

$$\mathscr{C}_{F_l F_r} = \mathscr{C}_{F_r F_l} = \mathscr{C}_{F_s F_s} = \{(F:G) | G \in F(\mathfrak{A})[F, A]\}.$$

Proof. By Lemma 5.8 it is obvious that $\{(F:G)|G \in F(\mathfrak{A})[F,A]\} = \{(F:G)_l | G \in F(\mathfrak{A})[F,A]\} \subseteq \mathscr{C}_{F_l F_r}$. Now, let $H = (F:X)_l$ for some $X \subseteq A$. By Proposition 4.3(1) we have $(F:(F:H)_r)_l = H$ and by Proposition 5.11 we have $(F:H)_r \in F(\mathfrak{A})[F,A]$. So by Lemma 5.8 we have $(F:(F:H)_r) = H$. It shows that $\mathscr{C}_{F_l F_r} \subseteq \{(F:G)|G \in F(\mathfrak{A})[F,A]\}$. Analogously, we can show that $\mathscr{C}_{F_r F_l} = \mathscr{C}_{F_s F_s} = \{(F:G)|G \in F(\mathfrak{A})[F,A]\}$. □

Theorem 5.10. *Let $\mathfrak{A}$ be a residuated lattice and F be a filter of $\mathfrak{A}$. Then $\mathfrak{A}$ is a RS_F-residuated lattice if and only if $(F:X)_r = (F:X)_l$ for any $X \subseteq A$.*

Proof. Let $\mathfrak{A}$ be a RS_F-residuated lattice and $a \in (F:X)_r$. We have $a \le (a \to x) \to x$ for any $x \in X$. Since $(F:X)_r$ is a filter so we obtain that $(a \to x) \to x \in (F:X)_r$. Also $x \le (a \to x) \to x$ for any $x \in X$ and it implies that $(a \to x) \to x \in Fi(X)$. Hence by Proposition 5.4(2) we conclude that $(a \to x) \to x \in (F:X)_r \cap Fi(X) \subseteq F$. It shows that for any $x \in X$ we have $(a \to x) \to x \in F$ and it states that $a \in (F:X)_l$. In a similar way we can show that $(F:X)_l \subseteq (F:X)_r$. It shows that $(F:X)_r = (F:X)_l$ for any $X \subseteq A$.

Conversely, if we have $(F:X)_r = (F:X)_l$ for any $X \subseteq A$ then $\mathfrak{A}$ is a RS_F-residuated lattice by Proposition 3.5. □

Definition 5.11. *Let $\mathfrak{A}$ be a residuated lattice. A filter F of $\mathfrak{A}$ will be called a right stabilizer filter of $\mathfrak{A}$ (RS filter) if $(F:x)_r = (F:x)_l$ for any $x \in A$.*

Proposition 5.12. *Let $\mathfrak{A}$ be a residuated lattice and F be a filter of $\mathfrak{A}$. The following assertions are equivalent:*

(1) *$\mathfrak{A}$ is a RS_F residuated lattice.*

(2) *F is a RS filter.*

(3) $(x \to y) \to y \in F$ implies $(y \to x) \to x \in F$, for any $x, y \in F$.

(4) $z, z \to ((y \to x) \to x) \in F$ implies $(x \to y) \to y \in F$, for any $x, y, z \in A$.

Proof.

(1)⇔(2): It is obvious by Theorem 5.10 and Definition 5.11.

(2)⇔(3): Let F be a RS filter of $\mathfrak{A}$ and $(x \to y) \to y \in F$ for arbitrary elements $x, y \in A$. So we have $x \in (F : y)_l = (F : y)_r$ and it implies that $(y \to x) \to x \in F$.

Conversely, let F satisfies (2) and $a \in (F : x)_r$. Then $(x \to a) \to a \in F$ and it implies that $(a \to x) \to x \in F$. It states that $a \in (F : x)_l$ and it shows that $(F : x)_r \subseteq (F : x)_l$. Analogously, we can show that $(F : x)_l \subseteq (F : x)_r$ and it means that F is a RS filter.

(3)⇔(4): See [1, Theorem 3.2].

□

REMARK 1. *According to [1, Theorem 3.10], each MV filter is a RS filter but the converse may be not true (See [1, Example 3.11]). Therefore each MV filter of a residuated lattice is a RS filter. Consequently, each Boolean filter of a residuated lattice is a RS filter, too.*

Corollary 5.13. *Let $\mathfrak{A}$ be a residuated lattice. The following assertions are equivalent:*

(1) $\mathfrak{A}$ *is a* RS_1 *residuated lattice.*

(2) *1 is a RS filter.*

(3) $x \to y = y$ *implies* $y \to x = x$, *for any* $x, y \in A$.

Proof. It is an immediate consequence of Proposition 5.12 by taking $F = 1$. □

Corollary 5.14. *The class of $\mathcal{RS} - \mathcal{RL}$ is a subquasivariety of the variety $\mathcal{RL}$.*

Proof. By Corollary 5.13, a residuated lattice $\mathfrak{A}$ is a RS residuated lattice if and only if it satisfies the quasi-identity $x \to y = y \Rightarrow y \to x = x$. It shows that $\mathcal{RS} - \mathcal{RL}$ can be be axiomatized by quasi-identities and it means that $\mathcal{RS} - \mathcal{RL}$ is a quasivariety. □

Theorem 5.15. *Let $\mathfrak{A}$ be a residuated lattice and F be a filter of $\mathfrak{A}$. If the complement of a filter G in the interval $\mathbf{Fi}(\mathfrak{A})[F, A] = (Fi(\mathfrak{A})[F, A]; \overline{\wedge}, \underline{\vee})$ is existed then $(F : G)_r \in Fi(\mathfrak{A})[F, A]$ and we have $G^c = (F : G)_r$.*

Proof. Let $G^c = H$. So we have $G \barwedge H = F$ and $G \veebar H = A$. By Proposition 4.10 follows that $H \subseteq (F : H)_r$. Let $a \in (F : H)_r$. So there are $g \in G$ and $h \in H$ such that $g \odot h \leq a$ and it means that $g \leq h \to a$. Since G is a filter so we have $h \to a \in G$. By hypothesis, it implies that $h \to a = ((h \to a) \to a) \to a \in F \subseteq H$ and it concludes that $a \in H$. It shows that $H = (F : G)_r$. $\square$

Corollary 5.16. *Let $\mathfrak{A}$ be a residuated lattice, F be a filter of $\mathfrak{A}$ and $X \subseteq A$. If the complement of F_X in the interval $\mathbf{Fi}(\mathfrak{A})[F, A]$ is existed then we have $F_X^c = (F : X)_r$.*

Proof. Let the complement of F_X in the interval $\mathbf{Fi}(\mathfrak{A})[F, A]$ is existed. By Theorem 5.15 we have $F_X^c = (F : F_X)_r$ and by Proposition 4.4(6) we have $(F : F_X)_r = (F : X)_r$. These show that $F_X^c = (F : X)_r$. $\square$

Corollary 5.17. *Let $\mathfrak{A}$ be a residuated lattice and F ba a filter of $\mathfrak{A}$. If the interval $\mathbf{Fi}(\mathfrak{A})[F, A]$ is a Boolean lattice, then F is a RS filter of $\mathfrak{A}$.*

Proof. Let $G \in Fi(\mathfrak{A})[F, A]$. By Theorem 5.15 we have $(F : G)_r \in Fi(\mathfrak{A})[F, A]$. So by Proposition 5.11 we conclude that $\mathfrak{A}$ is a RS_F residuated lattice and it states that F is a RS filter. $\square$

Theorem 5.18. *Let $\mathfrak{A}$ be a finite residuated lattice and F be a filter of $\mathfrak{A}$. Then F is a RS filter if and only if the interval $\mathbf{Fi}(\mathfrak{A})[F, A]$ is a Boolean lattice.*

Proof. Let F be a RS filter of $\mathfrak{A}$. Assume that $G \in \mathbf{Fi}(\mathfrak{A})[F, A]$. By Proposition 5.11 we obtain that $(F : G)_r$ is a filter of $\mathfrak{A}$ containing F and by Proposition 3.4(6) we have $G \barwedge (F : G)_r = G \cap (F : G)_r = F$. Hence, it is enough to prove that $G \veebar (F : G)_r = A$. Since $\mathfrak{A}$ is a finite residuated lattice, so $G = \{g_1, \cdots, g_n\}$ for some integer n. Let $a \in A$. By Proposition 2.2 we have $g_i \to a \leq g_i^2 \to a \leq \cdots \leq g_i^t \to a \leq \cdots$ for any $i = 1, \cdots, n$. Since A is a finite set so for any i there is an integer n_i such that $g_i^{n_i} \to a \leq \cdots \leq g_i^{n_i+k} \to a$ for any integer k. Set $x = (\odot_{i=1}^m g_i^{n_i}) \to a$. We have the following assertions:

$$\begin{aligned}
(g_j \to x) \to x &= (g_j \to ((\odot_{i=1}^m g_i^{n_i}) \to a)) \to x \\
&= ((g_j \odot (\odot_{i=1}^m g_i^{n_i})) \to a) \to x \\
&= ((g_j^{n_j+1} \odot (\odot_{i \neq j} g_i^{n_i})) \to a) \to x \\
&= ((\odot_{i \neq j} g_i^{n_i}) \to ((g_j^{n_j+1} \to a)) \to x \\
&= ((\odot_{i \neq j} g_i^{n_i}) \to ((g_j^{n_j} \to a)) \to x \\
&= ((g_j^{n_j} \odot (\odot_{i \neq j} g_i^{n_i}) \to a) \to x \\
&= ((\odot_{i=1}^m g_i^{n_i}) \to a) \to x \\
&= x \to x = 1 \in F
\end{aligned} \tag{1}$$

It shows that $(\odot_{i=1}^m g_i^{n_i}) \to a \in (F:G)_r$. We have $(\odot_{i=1}^m g_i^{n_i}) \odot ((\odot_{i=1}^m g_i^{n_i}) \to a) \leq a$ and it means that $a \in G \veebar (F:G)_r$. Hence the interval $\mathbf{Fi}(\mathfrak{A})[F,A]$ is a Boolean lattice. Conversely, it follows by Corollary 5.17. □

Theorem 5.19. *Let $\mathfrak{A}$ be a finite residuated lattice. Then the following statement are equivalent:*

(1) *$\mathfrak{A}$ is a RS_F residuated lattice.*

(2) *The interval $\mathbf{Fi}(\mathfrak{A})[F,A]$ is a Boolean lattice.*

Proof. It is straightforward by Theorem 5.15 and Theorem 5.18. □

References

[1] A. Ahadpanah, L. Torkzadeh, Normal filters in residuated filters, LE MATEMATICHE, Vol. LXX (2015), 81-92.

[2] A. Ahadpanah, L. Torkzadeh, Normal residuated lattices, Afr. Mat. (2015) 26:679Ű688 DOI 10.1007/s13370-014-0239-x.

[3] A. Borumand Saeid, N. Mohtashamnia, Stabilizer in residuated lattices, University Politehnica of Bucharest, Scientific Bulletin Series A - Applied Mathematics and Physics, 74(2), (2012), 65-74.

[4] R. A. Borzooei, A. Paad, Some new types of stabilizers in BL-algebras and their applications, Indian Journal of Science and Technology, 5(1) (2012) 1910-1915.

[5] W. J. Blok, D. Pigozzi, Algebraizable Logics, Mem. Am. Math. Soc., vol. 396, Amer. Math. Soc., Providence, 1989.

[6] S. Burris, H. P. Sankappanavar, 1981. A Course in Universal Algebra Springer-Verlag. ISBN 3-540-90578-2 Free online edition.

[7] D. Buşneag, D. Piciu, On the lattice of filters of a pseudo BL-algebra, Journal of Multiple Valued Logic and Soft Computing, vol. X (2006) 1-32.

[8] D. Buşneag, D. Piciu, Some types of filters in residuated lattices, Soft Comput. 18(5) (2014) 825-837.

[9] D. Buşneag, D. Piciu, A new approach for classification of filters in residuated lattices, Fuzzy Sets and Systems, 260 (2015) 121-130.

[10] F. Esteva, L. Godo, Monoidal t-norm based logic: towards a logic for left-continuous t-norms, Fuzzy Sets and Systems 124 (2001), 271-288.

[11] L. C. Ciungu, Classes of residuated lattices, Annals of University of Craiova. Math. Comp. Sci. Ser. 33 (2006) 189-207.

[12] L. C. Ciungu, Directly indecomposable residuated lattices, Iranian Journal of Fuzzy Systems Vol. 6, No. 2, (2009) 7-18.

[13] R. P. Dilworth, Abstract residuation over lattices, Bull. Amer. Math. Soc. 44 (1938) 262-268.

[14] R. P. Dilworth, Non-commutative residuated lattices, Trans. Amer. Math. Soc. 46 (1939) 426-444.

[15] A. Di Nola, G. Georgescu, A. Iorgulescu, Pseudo BL-algebras: Part I, Multiple Valued Logic, 8 (2002) 673-714.

[16] A. Dvurečenskij, J. Rachůnek, Probabilistic averaging in bounded $R\ell$-monoids, Semigroup Forum, 72 (2006) 190-206.

[17] P. Flondor, G. Georgescu, A. Iorgulescu, Pseudo t-norms and pseudo BL-algebras, Soft Comput. 5 (2001) 355-371.

[18] N. Galatos, P. Jipsen, T. Kowalski and H. Ono, Residuated lattices: an algebraic glimpse at substructural logics, Elsevier, 2007.

[19] G. Georgescu, A. Ioregulescu, Pseudo-MValgebras, Multiple Val. Logic, 6 (2001) 95-135.

[20] G. Grätzer, Lattice theory, W. H. Freeman and Company, San Francisco, (1979).

[21] M. Haveshki, M. Mohamadhasani, Stabilizer in BL-algebras and its properties. Int Math Forum 5(57) (2010), 2809-2816.

[22] M. Haveshki, Some Results on Stabilizers in Residuated Lattices, Çankaya University Journal of Science and Engineering 11(2) (2014) 7-17.

[23] U. Höhle, Commutative residuated monoids, in: U. Höhle, P. Klement (Eds.), Non-classical Logics and Their Aplications to Fuzzy Subsets, Kluwer Academic Publishers, 1995.

[24] P. Hájek, Metamathematics of fuzzy logic, Kluwer Acad. Publ., Dordrecht, 1998.

[25] M. Haveshki, Some results on stabilizers in residuated lattices, Çankaya University Journal of Science and Engineering Volume 11, 2 (2014) 7-17.

[26] P. M. Idziak, Lattice operations in BCK-algebras, Mathematica Japonica, 29(1984),

839-846.

[27] A. Iorgulescu, Classes of pseudo-BCK algebras *I*. Multiple-Valued Logic Soft Comput. 12 (2006) 71-130.

[28] K. Iséki, S. Tanaka, Ideal theory of BCK-algebras, Math. Jap. 21 (1976) 351-366.

[29] P. Jipsen, C. Tsinakis, A survey of residuated lattices, In: Ordered Algebraic Structures,(J.Martinez, ed) Kluwer Academic Publishers, Dordrecht, 2002, 19-56.

[30] T. Kowalski, H. Ono, Residuated lattices: an algebraic glimpse at logics without contraction, Japan Advanced Institute of Science and Technology, 2001.

[31] W. Krull, Axiomatische Begründung der allgemeinen Ideal theorie, Sitzungsberichte der physikalisch medizinischen Societĺad der Erlangen 56 (1924), 47-63.

[32] S. Motamed, L. Torkzadeh, A new class of BL-algebras, Soft Comput (2016). doi:10.1007/s00500-016-2043-z.

[33] M. Okada, K. Terui, The finite model property for various fragments of intuitionistic linear logic, Journal of Symbolic Logic, 64 (1999) 790-802.

[34] H. Ono, Y. Komori, Logics without the contraction rule, Journal of Symbolic Logic, 50 (1985), 169-201.

[35] S. Rasouli, Generalized co-annihilators in residuated lattices, submitted.

[36] B. Van Gasse, G. Deschrijver, C. Cornelis, E.E. Kerre, Filters of residuated lattices and triangle algebras, Inf. Sci. 180 (16) (2010) 3006-3020.

[37] M. Ward, Residuation in structures over which a multiplication is defined, Duke Math. Journal 3 (1937) 627-636.

[38] M. Ward, Structure Residuation, Annals of Mathematics, 2nd Ser. 39(3) (1938) 558-568.

[39] M. Ward, Residuated distributive lattices, Duke Math. J. 6 (1940) 641-651.

[40] M. Ward, R. P. Dilworth, Residuated Lattices, Proceedings of the National Academy of Sciences 24 (1938) 162-164.

[41] M. Ward, R. P. Dilworth, Residuated lattices, Transactions of the American Mathematical Society 45 (1939), 335-354.

[42] Y. Zhua, Y. Xu, On filter theory of residuated lattices, Information Sciences 180 (2010) 3614-3632.

Tuning the Program Transformers from LCC to PDL

Pere Pardo
Ruhr-Universität Bochum, Germany.
`pere.pardoventura@ruhr-uni-bochum.de`

Enrique Sarrión-Morillo, Fernando Soler-Toscano
Universidad de Sevilla, Spain.
`{esarrion,fsoler}@us.es`

Fernando R. Velázquez-Quesada
Universiteit van Amsterdam, The Netherlands.
`F.R.VelazquezQuesada@uva.nl`

Abstract

This work proposes an alternative definition of the so called program transformers used to obtain reduction axioms in the Logic of Communication and Change (LCC). Our proposal uses an elegant matrix treatment of Brzozowski's equational method instead of Kleene's translation from finite automata to regular expressions. The two alternatives are shown to be equivalent, with Brzozowski's method having the advantage of generating smaller expressions for models with average connectivity.

Keywords: Logic of communication and change, dynamic epistemic logic, propositional dynamic logic, action model, program transformer, reduction axiom

1 Introduction

Dynamic Epistemic Logic [1, 2] (*DEL*) encompasses several logical frameworks whose main aim is the study of different single- and multi-agent epistemic attitudes and the way they change due to diverse epistemic actions. These frameworks typically have two building blocks: a 'static' component, using some 'epistemic' model to represent the notion to be studied (e.g., knowledge or belief), and a 'dynamic' component, using model operations to represent actions that affect such notion (e.g., announcements or belief revision).[1]

Among the diverse existing *DEL* frameworks, the Logic of Communication and Change (LCC) of [6] stands as one of the most interesting. It consists of Propositional Dynamic Logic [7] (PDL), interpreted epistemically (its 'static' component), and the action models machinery [8, 9] for representing knowledge about actions (its 'dynamic' component). The LCC framework allows us to model not only diverse epistemic actions (as public, private or secret announcements) but also factual change.

A key feature of this logic is that it characterises the effect of an action model's execution via *reduction axioms*: valid formulas through which it is possible to rewrite a formula with action model (update) modalities as an equivalent one without them, thus reducing LCC to PDL and hence providing a compositional analysis for a wide range of informational events. For example, the reduction axiom for conjunction tells us that $\varphi \wedge \psi$ will be the case true after the pointed action model (U, e_i) is executed, $[U, e_i](\varphi \wedge \psi)$, if and only if both φ and ψ are true after executing such action, $[U, e_i]\varphi \wedge [U, e_i]\psi$. For another example, the reduction axiom for atoms p effectively reduces an LCC formula $[U, e_i]p$ into a formula about the conditions of the action e_i and its effect on p (see Table 1).

As one might expect, the crucial reduction axiom is the one characterising the effect of an action model over epistemic modalities π (i.e. over PDL programs):

$$[U, e_i][\pi]\varphi \leftrightarrow \bigwedge_{j=0}^{n-1} [T_{ij}^U(\pi)][U, e_j]\varphi$$

This axiom, presented in detail in what follows, characterises the epistemic change that the action model U brings about: after the pointed action model (U, e_i) is executed, every π-path in the resulting epistemic model leads to a φ-world, $[U, e_i][\pi]\varphi$, if and only if, for every action e_j in the action model U, every $T_{ij}^U(\pi)$-path in the original epistemic model ends in a world that, after the execution of (U, e_j), will satisfy φ. The axiom is based on the correspondence between action models and finite automata observed in [10]; its main component, the so-called program transformer function T_{ij}^U, follows Kleene's translation from finite automata to regular expressions [11].[2]

---

[1] This form of representing the dynamics is different from other approaches as, e.g., epistemic temporal logic [3, 4] (*ETL*), in which the static model already describes not only the relevant notion but also all the possible ways it can change due to the chosen epistemic action(s). See [5] for a comparison between *DEL* and *ETL*.

[2] See [12] for a deep discussion about the meaning of Kleene's theorem.

The present work proposes an alternative definition of program transformer, using instead a matrix treatment of Brzozowski's equational method for obtaining an expression representing the language accepted by a given finite automaton [13, 14].

Structure of the paper The paper starts in Section 2 by recalling the LCC framework together with its reduction axioms and its definition of program transformers. Section 3 explains how we can obtain, through Brzozowski's equational method, the corresponding expressions for Kleene closure, and then Section 4 introduces this paper's proposal, used to define an alternative translation from LCC to PDL. Section 5 comments on the computational complexity of this approach; the computational costs of the two methods are also compared using Prolog with different test-cases. Section 6 presents a summary and a discussion of further topics for research.

2 Logic of Communication and Change

This section recalls LCC's semantic structure, its language and semantic interpretation, and its axiom system. Throughout this paper, Var will denote a set of atoms (propositional variables), and Ag will denote a finite set of agents.

We start the definition of LCC by introducing the involved structures. First, the structure over which LCC formulas are interpreted.

Definition 1 (Epistemic model). An *epistemic model* M is a triple

$$(W, \langle R_a \rangle_{a \in \mathsf{Ag}}, V)$$

where $W \neq \varnothing$ is a set of worlds, $R_a \subseteq (W \times W)$ is an epistemic relation for each agent $a \in \mathsf{Ag}$ and $V : \mathsf{Var} \to \wp(W)$ is an atomic evaluation.

Note how the epistemic relations R_a are not required to satisfy any particular property. As usual, each possible world can be interpreted as a possible state of affairs (each one of them defined by the atomic valuation), and each relation R_a represents agent a's uncertainty about the situation: at world w, for agent a all worlds u such that wR_au are epistemically possible, i.e. are seen as possible by this agent. Figure 1 shows an example of an epistemic model.

Here is the structure for representing the knowledge about actions in the system.

Definition 2 (Action model). Let $\mathcal{L}$ be a language built upon Var and Ag that can be interpreted over epistemic models. An $\mathcal{L}$ *action model* U is a tuple

$$(\mathsf{E}, \langle \mathsf{R}_a \rangle_{a \in \mathsf{Ag}}, \mathsf{pre}, \mathsf{sub})$$

Figure 1: An epistemic model M with an actual p-world (gray) and a possible world with $\neg p$ (white). The arrows represent accessibility relations R_a, R_b and R_c, so only agent a knows that currently p, while b and c ignore whether p.

where $\mathsf{E} = \{\mathsf{e}_0, \ldots, \mathsf{e}_{n-1}\}$ is a *finite* non-empty set of actions, $\mathsf{R}_a \subseteq (\mathsf{E} \times \mathsf{E})$ is a relation for each $a \in \mathsf{Ag}$, $\mathsf{pre} : \mathsf{E} \to \mathcal{L}$ is a precondition map assigning a formula $\mathsf{pre}(\mathsf{e}) \in \mathcal{L}$ to each action $\mathsf{e} \in \mathsf{E}$, and $\mathsf{sub} : (\mathsf{E} \times \mathsf{Var}) \to \mathcal{L}$ is a postcondition map assigning a formula $\mathsf{sub}(\mathsf{e}, p) \in \mathcal{L}$ to each atom $p \in \mathsf{Var}$ at each action $\mathsf{e} \in \mathsf{E}$. The postcondition map should only change a finite number of atoms, so $\mathsf{sub}(\mathsf{e}, p) \neq p$ can hold only for a finite number of $p \in \mathsf{Var}$.[3] We emphasise that, in this definition, the language $\mathcal{L}$ is just a parameter.

Just as each relation R_a describes agent a's uncertainty about the situation, each relation R_a represents a's uncertainty about the executed action: $\mathsf{e}\mathsf{R}_a\mathsf{f}$ indicates a cannot distinguish f from e. Note, again, how the relation is not required to satisfy any particular property.

Example 1 (Announcements). Figure 2 illustrates three action models for announcements in a set of three agents $\mathsf{Ag} = \{a, b, c\}$. Each of the actions, say f, is purely epistemic (i.e., fact-preserving), so $\mathsf{sub}(\mathsf{f}, p) = p$ for any $p \in \mathsf{Var}$. Labeled arrows denote accessibility relations R_a, R_b or R_c; a gray circle denotes the action that is actually being executed, while other actions (wrongly believed by some agents to possibly take place) are represented by white circles. The preconditions are written below the corresponding actions.

As mentioned, action models represent both the actions and the knowledge agents have about these actions. Action models modify epistemic models in the following way.

Definition 3 (Update execution). Let $M = (W, \langle R_a \rangle_{a \in \mathsf{Ag}}, V)$ be an epistemic model and $\mathsf{U} = (\mathsf{E}, \langle \mathsf{R}_a \rangle_{a \in \mathsf{Ag}}, \mathsf{pre}, \mathsf{sub})$ an $\mathcal{L}$ action model, both over Var and Ag. Recall

---
[3]These 'finiteness' requirements (finite domain and only a finite number of atoms affected by the postcondition function) are needed to allow the pointed action model (U, e) —a pair with U an $\mathcal{L}$ action model and e a distinguished action in it— to be associated to a syntactic object and thus to be used within formulae. For details, the reader is referred to the discussion about action models in Section 6.1 of [1].

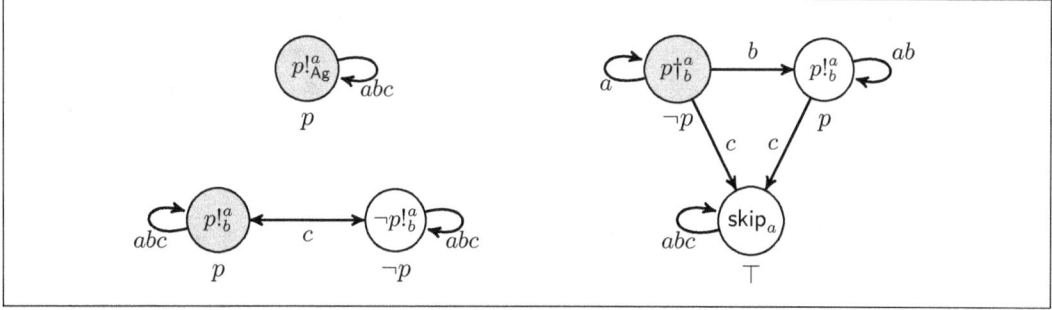

Figure 2: (Top left) A truthful public announcement by a that p, denoted $p!^a_{\text{Ag}}$. (Bottom left) A private announcement by a to b about p, denoted $p!^a_b$; here agent c only knows about the 'topic' of the message. (Right) A secret lie about p made by a to b, denoted $p\dagger^a_b$, is accepted by b as truthful, i.e. as if it was $p!^a_b$; agent c is not aware of any communication between a and b.

that $\mathcal{L}$ is any language built upon Var and Ag that can be interpreted over epistemic models, so we can assume the existence of a function $[\![\,\cdot\,]\!]^M$ returning those worlds in M in which each formula of $\mathcal{L}$ holds.

The update execution of U on M produces an epistemic model

$$(M \otimes \mathsf{U}) = (W^{M \otimes \mathsf{U}}, \langle R_a^{M \otimes \mathsf{U}} \rangle_{a \in \mathsf{Ag}}, V^{M \otimes \mathsf{U}})$$

given, for every $a \in \mathsf{Ag}$ and $p \in \mathsf{Var}$, by

$$W^{M \otimes \mathsf{U}} := \{\, (w, \mathsf{e}) \in W \times \mathsf{E} \mid w \in [\![\mathsf{pre}(\mathsf{e})]\!]^M \,\}$$

$$R_a^{M \otimes \mathsf{U}} := \{\, \langle (w, \mathsf{e}), (v, \mathsf{f}) \rangle \in W^{M \otimes \mathsf{U}} \times W^{M \otimes \mathsf{U}} \mid w R_a v \text{ and } \mathsf{e} R_a \mathsf{f} \,\}$$

$$V^{M \otimes \mathsf{U}}(p) := \{\, (w, \mathsf{e}) \in W^{M \otimes \mathsf{U}} \mid w \in [\![\mathsf{sub}(\mathsf{e}, p)]\!]^M \,\}$$

Thus, the update execution of U on M produces an epistemic model $M \otimes \mathsf{U}$ whose domain is the restricted cartesian product of the original models' domains.[4] In $M \otimes \mathsf{U}$, a world (w, e) satisfies an atom p if and only if w satisfied the formula $\mathsf{sub}(\mathsf{e}, p)$ in M; finally, an agent a sees a world (u, f) as possible from (w, e) if and only if she sees u from w (in M) and sees f from e (in U). If one works with a particular class of epistemic models in which the epistemic relations satisfy specific properties, then the chosen action models should be such that the update execution preserves these properties. This is straightforward in some cases as, e.g., reflexivity, transitivity and symmetry are preserved by update execution when the relations in

---

[4] If there is no world in M satisfying $\mathsf{pre}(\mathsf{e})$ for some action e in U, then the resulting structure is not an epistemic model, as its domain is empty.

the action models are reflexive, transitive and symmetric, respectively. But this is not always the case: for example, seriality is not preserved, even when the involved action models are serial. See Figure 3 for an illustration of different updates in an epistemic model.

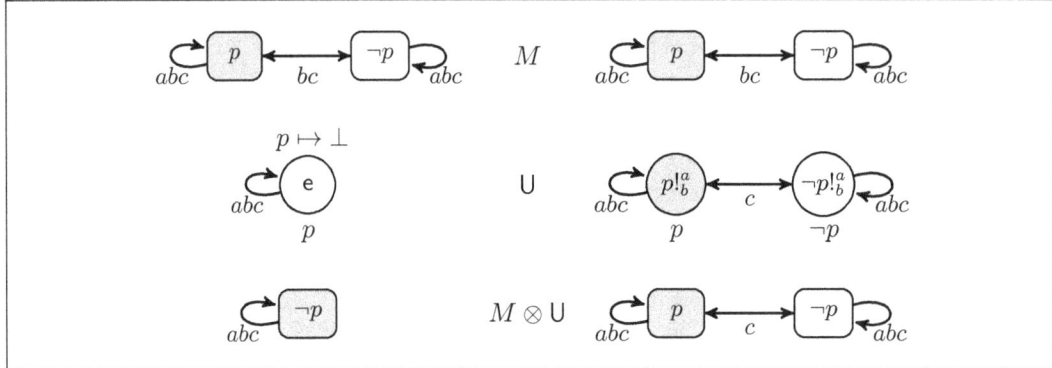

Figure 3: Two illustrations of update execution in the epistemic model M from Fig. 1 (topmost row here), where only agent a knows that p. (Left) Action e represents a public (i.e. publicly observable) change to $\neg p$; note that the postcondition is written on top of the action. After execution, it becomes common knowledge that $\neg p$. (Right) A private announcement by a to b about p results in a new model where it is public that b now knows whether p, and only c remains ignorant about p.

With the semantic structures already defined, it is time now to define the language that will be used to describe them. Note that the formulas (and programs) of the language $\mathcal{L}_{\mathsf{LCC}}$ are defined simultaneously with the notion of an $\mathcal{L}_{\mathsf{LCC}}$ action model (i.e. an action model using $\mathcal{L}_{\mathsf{LCC}}$ for its precondition and postcondition maps).

Definition 4 (Language $\mathcal{L}_{\mathsf{LCC}}$). The formulas φ and programs π of the language $\mathcal{L}_{\mathsf{LCC}}$ are given by, respectively:

$$\varphi ::= \top \mid p \mid \neg \varphi \mid \varphi \wedge \varphi \mid [\pi]\varphi \mid [\mathsf{U}, \mathsf{e}]\varphi$$
$$\pi ::= a \mid ?\varphi \mid \pi; \pi \mid \pi \cup \pi \mid \pi^*$$

where $p \in \mathsf{Var}$, $a \in \mathsf{Ag}$ and (U, e) is a pair with an $\mathcal{L}_{\mathsf{LCC}}$ action model U and an action e in this model.[5]

---
[5]More precisely, the language is defined by a double induction starting from the language $\mathsf{PDL}^0 = \mathsf{PDL}$, then defining $\mathcal{L}_{\mathsf{LCC}}^0$ as PDL^0 plus modalities of the form $[\mathsf{U}, \mathsf{e}]$ for U a PDL^0 action model, then defining PDL^1 as $\mathcal{L}_{\mathsf{LCC}}^0$ plus tests $?\varphi$ for $\varphi \in \mathcal{L}_{\mathsf{LCC}}^0$, then defining $\mathcal{L}_{\mathsf{LCC}}^1$ as PDL^1 plus modalities of the form $[\mathsf{U}, \mathsf{e}]$ for U a PDL^1 action model, and so on. The full language $\mathcal{L}_{\mathsf{LCC}}$ is then the union of all languages $\mathcal{L}_{\mathsf{LCC}}^i$ with i finite.

As the definition states, the set of LCC formulas contains the atomic propositions and $\top$, and it is closed under negation, conjunction, and modalities $[\pi]$ (for π a program) and $[\mathsf{U}, \mathsf{e}]$ (for U an $\mathcal{L}_{\mathsf{LCC}}$ action model and e an action in it).[6] On the other hand, the set of LCC programs contains basic programs for agents a and 'tests' $?\varphi$ (with φ a formula), and it is closed under sequential composition (;), non-deterministic choice ($\cup$) and Kleene closure (*).

It is only left to define the $[\![\cdot]\!]^M$ function associated to $\mathcal{L}_{\mathsf{LCC}}$ that collects the worlds of a given epistemic model M in which a given $\mathcal{L}_{\mathsf{LCC}}$ formula holds. In the case of LCC, this function also indicates which pairs of worlds are related by a given $\mathcal{L}_{\mathsf{LCC}}$ program.

Definition 5 (Semantics of $\mathcal{L}_{\mathsf{LCC}}$). Let $M = (W, \langle R_a \rangle_{a \in \mathsf{Ag}}, V)$ be an epistemic model and $\mathsf{U} = (\mathsf{E}, \langle R_a \rangle_{a \in \mathsf{Ag}}, \mathsf{pre}, \mathsf{sub})$ an action model. The function $[\![\cdot]\!]^M$, returning both those worlds in W in which an $\mathcal{L}_{\mathsf{LCC}}$ formula holds and those pairs in $W \times W$ in which an $\mathcal{L}_{\mathsf{LCC}}$ program holds, is given by

$$[\![\top]\!]^M := W \qquad\qquad [\![a]\!]^M := R_a$$
$$[\![p]\!]^M := V(p) \qquad\qquad [\![?\varphi]\!]^M := \mathsf{Id}_{[\![\varphi]\!]^M}$$
$$[\![\neg\varphi]\!]^M := W \setminus [\![\varphi]\!]^M \qquad\qquad [\![\pi_1; \pi_2]\!]^M := [\![\pi_1]\!]^M \circ [\![\pi_2]\!]^M$$
$$[\![\varphi_1 \wedge \varphi_2]\!]^M := [\![\varphi_1]\!]^M \cap [\![\varphi_2]\!]^M \qquad\qquad [\![\pi_1 \cup \pi_2]\!]^M := [\![\pi_1]\!]^M \cup [\![\pi_2]\!]^M$$
$$[\![[\pi]\varphi]\!]^M := \{w \in W \mid \forall v((w,v) \in [\![\pi]\!]^M \Rightarrow v \in [\![\varphi]\!]^M)\} \qquad [\![\pi^*]\!]^M := ([\![\pi]\!]^M)^*$$
$$[\![[\mathsf{U}, \mathsf{e}]\varphi]\!]^M := \{w \in W \mid w \in [\![\mathsf{pre}(\mathsf{e})]\!]^M \Rightarrow (w, \mathsf{e}) \in [\![\varphi]\!]^{M \otimes \mathsf{U}}\}$$

where $\circ$ and * are the composition and the reflexive transitive closure operator, respectively, and Id_U is the identity relation on $U \subseteq W$. Notice two special cases for test: $[\![?\bot]\!]^M = \varnothing$ and $[\![?\top]\!]^M = \mathsf{Id}_W$.

Even though LCC can be seen abstractly as the logic of regular programs (the PDL part) plus action models (modalities of the form $[\mathsf{U}, \mathsf{e}]$), it is also illustrative to discuss its epistemic interpretation, in particular, that of its PDL programs. Basic 'agent' programs $a \in \mathsf{Ag}$ produce formulas of the form $[a]\varphi$, read simply as *"agent a knows/believes φ"* as in standard Epistemic Logic. More complex programs also have epistemic readings. Formulas of the form $[\pi_1; \pi_2]\varphi$, relying on the sequential composition π_1 and then π_2, can be read as *"π_1 knows/believes that π_2 knows/believes φ"*, and thus can be used to express *nested* knowledge/belief; formulas of the form $[\pi_1 \cup \pi_2]\varphi$, relying on the union of the relations for π_1 and π_2, can be read as *"both π_1 and π_2 know/believe φ"*, and thus can be used to express *general* knowledge/belief among a group; finally, formulas of the form $[\pi^*]\varphi$, relying on the reflexive and

---

[6] From now on, all action models are assumed to be $\mathcal{L}_{\mathsf{LCC}}$ action models.

transitive closure of the relations for π, can be read as "φ *is the case,* π *knows it,* π *knows that she knows it, and so on*", and thus can be used to express *common knowledge* (or, if $\pi^+ := \pi; \pi^*$ is used instead of π^*, *common* belief). The modalities involving action models simply state the action's effects, with formulas of the form $[\mathsf{U}, \mathsf{e}]\varphi$ reading "φ *is the case after any execution of the pointed action* (U, e)".

Axiom system The axiom system for LCC, shown in Table 1, combines the known axiom system of its PDL fragment (the left column; [7]) with *recursion* axioms for its action model fragment (the right column). Intuitively, recursion axioms are valid formulae characterising a situation *after* an update execution in terms of a situation *before* such update, and thus indicating how to rewrite a formula with an action model modality as a provably equivalent one without them. Then, while soundness follows from the validity of these new axioms, completeness follows from the completeness of the basic system.[7]

(*taut*) propositional tautologies	(*top*) $[\mathsf{U}, \mathsf{e}]\top \leftrightarrow \top$
(*K*) $[\pi](\varphi_1 \to \varphi_2) \to ([\pi]\varphi_1 \to [\pi]\varphi_2)$	(*atm*) $[\mathsf{U}, \mathsf{e}]p \leftrightarrow (\mathsf{pre}(\mathsf{e}) \to \mathsf{sub}(\mathsf{e}, p))$
(*test*) $[?\varphi_1]\varphi_2 \leftrightarrow (\varphi_1 \to \varphi_2)$	(*neg*) $[\mathsf{U}, \mathsf{e}]\neg\varphi \leftrightarrow (\mathsf{pre}(\mathsf{e}) \to \neg[\mathsf{U}, \mathsf{e}]\varphi)$
(*seq*) $[\pi_1; \pi_2]\varphi \leftrightarrow [\pi_1][\pi_2]\varphi$	(*conj*) $[\mathsf{U}, \mathsf{e}](\varphi_1 \wedge \varphi_2) \leftrightarrow ([\mathsf{U}, \mathsf{e}]\varphi_1 \wedge [\mathsf{U}, \mathsf{e}]\varphi_2)$
(*choice*) $[\pi_1 \cup \pi_2]\varphi \leftrightarrow [\pi_1]\varphi \wedge [\pi_2]\varphi$	(K_U) $[\mathsf{U}, \mathsf{e}](\varphi_1 \to \varphi_2) \to ([\mathsf{U}, \mathsf{e}]\varphi_1 \to [\mathsf{U}, \mathsf{e}]\varphi_2)$
(*mix*) $[\pi^*]\varphi \leftrightarrow \varphi \wedge [\pi][\pi^*]\varphi$	(*prog*) $[\mathsf{U}, \mathsf{e}_i][\pi]\varphi \leftrightarrow \bigwedge_{j=0}^{n-1}[T_{ij}^\mathsf{U}(\pi)][\mathsf{U}, \mathsf{e}_j]\varphi$
(*ind*) $\varphi \wedge [\pi^*](\varphi \to [\pi]\varphi)) \to [\pi^*]\varphi$	(N_U) From $\vdash \varphi$ infer $\vdash [\mathsf{U}, \mathsf{e}]\varphi$
(*MP*) From $\vdash \varphi_1$ and $\vdash \varphi_1 \to \varphi_2$ infer $\vdash \varphi_2$	
(N_π) From $\vdash \varphi$ infer $\vdash [\pi]\varphi$	

Table 1: LCC calculus in [6] is that of PDL (left column) plus reduction axioms and necessitation rule for $[\mathsf{U}, \mathsf{e}]$ (right column).

In our particular case, recursion axioms for atomic propositions and boolean constants/operators are standard for action models with ontic (i.e., valuation) change [16]: while axiom (*atm*) states that an atom p will be the case *after* any update execution with action model U and action e, $[\mathsf{U}, \mathsf{e}]p$, if and only if, *before* the update, the formula $\mathsf{sub}(\mathsf{e}, p)$ holds whenever $\mathsf{pre}(\mathsf{e})$ holds, $\mathsf{pre}(\mathsf{e}) \to \mathsf{sub}(\mathsf{e}, p)$, axioms (*neg*) and (*conj*) state that update execution commutes with negation (modulo its precondition) and distributes over conjunction, respectively.

---

[7]The reader is referred to Chapter 7 of [1] (see also [15]) for an extensive explanation of this technique.

The most important recursion axiom, (*prog*), characterises the effect of an action model over LCC programs. It states that after any update execution with U on e_i every π-path in the resulting model will lead to a φ-world, $[U, e_i][\pi]\varphi$, if and only if, before the update, every $T_{ij}^U(\pi)$-path leads to a world that will satisfy φ after any update execution with U on e_j where e_j is any action on U, $\bigwedge_{j=0}^{n-1}[T_{ij}^U(\pi)][U, e_j]\varphi$. In this axiom, the program transformer T_{ij}^U is crucial, taking an LCC program π representing a path on $M \otimes U$ and returning an LCC program $T_{ij}^U(\pi)$ representing a 'matching' path on M, taking additional care that such path can be also reproduced in the action model U. A program transformer follows Kleene's translation from finite automata to regular expressions [11], and it is formally defined as follows.

Note also that the (valid) formula K_U is not listed among the LCC axioms in [6]. It has been added here not only because it cannot be derived from the rest of the system, but also because it allows the derivation of the crucial rule

$$\frac{\chi \leftrightarrow \psi}{[U, e]\chi \leftrightarrow [U, e]\psi} \; RE_U$$

This rule is needed for the *inside-out* translation of nested action model modalities (see footnote 10). The fact that K_U is not derivable from the rest of the system is stated in [15] (in particular, its Thm. 29), a paper which examines axiom systems for PAL, the logic of public announcements [φ!]. Their analysis of completeness proofs is based on a reduction from PAL, and thus it applies to LCC as well. (Of course, another alternative is to add RE_U directly since, following [15, Prop. 3], K_U is derivable from RE_U and the original LCC system, Table 1 minus K_U.) That the system on Table 1 is indeed sound and (weakly) complete w.r.t. the given semantic interpretation can be shown using the same technique as [15, Corollary 12].

Definition 6 (Program transformer [6]). Let $U = (E, \langle R_a \rangle_{a \in Ag}, \text{pre}, \text{sub})$ be an action model with $E = \{e_0, \ldots, e_{n-1}\}$. The *program transformer* T_{ij}^U ($i, j \in \{0, \ldots, n-1\}$) on the set of LCC programs is defined as:

$$T_{ij}^U(a) := \begin{cases} ?\text{pre}(e_i); a & \text{if } e_i R_a e_j \\ ?\bot & \text{otherwise} \end{cases} \qquad T_{ij}^U(?\varphi) := \begin{cases} ?(\text{pre}(e_i) \wedge [U, e_i]\varphi) & \text{if } i = j \\ ?\bot & \text{otherwise} \end{cases}$$

$$T_{ij}^U(\pi_1; \pi_2) := \bigcup_{k=0}^{n-1}(T_{ik}^U(\pi_1); T_{kj}^U(\pi_2)) \qquad T_{ij}^U(\pi_1 \cup \pi_2) := T_{ij}^U(\pi_1) \cup T_{ij}^U(\pi_2)$$

$$T_{ij}^U(\pi^*) := K_{ijn}^U(\pi)$$

with K_{ijn}^U inductively defined as follows:

$$K^{\mathsf{U}}_{ij0}(\pi) := \begin{cases} ?\top \cup T^{\mathsf{U}}_{ij}(\pi) & \text{if } i = j \\ T^{\mathsf{U}}_{ij}(\pi) & \text{otherwise} \end{cases}$$

$$K^{\mathsf{U}}_{ij(k+1)}(\pi) = \begin{cases} (K^{\mathsf{U}}_{kkk}(\pi))^* & \text{if } i = k = j \\ (K^{\mathsf{U}}_{kkk}(\pi))^*; K^{\mathsf{U}}_{kjk}(\pi) & \text{if } i = k \neq j \\ K^{\mathsf{U}}_{ikk}(\pi); (K^{\mathsf{U}}_{kkk}(\pi))^* & \text{if } i \neq k = j \\ K^{\mathsf{U}}_{ijk}(\pi) \cup (K^{\mathsf{U}}_{ikk}(\pi); (K^{\mathsf{U}}_{kkk}(\pi))^*; K^{\mathsf{U}}_{kjk}(\pi)) & \text{if } i \neq k \neq j \end{cases}$$

Example 2. In the action model of Fig. 3 (left), the axiom for the *public change to* $\neg p$ reduces an epistemic consequence $[\mathsf{U}, \mathsf{e}][a]\neg p$ to a claim before execution, namely $[?\mathsf{pre}(\mathsf{e}); a][\mathsf{U}, \mathsf{e}]\neg p$, which is necessarily true –see the left column below–. Similarly, in the action model of a private lying announcement Fig. 2 (right), enumerate the actions as $p\dagger^a_b = \mathsf{e}_0$ and $p!^a_b = \mathsf{e}_1$ and $\mathsf{skip}_a = \mathsf{e}_2$. Then, the axiom for the lying announcement $p\dagger^a_b$ turns the *believed lie* $[\mathsf{U}, p\dagger^a_b][b]p$ into a claim before the execution, also a tautology –see the right column–.

$$\begin{array}{rcl}
& & [\mathsf{U}, \mathsf{e}][a]\neg p \\
& \equiv & [?\mathsf{pre}(\mathsf{e}); a][\mathsf{U}, \mathsf{e}]\neg p \\
& \equiv & [?p; a]\big(\mathsf{pre}(\mathsf{e}) \to \neg[\mathsf{U}, \mathsf{e}]p\big) \\
& \equiv & p \to [a]\big(p \to \neg(\mathsf{pre}(\mathsf{e}) \to \mathsf{sub}(\mathsf{e}, p))\big) \\
& \equiv & p \to [a]\big(p \to \neg(p \to \bot)\big) \\
& \equiv & p \to [a]\big(p \to (p \wedge \top)\big) \\
& \equiv & p \to [a]\top \quad \equiv \quad \top
\end{array}$$

$$\begin{array}{rcl}
& & [\mathsf{U}, p\dagger^a_b][b]p \\
& \equiv & [T^{\mathsf{U}}_{01}(b)][\mathsf{U}, p!^a_b]p \\
& \equiv & [?\mathsf{pre}(p\dagger^a_b); b][\mathsf{U}, p!^a_b]p \\
& \equiv & [?\neg p; b]\big(\mathsf{pre}(p!^a_b) \to \mathsf{sub}(p!^a_b, p)\big) \\
& \equiv & \neg p \to [b]\big(p \to p\big) \\
& \equiv & \neg p \to [b]\top \\
& \equiv & \top
\end{array}$$

3 Program transformation through Brzozowski's equations

This paper proposes an alternative definition of program transformer, denoted $\mu^{\mathsf{U}}(\pi)[i,j]$, that differs from $T^{\mathsf{U}}_{ij}(\pi)$ mainly in the case for the Kleene closure operator. Before presenting the formal definitions in Section 4, we introduce the method in an informal way. In the action models of Figure 4, we tag every edge from e_i to e_j with a label $\pi \mid \mu^{\mathsf{U}}(\pi)[i,j]$ with π a program and and $\mu^{\mathsf{U}}(\pi)[i,j]$ its transformation. For example, in the agents' diagram below, the label from e_0 to e_1

$$a \mid ?\mathsf{pre}(\mathsf{e}_0); a \quad \text{means} \quad \mu^{\mathsf{U}}(a)[0,1] = ?\mathsf{pre}(\mathsf{e}_0); a.$$

i.e. $?\mathsf{pre}(\mathsf{e}_0); a$ is what we should test in (M, w) to ensure that, after executing $(\mathsf{U}, \mathsf{e}_0)$ over (M, w), an a-path from (w, e_0) to some state (w', e_1) will persist in $M \otimes \mathsf{U}$. (If no a-path from e_0 to e_1 exists, the transformation of a is $?\bot$.)

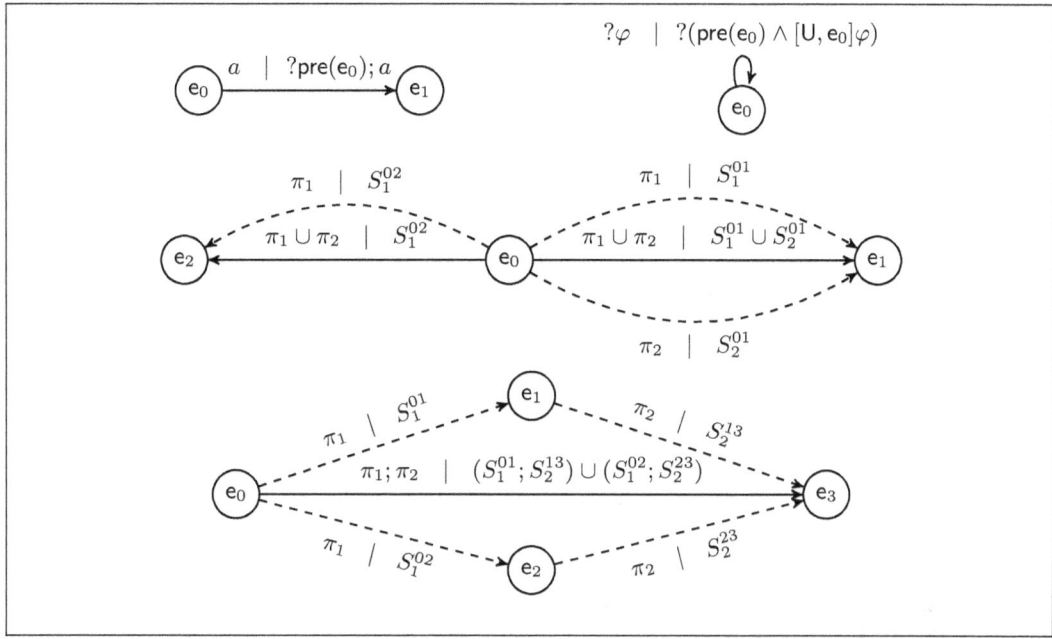

Figure 4: An illustration of action models and their program transformers for the following programs: *agent* (top left), *test* (top right), *choice* (mid) and *composition* (bottom). Dashed and solid lines represent, respectively, the original labels and those obtained after applying choice (mid) or product (bottom).

The construction of the diagrams on Figure 4 proceeds in a very similar way to that of Def. 6, just simplifying some trivial cases like $\pi \cup ?\bot$, which is reduced to π. The main novelty of our transformation is for the Kleene closure. We use a method proposed by Brzozowski [13], presented here in a matrix format (see [17, 18] for more an in-depth analysis about the improvements that we are applying to LCC language).

Kleene closure The following example will be used to illustrate the generation of the transformations of π^* from those of π. (The π^*-paths from e_i to e_j will be denoted by X^{ij}, while the corresponding π-paths are labeled as S^{ij}.)

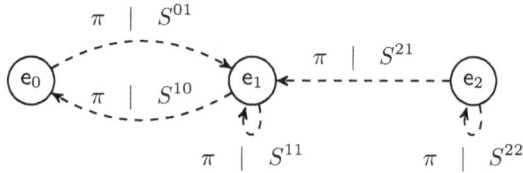

Generate an equation system[8] (Brzozowski [13]). E.g. for paths to e_0:

$$X^{00} = ?\mathsf{pre}(e_0) \cup (S^{01}; X^{10}) \tag{1}$$
$$X^{10} = (S^{10}; X^{00}) \cup (S^{11}; X^{10}) \tag{2}$$
$$X^{20} = (S^{22}; X^{20}) \cup (S^{21}; X^{10}) \tag{3}$$

Solve the system[9] using: substitution; associativity, distributivity [19], and Arden's Theorem [20] ($X = B \cup (A; X)$ implies $X = A^*; B$). E.g.

$$X^{00} = ?\mathsf{pre}(e_0) \cup (S^{01}; ((S^{10}; S^{01}) \cup S^{11})^*; S^{10}; ?\mathsf{pre}(e_0)) \tag{4}$$
$$X^{10} = ((S^{10}; S^{01}) \cup S^{11})^*; S^{10}; ?\mathsf{pre}(e_0) \tag{5}$$
$$X^{20} = (S^{22})^*; S^{21}; ((S^{10}; S^{01}) \cup S^{11})^*; S^{10}; ?\mathsf{pre}(e_0) \tag{6}$$

Similar processes produce labels for π^*-paths to e_1 and e_2, represented as:

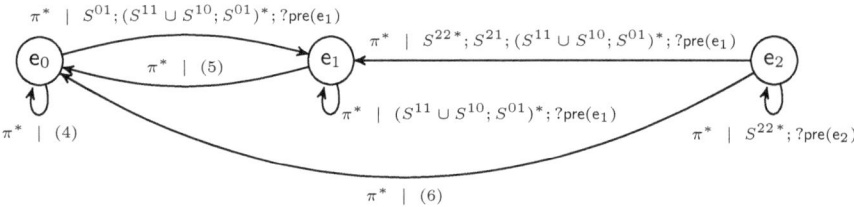

By using a matrix calculus similar to that in Chapter 3 of [14] we calculate all X^{ij} in parallel and thus avoid repeating the process for each destination node. The following section presents the formal definition of the matrix calculus; here we just illustrate the use of the matrix calculus. The equations (1)–(3) used above can be represented in the following matrix:

---

[8] For equation (2), observe how a π^*-path from e_1 to e_0 might start with S^{10} and then continue with X^{00} (an instance of π^* from e_0 to e_0), but it might also start with S^{11} and then continue with X^{10}. In equation (1), a π^*-path from e_0 to e_0 is to do nothing, but then the transformation should check $?\mathsf{pre}(e_0)$, i.e. whether e_0 is executable at the target state.

[9] We illustrate first how equation (2) is solved into (5):

$$\begin{aligned}
X^{10} &= (S^{10}; (?\mathsf{pre}(e_0) \cup (S^{01}; X^{10}))) \cup (S^{11}; X^{10}) &&\text{(substitute } X^{00} \text{ using (1))} \\
&= (S^{10}; ?\mathsf{pre}(e_0)) \cup (S^{10}; S^{01}; X^{10}) \cup (S^{11}; X^{10}) &&\text{(distributivity)} \\
&= (S^{10}; ?\mathsf{pre}(e_0)) \cup (((S^{10}; S^{01}) \cup S^{11}); X^{10}) &&\text{(associativity)} \\
&= ((S^{10}; S^{01}) \cup S^{11})^*; S^{10}; ?\mathsf{pre}(e_0) &&\text{(Arden's Theorem)}
\end{aligned}$$

Next, we use this to substitute X^{10} in (1) to obtain (4). Finally, we substitute X^{10} in (3) and apply Arden's Theorem to obtain (6).

	e_0	e_1	e_2	e_0	e_1	e_2
e_0	?⊥	S^{01}	?⊥	?pre(e_0)	?⊥	?⊥
e_1	S^{10}	S^{11}	?⊥	?⊥	?pre(e_1)	?⊥
e_2	?⊥	S^{21}	S^{22}	?⊥	?⊥	?pre(e_2)

The left part contains the π-paths from one node (row) to another one (column). It is an accessibility matrix for the π-graph above. Call $\mu^{\cup}(\pi)[i,j]$ the cell corresponding to row e_i and column e_j in this left part and $A^{\cup}[i,j]$ the cell with the same position at the right part. Observe that $A^{\cup}[i,j] =$?pre(e_i) if $i = j$ and ?⊥ otherwise. We may check that the equations for X^{ij} that we created above looking at the π-graph can be created now by:

$$X^{ij} = (\mu^{\cup}(\pi)[i,0]; X^{0j}) \cup (\mu^{\cup}(\pi)[i,1]; X^{1j}) \cup (\mu^{\cup}(\pi)[i,2]; X^{2j}) \cup A^{\cup}[i,j] \quad (7)$$

For example, the equations for X^{10} and X^{00} (equivalent to (2) and (1), resp.) are

$$X^{10} = (S^{10}; X^{00}) \cup (S^{11}; X^{10}) \cup (?\bot; X^{20}) \cup ?\bot \quad (8)$$
$$X^{00} = (?\bot; X^{00}) \cup (S^{01}; X^{10}) \cup (?\bot; X^{20}) \cup ?\text{pre}(e_0) \quad (9)$$

The greatest advantage of working with matrices is that we can perform several operations in parallel by working in a row. Applying Arden's Theorem to the e_1 row of the previous matrix gives:

	e_0	e_1	e_2	e_0	e_1	e_2
e_1	$(S^{11})^*; S^{10}$	?⊥	$(S^{11})^*; ?\bot$	$(S^{11})^*; ?\bot$	$(S^{11})^*; ?\text{pre}(e_1)$	$(S^{11})^*; ?\bot$

We replaced the left cell [e_1, e_1] with ?⊥ and concatenated its previous value $(S^{11})^*$ with the others cells in the row. After simplifying into ?⊥ cells we get:

	e_0	e_1	e_2	e_0	e_1	e_2
e_1	$(S^{11})^*; S^{10}$	?⊥	?⊥	?⊥	$(S^{11})^*; ?\text{pre}(e_1)$	?⊥

To check that we have applied Arden's Theorem, look at X^{10} (using (7) in the last matrix): $X^{10} = (S^{11})^*; S^{10}; X^{00}$. It is the result of applying Arden's Theorem to (8) (or (2)). Substitution can also be done in parallel:

	e_0	e_1	e_2	e_0	e_1	e_2
e_2	$(S^{21}; (S^{11})^*; S^{10})$ $\cup\, ?\bot$	?⊥	$(S^{21}; ?\bot)$ $\cup\, S^{22}$	$(S^{21}; ?\bot)$ $\cup\, ?\bot$	$(S^{21}; (S^{11})^*; ?\text{pre}(e_1))$ $\cup\, ?\bot$	$(S^{21}; ?\bot)$ $\cup\, ?\text{pre}(e_2)$

The above row for e_2 was obtained from the previous row by applying the following substitution into the original matrix: first, the left position $B = [e_2, e_1]$ (S^{21} in this case) is replaced with ?⊥; second, every other (left/right) position $D = [e_2, e_i]$ contains now a program with the form $(B; C) \cup D$, where C is the program in the (resp. left/right) $[e_1, e_i]$ position in the previous row for e_1. After simplifying into $?\bot$ cells (where appropriate) we obtain:

	e_0	e_1	e_2	e_0	e_1	e_2
e_2	$(S^{21};(S^{11})^*;S^{10})$	$?\bot$	S^{22}	$?\bot$	$(S^{21};(S^{11})^*;?\mathsf{pre}(e_1))$	$?\mathsf{pre}(e_2)$

To illustrate that we have done a substitution, consider the value of X^{21} in the matrix before the substitution (just as in the initial matrix):

$$X^{21} = (S^{21}; X^{11}) \cup (S^{22}; X^{21}) \tag{10}$$

And now consider the value of X^{11} after the application of Arden's Theorem:

$$X^{11} = ((S^{11})^*; S^{10}; X^{01}) \cup ((S^{11})^*; ?\mathsf{pre}(e_1)) \tag{11}$$

Using (11) to substitute X^{11} in (10) we get:

$$X^{21} = \left(S^{21}; \left(((S^{11})^*; S^{10}; X^{01}) \cup ((S^{11})^*; ?\mathsf{pre}(e_1))\right)\right) \cup (S^{22}; X^{21}) \tag{12}$$

that can be rewritten, using the distributive and associative properties, into:

$$X^{21} = (S^{21}; (S^{11})^*; S^{10}; X^{01}) \cup (S^{22}; X^{21}) \cup (S^{21}; (S^{11})^*; ?\mathsf{pre}(e_1)) \tag{13}$$

which is the equation obtained for X^{21} in the previous matrix, after the substitution.

In the following section we introduce the formal definitions of our matrix calculus to transform LCC programs.

4 A matrix calculus for program transformation

Definition 7 (Program transformation matrix). Let $\mathsf{U} = (\mathsf{E}, \mathsf{R}, \mathsf{pre}, \mathsf{sub})$ be an action model with $\mathsf{E} = \{e_0, \ldots, e_{n-1}\}$. The function $\mu^{\mathsf{U}} : \Pi \to \mathcal{M}_{n \times n}$, with Π the set of LCC programs and $\mathcal{M}_{n \times n}$ the class of n-square matrices, takes an LCC program π and returns a n-square matrix $\mu^{\mathsf{U}}(\pi)$ in which each cell $\mu^{\mathsf{U}}(\pi)[i, j]$ is an LCC program representing the transformation of π from e_i to e_j in the sense of the program transformers $T_{ij}^{\mathsf{U}}(\pi)$ of [6]. The recursive definition of $\mu^{\mathsf{U}}(\pi)$ is as follows.

- *Agents*:

$$\mu^{\mathsf{U}}(a)[i,j] := \begin{cases} ?\mathsf{pre}(e_i); a & \text{if } e_i \mathsf{R}_a e_j \\ ?\bot & \text{otherwise} \end{cases} \tag{14}$$

Tuning the Program Transformers from LCC to PDL

	e_0	e_1	e_2	e_0	e_1	e_2
e_0	$?\bot$	S^{01}	$?\bot$	$?\mathsf{pre}(e_0)$	$?\bot$	$?\bot$
e_1	S^{10}	S^{11}	$?\bot$	$?\bot$	$?\mathsf{pre}(e_1)$	$?\bot$
e_2	$?\bot$	S^{21}	S^{22}	$?\bot$	$?\bot$	$?\mathsf{pre}(e_2)$

The left part contains the π-paths from one node (row) to another one (column). It is an accessibility matrix for the π-graph above. Call $\mu^{\mathsf{U}}(\pi)[i,j]$ the cell corresponding to row e_i and column e_j in this left part and $A^{\mathsf{U}}[i,j]$ the cell with the same position at the right part. Observe that $A^{\mathsf{U}}[i,j] = ?\mathsf{pre}(e_i)$ if $i = j$ and $?\bot$ otherwise. We may check that the equations for X^{ij} that we created above looking at the π-graph can be created now by:

$$X^{ij} = (\mu^{\mathsf{U}}(\pi)[i,0]; X^{0j}) \cup (\mu^{\mathsf{U}}(\pi)[i,1]; X^{1j}) \cup (\mu^{\mathsf{U}}(\pi)[i,2]; X^{2j}) \cup A^{\mathsf{U}}[i,j] \qquad (7)$$

For example, the equations for X^{10} and X^{00} (equivalent to (2) and (1), resp.) are

$$X^{10} = (S^{10}; X^{00}) \cup (S^{11}; X^{10}) \cup (?\bot; X^{20}) \cup ?\bot \qquad (8)$$
$$X^{00} = (?\bot; X^{00}) \cup (S^{01}; X^{10}) \cup (?\bot; X^{20}) \cup ?\mathsf{pre}(e_0) \qquad (9)$$

The greatest advantage of working with matrices is that we can perform several operations in parallel by working in a row. Applying Arden's Theorem to the e_1 row of the previous matrix gives:

	e_0	e_1	e_2	e_0	e_1	e_2
e_1	$(S^{11})^*; S^{10}$	$?\bot$	$(S^{11})^*; ?\bot$	$(S^{11})^*; ?\bot$	$(S^{11})^*; ?\mathsf{pre}(e_1)$	$(S^{11})^*; ?\bot$

We replaced the left cell $[e_1, e_1]$ with $?\bot$ and concatenated its previous value $(S^{11})^*$ with the others cells in the row. After simplifying into $?\bot$ cells we get:

	e_0	e_1	e_2	e_0	e_1	e_2
e_1	$(S^{11})^*; S^{10}$	$?\bot$	$?\bot$	$?\bot$	$(S^{11})^*; ?\mathsf{pre}(e_1)$	$?\bot$

To check that we have applied Arden's Theorem, look at X^{10} (using (7) in the last matrix): $X^{10} = (S^{11})^*; S^{10}; X^{00}$. It is the result of applying Arden's Theorem to (8) (or (2)). Substitution can also be done in parallel:

	e_0	e_1	e_2	e_0	e_1	e_2
e_2	$(S^{21}; (S^{11})^*; S^{10}) \cup ?\bot$	$?\bot$	$(S^{21}; ?\bot) \cup S^{22}$	$(S^{21}; ?\bot) \cup ?\bot$	$(S^{21}; (S^{11})^*; ?\mathsf{pre}(e_1)) \cup ?\bot$	$(S^{21}; ?\bot) \cup ?\mathsf{pre}(e_2)$

The above row for e_2 was obtained from the previous row by applying the following substitution into the original matrix: first, the left position $B = [e_2, e_1]$ (S^{21} in this case) is replaced with $?\bot$; second, every other (left/right) position $D = [e_2, e_i]$ contains now a program with the form $(B; C) \cup D$, where C is the program in the

83

(resp. left/right) $[e_1, e_i]$ position in the previous row for e_1. After simplifying into $?\bot$ cells (where appropriate) we obtain:

	e_0	e_1	e_2	e_0	e_1	e_2
e_2	$(S^{21};(S^{11})^*;S^{10})$	$?\bot$	S^{22}	$?\bot$	$(S^{21};(S^{11})^*;?\mathsf{pre}(e_1))$	$?\mathsf{pre}(e_2)$

To illustrate that we have done a substitution, consider the value of X^{21} in the matrix before the substitution (just as in the initial matrix):

$$X^{21} = (S^{21}; X^{11}) \cup (S^{22}; X^{21}) \qquad (10)$$

And now consider the value of X^{11} after the application of Arden's Theorem:

$$X^{11} = ((S^{11})^*; S^{10}; X^{01}) \cup ((S^{11})^*; ?\mathsf{pre}(e_1)) \qquad (11)$$

Using (11) to substitute X^{11} in (10) we get:

$$X^{21} = \left(S^{21}; \left(((S^{11})^*; S^{10}; X^{01}) \cup ((S^{11})^*; ?\mathsf{pre}(e_1))\right)\right) \cup (S^{22}; X^{21}) \qquad (12)$$

that can be rewritten, using the distributive and associative properties, into:

$$X^{21} = (S^{21}; (S^{11})^*; S^{10}; X^{01}) \cup (S^{22}; X^{21}) \cup (S^{21}; (S^{11})^*; ?\mathsf{pre}(e_1)) \qquad (13)$$

which is the equation obtained for X^{21} in the previous matrix, after the substitution.

In the following section we introduce the formal definitions of our matrix calculus to transform LCC programs.

4 A matrix calculus for program transformation

Definition 7 (Program transformation matrix). Let $\mathsf{U} = (\mathsf{E}, \mathsf{R}, \mathsf{pre}, \mathsf{sub})$ be an action model with $\mathsf{E} = \{e_0, \ldots, e_{n-1}\}$. The function $\mu^\mathsf{U} : \Pi \to \mathcal{M}_{n \times n}$, with Π the set of LCC programs and $\mathcal{M}_{n \times n}$ the class of n-square matrices, takes an LCC program π and returns a n-square matrix $\mu^\mathsf{U}(\pi)$ in which each cell $\mu^\mathsf{U}(\pi)[i,j]$ is an LCC program representing the transformation of π from e_i to e_j in the sense of the program transformers $T^\mathsf{U}_{ij}(\pi)$ of [6]. The recursive definition of $\mu^\mathsf{U}(\pi)$ is as follows.

- *Agents*:

$$\mu^\mathsf{U}(a)[i,j] := \begin{cases} ?\mathsf{pre}(e_i); a & \text{if } e_i \mathsf{R}_a e_j \\ ?\bot & \text{otherwise} \end{cases} \qquad (14)$$

- *Test*:
$$\mu^{\mathsf{U}}(?\varphi)[i,j] := \begin{cases} ?(\mathsf{pre}(\mathsf{e}_i) \wedge [\mathsf{U},\mathsf{e}_i]\varphi) & \text{if } i = j \\ ?\bot & \text{otherwise} \end{cases} \quad (15)$$

- *Non-deterministic choice*:
$$\mu^{\mathsf{U}}(\pi_1 \cup \pi_2)[i,j] := \oplus \left\{ \mu^{\mathsf{U}}(\pi_1)[i,j],\ \mu^{\mathsf{U}}(\pi_2)[i,j] \right\} \quad (16)$$

where $\oplus \Gamma$ is the non-deterministic choice of the programs in Γ set after removing occurrences of $?\bot$, that is,

$$\oplus \Gamma := \begin{cases} \bigcup (\Gamma \setminus \{?\bot\}) & \text{if } \varnothing \neq \Gamma \neq \{?\bot\} \\ ?\bot & \text{otherwise} \end{cases} \quad (17)$$

being $\bigcup$ the generalised non-deterministic choice of a non-empty set of programs.

- *Sequential composition*:
$$\mu^{\mathsf{U}}(\pi_1; \pi_2)[i,j] := \oplus \left\{ \mu^{\mathsf{U}}(\pi_1)[i,k] \odot \mu^{\mathsf{U}}(\pi_2)[k,j] \ \middle| \ 0 \leq k \leq n-1 \right\} \quad (18)$$

where $\sigma \odot \rho$ is the sequential composition of σ and ρ after removing superfluous occurrences of $?\bot$ and $?\top$, that is,

$$\sigma \odot \rho := \begin{cases} \sigma; \rho & \text{if } \sigma \neq ?\bot \neq \rho \text{ and } \sigma \neq ?\top \neq \rho \\ \sigma & \text{if } \sigma \neq ?\top = \rho \\ \rho & \text{if } \sigma = ?\top \\ ?\bot & \text{otherwise} \end{cases} \quad (19)$$

- *Kleene closure*:
$$\mu^{\mathsf{U}}(\pi^*) := S_0^{\mathsf{U}} \left(\mu^{\mathsf{U}}(\pi) \mid A^{\mathsf{U}} \right) \quad (20)$$

where $\mu^{\mathsf{U}}(\pi) \mid A^{\mathsf{U}}$ is the $n \times 2n$ matrix obtained by augmenting $\mu^{\mathsf{U}}(\pi)$ with A^{U}, an $n \times n$ matrix defined as

$$A^{\mathsf{U}}[i,j] := \begin{cases} ?\mathsf{pre}(\mathsf{e}_i) & \text{if } i = j \\ ?\bot & \text{otherwise} \end{cases} \quad (21)$$

The function S_k^{U} (with $0 \leq k \leq n$), defined as

$$S_k^{\mathsf{U}}(M \mid A) := \begin{cases} A & \text{if } k = n \\ S_{k+1}^{\mathsf{U}}(Subs_k(Ard_k(M \mid A))) & \text{otherwise} \end{cases} \qquad (22)$$

receives an argument $M \mid A$ and performs an iterative process applying Arden's Theorem to row k (via function $Ard_k : \mathcal{M}_{n \times 2n} \to \mathcal{M}_{n \times 2n}$) and substituting rows different from k (via function $Subs_k : \mathcal{M}_{n \times 2n} \to \mathcal{M}_{n \times 2n}$) until a $k = n$, then returning the right part of the augmented matrix. The two auxiliary functions, Ard_k and $Subs_k$, are given by

$$Ard_k(N)[i,j] := \begin{cases} N[i,j] & \text{if } i \neq k \\ ?\bot & \text{if } i = k = j \\ N[i,j] & \text{if } i = k \neq j \text{ and } N[k,k] = ?\bot \\ N[k,k]^* \odot N[i,j] & \text{otherwise} \end{cases} \qquad (23)$$

$$Subs_k(N)[i,j] := \begin{cases} N[i,j] & \text{if } i = k \\ ?\bot & \text{if } i \neq k = j \\ \oplus\{N[i,k] \odot N[k,j], N[i,j]\} & \text{otherwise} \end{cases} \qquad (24)$$

The operators '$\oplus$' and '$\odot$' used in the previous definition are versions of non-deterministic choice and sequential composition that remove unnecessary occurrences of $?\bot$ and $?\top$; thus returning programs that are (potentially) syntactically shorter but nevertheless semantically equivalent to their PDL counterparts '$\cup$' and ';', as the following propositions show.

Proposition 1. *Let M be an epistemic model and Γ a set of LCC programs. Then,*

$$[\![\oplus \Gamma]\!]^M = [\![\bigcup \Gamma]\!]^M$$

Proof. Take any epistemic model M. Equation (17) states that $\oplus \Gamma$ is a non-deterministic choice of the LCC programs in Γ that returns $\bigcup(\Gamma \setminus \{?\bot\})$ when Γ is different from both $\varnothing$ and $\{?\bot\}$, and $?\bot$ otherwise. In the first case, $[\![\oplus \Gamma]\!]^M = [\![\bigcup \Gamma]\!]^M$ because $[\![\bigcup \Gamma]\!]^M = [\![\bigcup(\Gamma \setminus \{?\bot\})]\!]^M$; in the second, $[\![\oplus \Gamma]\!]^M = [\![\bigcup \Gamma]\!]^M$ because $[\![\bigcup \varnothing]\!]^M = [\![\bigcup \{?\bot\}]\!]^M = [\![?\bot]\!]^M = \varnothing$. □

Proposition 2. *Let M be an epistemic model and σ, ρ two LCC programs. Then,*

$$[\![\sigma; \rho]\!]^M = [\![\sigma \odot \rho]\!]^M$$

Proof. Take any epistemic model M. Equation (19) states that $\sigma \odot \rho$ differs from $\sigma;\rho$ only when either σ or else ρ is $?\bot$ or $?\top$. But, in such cases:

- $[\![\sigma;?\bot]\!]^M = [\![?\bot;\sigma]\!]^M = [\![?\bot]\!]^M$; hence, $[\![\sigma;\rho]\!]^M = [\![\sigma \odot \rho]\!]^M$.

- $[\![\sigma;?\top]\!]^M = [\![?\top;\sigma]\!]^M = [\![\sigma]\!]^M$; hence, $[\![\sigma;\rho]\!]^M = [\![\sigma \odot \rho]\!]^M$.

$\square$

The rest of this section is devoted to prove that the function μ^{U} returns an LCC program that is semantically equivalent to the one returned by the program transformer T^{U} of [6].

Lemma 1. *Let* $\mathsf{U} = (\mathsf{E}, \mathsf{R}, \mathsf{pre}, \mathsf{sub})$ *be an action model with* $\mathsf{e}_i, \mathsf{e}_j \in \mathsf{E}$; *let* π *be an LCC program. For any epistemic model M,*

$$[\![T^{\mathsf{U}}_{ij}(\pi)]\!]^M = [\![\mu^{\mathsf{U}}(\pi)[i,j]]\!]^M$$

Proof. By induction on the complexity of π. Let M be an epistemic model; then

(Base Cases: a and $?\varphi$) Trivial, as the definitions of T^{U}_{ij} and $\mu^{\mathsf{U}}(\pi)[i,j]$ are identical for both a and $?\varphi$.

(Ind. Case $\pi_1 \cup \pi_2$) Suppose (Ind. Hyp.) the claim holds for π_1 and π_2. Then

$$\begin{aligned}
[\![T^{\mathsf{U}}_{ij}(\pi_1 \cup \pi_2)]\!]^M &= [\![T^{\mathsf{U}}_{ij}(\pi_1) \cup T^{\mathsf{U}}_{ij}(\pi_2)]\!]^M & \text{(Def. 6)}\\
&= [\![T^{\mathsf{U}}_{ij}(\pi_1)]\!]^M \cup [\![T^{\mathsf{U}}_{ij}(\pi_2)]\!]^M & (\text{Def. of } [\![\cdot]\!]^M)\\
&= [\![\mu^{\mathsf{U}}(\pi_1)[i,j]]\!]^M \cup [\![\mu^{\mathsf{U}}(\pi_2)[i,j]]\!]^M & \text{(Ind. Hyp.)}\\
&= [\![\mu^{\mathsf{U}}(\pi_1)[i,j] \cup \mu^{\mathsf{U}}(\pi_2)[i,j]]\!]^M & (\text{Def. of } [\![\cdot]\!]^M)\\
&= [\![\oplus\{\mu^{\mathsf{U}}(\pi_1)[i,j], \mu^{\mathsf{U}}(\pi_2)[i,j]\}]\!]^M & \text{(Prop. 1)}\\
&= [\![\mu^{\mathsf{U}}(\pi_1 \cup \pi_2)[i,j]]\!]^M & (\text{Def. of } \mu^{\mathsf{U}}(\pi_1 \cup \pi_2) \text{ in (16))}
\end{aligned}$$

(Ind. Case $\pi_1;\pi_2$) Suppose (Ind. Hyp.) the claim holds for π_1 and π_2. Then

$$[\![T^{\mathsf{U}}_{ij}(\pi_1;\pi_2)]\!]^M = [\![\bigcup_{k=0}^{n-1}(T^{\mathsf{U}}_{ik}(\pi_1); T^{\mathsf{U}}_{kj}(\pi_2))]\!]^M \quad \text{(Def. 6)}$$

$$= \bigcup_{k=0}^{n-1}\left([\![T^{\mathsf{U}}_{ik}(\pi_1)]\!]^M \circ [\![T^{\mathsf{U}}_{kj}(\pi_2)]\!]^M\right) \quad (\text{Def. of } [\![\cdot]\!]^M)$$

$$= \bigcup_{k=0}^{n-1}\left([\![\mu^{\mathsf{U}}(\pi_1)[i,k]]\!]^M \circ [\![\mu^{\mathsf{U}}(\pi_2)[k,j]]\!]^M\right) \quad \text{(Ind. Hyp.)}$$

$$= [\![\bigcup_{k=0}^{n-1}\left(\mu^{\mathsf{U}}(\pi_1)[i,k]; \mu^{\mathsf{U}}(\pi_2)[k,j]\right)]\!]^M \quad (\text{Def. of } [\![\cdot]\!]^M)$$

$$= [\![\bigcup_{k=0}^{n-1}\left(\mu^{\mathsf{U}}(\pi_1)[i,k] \odot \mu^{\mathsf{U}}(\pi_2)[k,j]\right)]\!]^M \quad \text{(Prop. 2)}$$

$$= [\![\oplus\{\mu^{\mathsf{U}}(\pi_1)[i,k] \odot \mu^{\mathsf{U}}(\pi_2)[k,j] \mid 0 \leq k \leq n-1\}]\!]^M \quad \text{(Prop. 1)}$$

$$= [\![\mu^{\mathsf{U}}(\pi_1;\pi_2)[i,j]]\!]^M \quad (\text{Def. of } \mu^{\mathsf{U}}(\pi_1;\pi_2) \text{ in (18)})$$

(Ind. Case π^*) Suppose (Ind. Hyp.) the claim holds for π and observe how $[\![\pi^*]\!]^M = [\![?\top \cup (\pi;\pi^*)]\!]^M$. Now,

$$[\![T^{\mathsf{U}}_{ij}(\pi^*)]\!]^M = [\![T^{\mathsf{U}}_{ij}(?\top \cup \pi;\pi^*)]\!]^M$$

$$= [\![T^{\mathsf{U}}_{ij}(?\top)]\!]^M \cup [\![\bigcup_{k=0}^{n-1}(T^{\mathsf{U}}_{ik}(\pi); T^{\mathsf{U}}_{kj}(\pi^*))]\!]^M \quad \text{(Def. 6)}$$

$$= [\![T^{\mathsf{U}}_{ij}(?\top)]\!]^M \cup \bigcup_{k=0}^{n-1}\left([\![T^{\mathsf{U}}_{ik}(\pi)]\!]^M \circ [\![T^{\mathsf{U}}_{kj}(\pi^*)]\!]^M\right) \quad (\text{Def. of } [\![\cdot]\!]^M)$$

$$= [\![T^{\mathsf{U}}_{ij}(?\top)]\!]^M \cup \bigcup_{k=0}^{n-1}\left([\![\mu^{\mathsf{U}}(\pi)[i,k]]\!]^M \circ [\![T^{\mathsf{U}}_{kj}(\pi^*)]\!]^M\right) \quad \text{(Ind. Hyp.)}$$

The last equality produces n^2 relational equations. By abbreviating $[\![T^{\mathsf{U}}_{ij}(\pi^*)]\!]^M$ as X^{ij} for every $0 \leq i,j \leq n-1$, we get

$$X^{ij} = [\![T^{\mathsf{U}}_{ij}(?\top)]\!]^M \cup \bigcup_{k=0}^{n-1}\left([\![\mu^{\mathsf{U}}(\pi)[i,k]]\!]^M \circ X^{kj}\right) \qquad (25)$$

Thus, it is enough to prove that $[\![\mu^{\mathsf{U}}(\pi^*)[i,j]]\!]^M$ is a solution for X^{ij}. This is shown in the following three propositions about the functions building $\mu^{\mathsf{U}}(\pi^*)$.

Proposition 3. *Take* $\Omega = (\mu^{\mathsf{U}}(\pi) \mid A^{\mathsf{U}})$ *(see (20)). Then,*

$$X^{ij} = [\![\Omega[i,j+n]]\!]^M \cup \bigcup_{k=0}^{n-1}\left([\![\Omega[i,k]]\!]^M \circ X^{kj}\right) \qquad (26)$$

Proof. It will be shown that the right-hand side (r.h.s.) of (25) and (26) coincide. Their respective rightmost parts are equivalent since, for $0 \leq k \leq n-1$, $\Omega[i,k] = \mu^{\mathsf{U}}(\pi)[i,k]$ (recall that Ω is built by adding additional columns at the right of the n first columns of $\mu^{\mathsf{U}}(\pi)$, and the matrix's indexes start from 0). For the leftmost parts,

$$[\![T^{\mathsf{U}}_{ij}(?\top)]\!]^M = \begin{cases} [\![?(\mathsf{pre}(e_i) \wedge [\mathsf{U},e_i]\top)]\!]^M & \text{if } i = j \\ [\![?\bot]\!]^M & \text{otherwise} \end{cases} \quad \text{(Def. 6)}$$

$$= \begin{cases} [\![?\mathsf{pre}(e_i)]\!]^M & \text{if } i = j \\ [\![?\bot]\!]^M & \text{otherwise} \end{cases} \quad \text{(as } [\mathsf{U},e_i]\top \text{ is trivially true)}$$

$$= [\![A^{\mathsf{U}}[i,j]]\!]^M = [\![\Omega[i,j+n]]\!]^M \qquad ((21) \text{ and Def. of } \Omega)$$

□

Proposition 4. *For $0 \leq k \leq n-1$, if N is a matrix of size $n \times 2n$ with all cells in columns $0,\ldots,k-1$ equal to $?\bot$, then $Subs_k(Ard_k(N))$ contains all cells in columns $0,\ldots,k$ equal to $?\bot$.*

Proof. Start with $Ard_k(N)$. Observe in (23) that the only modified cells are in the kth row. Cell $Ard_k(N)[k,k]$ in the kth column is converted into $?\bot$. With respect to cells in columns from 0 to $k-1$, if they were $?\bot$, they continue being $?\bot$: those cells $N[i,j]$ do not change, if $N[k,k] = ?\bot$, or otherwise are converted by (23) into $N[k,k]^* \odot N[i,j]$ and, by (19), if $N[i,j] = ?\bot$, then $N[k,k]^* \odot N[i,j] = ?\bot$.

Now, call N' the output of $Ard_k(N)$ and observe $Subs_k(N')$'s definition (24): the only cells that change are in rows different to k. With respect to any such row i, the position in the kth column is made $?\bot$. For cells in previous columns, $j < k$, the last case in the definition returns $\oplus\{N'[i,k] \odot N'[k,j], N'[i,j]\}$. But as N' is the result of $Ard_k(N)$, $N'[k,j]$ is $?\bot$ (because, as argued above, $Ard_k(N)$ works over the kth row and keeps the $?\bot$ in columns before k). Also, $N'[i,j] = ?\bot$, as columns $j < k$ are filled with $?\bot$. So $\oplus\{N'[i,k] \odot N'[k,j], N'[i,j]\}$ becomes $\oplus\{N'[i,k] \odot ?\bot, ?\bot\}$ and, by (17) and (19), it is $?\bot$. □

Proposition 5. *Given an $n \times 2n$ matrix N of LCC programs, the equations built using (26), with $\Omega = Subs_k(Ard_k(N))$, $0 \leq k \leq n-1$, are correct transformations of the equations built in the same way with $\Omega = N$.*

Proof. As argued in the proof of Proposition 4, $Ard_k(N)$ works only on the kth row. If $N[k,k] = ?\bot$, nothing is done, so according to (26) the equations for X^{kj} ($0 \leq j \leq n-1$) do not change. Otherwise, the kth row of N changes: all cells $N[k,j]$ with $j \neq k$ become $N[k,k]^* \odot N[k,j]$, except $N[k,k]$ which becomes $?\bot$. Then, for every $0 \leq j \leq n-1$, the equation for X^{kj} becomes (using index t instead of k and removing $[\![?\bot]\!]^M \circ X^{kj}$ from the union):

$$X^{kj} = [\![N[k,k]^* \odot N[k,j+n]]\!]^M \cup \bigcup_{\substack{0 \leq t \leq n-1 \\ t \neq k}} \left([\![N[k,k]^* \odot N[k,t]]\!]^M \circ X^{tj} \right)$$

By Proposition 2 and $[\![\,\cdot\,]\!]^M$'s definition, this can be rewritten as

$$X^{kj} = ([\![N[k,k]]\!]^M)^* \circ [\![N[k,j+n]]\!]^M \cup \\ \bigcup_{\substack{0 \le t \le n-1 \\ t \ne k}} \left(([\![N[k,k]]\!]^M)^* \circ [\![N[k,t]]\!]^M \circ X^{tj}\right) \qquad (27)$$

which is an application of Arden's Theorem [20] to the corresponding equation for the original row in N:

$$X^{kj} = [\![N[k,j+n]]\!]^M \cup \bigcup_{0 \le t \le n-1} \left([\![N[k,t]]\!]^M \circ X^{tj}\right) \qquad (28)$$

Arden's Theorem (which works on regular algebras, such as LCC programs) gives $X = A^* \circ B$ as a solution for $X = (A \circ X) \cup B$. In (28), X is X^{kj}, A is $[\![N[k,k]]\!]^M$, and B is the union of all terms in the r.h.s. of (28) except $[\![N[k,k]]\!]^M \circ X^{kj}$. Besides Arden's Theorem, from (28) to (27) we use $\circ$'s distribution over $\cup$, $A \circ (B \cup C) = (A \circ B) \cup (A \circ C)$.

Now denote by N' the output of $Ard_k(N)$. We move to $Subs_k(N')$ to show that the equations obtained from it with (26) are correct transformations of the equations built from N'. The only modified cells in $Subs_k(N')$ are in rows different to k, so it only affects equations for X^{ij} with $i \ne k$. According to (26), if $\Omega = N'$, these equations are (using t instead of k):

$$X^{ij} = [\![N'[i,j+n]]\!]^M \cup \bigcup_{t=0}^{n-1} \left([\![N'[i,t]]\!]^M \circ X^{tj}\right) \qquad (29)$$

The same equation for $\Omega = Subs_k(N')$ becomes the following (we remove from the union the term $[\![?\bot]\!]^M \circ X^{kj}$, as it is equivalent to $\varnothing$):

$$X^{ij} = [\![\oplus\{N'[i,k] \odot N'[k,j+n], N'[i,j+n]\}]\!]^M \cup \\ \bigcup_{\substack{0 \le t \le n-1 \\ t \ne k}} \left([\![\oplus\{N'[i,k] \odot N'[k,t], N'[i,t]\}]\!]^M \circ X^{tj}\right) \qquad (30)$$

By using Propositions 1 and 2 and the properties of $[\![\,\cdot\,]\!]^M$, equation (30) becomes

$$X^{ij} = ([\![N'[i,k]]\!]^M \circ [\![N'[k,j+n]]\!]^M) \cup [\![N'[i,j+n]]\!]^M \cup \\ \bigcup_{\substack{0 \le t \le n-1 \\ t \ne k}} \left((([\![N'[i,k]]\!]^M \circ [\![N'[k,t]]\!]^M) \cup [\![N'[i,t]]\!]^M) \circ X^{tj}\right) \qquad (31)$$

But note that in the equation for X^{kj}, which is the same at N' and $Subs_k(N')$, the kth row of N' is not changed by $Subs_k(N')$:

$$X^{kj} = [\![N'[k, j+n]]\!]^M \cup \bigcup_{\substack{0 \leq t \leq n-1 \\ t \neq k}} \left([\![N'[k, t]]\!]^M \circ X^{tj}\right) \tag{32}$$

We have eliminated the term $[\![N'[k, k]]\!]^M \circ X^{kj}$ in (32) because $N' = Ard_k(N)$ and by (23), $N'[k, k] = ?\bot$, which produces $[\![N'[k, k]]\!]^M \circ X^{kj} = \varnothing$.

Observe that (31) can be obtained from (29) by replacing X^{kj} by the r.h.s. of (32) and applying the distribution of $\circ$ over $\cup$. So the modified equation (30) is equivalent to correct transformations of the original one (29). □

The proof of the case π^* in Lemma 1 can be finished now. Take the set of relational equations given by (25). By (20), $\mu^{\cup}(\pi^*)$ operates by iterating calls to $S_k^{\cup}$ (with k from 0 to n) with $\Omega = (\mu^{\cup}(\pi) \mid A^{\cup})$ as the initial argument. Let M_{-1} be Ω and M_k the output of $S_k^{\cup}(M_{k-1})$. By Proposition 3, (26) gives equations equivalent to (25). By Proposition 5, the equations are correct for each successive M_k ($0 \leq k \leq n-1$). As the calls to $S_k^{\cup}$ are done iteratively with k from 0 to $n-1$, Proposition 4 guarantees that, in M_{n-1}, all cells in columns for 0 to $n-1$ are equal to $?\bot$. Thus, equations (26) for M_{n-1} are:

$$X^{ij} = [\![M_{n-1}[i, j+n]]\!]^M \tag{33}$$

The rightmost union in (26) has disappeared ($M[i, k] = ?\bot$ for $0 \leq k \leq n-1$, and $[\![?\bot]\!]^M = \varnothing$). Now, by $S_k^{\cup}$'s definition in (22), $M_{n-1}[i, j+n] = M_n[i, j] = \mu^{\cup}(\pi^*)[i, j]$, so $X^{ij} = [\![\mu^{\cup}(\pi^*)[i, j]]\!]^M$. Then, since X^{ij} represents $[\![T_{ij}^{\cup}(\pi^*)]\!]^M$,

$$[\![T_{ij}^{\cup}(\pi^*)]\!]^M = [\![\mu^{\cup}(\pi^*)[i, j]]\!]^M$$

which completes the proof. □

We can now define new translation functions t', r' as follows. Note that t' and r' are defined as the translation functions t, r for formulas φ and programs π proposed in [6], with the only exception of formulas of the form $[\mathsf{U}, \mathsf{e}_i][\pi]\varphi$.[10] Note also the *inside-out* approach in the case $t([\mathsf{U}, \mathsf{e}][\mathsf{U}', \mathsf{f}]\varphi) = t([\mathsf{U}, \mathsf{e}]t([\mathsf{U}', \mathsf{f}]\varphi))$, which requires rule RE$_{\mathsf{U}}$ (with $\chi = [\mathsf{U}', \mathsf{f}]\varphi$ and $\psi = t(\chi)$) in order to prove that the translation is indeed provably equivalent (i.e. $\vdash \phi \leftrightarrow t(\phi)$).

---

[10]Two minor typos for the cases $[\mathsf{U}, \mathsf{e}]p$ and $[\mathsf{U}, \mathsf{e}_i][\pi]\varphi$ are also corrected here w.r.t. [6] (the first was given by $t(\mathsf{pre}(\mathsf{e})) \to \mathsf{sub}(\mathsf{e}, p)$, and the second by $\bigwedge_{j=0}^{n-1}[T_{ij}^{\cup}(r(\pi))]t([\mathsf{U}, \mathsf{e}_j]\varphi)$).

$$
\begin{aligned}
t'(\top) &= \top & r'(a) &= a \\
t'(p) &= p & r'(B) &= B \\
t'(\neg\varphi) &= \neg t'(\varphi) & r'(?\varphi) &= ?t'(\varphi) \\
t'(\varphi_1 \wedge \varphi_2) &= t'(\varphi_1) \wedge t'(\varphi_2) & r'(\pi_1;\pi_2) &= r'(\pi_1);r'(\pi_2) \\
t'([\pi]\varphi) &= [r'(\pi)]t'(\varphi) & r'(\pi_1 \cup \pi_2) &= r'(\pi_1) \cup r'(\pi_2) \\
t'([\mathsf{U},\mathsf{e}]\top) &= \top & r'(\pi^*) &= (r'(\pi))^* \\
t'([\mathsf{U},\mathsf{e}]p) &= t'(\mathsf{pre}(\mathsf{e})) \to t'(\mathsf{sub}(\mathsf{e},p)) \\
t'([\mathsf{U},\mathsf{e}]\neg\varphi) &= t'(\mathsf{pre}(\mathsf{e})) \to \neg t'([\mathsf{U},\mathsf{e}]\varphi) \\
t'([\mathsf{U},\mathsf{e}](\varphi_1 \wedge \varphi_2)) &= t'([\mathsf{U},\mathsf{e}]\varphi) \wedge t'([\mathsf{U},\mathsf{e}]\varphi_2) \\
t'([\mathsf{U},\mathsf{e}_i][\pi]\varphi) &= \bigwedge_{\substack{0 \leq j \leq n-1 \\ \mu^{\mathsf{U}}(\pi)[i,j] \neq ?\bot}} [r'(\mu^{\mathsf{U}}(\pi)[i,j])]t'([\mathsf{U},\mathsf{e}_j]\varphi) \\
t'([\mathsf{U},\mathsf{e}][\mathsf{U}',\mathsf{e}']\varphi) &= t'([\mathsf{U},\mathsf{e}]t'([\mathsf{U}',\mathsf{e}']\varphi))
\end{aligned}
$$

Corollary 1. *The translation functions t', r' reduce the language of LCC to that of PDL. This translation is correct.*

Proof. The effective reduction from LCC to PDL is immediate by inspection. Its correctness follows from that in [6], with Lemma 1 for the case $[\mathsf{U},\mathsf{e}_i][\pi]\varphi$. □

Definition 8. We define a new axiom system for LCC by replacing the reduction axiom for PDL programs with the following

$$[\mathsf{U},\mathsf{e}_i][\pi]\varphi \leftrightarrow \bigwedge_{\substack{0 \leq j \leq n-1 \\ \mu^{\mathsf{U}}(\pi)[i,j] \neq ?\bot}} [\mu^{\mathsf{U}}(\pi)[i,j]][\mathsf{U},\mathsf{e}_j]\varphi \quad \text{(prog)}$$

Corollary 2. *The axiom system for LCC from Def. 8 is sound and complete.*

Proof. The only new axiom, that for PDL-programs, is sound by Lemma 1. For completeness, the proof system for PDL is complete, and every LCC formula is provably equivalent to a PDL formula using Corollary 1. □

5 Complexity of the new transformers

The original program transformers in [6] require exponential time due to the use of Kleene's method [11]. Moreover, the size of the transformed formulas of type π^* is also exponential because of the definition of K^{U}_{ijn} (Def. 6).

In order to study the complexity of our program transformers, we first implemented in Prolog both the original program transformers and our matrix calculus. Figure 5 shows the result for our transformers for two kinds of models, *complete* and *chain* models, from 1 to 20 states. The graph's vertical axis, which is shown in logarithmic scale, presents the number of PDL operators in the transformed program $\mu^{\cup}(\pi^*)[n-1,0]$ for n the number of states in the model.

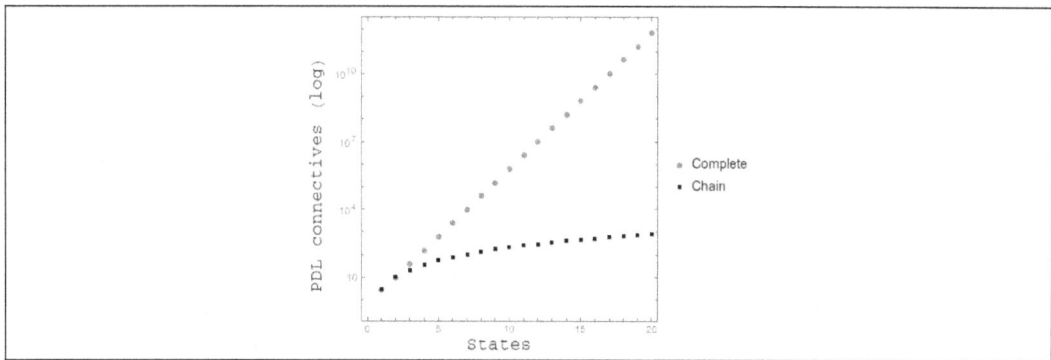

Figure 5: Number of PDL connectives in $\mu^{\cup}(\pi^*)$ for different action models

A model is *complete* when it is fully connected, i.e., when each $\mu^{\cup}(\pi)[i,j] = s(i,j)$ is an atomic expression (which is not further analysed by the implementations). All values $s(i,j)$ in $\mu^{\cup}(\pi)$ are assumed to be different, avoiding simplifications of repeated patterns. The number of operators in $\mu^{\cup}(\pi^*)[n-1,0]$ is in the order of 2^{2n}. In the worst case our transformers produce an exponential output, which implies that the required time is also exponential. In a *chain model*, each state is connected with itself, the previous and next one. Thus, $\mu^{\cup}(\pi)[i,j]$ is $s(i,j)$ when $i = j$ or $|i - j| = 1$, and $?\bot$ otherwise. Now the number of operators in $\mu^{\cup}(\pi^*)[n-1,0]$ is in the order of $2n^2$, so in this case the length of the output is polynomial. We chose models with chain-like structure because it makes it easier to generate models of increasing size with limited connectivity. Similar results can be obtained for other kinds of models with similar average connectivity, as the key is the number of instances of $?\bot$ spread along the matrices.

The results for the original program transformers are not shown in Figure 5 as they are, for both the complete and chain models, as our worst case. (The reason is that they do not benefit from removing superfluous $?\bot$.)

An advantage of our transformers is that they do not require exponential space in cases other than the worst one, in contrast with the original transformers which always perform in the same (exponential) way. An additional advantage can be found in the reusability of the information produced during program transformations. As we argued, working with matrices allows to perform several operations in parallel. Indeed, matrix $\mu^{\cup}(\pi)$ contains the transformation of program π within all states in the action model U. As building the matrix with the transformations of a given program involves building the matrices for its subprograms, the information of each generated matrix is properly stored, and then it can be reused in the same or subsequent transformations within the same action model. This is a practical way to reduce the computation time required for program transformation.

6 Summary and Future Work

In this work, we presented an alternative definition of the program transformers used to obtain reduction axioms in LCC. The proposal uses a matrix treatment of Brzozowski's equational method in order to obtain a regular expression representing the language accepted by a finite automaton. While Brzozowski's method and that used in the original LCC paper [6] are equivalent, the first is computationally more efficient in cases different to the worst one; moreover, the matrix treatment presented here is more synthetic, simple and elegant, thus allowing a simpler implementation.

Towards future work, some definitions used by program transformers (particularly the $\odot$ operation) can be modified to obtain even simpler expressions. For example, $\sigma \odot \rho$ might be defined as σ if $\sigma \neq ?\top = \rho$ and as ρ if $\sigma = ?\top$. Moreover, the algorithm implementing Ard_k and $Subs_k$ functions can be improved by disregarding the $N[i,j]$ elements with $j < k$ or $j > n+k$ (being $N[i,j]$ a $n \times 2n$ matrix), since those are necessarily equal to $?\bot$. These changes, despite not lowering the translation's complexity order, would nevertheless make it more efficient.

Acknowledgements

We would like to thank two anonymous reviewers for their helpful comments. This work is supported by the Spanish Ministry of Economy and Competitiveness, under Research Project FFI2014-56219-P. The research of P. Pardo was supported by a Sofja Kovalevkaja award of the Alexander von Humboldt-Foundation, funded by the German Ministry for Education and Research. This paper is an extended version of a previous conference paper [21] at JELIA 2014.

References

[1] H. van Ditmarsch, W. van der Hoek, B. Kooi, Dynamic Epistemic Logic, Vol. 337 of Synthese Library Series, Springer, 2007.

[2] J. van Benthem, Logical Dynamics of Information and Interaction, Cambridge University Press, 2011.

[3] R. Fagin, J. Y. Halpern, Y. Moses, M. Y. Vardi, Reasoning about knowledge, The MIT Press, Cambridge, Mass., 1995.

[4] R. Parikh, R. Ramanujam, A knowledge based semantics of messages, Journal of Logic, Language and Information 12 (4) (2003) 453–467.

[5] J. van Benthem, J. Gerbrandy, T. Hoshi, E. Pacuit, Merging frameworks for interaction, Journal of Philosophical Logic 38 (5) (2009) 491–526. doi:10.1007/s10992-008-9099-x.

[6] J. van Benthem, J. van Eijck, B. Kooi, Logics of communication and change, Information and Computation 204 (11) (2006) 1620–1662. doi:10.1016/j.ic.2006.04.006.

[7] D. Harel, D. Kozen, J. Tiuryn, Dynamic Logic, MIT Press, Cambridge, MA, 2000.

[8] A. Baltag, L. S. Moss, S. Solecki, The logic of public announcements and common knowledge and private suspicions, in: I. Gilboa (Ed.), TARK, Morgan Kaufmann, San Francisco, CA, USA, 1998, pp. 43–56.

[9] A. Baltag, L. S. Moss, Logics for epistemic programs, Synthese 139 (2) (2004) 165–224.

[10] J. van Benthem, B. Kooi, Reduction axioms for epistemic actions, in: R. Schmidt, I. Pratt-Hartmann, M. Reynolds, H. Wansing (Eds.), Advances in Modal Logic (Number UMCS-04-09-01 in Technical Report Series), Department of Computer Science, University of Manchester, 2004, pp. 197–211.

[11] S. Kleene, Representation of events in nerve nets and finite automata, in: C. E. Shannon, J. McCarthy (Eds.), Automata Studies, Princeton University Press, Princeton, NJ, 1956, pp. 3–41.

[12] J. Sakarovitch, Automata and rational expressions, CoRR abs/1502.03573.
URL http://arxiv.org/abs/1502.03573

[13] J. A. Brzozowski, Derivatives of regular expressions, Journal of the ACM 11 (4) (1964) 481–494.

[14] J. H. Conway, Regular Algebra and Finite Machines, Chapman and Hall, 1971.

[15] Y. Wang, Q. Cao, On axiomatizations of public announcement logic, Synthese 190 (1) (2013) 103–134. doi:10.1007/s11229-012-0233-5.

[16] H. van Ditmarsch, B. Kooi, Semantic results for ontic and epistemic change, in: G. Bonanno, W. van der Hoek, M. Wooldridge (Eds.), Logic and the Foundations of Game and Decision Theory (LOFT7), Vol. 3 of Texts in Logic and Games, Amsterdam University Press, Amsterdam, The Netherlands, 2008, pp. 87–117.

[17] H. Gruber, M. Holzer, Finite automata, digraph connectivity, and regular expression size, in: L. Aceto, I. Damgård, L. A. Goldberg, M. M. Halldórsson, A. Ingólfsdóttir, I. Walukiewicz (Eds.), Automata, Languages and Programming, 35th International Colloquium, ICALP 2008, Reykjavik, Iceland, July 7-11, 2008, Proceedings, Part II - Track B: Logic, Semantics, and Theory of Programming & Track C: Security and Cryptography Foundations, Vol. 5126 of Lecture Notes in Computer Science, Springer, 2008, pp. 39–50. doi:10.1007/978-3-540-70583-3_4.

[18] H. Gruber, M. Holzer, Provably shorter regular expressions from finite automata, International Journal of Foundations of Computer Science 24 (8) (2013) 1255–1279. doi:10.1142/S0129054113500330.

[19] D. Kozen, On kleene algebras and closed semirings, in: B. Rovan (Ed.), MFCS, Vol.

452 of Lecture Notes in Computer Science, Springer, 1990, pp. 26–47.

[20] D. N. Arden, Delayed-logic and finite-state machines, in: SWCT (FOCS), IEEE Computer Society, 1961, pp. 133–151.

[21] P. Pardo, E. Sarrión-Morillo, F. Soler-Toscano, F. R. Velázquez-Quesada, Efficient program transformers for translating LCC to PDL, in: E. Fermé, J. Leite (Eds.), Logics in Artificial Intelligence - 14th European Conference, JELIA 2014, Funchal, Madeira, Portugal, September 24-26, 2014. Proceedings, Vol. 8761 of LNCS, Springer, 2014, pp. 253–266.

Lighthouse Principle for Diffusion in Social Networks

Sanaz Azimipour
University of Tehran, Tehran, Iran.
Sanaz.a234@gmail.com

Pavel Naumov
Vassar College, Poughkeepsie, New York, USA.
pnaumov@vassar.edu

Abstract

The article investigates an influence relation between two sets of agents in a social network. It proposes a logical system that captures propositional properties of this relation valid in all threshold models of social networks with the same structure. The logical system consists of Armstrong axioms for functional dependence and an additional Lighthouse axiom. The main results are soundness, completeness, and decidability theorems for this logical system.

1 Introduction

1.1 Social Networks

In this article we study influence in social networks. When a new product is introduced to the market, it is usually first adopted by a few users that are called "early adopters". These users might adopt the product because they are fans of the company introducing the product, as a result of the marketing campaign conducted by the company, or because they have a genuine need for this type of product. Once the early adopters start using the product, they put peer pressure on their friends and acquaintances in the social network, who might eventually follow them in adopting the product. The friends of the early adopters might eventually influence their own friends and so on, until the product is potentially adopted by a significant part of the network.

A similar phenomenon could be observed with the diffusion of certain behaviours, like smoking, the adoption of new words and technical innovations, and the propagation of beliefs.

There are two most widely used models that formally capture diffusion process in social networks. One of them is the *stochastic* model [20, 12]. This model distinguishes active and inactive vertices of the network. Once a vertex v becomes active, it gets a single chance to activate each neighbour u with a given probability $p_{v,u}$. This process continues until no more activations can happen.

In this article we focus on the second model, called *threshold* model [26, 14, 11, 1], originally introduced by Granovetter [8] and Schelling [21]. In this model each agent has a non-negative threshold value representing the agent's resistance to adoption of a given product. If the pressure from those peers of the agent who already adopted the product reaches the threshold value, then the agent also adopts the product. We assume that each of the other agents has a non-negative, but possibly zero, influence on the given agent. The peer pressure on an agent to adopt a product is the sum of influences on the agent of all agents who have already adopted the product. It is assumed in this model that, once the product is adopted, the agent keeps using the product and putting pressure on her peers indefinitely.

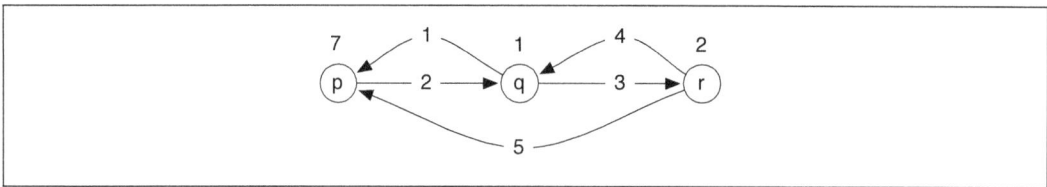

Figure 1: Social Network N_1

Consider, for example, social network N_1 depicted in Figure 1. This network consists of three agents: p, q, and r that have threshold values 7, 1, and 2 respectively. The threshold value of a node is shown on the diagram above the node representing the agent. The influence of one agent on another is shown in this figure by the label on the directed edge connecting the two agents. For instance, the influence of agent r on agent p is 5. If an agent has zero influence on another agent, no edge is shown. Thus, influence of agent p on agent r is zero.

Suppose that a marketing company gives agent p a free sample of the product and the agent starts using it. Since agent p has influence 2 on agent q and threshold value of agent q is only 1, she will eventually also adopt the product. In turn, the adoption of the product by agent q will eventually lead to an adoption by agent r because threshold value of agent r is only 2 and the influence of agent q on agent r is 3. Thus, the adoption by agent p eventually leads to an adoption of this product by agent r. We denote this fact by $N_1 \vDash p \triangleright r$.

In this article we study relation $A \triangleright B$ between group of agents A and B that could be informally described[1] as "if all agents in set A use the product, then all agents in set B will eventually adopt the product". For example, for the above discussed social network N_1, we have $N_1 \vDash \{p\} \triangleright \{q, r\}$, which we usually write as just $N_1 \vDash p \triangleright q, r$.

At the same time, if a free sample of the product is given to agent r, then agent q will eventually adopt it because her threshold value is 1 and the influence of agent r on her is 4. Once agent q adopts the product, however, the product diffusion stops and the product will never be adopted by agent p because her threshold value is 7 and the total peer pressure from agents q and r on p will be only $1 + 5 = 6$. Therefore, for example, $N_1 \vDash \neg(r \triangleright p)$.

The properties of relation $A \triangleright B$ that we have discussed so far were specific to social network N_1. Let us now consider social network N_2 depicted in Figure 2. If

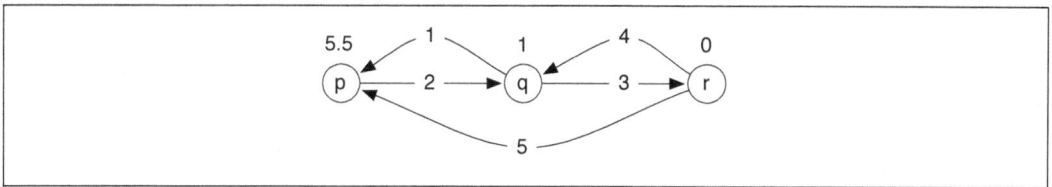

Figure 2: Social Network N_2

a free sample of the product is given in network N_2 to agent r and she starts using it, then, like it was for the network N_1, agent q will eventually adopt the product because her threshold value is only 1 and influence of agent r on agent q is 4. Unlike network N_1, however, the product diffusion does not stop at this point because now the total peer pressure of agents q and r on agent p is still $1 + 5 = 6$, but the threshold value of agent p in this network is only 5.5. Thus, agent p eventually will adopt the product. In other words, $N_2 \vDash r \triangleright p$.

An interesting property of network N_2 is that agent r has threshold value 0. Thus, she will eventually adopt the product even if no free product samples are given to any of the agents: $N_2 \vDash \varnothing \triangleright r$.

1.2 Sociograms

So far, we have discussed properties of specific social networks. In this article we study properties common to a class of networks. The classes of networks can be defined on different levels of abstraction. Perhaps the most natural approach is to

---

[1] We formally specify this relation in Definition 7.

study common properties of social networks that have the same topological structure. In other words, to study properties that do not depend on a specific choice of influence and threshold values, but only on the (unlabeled) graph of the network. Although such an approach appears to be the most natural, it unexpectedly results in a very complicated principles that seems to capture more properties of real numbers than properties of the influence relation.

We adopt a different level of abstraction in which we assume that the graph and the distribution of influences is fixed. We study all properties that are universal no matter what the threshold values are. This level of abstraction results in a simple set of properties that can be captured by the complete logic system presented in this paper. In the conclusion we discuss examples of properties of influence that are true for all graphs without fixing distribution of influences *and* distribution of the thresholds. To distinguish graphs labeled with influences and thresholds from those labeled with influences only, we call the former *social networks* and the latter *sociograms*. To some degree, the threshold values characterize the relation that exists between the product and the individual agents and the sociogram describes the influence relation between the agents. The term sociogram has been first introduced by psychosociologist Jacob Levy Moreno [16]. The sociograms, as defined in this article, are directed labeled graphs. The original Moreno's sociograms were neither directed nor labeled.

For example, the above discussed social networks N_1 and N_2 are different only by the threshold values that the agents have. Thus, we say that social networks N_1 and N_2 have the same sociogram. This common sociogram S_1 for networks N_1 and N_2 is depicted in Figure 3.

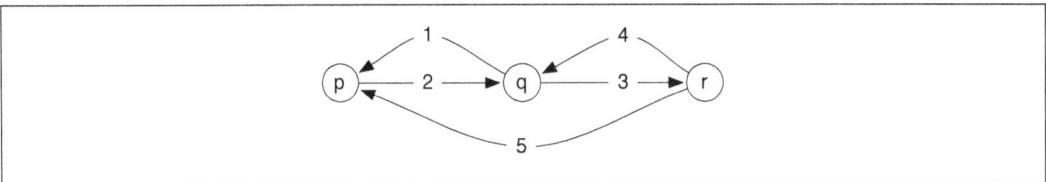

Figure 3: Sociogram S_1

We write $S \vDash \varphi$ if property φ is true for all social networks with sociogram S. For example, as we show in Proposition 1,

$$S_1 \vDash p \triangleright r \to q \triangleright r. \tag{1}$$

In other words, under any assignment of threshold values on sociogram S_1, if giving a free sample of the product to agent p will eventually lead to agent r adopting the product, then giving a free sample of the product to agent q would have the same effect.

1.3 Lighthouse Axiom

The main result of this article is a complete axiomatization of the propositional properties of relation $A \triangleright B$ for any given sociogram. Such an axiomatization consists of three axioms common to all sociograms and a sociogram-specific fourth axiom. The first three axioms are

1. Reflexivity: $A \triangleright B$ if $B \subseteq A$,

2. Transitivity: $A \triangleright B \to (B \triangleright C \to A \triangleright C)$,

3. Augmentation: $A \triangleright B \to (A, C \triangleright B, C)$,

where A, B denotes the union of sets A and B. These axioms were originally proposed by Armstrong [2] to describe functional dependence relation in database theory. They became known in database literature as Armstrong's axioms [7, p. 81]. Väänänen proposed a first order version of these principles [24] and their generalization for reasoning about approximate dependency [25]. Beeri, Fagin, and Howard [4] suggested a variation of Armstrong's axioms that describes properties of multi-valued dependence. Naumov and Nicholls [17] proposed another variation of these axioms that describes rationally functional dependence. The influence semantics of these axioms that we introduce in this article does not appear to be connected to the functional dependency semantics.

The sociogram-dependent fourth axiom captures the fact that in every group of agents in which at least one agent eventually adopts the product there is always an agent (or a nonempty subgroup of agents) who adopts the product first. In marketing such agents are sometimes called *lighthouse customers*. In any given group of agents, the distinctive property of lighthouse customers is that they adopt the product without any peer pressure coming from other agents in this group. The lighthouse customers adopt the product as a result of the peer pressure from the outside of the group. Our fourth axiom postulates the existence of lighthouse customers in any group of agents in which at least one agent eventually will adopt the product. Thus, we call this postulate *Lighthouse axiom*.

One possible way to state Lighthouse axiom is to say that if all agents in network N are partitioned into disjoint sets A and B, see Figure 4, and there is an agent $a \in A$ such that $N \vDash B \triangleright a$, then there must exist a "lighthouse" agent $\ell \in A$ such that the total peer pressure of all agents in set B on agent ℓ is no less than the

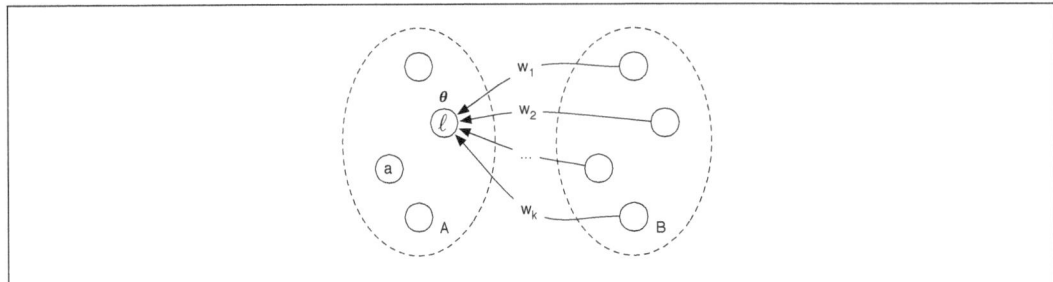

Figure 4: Lighthouse Axiom

threshold value of agent ℓ:

$$\theta \le w_1 + w_2 + \cdots + w_k.$$

Unfortunately, when stated this way, Lighthouse axiom refers to threshold value θ of agent ℓ. Thus, in this form, it is a property of the social network, rather than the corresponding sociogram.

It turns out, however, that there is a way to re-word the axiom so that it does not refer to threshold values. Namely, let us assume that for every agent $a \in A$ we choose a set of agents $C_a \subseteq A \cup B$ such that peer pressure of set C_a on agent a is no less than peer pressure of set B on agent a. The new form of Lighthouse axiom states that, under the above condition, if $N \vDash B \rhd a$, then there exists a "lighthouse" agent $\ell \in A$ such that $N \vDash C_\ell \rhd \ell$. The main result of this article is the completeness theorem for the logical system consisting of this form of Lighthouse axiom and the three Armstrong axioms.

1.4 Related Literature

Several logical frameworks for reasoning about diffusion in social networks have been studied before. Seligman, Liu, and Girard [22] proposed Facebook Logic for capturing properties of epistemic social networks in modal language, but did not give any axiomatization for this logic. They further developed this approach in papers [23, 13] where they introduced dynamic friendship relations. Christoff and Hansen [5] simplified Seligman, Liu, and Girard setting and gave a complete axiomatization of the logical system for this new setting. Christoff and Rendsvig proposed Minimal Threshold Influence Logic [6] that uses modal language to capture dynamic of diffusion in a threshold model and gave a complete axiomatization of this logic. Baltag, Christoff, Rendsvig, and Smets [3] discussed logics for informed update and prediction update. Informally, the languages of the described above systems feel significantly richer than the more succinct language of our system. However, neither of these systems capture principles similar to our Lighthouse axiom. Naumov and Tao [19, 18] used Armstrong's axioms to describe influence in social networks. They considered relation $A \rhd_b B$ that stands for "given marketing budget b, group of agents A can influence group of agents B". They gave modified versions of Armstrong axioms that capture properties of this relation for preventive and promotional marketing. Since they do not assume a fixed sociogram of the network, their approach does not capture any properties similar to our Lighthouse principle.

Diffusion in social networks is a special case of information flow on graphs. Logical systems for reasoning about various types of graph information flow has been studied before. Lighthouse axiom has certain resemblance with Gateway axiom for functional dependence on hypergraphs of secrets [15], Contiguity axiom [9] for graphical games, and Shield Wall axiom for fault tolerance in belief formation networks [10].

1.5 Outline

This article is organized as following. In Section 2 we introduce formal syntax and semantics of our logical system. Section 3 list the four axioms of the system. In Section 4, we give several examples of formal proofs in our system. In Section 5 we show some auxiliary results that are used later. Section 6 and Section 7 prove soundness and completeness theorems respectively. Section 9 concludes with a discussion of logical properties of unlabeled sociograms.

2 Syntax and Semantics

In this section we formally define a social network, a sociogram, and the influence relation.

Definition 1. *For any finite set $\mathcal{A}$, let $\Phi(\mathcal{A})$ be the minimal set of formulas such that*

1. $\bot \in \Phi(\mathcal{A})$,

2. $A \rhd B \in \Phi(\mathcal{A})$, *for each subsets $A, B \subseteq \mathcal{A}$,*

3. $\varphi \to \psi \in \Phi(\mathcal{A})$ *for each $\varphi, \psi \in \Phi(\mathcal{A})$.*

We assume that disjunction $\vee$ is defined through implication $\to$ and false constant $\bot$ in the standard way.

Definition 2. *A sociogram is pair $(\mathcal{A}, w)$, where*

1. *$\mathcal{A}$ is an arbitrary finite set (of agents),*

2. *w is a function that maps $\mathcal{A}^2$ into non-negative real numbers. Value $w(a,b)$ represents influence of agent a on agent b.*

Definition 3. *A social network is triple $(\mathcal{A}, w, \theta)$, where*

1. *$(\mathcal{A}, w)$ is a sociogram,*

2. *θ is a function that maps $\mathcal{A}$ into non-negative real numbers. Value $\theta(a)$ represents threshold value of agent $a \in \mathcal{A}$.*

We say that social network $(\mathcal{A}, w, \theta)$ is based on sociogram $(\mathcal{A}, w)$. We now proceed to define peer pressure on an agent by a group of agents in a given sociogram.

Definition 4. *For any sociogram $(\mathcal{A}, w)$ and any subset of agents $A \subseteq \mathcal{A}$, let $\|A\|_b = \sum_{a \in A} w(a, b)$.*

In the introduction we said that if, at some moment in time, an agent experiences peer pressure higher than her threshold value, then at some point in the future she will adopt the product. For the sake of simplicity, in our formal model we assume that time is discrete and that if at moment k an agent experiences sufficient peer pressure, then she adopts the product at moment $k+1$. Although this assumption, generally speaking, affects the "time dynamics" of product diffusion, it does not affect the final outcome of diffusion. Thus, this assumption, while simplifying the formal setting, does not change the properties of influence relation $A \triangleright B$. Given this assumption, if free samples of the product are given to all agents in set A at moment 0, then by A^k we mean the set of all agents who will adopt the product by moment k. The formal definition of A^k is below.

Definition 5. *For any $A \subseteq \mathcal{A}$ and any $k \in \mathbb{N}$, let subset $A^k \subseteq \mathcal{A}$ be defined recursively as follows:*

1. *$A^0 = A$,*

2. *$A^{k+1} = A^k \cup \{x \in \mathcal{A} \mid \|A^k\|_x \geq \theta(x)\}$.*

Corollary 1. *$(A^n)^k = A^{n+k}$.*

If free samples of the product are given to all agents in set A, then by A^* we mean the set of all agents who will eventually adopt the product. The formal definition of A^* is below.

Definition 6.
$$A^* = \bigcup_{k \geq 0} A^k.$$

The next definition specifies the formal semantics of our logical system. In particular, item 2 in this definition specifies the formal meaning of the influence relation.

Definition 7. *For any social network $N = (\mathcal{A}, w, \theta)$ and any $\varphi \in \Phi(\mathcal{A})$, let satisfiability relation $N \vDash \varphi$ be defined as follows*

1. $N \nvDash \bot$,

2. $N \vDash A \rhd B$ *if* $B \subseteq A^*$,

3. $N \vDash \psi \to \chi$ *if* $N \nvDash \psi$ *or* $N \vDash \chi$.

3 Axioms

Our logical system for an arbitrary sociogram $S = (\mathcal{A}, w)$ consists of propositional tautologies in language $\Phi(\mathcal{A})$ and the following additional axioms:

1. Reflexivity: $A \rhd B$ if $B \subseteq A$,

2. Transitivity: $A \rhd B \to (B \rhd C \to A \rhd C)$,

3. Augmentation: $A \rhd B \to (A, C \rhd B, C)$,

4. Lighthouse: if $A \sqcup B$ is a partition of the set of all agents $\mathcal{A}$ and $\{C_a\}_{a \in A}$ is a family of sets of agents such that $\|B\|_a \leq \|C_a\|_a$ for each $a \in A$, then
$$\bigvee_{a \in A} B \rhd a \to \bigvee_{a \in A} C_a \rhd a.$$

We write $\vdash_S \varphi$ if formula φ can be derived in our system using Modus Ponens inference rule. We sometimes write just $\vdash \varphi$ if the value of subscript S is clear from the context. We also write $X \vdash_S \varphi$ if formula φ could be derived in our system extended by a set of additional axioms X.

4 Examples

In this section we give three examples of formal proofs in our logical system to illustrate how the system works. Soundness of the system is shown in Section 6. We start by proving statement (1) from the introduction.

Proposition 1. $\vdash_{S_1} p \triangleright r \to q \triangleright r$, where S_1 is the sociogram depicted in Figure 3.

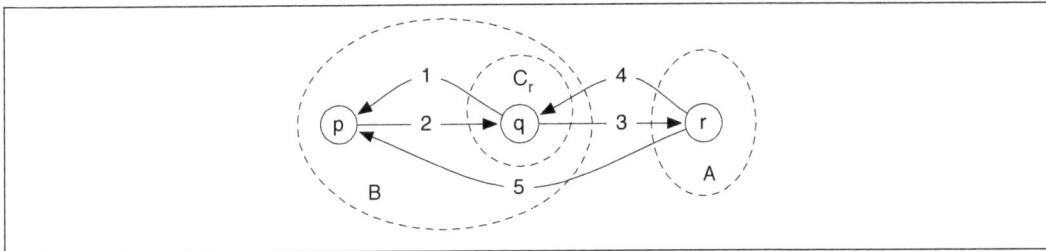

Figure 5: Towards Proof of Proposition 1

Proof. Let $A = \{r\}$, $B = \{p, q\}$, and $C_r = \{q\}$, see Figure 5. Note that

$$\|B\|_r = w(p, r) + w(q, r) = 0 + 3 = 3 = w(q, r) = \|C_r\|_r.$$

Hence, by Lighthouse axiom,

$$\vdash p, q \triangleright r \to q \triangleright r. \tag{2}$$

At the same time, by Transitivity axiom,

$$\vdash p, q \triangleright p \to (p \triangleright r \to p, q \triangleright r).$$

By Reflexivity axiom, $\vdash p, q \triangleright p$. Thus, by Modus Ponens inference rule,

$$\vdash p \triangleright r \to p, q \triangleright r.$$

Therefore, $\vdash p \triangleright r \to q \triangleright r$ using statement (2) and propositional logic reasoning. □

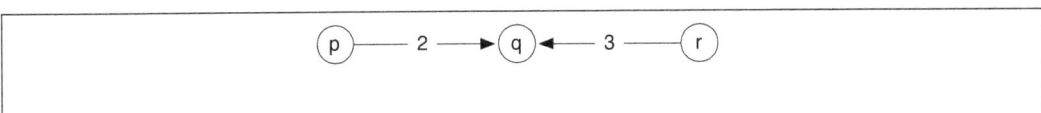

Figure 6: Sociogram S_2

Let us now consider sociogram S_2 depicted in Figure 6. Since in this sociogram agent r has higher influence on agent q than agent p, one might expect the following statement to be true for all social networks over sociogram S_2:

$$p \triangleright q \to r \triangleright q. \qquad (3)$$

Surprisingly, this is false. Namely, this statement is false for the social network depicted in Figure 7. This happens because agent r in this social network has

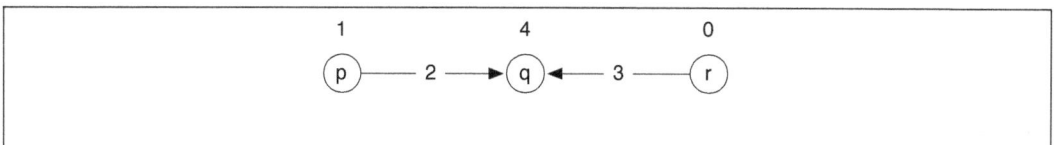

Figure 7: Social Network

threshold value 0. In other words, agent r is an "early adopter" who does not need any external peer pressure in order to buy the product. As a result, see Figure 8, we have $\{p\}^1 = \{p, r\}$. Once agent r adopts the product, the total peer pressure on agent q becomes $2 + 3 = 5$ and she will adopt the product as well. On the other hand, if the free sample is given to agent r, then neither agent p nor agent q ever adopt the product.

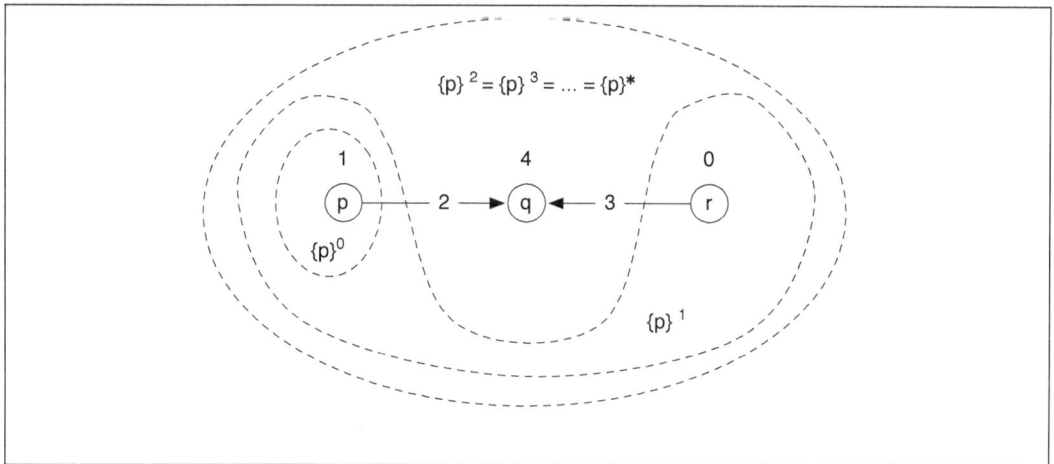

Figure 8: Social Network

Although statement (3) does not hold for some social networks over sociogram S_2, in the next proposition we show that a slightly modified version of this statement does hold for all such networks.

Proposition 2. $\vdash_{S_2} p \triangleright q \to (r \triangleright q \vee \varnothing \triangleright r)$, where S_2 is the sociogram depicted in Figure 6.

Proof. Let $A = \{q, r\}$, $B = \{p\}$, $C_q = \{r\}$, and $C_r = \varnothing$, see Figure 9. Note that

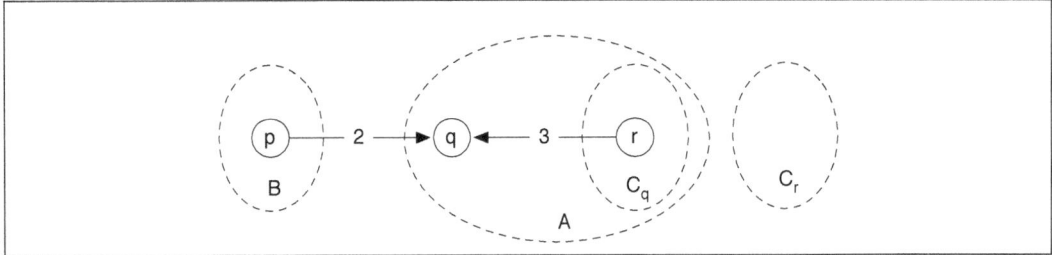

Figure 9: Towards Proof of Proposition 2

$$\|B\|_q = w(p, q) = 2 < 3 = w(r, q) = \|C_q\|_q$$

and

$$\|B\|_r = w(p, r) = 0 = \|\varnothing\|_r = \|C_r\|_r.$$

Thus, by Lighthouse axiom,

$$\vdash p \triangleright q \vee p \triangleright r \to r \triangleright q \vee \varnothing \triangleright r.$$

Therefore, $\vdash p \triangleright r \to r \triangleright q \vee \varnothing \triangleright r$. □

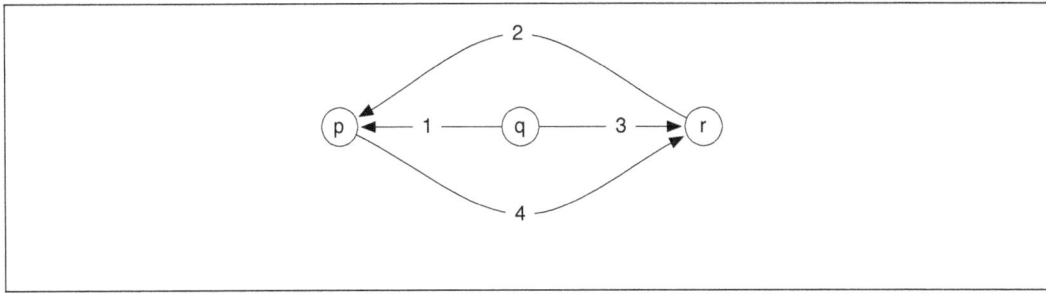

Figure 10: Sociogram S_3

Proposition 3. $\vdash_{S_3} q \triangleright p \vee q \triangleright r \to p \triangleright r \vee r \triangleright p$, where S_3 is the sociogram depicted in Figure 10.

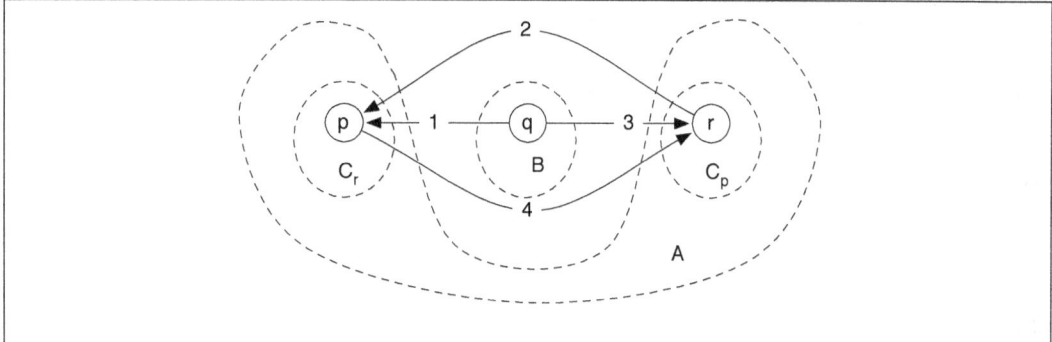

Figure 11: Towards Proof of Proposition 3

Proof. Let $A = \{p, r\}$, $B = \{q\}$, $C_p = \{r\}$, and $C_r = \{p\}$, see Figure 11. Note that

$$\|B\|_p = w(q, p) = 1 < 2 = w(r, p) = \|C_p\|_p$$

and

$$\|B\|_r = w(q, r) = 3 < 4 = w(p, r) = \|C_r\|_r.$$

Therefore, by Lighthouse axiom, $\vdash q \triangleright p \vee q \triangleright r \to p \triangleright r \vee r \triangleright p$. □

5 Properties of Star Closure

In this section we prove several technical properties of A^* that are used later in the proofs of soundness and completeness.

Lemma 1. *If $A^1 = A$, then $A^k = A$ for each $k \geq 0$.*

Proof. We prove this lemma by induction on k. If $k = 0$, then $A^0 = A$ by Definition 5. If $k > 0$, then by Corollary 1, assumption $A^1 = A$, and the induction hypothesis, $A^k = (A^1)^{k-1} = A^{k-1} = A$. □

Lemma 2. $A^* = A^k$ *for some $k \geq 0$.*

Proof. The statement of the lemma follows from the assumption in Definition 3 that set $\mathcal{A}$ is finite. □

Lemma 3. *If $x \notin A^*$, then $\theta(x) > \|A^*\|_x$, for each subset $A \subseteq \mathcal{A}$ and each agent $x \in \mathcal{A}$.*

Proof. By Lemma 2, there is $k \geq 0$ such that $A^* = A^k$. Suppose that $\|A^*\|_x \geq \theta(x)$. Thus, $\|A^k\|_x \geq \theta(x)$. Hence, $x \in A^{k+1}$, by Definition 5. Thus, $x \in A^*$ by Definition 6, which is a contradiction to the assumption of the lemma. □

Lemma 4. $A \subseteq A^*$.

Proof. By Definition 5 and Definition 6, $A = A^0 \subseteq \bigcup_{k \geq 0} A^k = A^*$. □

Lemma 5. $(A^*)^* \subseteq A^*$.

Proof. By Lemma 2, there are $n, k \geq 0$ such that $A^* = A^n$ and $(A^*)^* = (A^*)^k$. Thus, by Corollary 1 and Definition 6,

$$(A^*)^* = (A^*)^k = (A^n)^k = A^{n+k} \subseteq \bigcup_{m \geq 0} A^m = A^*.$$

□

Lemma 6. *If* $A \subseteq B$, *then* $A^k \subseteq B^k$, *for each* $k \geq 0$.

Proof. We prove the statement of the lemma by induction on k. If $k = 0$, then $A^0 = A \subseteq B = B^0$ by Definition 5.

Suppose that $A^k \subseteq B^k$. Let $x \in A^{k+1}$. It suffices to show that $x \in B^{k+1}$. Indeed, by Definition 5, assumption $x \in A^{k+1}$ implies that either $x \in A^k$ or $\|A^k\|_x \geq \theta(x)$. In the first case, by the induction hypothesis, $x \in A^k \subseteq B^k$. Thus, $x \in B^k$. Therefore, $x \in B^{k+1}$ by Definition 5.

In the second case, by Definition 4 and assumption $A^k \subseteq B^k$,

$$\|B^k\|_x = \sum_{b \in B^k} w(b, x) \geq \sum_{a \in A^k} w(a, x) = \|A^k\|_x \geq \theta(x).$$

Therefore, $x \in B^{k+1}$ by Definition 5. □

Corollary 2. *If* $A \subseteq B$, *then* $A^* \subseteq B^*$.

Lemma 7. $A^* \cup B^* \subseteq (A \cup B)^*$.

Proof. Note that $A \subseteq A \cup B$ and $B \subseteq A \cup B$. Thus, $A^* \subseteq (A \cup B)^*$ and $B^* \subseteq (A \cup B)^*$ by Corollary 2. Therefore, $A^* \cup B^* \subseteq (A \cup B)^*$. □

6 Soundness

In this section we prove the soundness of our logical system with respect to the semantics given in Definition 7. The soundness of propositional tautologies and Modus Ponens inference rule is straightforward. Below we show the soundness of each of the remaining four axioms as separate lemmas. In the lemmas that follow we assume that $S = (\mathcal{A}, w, \theta)$ is a social network and A, B, and C are subsets of $\mathcal{A}$.

Lemma 8. *If $B \subseteq A$, then $S \vDash A \triangleright B$.*

Proof. By Lemma 4, $A \subseteq A^*$. Thus, $B \subseteq A^*$ by the assumption of the lemma. Therefore, $S \vDash A \triangleright B$, by Definition 7. □

Lemma 9. *If $S \vDash A \triangleright B$ and $S \vDash B \triangleright C$, then $S \vDash A \triangleright C$.*

Proof. By Definition 7, assumption $S \vDash A \triangleright B$ implies that $B \subseteq A^*$. Hence, $B^* \subseteq (A^*)^*$ by Corollary 2. Thus, $B^* \subseteq A^*$ by Lemma 5. At the same time, $C \subseteq B^*$ by assumption $S \vDash B \triangleright C$ and Definition 7. Thus, $C \subseteq A^*$. Therefore, $S \vDash A \triangleright C$ by Definition 7. □

Lemma 10. *If $S \vDash A \triangleright B$, then $S \vDash A, C \triangleright B, C$.*

Proof. Suppose that $S \vDash A \triangleright B$. Thus, $B \subseteq A^*$ by Definition 7. Note that $C \subseteq C^*$ by Lemma 4. Thus, $B \cup C \subseteq A^* \cup C^* \subseteq (A \cup C)^*$, by Lemma 7. Therefore, $S \vDash A, C \triangleright B, C$, by Definition 7. □

Lemma 11. *If $S \vDash B \triangleright a_0$ for some $a_0 \in A$, then there is $\ell \in A$ such that $S \vDash C_\ell \triangleright \ell$, where $A \sqcup B$ is a partition of the set of all agents $\mathcal{A}$ and $\{C_a\}_{a \in A}$ is a family of sets of agents such that $\|B\|_a \leq \|C_a\|_a$ for each $a \in A$.*

Proof. Note that assumption $S \vDash B \triangleright a_0$ by Definition 7 implies that $a_0 \in B^*$. On the other hand, assumption $a_0 \in A$ implies that $a_0 \notin B$ because $A \sqcup B$ is a partition of set $\mathcal{A}$. Thus, $B^* \neq B$. Hence, by Definition 6, there must exist k such that $B^k \neq B$. Then, $B^1 \neq B$ by Lemma 1. Thus, there must exist $\ell \in B^1 \setminus B$. Hence, $\|B\|_\ell \geq \theta(\ell)$ by Definition 5. Then, by the assumption of the lemma, $\|C_\ell\|_\ell \geq \|B\|_\ell \geq \theta(\ell)$. Thus, $\ell \in C_\ell^1$, by Definition 5. Hence, $\ell \in C_\ell^*$ by Definition 6. Therefore, $S \vDash C_\ell \triangleright \ell$ by Definition 7. Finally, note that $\ell \in A$ because $\ell \in B^1 \setminus B$ and $A \sqcup B$ is a partition of the set $\mathcal{A}$. □

This concludes the proof of the soundness of our logical system.

7 Completeness

In this section we prove the completeness of our logical system with respect to the semantics given in Definition 7. This result is formally stated as Theorem 1 in the end of this section. The proof of completeness consists in the construction of a "canonical" social network. We start, however, we a few technical lemmas and definitions.

7.1 Preliminaries

Let us first prove a useful property of real numbers.

Lemma 12. *If $\varepsilon > 0$ is a real number and x and y are any real numbers such that either $x = y$ or $|x - y| > \varepsilon$. Then, $x + \varepsilon > y$ implies $x \geq y$.*

Proof. Suppose $y > x$. Hence, $x \neq y$. Thus, $|x - y| > \varepsilon$, by the assumption of the lemma. Then, $y - x > \varepsilon$, because $y > x$. Therefore, $x + \varepsilon < y$. □

We now assume a fixed sociogram $(\mathcal{A}, w)$ and a fixed maximal consistent subset X of $\Phi(\mathcal{A})$.

Definition 8. $\hat{A} = \{a \in \mathcal{A} \mid X \vdash A \triangleright a\}$ *for each subset $A \subseteq \mathcal{A}$.*

Choose ε to be any positive real number such that $\varepsilon < \|A\|_a - \|B\|_a$ for each agent $a \in \mathcal{A}$ and each subsets $A, B \subseteq \mathcal{A}$, such that $\|A\|_a > \|B\|_a$. This could be achieved because set $\mathcal{A}$ is finite.

Lemma 13. *For any subsets $A, B \subseteq \mathcal{A}$ and any agent $a \in \mathcal{A}$ if $\|A\|_a + \varepsilon > \|B\|_a$, then $\|A\|_a \geq \|B\|_a$.*

Proof. By the choice of ε, we have either $\|A\|_a = \|B\|_a$ or $|(\|A\|_a - \|B\|_a)| > \varepsilon$. Thus, $\|A\|_a \geq \|B\|_a$ by Lemma 12. □

Lemma 14. $A \subseteq \hat{A}$ *for each subset $A \subseteq \mathcal{A}$.*

Proof. Suppose that $a \in A$. Thus, $\vdash A \triangleright a$ by Reflexivity axiom. Therefore, $a \in \hat{A}$ by Definition 8. □

Lemma 15. $X \vdash A \triangleright \hat{A}$, *for each subset $A \subseteq \mathcal{A}$.*

Proof. Let $\hat{A} = \{a_1, \ldots, a_n\}$. By the definition of $\hat{A}$, $X \vdash A \triangleright a_i$, for any $i \leq n$. We prove, by induction on k, that $X \vdash A \triangleright a_1, \ldots, a_k$ for each $0 \leq k \leq n$.
Base Case: $X \vdash A \triangleright \varnothing$ by Reflexivity axiom.

Induction Step: Assume that $X \vdash A \rhd a_1, \ldots, a_k$. By Augmentation axiom,

$$X \vdash A, a_{k+1} \rhd a_1, \ldots, a_k, a_{k+1}. \tag{4}$$

Recall that $X \vdash A \rhd a_{k+1}$. Again by Augmentation axiom, $X \vdash A \rhd A, a_{k+1}$. Hence, $X \vdash A \rhd a_1, \ldots, a_k, a_{k+1}$, by (4) and Transitivity axiom. □

7.2 Canonical Social Network

Next, based on the sociogram $(\mathcal{A}, w)$ and the maximal consistent set X, we define the "canonical" social network $N_X = (\mathcal{A}, w, \theta)$. We then proceed to prove the core properties of this network.

Definition 9.
$$\theta(a) = \begin{cases} 0, & \text{if } X \vdash \varnothing \rhd a, \\ \max_{a \notin \widehat{B}} \|\widehat{B}\|_a + \varepsilon, & \text{otherwise.} \end{cases}$$

The maximum in the above definition is taken over all subsets B of $\mathcal{A}$ such that $\widehat{B}$ does not contain agent a.

Lemma 16. *Function $\theta(a)$ is well-defined for each $a \in \mathcal{A}$.*

Proof. We need to show that if $X \nvdash \varnothing \rhd a$, then there is at least one subset $B \subseteq \mathcal{A}$ such that $a \notin \widehat{B}$. It suffices to show that $a \notin \widehat{\varnothing}$, which is true due to assumption $X \nvdash \varnothing \rhd a$ and Definition 8. □

Lemma 17. *For any subset $B \subseteq \mathcal{A}$, if $a \in \mathcal{A} \setminus B^*$, then there is $C \subseteq \mathcal{A}$ such that $a \notin \widehat{C}$ and $\theta(a) = \|\widehat{C}\|_a + \varepsilon$.*

Proof. If $\theta(a) = 0$, then $a \in B^1$ due to Definition 5. Thus, $a \in B^*$ by Definition 6, which is a contradiction to the assumption $a \in \mathcal{A} \setminus B^*$. Suppose now that $\theta(a) > 0$, thus, by Definition 9, there is at least one $C \subseteq \mathcal{A}$ such that $a \notin \widehat{C}$ and $\theta(a) = \|\widehat{C}\|_a + \varepsilon$. □

Lemma 18. *If $B \subseteq \mathcal{A}$ and $a \in \mathcal{A} \setminus \widehat{B}$, then $\theta(a) > \|\widehat{B}\|_a$.*

Proof. Case I: $X \vdash \varnothing \rhd a$. Note that $X \vdash B \rhd \varnothing$ by Reflexivity axiom. Thus, $X \vdash B \rhd a$ by Transitivity axiom. Hence, $a \in \widehat{B}$ by Definition 8, which is a contradiction to the assumption of the lemma.

Case II: $X \nvdash \varnothing \rhd a$. Thus, $\theta(a) > \|\widehat{B}\|_a$ by Definition 9. □

Lemma 19. $(\widehat{B})^k = \widehat{B}$ *for each $B \subseteq \mathcal{A}$ and each $k \geq 0$.*

Proof. We prove this statement by induction on k. If $k = 0$, then $(\widehat{B})^k = \widehat{B}$, by Definition 5. Note next that by Definition 5, the induction hypothesis, and Lemma 18,

$$\begin{aligned}(\widehat{B})^{k+1} &= (\widehat{B})^k \cup \{a \in \mathcal{A} \mid \|(\widehat{B})^k\|_a \geq \theta(a)\} \\ &= \widehat{B} \cup \{a \in \mathcal{A} \mid \|\widehat{B}\|_a \geq \theta(a)\} \\ &= \widehat{B} \cup \{a \in \mathcal{A} \setminus \widehat{B} \mid \|\widehat{B}\|_a \geq \theta(a)\} = \widehat{B} \cup \varnothing = \widehat{B}.\end{aligned}$$

□

Lemma 20. $(\widehat{B})^* = \widehat{B}$ for each $B \subseteq \mathcal{A}$.

Proof. By Definition 6 and Lemma 19, $(\widehat{B})^* = \bigcup_{k \geq 0}(\widehat{B})^k = \bigcup_{k \geq 0}\widehat{B} = \widehat{B}$. □

Lemma 21. For each $B \subseteq \mathcal{A}$, if $a \in B^*$, then $X \vdash B \rhd a$.

Proof. Suppose $a \in B^*$. By Lemma 14, $B \subseteq \widehat{B}$. Then, $B^* \subseteq (\widehat{B})^*$ by Corollary 2. Thus, $a \in (\widehat{B})^*$. Hence, $a \in \widehat{B}$ by Lemma 20. Therefore, $X \vdash B \rhd a$ by Definition 8.

□

Lemma 22. For each $B \subseteq \mathcal{A}$ and each $a \in \mathcal{A}$, if $X \vdash B \rhd a$, then $a \in B^*$.

Proof. By Lemma 3, $\theta(x) > \|B^*\|_x$ for each $x \in \mathcal{A} \setminus B^*$. At the same time, by Lemma 17, for each $x \in \mathcal{A} \setminus B^*$ there is C_x such that $x \notin \widehat{C_x}$ and $\theta(x) = \|\widehat{C_x}\|_x + \varepsilon$. Hence, $\|\widehat{C_x}\|_x + \varepsilon > \|B^*\|_x$ for each $x \in \mathcal{A} \setminus B^*$. Thus, by Lemma 13, $\|\widehat{C_x}\|_x \geq \|B^*\|_x$ for each $x \in \mathcal{A} \setminus B^*$.

Consider partition $(\mathcal{A} \setminus B^*) \sqcup B^*$ of $\mathcal{A}$. By Lighthouse axiom,

$$\vdash \bigvee_{x \in \mathcal{A} \setminus B^*} B^* \rhd x \to \bigvee_{x \in \mathcal{A} \setminus B^*} \widehat{C_x} \rhd x. \tag{5}$$

Suppose that $a \notin B^*$, Lemma 4 and Reflexivity axiom imply that $\vdash B^* \rhd B$. Thus, by assumption $X \vdash B \rhd a$ and Transitivity axiom, $X \vdash B^* \rhd a$. Hence, statement (5) implies that

$$X \vdash \bigvee_{x \in \mathcal{A} \setminus B^*} \widehat{C_x} \rhd x.$$

Then, due to the maximality of set X, there must exist $x_0 \in \mathcal{A} \setminus B^*$ such that $X \vdash \widehat{C_{x_0}} \rhd x_0$. Thus, $X \vdash C_{x_0} \rhd x_0$, due to Lemma 15 and Transitivity axiom: $\vdash C_{x_0} \rhd \widehat{C_{x_0}} \to (\widehat{C_{x_0}} \rhd x_0 \to C_{x_0} \rhd x_0)$. Hence, $x_0 \in \widehat{C_{x_0}}$ by Definition 8, which is a contradiction with the choice of set C_x. □

Lemma 23. $N_X \vDash \varphi$ if and only if $\varphi \in X$, for each formula $\varphi \in \Phi(\mathcal{A})$.

Proof. We prove this lemma by induction on structural complexity of formula φ. Cases when formula φ is $\bot$ or has form $\psi_1 \to \psi_2$ follow in the standard way from Definition 7 and the assumptions of maximality and consistency of set X. Suppose that φ has form $A \triangleright B$.

($\Rightarrow$): Suppose that $N_X \vDash A \triangleright B$. Then $B \subseteq A^*$ by Definition 7. Hence, $b \in A^*$ for each $b \in B$. Thus, $X \vdash A \triangleright b$ for each $b \in B$ by Lemma 21. Hence, $b \in \hat{A}$ for each $b \in B$ by Definition 8. In other words, $B \subseteq \hat{A}$. Thus, by Reflexivity axiom, $\vdash \hat{A} \triangleright B$. On the other hand, $X \vdash A \triangleright \hat{A}$ by Lemma 15. Therefore, $X \vdash A \triangleright B$ by Transitivity axiom.

($\Leftarrow$): Assume $X \vdash A \triangleright B$. By Reflexivity axiom, $\vdash B \triangleright b$ for every $b \in B$. Hence, $X \vdash A \triangleright b$ for each $b \in B$ by Transitivity axiom. Thus, $b \in A^*$ for each $b \in B$, by Lemma 22. In other words, $B \subseteq A^*$. Therefore, $N_X \vDash A \triangleright B$ by Definition 7. $\square$

7.3 Main Result

We are now ready to state and prove the completeness theorem for our logical system with respect to the semantics given in Definition 7.

Theorem 1. *For any sociogram $(\mathcal{A}, w)$ and any formula $\varphi \in \Phi(\mathcal{A})$, if $N \vDash \varphi$ for each social network N based on sociogram $(\mathcal{A}, w)$, then $\vdash \varphi$.*

Proof. Suppose that $\nvdash \varphi$. Let X be a maximal consistent subset of $\Phi(\mathcal{A})$ such that $\varphi \notin X$. By Lemma 23, $N_X \nvDash \varphi$. $\square$

8 Decidability

In this section we discuss decidability of our logical system for any fixed sociogram $(\mathcal{A}, w)$. Note that we allow arbitrary real numbers as subscripts in formula $A \triangleright_c B$. Thus, the set of all formulas $\Phi(\mathcal{A})$ is uncountable and its elements can not be used as inputs of a Turing machine. In order to avoid this issue, in this section we modify Definition 1, Definition 3, and Definition 2 by assuming that only rational numbers could be used as subscripts in our atomic formulas $A \triangleright_c B$, as influence values, and as threshold values. It is easy to see that the above proof of completeness is still valid. From this change point of view, the only non-trivial place is the choice of ε for the given sociogram $(\mathcal{A}, w)$ that we have made right after Definition 8. Note, however, that the required ε could always be choose to be a rational number because 0 is a limit point of the set of positive rational numbers.

Theorem 2. *For any given sociogram $S = (\mathcal{A}, w)$, set $\{\varphi \in \Phi(\mathcal{A}) \mid \vdash_S \varphi\}$ is decidable.*

Proof. According to Theorem 1, $\vdash_S \varphi$ if and only if formula φ is true for each social network $(\mathcal{A}, w, \theta)$ based on sociogram S. This, of course, does not imply the decidability because there are infinitely many social networks based on sociogram S. However, it turns out that the proof of Theorem 1 that we gave above actually shows a stronger result: $\vdash_S \varphi$ if and only if formula φ is true for each social network from a specific finite class $C(S)$ of networks based on sociogram S.

Once existence of such *finite* class of social networks $C(S)$ is establish, we should be able to claim the decidability result because one can always verify if a formula φ is true for each out of *finitely* many given networks.

We are now ready to describe the finite class of social networks $C(S)$. The social network over sociogram S is completely defined by specifying threshold function θ. In the proof of Theorem 1, this is done in Definition 9. This definition depends on ε and maximal consistent set of formulas X. Note however that the choice of ε does not depend on X and could be made based on sociogram S alone. Once ε is fixed, the set of all values of function θ, as specified in Definition 9, belongs to *finite* set

$$\{0\} \cup \{\|A\|_a + \varepsilon \mid a \in \mathcal{A}, A \subseteq \mathcal{A}\}.$$

The set of all social networks over sociogram S whose threshold functions use only values from the above set is the desired finite class of social networks $C(S)$. □

9 Conclusion

In this article we have studied properties of influence common to all social networks with the same weighted sociogram. We introduced a logical system for reasoning about these properties and proved soundness and completeness of this system. We have established that the logical system is decidable if its syntax and semantics are restricted to rational numbers.

As has been mentioned in that introduction, perhaps more natural question to consider is axiomatization of all common influence properties of social networks with the same graph, without fixing distribution of either weights or thresholds. Surprisingly, such setting yields a much more complicated set of properties. We discuss some of these properties below.

Consider, for example, unweighted sociogram U_1 depicted in Figure 12. Let $N = (\mathcal{A}, w, \theta)$ be a social network based on U_1. Furthermore, assume that in social network N (i) neither of the agents p_1, p_2, q_1, q_2 is an early adopter, (ii) $N \vDash p_1, p_2 \triangleright r$, and (iii) $N \vDash q_1, q_2 \triangleright r$. Thus, $w(p_1, r) + w(p_2, r) \geq \theta(r)$ and $w(q_1, r) + w(q_2, r) \geq \theta(r)$. The first inequality implies that at least one out of $w(p_1, r)$ and $w(p_2, r)$ is greater or equal than $\theta(r)/2$. In other words, there is $i \in \{1, 2\}$ such that $w(p_i, r) \geq \theta/2$.

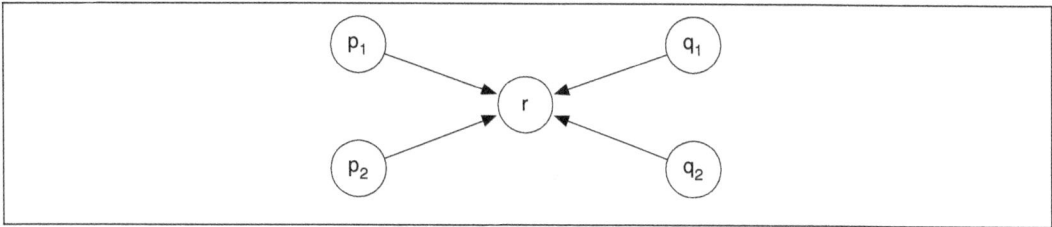

Figure 12: Unweighted Sociogram U_1

Similarly, the second inequality implies that there is $j \in \{1,2\}$ such that $w(q_j, r) \geq \theta/2$. Thus,

$$\|\{p_i, q_j\}\|_r = w(p_i, r) + w(q_j,) \geq \theta/2 + \theta/2 = \theta.$$

Hence, $r \in \{p_i, q_j\}^1 \subseteq \{p_i, q_j\}^*$. Then, $N \vDash p_i, q_j \triangleright r$. So, we have shown that for any social network N based on unweighted sociogram U_1 and satisfying the conditions (i), (ii), (iii), there are $i, j \in \{1, 2\}$ such that $N \vDash p_i, q_j \triangleright r$. This could be formally stated as

$$U_2 \vDash p_1, p_2 \triangleright r \wedge q_1, q_2 \triangleright r \to \bigvee_{i=1}^{2} \bigvee_{j=1}^{2} p_i, q_j \triangleright r \vee \bigvee_{x \in \{p_1, p_2, q_1, q_2\}} \varnothing \triangleright x,$$

where disjunction $\bigvee_{x \in \{p_1, p_2, q_1, q_2\}} \varnothing \triangleright x$ captures the statement that one of agents p_1, p_2, q_1, q_2 is an early adopter. The above principle is just an example of a nontrivial property of diffusion common to all social networks with the same unweighted sociogram. This example can be stated in a more general form as

$$U_2 \vDash \bigwedge_{i=1}^{n} p_{i1}, p_{i2}, \ldots, p_{in} \triangleright q$$

$$\to \bigvee_{j_1=1}^{n} \bigvee_{j_2=1}^{n} \cdots \bigvee_{j_n=1}^{n} p_{1j_1}, p_{2j_2}, \ldots, p_{nj_n} \triangleright q \vee \bigvee_{i=1}^{n} \bigvee_{j=1}^{n} \varnothing \triangleright p_{ij},$$

where U_2 is unweighted sociogram depicted in Figure 13. Complete axiomatization of properties of influence common to all social networks with a given graph remains an open problem.

Another possible extension of our work, suggested by an anonymous reviewer, is to consider common logical principles of all social networks in which all agents have the same threshold values. Such more narrow class of models would results in a larger set of universally true principles, some of which will not be provable from the axioms of our logical system. Formula $p \triangleright q \to p \triangleright r$ is an example of such

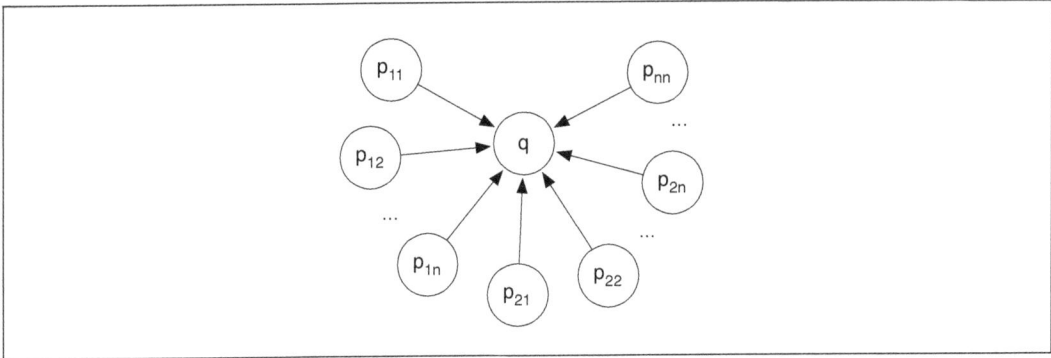

Figure 13: Unweighted Sociogram U_2

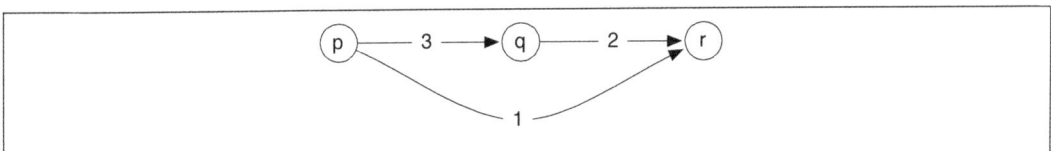

Figure 14: Sociogram S_4

principle for the sociogram depicted in Figure 14. To see how much different this new setting is from the one discussed earlier in the article, note that in all models from this class, either all agents are early adopters or none is.

References

[1] Krzysztof R Apt and Evangelos Markakis. Social networks with competing products. *Fundamenta Informaticae*, 129(3):225–250, 2014.

[2] W. W. Armstrong. Dependency structures of data base relationships. In *Information Processing 74 (Proc. IFIP Congress, Stockholm, 1974)*, pages 580–583. North-Holland, Amsterdam, 1974.

[3] Alexandru Baltag, Zoé Christoff, Rasmus K. Rendsvig, and Sonja Smets. Dynamic epistemic logics of diffusion and prediction in social networks. In *12th Conference on Logic and the Foundations of Game and Decision Theory (LOFT), Maastricht, the Netherlands*, 2016.

[4] Catriel Beeri, Ronald Fagin, and John H. Howard. A complete axiomatization for functional and multivalued dependencies in database relations. In *SIGMOD '77: Proceedings of the 1977 ACM SIGMOD international conference on Management of data*, pages 47–61, New York, NY, USA, 1977. ACM.

[5] Zoé Christoff and Jens Ulrik Hansen. A logic for diffusion in social networks. *Journal of Applied Logic*, 13(1):48 – 77, 2015.

[6] Zoé Christoff and Rasmus K. Rendsvig. Dynamic logics for threshold models and their epistemic extension. In *Epistemic Logic for Individual, Social, and Interactive Epistemology workshop*, 2014.

[7] Hector Garcia-Molina, Jeffrey Ullman, and Jennifer Widom. *Database Systems: The Complete Book*. Prentice-Hall, second edition, 2009.

[8] Mark Granovetter. Threshold models of collective behavior. *American journal of sociology*, pages 1420–1443, 1978.

[9] Kristine Harjes and Pavel Naumov. Functional dependence in strategic games. In *1st International Workshop on Strategic Reasoning, March 2013, Rome, Italy, Electronic Proceedings in Theoretical Computer Science 112*, pages 9–15, 2013. Full version to appear in Notre Dame Journal of Formal Logic.

[10] Sarah Holbrook and Pavel Naumov. Fault tolerance in belief formation networks. In Luis Fariñas del Cerro, Andreas Herzig, and Jérôme Mengin, editors, *JELIA*, volume 7519 of *Lecture Notes in Computer Science*, pages 267–280. Springer, 2012.

[11] David Kempe, Jon Kleinberg, and Éva Tardos. Maximizing the spread of influence through a social network. In *Proceedings of the ninth ACM SIGKDD international conference on Knowledge discovery and data mining*, pages 137–146. ACM, 2003.

[12] David Kempe, Jon Kleinberg, and Éva Tardos. Influential nodes in a diffusion model for social networks. In *Automata, languages and programming*, pages 1127–1138. Springer, 2005.

[13] Fenrong Liu, Jeremy Seligman, and Patrick Girard. Logical dynamics of belief change in the community. *Synthese*, 191(11):2403–2431, 2014.

[14] Michael W Macy. Chains of cooperation: Threshold effects in collective action. *American Sociological Review*, pages 730–747, 1991.

[15] Sara Miner More and Pavel Naumov. The functional dependence relation on hypergraphs of secrets. In João Leite, Paolo Torroni, Thomas Ågotnes, Guido Boella, and Leon van der Torre, editors, *CLIMA*, volume 6814 of *Lecture Notes in Computer Science*, pages 29–40. Springer, 2011.

[16] Jacob Levy Moreno. *Who shall survive?: A new approach to the problem of human interrelations*. Nervous and Mental Disease Publishing Co, 1934.

[17] Pavel Naumov and Brittany Nicholls. Rationally functional dependence. *Journal of Philosophical Logic*, 43(2-3):603–616, 2014.

[18] Pavel Naumov and Jia Tao. Marketing impact on diffusion in social networks. *Journal of Applied Logic*. (to appear).

[19] Pavel Naumov and Jia Tao. Marketing impact on diffusion in social networks. In *12th Conference on Logic and the Foundations of Game and Decision Theory (LOFT), Maastricht, the Netherlands*, 2016.

[20] Kazumi Saito, Ryohei Nakano, and Masahiro Kimura. Prediction of information diffusion probabilities for independent cascade model. In Ignac Lovrek, Robert J. Howlett, and Lakhmi C. Jain, editors, *Knowledge-Based Intelligent Information and Engineering Systems*, volume 5179 of *Lecture Notes in Computer Science*, pages 67–75. Springer

Berlin Heidelberg, 2008.

[21] Thomas C. Schelling. Dynamic models of segregation. *Journal of Mathematical Sociology*, 1:143–186, 1971.

[22] Jeremy Seligman, Fenrong Liu, and Patrick Girard. Logic in the community. In *Logic and Its Applications*, pages 178–188. Springer, 2011.

[23] Jeremy Seligman, Fenrong Liu, and Patrick Girard. Facebook and the epistemic logic of friendship. In *14th conference on Theoretical Aspects of Rationality and Knowledge (TARK '13), January 2013, Chennai, India*, pages 229–238, 2013.

[24] Jouko Väänänen. *Dependence logic: A new approach to independence friendly logic*, volume 70. Cambridge University Press, 2007.

[25] Jouko Väänänen. The logic of approximate dependence. *arXiv preprint arXiv:1408.4437*, 2014.

[26] Thomas W Valente. Social network thresholds in the diffusion of innovations. *Social networks*, 18(1):69–89, 1996.

A Labelled Sequent Calculus for Half-Order Modal Logic

Romas Alonderis
*Vilnius University Institute of Data Science and Digital Technologies
Akademijos 4, Vilnius 2600, LITHUANIA*
r.alonderis@post.penki.lt

Jūratė Sakauskaitė
*Vilnius University Institute of Data Science and Digital Technologies
Akademijos 4, Vilnius 2600, LITHUANIA*
jurate.sakalauskaite@mii.vu.lt

Abstract

We introduce the labelled Gentzen-type structural rules and cut-free sequent calculus **GHOML** for the half-order modal logic without function symbols and prove that the calculus is sound and complete for the logic. Using syntactic methods, we prove that the structural and cut rules are admissible in **GHOML**. The obtained calculus enables us to present a decision procedure for the half-order modal logic considered.

Key words: half-order modal logic, sequent calculus, admissibility of cut and structural rules, decidability

Classification codes: 03B44, 03F03

1 Introduction

In [2], a novel extension of normal propositional modal logic is introduced. The freeze quantifier is used in the obtained logic (called half-order modal logic) instead of the traditional universal and existential quantifiers. The freeze quantifier "$x.$" binds ("freezes") the variable x to the unique state-value of the current state of a Kripke structure. For example, if the unique state-values of the states s_1 and s_2 are

---

We thank the anonymous referees for their thorough review, which significantly contributed to improving the quality of the publication.

5 and 7, respectively, then the formula $x.(3 \leq x)$ is equivalent to $3 \leq 5$ at the state s_1 and to $3 \leq 7$ at the state s_2.

Addition of variables to the language that range over a aet of values and which can then be quantified over, increases the expressivity of the language without requiring to adopt a full first-order modal logic. This allows us to avoid the complications, both technical and motivational, inherent in such a move. In this respect, half-order modal logics are interesting, and, one could argue, understudied.

Half-order modal logic (**HOML**) is based on the normal modal logic **K**. The extensions of this logic are obtained by introducing additional modal operators and properties for the set of states and the accessibility relation. The corresponding Hilbert-type proof systems are obtained by adding appropriate axioms and derivation rules so that the new semantical features are captured.

Modal logics are used to express the properties of e. g., knowledge, belief, desire, intention, [3, 5]. For example, if the accessibility relation between states of interpretations is the equivalence relation, then we get the modal logic **S5** used to describe the perfect knowledge. The normal modal logics extended to the corresponding half-order logics provides us additional expression means. For instance, let the half-order logic of knowledge (**HOLK**) be obtained from the half-order modal logic **S5** or from some its sub-logic, where the set of values is a domain of agents. The formula

$$x.Ky.x = y$$

of **HOLK**, where "x." denotes the personal pronoun "I", which cannot be modeled adequuately in propositional epistemic logics, captures the assertion "I know who I am". If we have two modal operators, K for knowledege and $\Box$ to change the agent that is reasoning, then we can express properties such as "I know that everyone who I know knows that I am smarter than (s)he is":

$$x.K\Box y.K(x \; smarter\text{-}than \; y),$$

[2]. The formula

$$x.K\Box y.B(x \; smarter\text{-}than \; y),$$

where B is the modal operator of belief, expresses the property "I know that everyone who I know believes that I am smarter than (s)he is".

The half-order real time temporal logic **TPTL**, which is the temporal extension of **HOML**, is used in verification of real-time systems, [2].

Cut-free sequent calculi are comparatively convenient means for backward proof search of sequents/formulas. To check if an arbitrary sequent S is derivable in such a calculus, the derivation rules of the calculus are applied backward to S and the resulting sequents. In the cases where the premises of the derivation rules are simpler than the conclusions, the process is finite and each time the same, routine: determine the outermost symbol of a formula and apply the corresponding derivation rule. (If the calculus contain rules the premises of which are not simpler than the corresponding conclusions, then some additional means is required to ensure termination of backward proof-search.) The rule of cut destroys such a procedure, since this rule is not deterministic from the backward perspective. It contains the formula (called the cut formula) in the premises which is not specified in the conclusion. Therefore we do not know how many times this rule should be applied and what cut formula should be chosen each time so that to derive the considered sequent or to prove that it is not derivable.

Hilbert-type calculi such as **HHOML** given in Section 3 contain the Modus Ponens rule which is an analogue of the rule of cut. It would be hard (if possible) to describe effective proof-search by means of such calculi suitable for all formulas of a logic.

In the present paper, we consider the proof-theory of function symbols free version of **HOML**, (**HOML**$^{-f}$ in notation) by introducing the cut-free Gentzen-type labelled sequent calculus **GHOML** and by investigating the properties of the calculus. The sequent calculus is obtained by adding new quantifier and equality rules to the labelled sequent calculus for logic **K**, introduced in [7]. The shape of the rules for the freeze quantifier are related to the fact that the quantifier is its own dual, and thus it makes sense to have a universal-like succedent rule coupled with an existential-like antecedent one.

We prove that all the rules of **GHOML** are invertible, that the structural rules of weakening, contraction, and the rule of cut are admissible in the calculus, and it is sound and complete for **HOML**$^{-f}$. These properties enable us to present the decision procedure for **HOML**$^{-f}$. Given a formula ϕ, the procedure allows us to determine effectively if ϕ is valid in **HOML**$^{-f}$ (or, in other words, if ϕ is a theorem of **HOML**$^{-f}$). The decision procedure is based on the backward sequent proof-search by means of the calculus **GHOML**. Since the calculus is sound and complete for **HOML**$^{-f}$, we get that a formula is derivable in the calculus, iff it is valid in **HOML**$^{-f}$.

We also show that the calculus **GHOML** can be extended to the sequent calculi of the other half-order modal logic such as t based on **S5** and its sub-logics, e. g., **T**.

The fact that function symbols are dropped simplifies the term unification problem and allows us to focus on the introduced ($a = b \Rightarrow$) and quantifier rules.

Cut-free sequent calculi were introduced and considered for modal logics. For example, the Gentzen-type sequent calculi for various propositional modal logics are considered in [10, 5, 4]. The sequent calculus for the temporal logic with time gaps is investigated in [1]. The labelled Gentzen-type sequent calculi for modal logics are presented in [7]. The Gentzen-type sequent calculus with marks for logic **S4** is considered in [6].

To our knowledge, sequent calculi for half-order logics have not yet been investigated before in the literature.

This paper is organized as follows. In Section 2, we present the syntax and semantics of **HOML**$^{-f}$. The calculus **GHOML** is introduced in Section 3. The properties of the calculus are considered in Section 4. In Section 5, we describe the decision procedure for the half-order modal logic without function symbols, using the calculus **GHOML**. Concluding remarks are in Section 6.

2 Syntax and semantics of HOML^{-f}

The logic **HOML**$^{-f}$ is obtained from the half-order modal logic introduced in [2] by removing function symbols.

In this section, we introduce the syntax and semantics of **HOML**$^{-f}$. The definitions are taken from [2] except that:

(1) the function symbols are dropped;

(2) following [8], we separate free and bounded variables by introducing the set V_y of bound variables and the set V_a of free variables such that $V_y \cap V_a = \emptyset$. Such separation of free and bound variables assures us that each variable substitution is safe, where safe substitution of the variable v_1 by v_2 means, as usual, that no free occurrence of v_1 is within the scope of a quantifier binding v_2.

The formulas of **HOML**$^{-f}$ are constructed from predicate symbols, including equality, using implication ('$\rightarrow$'), the constant **false** ('$\bot$'), the modal operator $\Box$, and the freeze quantifier.

Let V_y, V_a, and P be infinite, effectively enumerable sets of bound variables, free variables, and predicate symbols, respectively. Since function symbols are absent, the notions of 'term' and 'free variable' coincide. The atomic formulas α, and formulas ϕ of **HOML**$^{-f}$ are inductively defined as follows:

$$\alpha := a = b \,|\, p(b_1, \ldots, b_m) \,|\, \bot,$$
$$\phi := \alpha \,|\, \phi_1 \rightarrow \phi_2 \,|\, \Box\phi \,|\, x.\phi'.$$

where $m \geq 0$; $a, b, b_1, \ldots, b_m \in V_a$; $x \in V_y$; $p \in P$, and ϕ' is obtained from a formula ϕ by replacing some, if any, occurrences of a free variable by x. To avoid confusion, we require that no quantifier '$x.$' occur within the scope of '$x.$'; the generality is not lost, since $x.x.\phi$ and $x.\phi$ are intended to express the same assertion, and the same is true for, e. g., $x.(\phi \to x.\psi_x)$ and $x.(\phi \to y.\psi_y)$ by renaming variables (here y does not occur in $x.(\phi \to x.\psi_x)$ and ψ_y is obtained from ψ_x by substituting y for x); these formulas are equivalent in [2].

In the paper, we use: 1) the letters a, b, c, d, e to denote free variables, 2) x, y, z, to denote bound variables, 3) ϕ, ψ, φ to denote arbitrary formulas. The following abbreviations are used here:
1) $\neg\phi$ for $\phi \to \bot$, 2) $\phi \vee \psi$ for $\neg\phi \to \psi$, 2) $\phi \wedge \psi$ for $\neg(\neg\phi \vee \neg\psi)$, 3) $\phi \leftrightarrow \psi$ for $(\neg\phi \vee \psi) \wedge (\neg\psi \vee \phi)$, 4) and $\Diamond$ for $\neg\Box\neg$.

The complexity of a formula ϕ is the number of occurrences of predicate symbols, propositional connectives, the modal operator, and of the freeze quantifier in ϕ.

The expression obtained from a formula by dropping some quantifiers and containing some bound variables not bound by quantifiers is called a quasi-formula. For example, $\alpha_1(x) \to y.\alpha_2(y)$ is a quasi-formula obtained from, e. g., the formula $x.\alpha_1(x) \to y.\alpha_2(y)$.

The expression $\xi[\chi := b]$, where $\chi \in V_a \cup V_x$ and ξ is a formula or a quasi-formula, is obtained from ξ by substituting b for each non-bound by a quantifier occurrence of χ in ξ. If each occurrence of χ is bound by a quantifier or χ does not occur in ξ, then $\xi[\chi := b]$ is equal to ξ.

The expression $\xi\langle\chi := b\rangle$ is obtained from ξ by replacing zero, one, or more (all, respectively) non-bound by a quantifier occurrences of χ by b.

In most cases, we use the symbol '$\rightleftharpoons$' to denote equality outside the formulas and quasi-formulas, e. g., the expression $S \rightleftharpoons \{\phi_1, \phi_2\}$ denotes that S is equal to the set $\{\phi_1, \phi_2\}$.

If $\Gamma \rightleftharpoons (\phi_1, \ldots, \phi_n)$, then $\Gamma[a := b] \rightleftharpoons (\phi_1[a := b], \ldots, \phi_n[a := b])$.

An interpretation
$$(\mathcal{S}, \to_\Box, \mathcal{U}, \|, [\![a]\!]_{a \in V_b}, [\![p]\!]_{p \in P}, s_0)$$
for half-order modal logic consists of:

- a non-empty set $\mathcal{S}$ of states;

- an accessibility relation $\to_\Box \subseteq \mathcal{S}^2$ on the states;

- a non-empty set $\mathcal{U}$ of values;

- a value function $\|: \mathcal{S} \to \mathcal{U}$ that associates a value $|s|$ with every state s;

- a rigid assignment function $[\![a]\!] \in \mathcal{U}$ for all variables $a \in V_b$;

- a flexible assignment function $[\![p]\!] : \mathcal{S} \to 2^{\mathcal{U} \times \cdots \times \mathcal{U}}$ for all predicate symbols $p \in P$;

- an initial state s_0.

The interpretation $\mathcal{M}$ is a model of the formula ϕ, iff $\mathcal{M} \models \phi$, for the following inductive definition of the truth predicate $\models$:

$$
\begin{array}{lll}
\mathcal{M} \models a = b & \text{iff} & [\![a]\!] = [\![b]\!], \\
\mathcal{M} \models p(b_1, \ldots, b_n) & \text{iff} & ([\![b_1]\!], \ldots, [\![b_n]\!]) \in [\![p]\!](s_0), \\
\mathcal{M} \not\models \bot, & & \\
\mathcal{M} \models \phi_1 \to \phi_2 & \text{iff} & \mathcal{M} \not\models \phi_1 \text{ or } \mathcal{M} \models \phi_2, \\
\mathcal{M} \models \Box \phi & \text{iff} & \mathcal{M}[s_0 := t] \models \phi \text{ for all } t \in \mathcal{S} \text{ with } s_0 \to_\Box t, \\
\mathcal{M} \models x.\phi & \text{iff} & \mathcal{M}[[\![a]\!] := |s_0|] \models \phi[x := a], \text{ where } a \text{ does not occur in } \phi
\end{array}
$$

Here $\mathcal{M}[s_0 := t]$ denotes the interpretation that differs, if any, from $\mathcal{M}$ only in its initial state, t; the interpretation $\mathcal{M}[[\![a]\!] := |s_0|]$ differs, if any, from $\mathcal{M}$ only in its assignment function for a. For example,

$$(\{s,t\}, s \to_\Box t, [\![p]\!](t) \rightleftharpoons \{|s|,|t|\}, [\![p]\!](s) \rightleftharpoons \emptyset, s)$$

is the model of the formula $x.\Diamond y.p(x,y)$.

The formula ϕ is satisfiable (valid, $\models \phi$ in notation), iff some (every) interpretation is a model for ϕ.

3 Deductive systems

The Hilbert-type calculus **HHOML** for **HOML**$^{-f}$ is taken from [2], except that the function symbols are dropped and different letters are used to denote free and bound variables. The calculus is defined by the following postulates:

PROP1 Propositional tautologies are axioms.

PROP2 Modus Ponens rule

$$\frac{\phi_1; \; \phi_1 \to \phi_2}{\phi_2} \; (MP).$$

K1   Modal axiom schema

$$\Box(\phi_1 \to \phi_2) \to (\Box\phi_1 \to \Box\phi_2).$$

This axiom schema expresses the distributivity of the modal operator '$\Box$' over implication.

K2   Modal rule

$$\frac{\phi}{\Box\phi}\ (\Box).$$

This rule allows us to infer the formula $\Box\phi$ from ϕ.

Q1   Quantifier axiom schema

$$x.(\phi_1 \to \phi_2) \leftrightarrow (x.\phi_1 \to x.\phi_2).$$

The quantifier axiom schema expresses the distributivity of the freeze quantifier over implication.

Q2   Quantifier rule

$$\frac{\phi_1 \hookrightarrow \cdots \hookrightarrow \phi_n \hookrightarrow \psi[x := b]}{\phi_1 \hookrightarrow \cdots \hookrightarrow \phi_n \hookrightarrow x.\psi},$$

where 1) $\phi \hookrightarrow \psi$ stands for $\phi \to \Box\psi$ and $\hookrightarrow$ associates to the right and 2) the variable b does not occur in ϕ_i for all $1 \leq i \leq n$. This rule allows us to infer the formula $\phi_1 \hookrightarrow \cdots \hookrightarrow \phi_n \hookrightarrow x.\psi$ from the formula $\phi_1 \hookrightarrow \cdots \hookrightarrow \phi_n \hookrightarrow \psi[x := b]$.

Q2*   Quantifier rule

$$\frac{\phi[x := b]}{x.\phi}.$$

This rule is the instance of the rule **Q2** with $n = 0$. It allows us to infer the formula $x.\phi$ from $\phi[x := b]$.

Q3   Quantifier axiom schema

$$x.\phi \leftrightarrow \phi,$$

where x does not occur in ϕ. This axiom schema expresses the fact that the formulas $x.\phi$ and ϕ are equivalent if x does not occur in ϕ.

EQ1   Equality axiom schema

$$b = b.$$

This simple axiom schema states that the value assigned to any variable b at a state of an interpretation is equal to itself.

EQ2   Equality axiom schema

$$(a = b) \to (\alpha \to \alpha \langle a := b \rangle).$$

According this axiom schema, if the values assigned to the variables a and b coincide, then the implication formula $(\alpha \to \alpha \langle a := b \rangle)$ is true.

RIG1   Term rigidity axiom schema

$$(a = b) \to \Box(a = b).$$

This axiom schema states that if any two variables a and b are assigned the same value at a state s, then the variables are assigned the same value at each state accessible from s.

RIG2   Term rigidity axiom schema

$$(a \neq b) \to \Box(a \neq b).$$

The statement of this axiom schema is complementary to the previous one: if the values assigned to any variables a and b differ at a state s, then the values assigned to the variables differ at each state accessible from s.

QEQ   Axiom schema
$$x.y.(x = y).$$

This axiom schema states that any two variables x and y bound by the freeze quantifier are assigned the same value at every state, i. e., the value associated with every state is unique.

Remark 3.1. *It follows from [2] that **HHOML** is sound and complete for the logic $\mathbf{HOML}^{-f}$: a formula without function symbols is valid in $\mathbf{HOML}^{-f}$, iff it is* derivable in **HHOML**.

The expression $a <> b$ denotes that a and b are different free variables, i. e., the formula $a = b$ is not of the shape $c = c$.

3.1 Gentzen-type labelled sequent calculus

For the sequent calculus we introduce additionally:

1. The formulas of the type $x.x \stackrel{\circ}{=} b$, which are used in the quantifier rules and allow us to restrict the number of backward applications of the rule $(x. \Rightarrow)$ in backward proof-search. The sign '$\stackrel{\circ}{=}$' occurs only in the formulas of this type, e. g., we do not consider formulas of the type $a \stackrel{\circ}{=} b$ or $(x.x \stackrel{\circ}{=} b) \to \alpha$.

2. Labels, denoted by the letters i, j, and l.

3. Formulas of the type $i \mapsto j$, called relation atoms. Formulas of this type do not occur within the scope of quantifiers, propositional or modal operators, e. g., we do not consider formulas of the type $x.(i \mapsto j)$, $\Box(i \mapsto j)$, or $(i_1 \mapsto j_1) \to (i_2 \mapsto j_2)$.

4. Labelled formulas $^i\phi$, where ϕ is an unlabelled formula, except a relation atom.

A multiset is a generalization of the concept of a set that, unlike a set, allows multiple instances of the multiset's elements. For example, $\{a, a, b\}$ and $\{a, b\}$ are different multisets although they are the same set. However, order does not matter, thus $\{a, a, b\}$ and $\{a, b, a\}$ are the same multiset.

As usual, sequents are objects of the shape $\Gamma \Rightarrow \Delta$, where Γ and Δ are finite, possibly empty, multisets consisting of relation atoms and labelled formulas, except that '$\stackrel{\circ}{=}$' and relation atoms do not occur in Δ. The letter S (possibly subscripted) is used in the paper to denote sequents. Any sequent

$$\phi_1, \ldots, \phi_m \Rightarrow \psi_1, \ldots, \psi_n$$

is understood informally as the formula

$$(\phi_1 \wedge \ldots \wedge \phi_m) \to (\psi_1 \vee \ldots \vee \psi_n).$$

The Gentzen-type labelled sequent calculus **GHOML** for the half-order modal logic **HOML**$^{-f}$ without function symbols is defined by the following axiom schemata and derivation rules (the modal rules are taken from [7]):

1. Axiom schemata:

$$\Gamma, {}^i\bot \Rightarrow \Delta,$$

$$\Gamma, {}^i\alpha \Rightarrow \Delta, {}^i\alpha,$$

$$\Gamma \Rightarrow \Delta, {}^ib = b.$$

2. Propositional rules:

$$\frac{\Gamma \Rightarrow \Delta, {}^i\phi \quad {}^i\psi, \Gamma \Rightarrow \Delta}{{}^i\phi \to \psi, \Gamma \Rightarrow \Delta} \; (\to\Rightarrow), \qquad \frac{\Gamma, {}^i\phi \Rightarrow {}^i\psi, \Delta}{\Gamma \Rightarrow {}^i\phi \to \psi, \Delta} \; (\Rightarrow\to).$$

3. Modal rules:

$$\frac{{}^j\phi, {}^i\Box\phi, i\mapsto j, \Gamma \Rightarrow \Delta}{{}^i\Box\phi, i\mapsto j, \Gamma \Rightarrow \Delta} \; (\Box\Rightarrow), \qquad \frac{i\mapsto j, \Gamma \Rightarrow {}^j\phi, \Delta}{\Gamma \Rightarrow {}^i\Box\phi, \Delta} \; (\Rightarrow\Box),$$

where j does not occur in the conclusion in the rule $(\Rightarrow\Box)$.

4. Quantifier rules:

$$\frac{{}^ix.x \stackrel{\circ}{=} b, {}^i\phi[x:=b], \Gamma \Rightarrow \Delta}{{}^ix.\phi, \Gamma \Rightarrow \Delta} \; (x.\Rightarrow), \qquad \frac{{}^ix.x \stackrel{\circ}{=} b, \Gamma \Rightarrow \Delta, {}^i\phi[x:=b]}{\Gamma \Rightarrow \Delta, {}^ix.\phi} \; (\Rightarrow x.).$$

The rules $(x.\Rightarrow)$ and $(\Rightarrow x.)$ require that b not occur in the conclusion. The rule $(x.\Rightarrow)$ requires additionally that ϕ is not of the shape $x.x \stackrel{\circ}{=} a$.

5. Equality rules:

$$\frac{\Gamma[a:=b] \Rightarrow \Delta[a:=b]}{{}^ia = b, \Gamma \Rightarrow \Delta} \; (a:=b),$$

$$\frac{{}^ia = b, ({}^ix.\sigma_1 \; \theta \; \sigma_2), ({}^iy.\sigma_3 \; \xi \; \sigma_4), \Gamma \Rightarrow \Delta}{({}^ix.\sigma_1 \; \theta \; \sigma_2), ({}^iy.\sigma_3 \; \xi \; \sigma_4), \Gamma \Rightarrow \Delta} \; (a=b\Rightarrow).$$

In the rule $(a=b\Rightarrow)$:

$$\theta, \xi \in \{=, \stackrel{\circ}{=}\}, \quad \{\sigma_1, \sigma_2\} \leftrightharpoons \{x, a\}, \quad \{\sigma_3, \sigma_4\} \leftrightharpoons \{y, b\},$$

and if θ (ξ) is $\stackrel{\circ}{=}$, then σ_1 (σ_3) is x (y). This rule requires the following side conditions to be met: 1) $a <> b$, 2) $\Gamma \neq (\Gamma', {}^ia = b)$, and 3) both a and b

to occur in (Γ, Δ). The side conditions enable us to restrict the number of applications of the rule in proof-search by prohibiting the useless applications. From the bottom-up perspective, the rule $(a = b \Rightarrow)$ states that, if a and b are equal to the value associated with the current state, then a and b are equal between themselves.

The additional active formulas in the quantifier rules together with the rule $(a = b \Rightarrow)$ allow us to derive formulas of the type $x.\alpha(x) \to y.\alpha(y)$; a special equality sign in the active formulas is used to restrict the number of backward rule $(x. \Rightarrow)$ applications in proof-search.

Applied backwards, the rule $(a = b \Rightarrow)$ introduces the equality formula $a = b$ in the antecedent of the premise, if the free variables a and b are equal to the value associated with the current state. It is used to derive formulas of the type

$$(x.x = a \wedge y.b = y) \to a = b \quad \text{or} \quad (x.x = a \wedge y.b = y) \to (\alpha(a) \to \alpha(b)).$$

Now we recall some definitions. In the rule $(\Rightarrow \to)$: the explicit formula ${}^i\phi \to \psi$ in the conclusion is the *principal* formula; ${}^i\phi$ and ${}^i\psi$ in the premise are *active* formulas; the formulas in Γ and Δ are *context* formulas. The principal, active, and context formulas for the remaining propositional, quantifier, and modal rules are defined in the same way; the rule $(\Box \Rightarrow)$ has two principal formulas ${}^i\Box\phi$ and $i \mapsto j$; the rules $(x. \Rightarrow)$ and $(\Rightarrow x.)$ have two active formulas ${}^ix.x \stackrel{\circ}{=} b$ and ${}^i\phi[x := b]$. The explicit quantifier formulas in the conclusion of $(a = b \Rightarrow)$ are principal, and the formula ${}^ia = b$ is active. All the formulas in the conclusion of the rule $(a := b)$ are principal.

Given a sequent S, a **GHOML** proof-search tree with the sequent S at the root is obtained in a usual way by subsequently applying backwards the **GHOML** derivation rules to S and the sequents obtained in the course of the tree construction.

The *height* of a proof-search tree is the length of the longest branch in it. The *length* of a branch is measured by the number of rule applications present on it.

A proof-search tree, all the branches of which end up in axioms, is called a *derivation tree*. Any sequent S at the root of a derivation tree, generated by the rules of the calculus **GHOML**, is called *derivable* in **GHOML** ($GHOML \vdash S$ in notation; for the sake of readability, we omit $GHOML$ when it is clear which calculus is meant).

Let us consider the example of the backward proof-search tree demonstrating the above concepts:

$$\dfrac{\dfrac{\dfrac{\dfrac{^j\alpha_1, i\mapsto j, {}^i\Box\alpha_1 \Rightarrow {}^i\alpha_3 \to \alpha_2, {}^j\alpha_1}{i\mapsto j, {}^i\Box\alpha_1 \Rightarrow {}^i\alpha_3 \to \alpha_2, {}^j\alpha_1}(\Box\Rightarrow)}{{}^i\Box\alpha_1 \Rightarrow {}^i\alpha_3 \to \alpha_2, {}^i\Box\alpha_1}(\Rightarrow\Box)}{{}^i\Box\alpha_1 \to \alpha_2, {}^i\Box\alpha_1 \Rightarrow {}^i\alpha_3 \to \alpha_2} \quad \dfrac{\dfrac{{}^i\alpha_2, {}^i\Box\alpha_1, {}^i\alpha_3 \Rightarrow {}^i\alpha_2}{{}^i\alpha_2, {}^i\Box\alpha_1 \Rightarrow {}^i\alpha_3 \to \alpha_2}(\Rightarrow\to)}{}(\Rightarrow\to)}{{}^i\Box\alpha_1 \to \alpha_2 \Rightarrow {}^i\Box\alpha_1 \to (\alpha_3 \to \alpha_2)}(\to\Rightarrow)$$

This tree is generated by backward applying one by one derivation rules to the corresponding sequents, starting from the root, i. e., the sequent

$${}^i\Box\alpha_1 \to \alpha_2 \Rightarrow {}^i\Box\alpha_1 \to (\alpha_3 \to \alpha_2).$$

The topmost sequents

$${}^j\alpha_1, i\mapsto j, {}^i\Box\alpha_1 \Rightarrow {}^i\alpha_3 \to \alpha_2, {}^j\alpha_1 \quad \text{and} \quad {}^i\alpha_2, {}^i\Box\alpha_1, {}^i\alpha_3 \Rightarrow {}^i\alpha_2$$

are the leaves of the tree. A branch of a tree is the path from the root to a leaf inclusive. This tree has two branches; the height of the left branch is 5 and the height of the right one is 4. Hence the height of the tree is 5. Since all the branches end up in axioms, this tree is a derivation tree. Hence the root sequent is derivable in **GHOML**, denoted by

$$GHOML \vdash {}^i\Box\alpha_1 \to \alpha_2 \Rightarrow {}^i\Box\alpha_1 \to (\alpha_3 \to \alpha_2)$$

or

$$\vdash {}^i\Box\alpha_1 \to \alpha_2 \Rightarrow {}^i\Box\alpha_1 \to (\alpha_3 \to \alpha_2).$$

Since the derivation rules can be applied in any order, we can construct another derivation tree with the same root sequent:

$$\dfrac{\dfrac{\dfrac{\dfrac{^j\alpha_1, i\mapsto j, {}^i\Box\alpha_1 \Rightarrow {}^i\alpha_3 \to \alpha_2, {}^j\alpha_1}{i\mapsto j, {}^i\Box\alpha_1 \Rightarrow {}^i\alpha_3 \to \alpha_2, {}^j\alpha_1}(\Box\Rightarrow)}{{}^i\Box\alpha_1 \Rightarrow {}^i\alpha_3 \to \alpha_2, {}^i\Box\alpha_1}(\Rightarrow\Box)}{\Rightarrow {}^i\Box\alpha_1 \to (\alpha_3 \to \alpha_2), {}^i\Box\alpha_1}(\Rightarrow\to) \quad \dfrac{\dfrac{\dfrac{{}^i\alpha_2, {}^i\Box\alpha_1, {}^i\alpha_3 \Rightarrow {}^i\alpha_2}{{}^i\alpha_2, {}^i\Box\alpha_1 \Rightarrow {}^i\alpha_3 \to \alpha_2}(\Rightarrow\to)}{{}^i\alpha_2 \Rightarrow {}^i\Box\alpha_1 \to (\alpha_3 \to \alpha_2)}(\Rightarrow\to)}{}}{{}^i\Box\alpha_1 \to \alpha_2 \Rightarrow {}^i\Box\alpha_1 \to (\alpha_3 \to \alpha_2)}(\to\Rightarrow)$$

A derivation rule is called *height-preserving admissible*, iff derivability of its premise(es) implies derivability of its conclusion, the height of the conclusion derivation being not greater than that of any premise.

A derivation rule is called *height-preserving invertible*, iff derivability of its conclusion implies derivability of its premise(es), the height of the derivation of any premise being not greater than that of the conclusion.

Remark 3.2. *The fact that the sequents are of multiset type implies that the structural rule of permutation is admissible in* **GHOML**.

Since the labels are relevant mainly in the modal rules, we often omit them for the sake of readability when application of the modal rules is not explicitly involved into the consideration.

Remark 3.3. *If the variable b is allowed to occur in the conclusion of the rules $(\Rightarrow x.)$ and $(x. \Rightarrow)$, then these rules are not sound: the non-valid sequents $(\Rightarrow {}^i x.x = b)$ and $({}^i x. \neg (x = b) \Rightarrow)$ are derivable in such a calculus.*

Remark 3.4. *Without '$\doteq$', the backward proof-search of, e. g., the sequent ${}^i x.x = b \Rightarrow$, does not terminate (the labels are omitted):*

$$\cfrac{\cfrac{\cfrac{\cdots}{x.x = d, d = c, c = b \Rightarrow}\,(x.\Rightarrow)}{x.x = c, c = b \Rightarrow}\,(x.\Rightarrow)}{x.x = b \Rightarrow}\,(x.\Rightarrow)$$

4 Some properties of GHOML

Lemma 4.1. *Any sequent $S \leftrightharpoons (\Gamma, {}^i\phi \Rightarrow {}^i\phi, \Delta)$ is derivable in* **GHOML**.

Proof. The lemma is proved by induction on the complexity C of ϕ. If $C = 1$, then S is an axiom. Let $C > 1$:

1. If $\phi \leftrightharpoons x.\psi$, then, starting from the bottom, we generate the following backward proof-search tree (the labels are omitted):

$$\cfrac{\cfrac{\cfrac{\cfrac{S_1 \leftrightharpoons (x.x \doteq a, x.x \doteq a, \Gamma, \psi[x := a] \Rightarrow \psi[x := a], \Delta)}{b = a, \ x.x \doteq b, x.x \doteq a, \Gamma, \psi[x := a] \Rightarrow \psi[x := b], \Delta}\,(b := a)}{x.x \doteq b, x.x \doteq a, \Gamma, \psi[x := a] \Rightarrow \psi[x := b], \Delta}\,(b = a \Rightarrow)}{x.x \doteq a, \Gamma, \psi[x := a] \Rightarrow x.\psi, \Delta}\,(\Rightarrow x.)}{S \leftrightharpoons (\Gamma, x.\psi \Rightarrow x.\psi, \Delta)}\,(x.\Rightarrow)$$

By inductive hypothesis, the sequent S_1 is derivable; hence S is derivable as well.

2. If $\phi \leftrightharpoons \psi \to \varphi$, then we generate the following backward proof-search tree (the labels are omitted):

$$\frac{\dfrac{\Gamma,\psi \Rightarrow \psi,\varphi,\Delta \quad \Gamma,\varphi,\psi \Rightarrow \varphi,\Delta}{\Gamma,\psi \to \varphi,\psi \Rightarrow \varphi,\Delta}(\to\Rightarrow)}{S \leftrightharpoons (\Gamma,\psi \to \varphi \Rightarrow \psi \to \varphi,\Delta)}(\Rightarrow\to)$$

By inductive hypothesis, the both topmost sequents are derivable; this fact yields that S is derivable as well.

3. If $\phi \leftrightharpoons \Box\psi$, then we generate the following backward proof-search tree:

$$\frac{\dfrac{i \mapsto j,{}^j\psi,{}^i\Box\psi,\Gamma \Rightarrow {}^j\psi,\Delta}{i \mapsto j,{}^i\Box\psi,\Gamma \Rightarrow {}^j\psi,\Delta}(\Box\Rightarrow)}{S \leftrightharpoons ({}^i\Box\psi,\Gamma \Rightarrow {}^i\Box\psi,\Delta)}(\Rightarrow\Box)$$

The topmost sequent is derivable, according to inductive hypothesis; hence S is derivable as well.

All possible types of ϕ has been considered and the lemma is proved. □

Proposition 4.2. *The rule*
$$\frac{{}^ib = b, \Gamma \Rightarrow \Delta}{\Gamma \Rightarrow \Delta},$$
is height-preserving admissible in **GHOML**.

Proof. The proposition is proved using induction on the height H of the derivation of the premise. If $H = 0$, then the premise is an axiom. This fact implies that the conclusion $\Gamma \Rightarrow \Delta$ is an axiom as well and the proof is obtained. Let $H > 0$ (the labels are omitted):

$$\frac{\cdots}{b = b, \Gamma \Rightarrow \Delta}(r).$$

1. Let (r) be $(\to\Rightarrow)$:

$$\frac{b = b, \Gamma \Rightarrow \Delta, \phi \quad b = b, \psi, \Gamma \Rightarrow \Delta}{b = b, \phi \to \psi, \Gamma \Rightarrow \Delta}(\to\Rightarrow).$$

According to inductive hypothesis, $\vdash \Gamma \Rightarrow \Delta, \phi$ and $\vdash \psi, \Gamma \Rightarrow \Delta$. We apply the rule $(\to\Rightarrow)$ to these sequents and infer $\phi \to \psi, \Gamma \Rightarrow \Delta$.

2. Let (r) be $(c := d)$:

$$\frac{b = b[c := d], \Gamma[c := d] \Rightarrow \Delta[c := d]}{b = b, c = d, \Gamma \Rightarrow \Delta}(c := d).$$

According to inductive hypothesis, $\vdash \Gamma[c := d] \Rightarrow \Delta[c := d]$. We apply the rule $(c := d)$ to this sequent and infer $c = d, \Gamma \Rightarrow \Delta$.

3. The remaining cases when (r) is another rule of **GHOML** are dealt with similarly as the previous one. We get $\vdash S \leftrightharpoons (\Gamma' \Rightarrow \Delta')$ from the premise $b = b, \Gamma' \Rightarrow \Delta'$ of (r), using inductive hypothesis. The required sequent is obtained by applying (r) to S.

$\square$

Proposition 4.3. *The rule*
$$\frac{{}^i\alpha, {}^i\alpha, \Gamma \Rightarrow \Delta}{{}^i\alpha, \Gamma \Rightarrow \Delta}$$
is height-preserving admissible in **GHOML**.

Proof. The proposition is proved using induction on the height H of derivation of the premise and Proposition 4.2. If $H = 0$, then the premise is an axiom. This fact implies that the conclusion is an axiom as well and the proof is obtained. Let $H > 0$:
$$\frac{\cdots}{{}^i\alpha, {}^i\alpha, \Gamma \Rightarrow \Delta}\ (r).$$

We consider some characteristic cases of (r) (the labels are omitted).

1. Let (r) be $(a := b)$ and α be $a = b$.
$$\frac{a = b[a := b], \Gamma[a := b] \Rightarrow \Delta[a := b]}{a = b, a = b, \Gamma \Rightarrow \Delta}\ (a := b).$$

The facts that $a = b[a := b]$ gives $b = b$ and the premise is derivable yields $\vdash \Gamma[a := b] \Rightarrow \Delta[a := b]$, using Proposition 4.2. We apply the rule $(a := b)$ to this sequent and infer $a = b, \Gamma \Rightarrow \Delta$.

2. Let (r) be $(\Rightarrow\rightarrow)$:
$$\frac{\alpha, \alpha, \Gamma, \phi \Rightarrow \Delta, \psi}{\alpha, \alpha, \Gamma \Rightarrow \Delta, \phi \rightarrow \psi}\ (\Rightarrow\rightarrow).$$

According to inductive hypothesis, $\vdash \alpha, \Gamma, \phi \Rightarrow \Delta, \psi$. We apply the rule $(\Rightarrow\rightarrow)$ to this sequent and infer $\alpha, \Gamma \Rightarrow \Delta, \phi \rightarrow \psi$.

3. The cases when (r) is one of the remaining rules of **GHOML** are considered in the same way as the previous one: we get $\vdash S \leftrightharpoons (\alpha, \Gamma' \Rightarrow \Delta')$ from the premise $\alpha, \alpha, \Gamma' \Rightarrow \Delta'$ of (r), using inductive hypothesis; the required sequent is obtained by applying (r) to S.

The sequent $\Gamma(i/j) \Rightarrow \Delta(i/j)$ is obtained from the sequent $\Gamma \Rightarrow \Delta$ by substituting the label i for j.

Lemma 4.4 (Label substitution). *The rule*

$$\frac{\Gamma \Rightarrow \Delta}{\Gamma(i/j) \Rightarrow \Delta(i/j)}$$

is height-preserving admissible in **GHOML**.

Proof. The lemma is proved by induction on the height H of the derivation of the premise. If $H = 0$, then the premise is an axiom. This fact implies that the conclusion is an axiom as well. Let $H > 0$. We consider some typical cases of the inductive step.

1. Let the derivation of the premise be concluded by

$$\frac{\Gamma \Rightarrow \Delta, {}^j\phi \quad {}^j\psi, \Gamma \Rightarrow \Delta}{{}^j\phi \to \psi, \Gamma \Rightarrow \Delta} \ (\to\Rightarrow).$$

According to inductive hypothesis, $\vdash \Gamma(i/j) \Rightarrow \Delta(i/j), {}^i\phi$ and $\vdash {}^i\psi, \Gamma(i/j) \Rightarrow \Delta(i/j)$. We apply the rule $(\to\Rightarrow)$ to these sequents and infer the required sequent ${}^i\phi \to \psi, \Gamma(i/j) \Rightarrow \Delta(i/j)$.

2. Let the derivation of the premise be concluded by

$$\frac{S \leftrightharpoons (j \mapsto i, \Gamma \Rightarrow \Delta, {}^i\phi)}{\Gamma \Rightarrow \Delta, {}^j\square\phi} \ (\Rightarrow\square).$$

According to inductive hypothesis, $\vdash S(k/i) \leftrightharpoons (j \mapsto k, \Gamma \Rightarrow \Delta, {}^k\phi)$, where k does not occur in S. This fact yields

$$\vdash S(k/i)(i/j) \leftrightharpoons (i \mapsto k, \Gamma(i/j) \Rightarrow \Delta(i/j), {}^k\phi),$$

by inductive hypothesis. We apply the rule $(\Rightarrow\square)$ to this sequent and infer the required one $\Gamma(i/j) \Rightarrow \Delta(i/j), {}^i\square\phi$.

We had to use the inductive hypothesis twice here, since the label introduced by (i/j) and the new label introduced in the premise of $(\Rightarrow\square)$ coincide.

3. Let the derivation of the premise be concluded by

$$\frac{S \leftrightharpoons (i \mapsto j, {}^j\phi, {}^i\Box\phi, \Gamma \Rightarrow \Delta)}{i \mapsto j, {}^i\Box\phi, \Gamma \Rightarrow \Delta} \ (\Box \Rightarrow).$$

According to inductive hypothesis,

$$\vdash S(i/j) \leftrightharpoons (i \mapsto i, {}^i\phi, {}^i\Box\phi, \Gamma(i/j) \Rightarrow \Delta(i/j)).$$

We apply the rule $(\Box \Rightarrow)$ to $S(i/j)$ and infer the required one ${}^i\Box\phi, \Gamma(i/j) \Rightarrow \Delta(i/j)$.

4. The cases when the derivation of the premise is concluded by applying one of the remaining rules of **GHOML** are considered in the same way as the previous one: from the premise S of the considered rule (r), we get $\vdash S(i/j)$, using inductive hypothesis; the required sequent is obtained by applying (r) to $S(i/j)$.

$\square$

Lemma 4.5 (Variable substitution). *The rule*

$$\frac{\Gamma \Rightarrow \Delta}{\Gamma[a := b] \Rightarrow \Delta[a := b]} \ (Sub)$$

is height-preserving admissible in **GHOML**.

Proof. The lemma is proved by induction on the height h of derivation of the premise. The proof is obvious if $h = 0$, since the conclusion is an axiom in this case. The proof is obtained if the substitution $(\Gamma, \Delta)[a := b]$ is void. Let the substitution $(\Gamma, \Delta)[a := b]$ be non-void, $h > 0$, and the derivation of the premise be concluded by:

1.
$$\frac{S \leftrightharpoons ({}^ix.x \overset{\circ}{=} c, \Gamma, {}^i\phi[x := c] \Rightarrow \Delta)}{\Gamma, {}^ix.\phi \Rightarrow \Delta} \ (x. \Rightarrow),$$

where c does not occur in the conclusion.

We consider two sub-cases here (the label 'i' is omitted):

(a) the variable c does not occur in $\{a, b\}$. According to the inductive hypothesis,

$$GHOML \vdash x.x \overset{\circ}{=} c, \Gamma[a := b], \phi[x := b][a := b] \Rightarrow \Delta[a := b].$$

(The vacuous substitutions are omitted here and below.) We apply the rule $(x. \Rightarrow)$ to this sequent and infer

$$\Gamma[a := b], x.\phi[a := b] \Rightarrow \Delta[a := b].$$

(b) the variable c occurs in $\{a, b\}$. We choose the variable d which does not occur either in the conclusion or in $\{a, b\}$. According to the inductive hypothesis,

$$GHOML \vdash S[c := d] \leftrightharpoons (x.x \stackrel{\circ}{=} d, \Gamma, \phi[x := d] \Rightarrow \Delta).$$

Since the derivation height is not increased, we still can apply the inductive hypothesis to S; hence

$$GHOML \vdash x.x \stackrel{\circ}{=} d, \Gamma[a := b], \phi[x := d][a := b] \Rightarrow \Delta[a := b].$$

We apply the rule $(x. \Rightarrow)$ to this sequent and infer

$$\Gamma[a := b], x.\phi[a := b] \Rightarrow \Delta[a := b].$$

2. let the derivation of the premise of (Sub) be concluded by:

$$\frac{S_1 \leftrightharpoons (c = d, \ x.x = c, y.y = d, \Gamma \Rightarrow \Delta)}{S \leftrightharpoons (x.x = c, y.y = d, \Gamma \Rightarrow \Delta)} \ (c = d \Rightarrow).$$

We consider two sub-cases here:

(a) let $a = c$ and $b = d$. The sequent

$$S_1[c := d] \leftrightharpoons (d = d, \ x.x = d, y.y = d, \Gamma[c := d] \Rightarrow \Delta[c := d])$$

is derivable by inductive hypothesis. The proof is obtained by eliminating $d = d$ from this sequent, using Proposition 4.2.

(b) let $a = c$ and $b = e$. The sequent

$$S_1[c := e] \leftrightharpoons (e = d, \ x.x = e, y.y = d, \Gamma[c := e] \Rightarrow \Delta[c := e])$$

is derivable, according to the inductive hypothesis. If $\Gamma[c := e]$ contains the member $e = d$, then the proof is obtained by eliminating one occurrence of $e = d$, using Proposition 4.3. Otherwise, we apply the rule $(c = e \Rightarrow)$ to $S_1[c := e]$ and infer $S[c := e]$.

3. Let the derivation of the premise of (Sub) be concluded by

$$\frac{S_1 \leftrightharpoons (\Gamma[c:=d] \Rightarrow \Delta[c:=d])}{S \leftrightharpoons (c=d, \Gamma \Rightarrow \Delta)} \ (c:=d).$$

The sequent

$$S_1[a:=b] \leftrightharpoons (\Gamma[c:=d][a:=b] \Rightarrow \Delta[c:=d][a:=b])$$

is derivable in **GHOML**, based on the inductive hypothesis (if the substitution $[a:=b]$ is void in $S_1[a:=b]$, e. g., if a and c coincide, then the inductive hypothesis is not needed). The proof is obtained by applying the rule $(c:=d[a:=b])$ to this sequent and inferring $S[a:=b]$.

4. The cases where the derivation of the premise of (Sub) is concluded by applying one of the remaining rules of **GHOML** are easier, since these rules do not involve manipulation of free variables: from the premise S of the considered rule (r) application, we get $\vdash S[a:=b]$, using inductive hypothesis; the required sequent is obtained by applying (r) to $S[a:=b]$.

□

Lemma 4.6. *The rule of weakening*

$$\frac{\Gamma \Rightarrow \Delta}{\Pi, \Gamma \Rightarrow \Delta, \Lambda} \ (W),$$

where Π and Λ are any finite multisets of formulas, is height-preserving admissible in **GHOML**.

Proof. The lemma is proved using the inductive hypothesis on height h of the derivation of the premise.

The base case is obvious, since the premise is an axiom when $h=0$.

Let $h>0$. We consider the following cases:

1. Let the derivation of the premise be concluded by

$$\frac{(^i a = b), (^i x.x = a), (^i y.y = b), \Gamma \Rightarrow \Delta}{(^i x.x = a), (^i y.y = b), \Gamma \Rightarrow \Delta} \ (a = b \Rightarrow).$$

We obtain $\vdash \Pi, (^i a = b), (^i x.x = a), (^i y.y = b), \Gamma \Rightarrow \Delta, \Lambda$ from the premise, using the inductive hypothesis. If $\Pi \neq (^i a = b, \Pi')$, then we apply the rule

$(a = b \Rightarrow)$ to this sequent and infer

$$\Pi, (^i x.x = a), (^i y.y = b), \Gamma \Rightarrow \Delta, \Lambda.$$

Otherwise, we have $\vdash \Pi', (^i a = b), (^i a = b), (^i x.x = a), (^i y.y = b), \Gamma \Rightarrow \Delta, \Lambda$. The latter fact yields

$$\vdash \Pi', (^i a = b), (^i x.x = a), (^i y.y = b), \Gamma \Rightarrow \Delta, \Lambda,$$

according to Lemma 4.3, and the lemma is proved in this case.

2. Let the derivation of the premise be concluded by

$$\frac{\Gamma[a := b] \Rightarrow \Delta[a := b]}{a = b, \Gamma \Rightarrow \Delta} \; (a := b).$$

According to Lemma 4.5, $\vdash \Gamma[a := b][b := c] \Rightarrow \Delta[a := b][b := c]$, where c occurs neither in the conclusion nor in Π, Λ. Hence

$$\vdash \Pi, \Gamma[a := b][b := c] \Rightarrow \Delta[a := b][b := c], \Lambda,$$

based on inductive hypothesis. The latter fact yields

$$\vdash S_1 \leftrightharpoons (\Pi[a := c], \Gamma[a := b][b := c] \Rightarrow \Delta[a := b][b := c], \Lambda[a := c]),$$

using Lemma 4.5. We have

$$\frac{S_1}{S \leftrightharpoons (a = c, \Pi, \Gamma[b := c] \Rightarrow \Delta[b := c], \Lambda)} \; (a := c).$$

Hence $\vdash S[c := b] \leftrightharpoons (a = b, \Pi, \Gamma \Rightarrow \Delta, \Lambda)$, using Lemma 4.5.

3. Let the derivation of the premise be concluded by

$$\frac{^i x.x \stackrel{\circ}{=} b, {}^i\phi[x := b], \Gamma \Rightarrow \Delta}{^i x.\phi, \Gamma \Rightarrow \Delta} \; (x. \Rightarrow).$$

We obtain $\vdash {}^i x.x \stackrel{\circ}{=} c, {}^i\phi[x := c], \Gamma \Rightarrow \Delta$ from the premise, according to Lemma 4.5; here c occurs neither in the premise nor in (Π, Λ). Hence

$$\vdash \Pi, {}^i x.x \stackrel{\circ}{=} c, {}^i\phi[x := c], \Gamma \Rightarrow \Delta, \Lambda,$$

according to the inductive hypothesis. We apply the rule $(x. \Rightarrow)$ to this sequent

and infer
$$\Pi, {}^i x.\phi, \Gamma \Rightarrow \Delta, \Lambda.$$

4. Let the derivation of the premise be concluded by
$$\frac{\Gamma \Rightarrow \Delta, {}^j\phi \quad {}^j\psi, \Gamma \Rightarrow \Delta}{{}^j\phi \to \psi, \Gamma \Rightarrow \Delta} \; (\to\Rightarrow).$$

According to inductive hypothesis, $\vdash \Gamma, \Pi \Rightarrow \Delta, \Lambda, {}^j\phi$ and $\vdash {}^j\psi, \Gamma, \Pi \Rightarrow \Delta, \Lambda$. We apply the rule ($\to\Rightarrow$) to these sequents and infer the required one ${}^j\phi \to \psi, \Gamma, \Pi \Rightarrow \Delta, \Lambda$.

5. The remaining cases are considered in the usual way: from the premise $\Gamma' \Rightarrow \Delta'$ of the considered rule (r) application, we get $\vdash S \leftrightharpoons (\Pi, \Gamma' \Rightarrow \Delta', \Lambda)$. The required sequent is obtained by applying (r) to S.

$\square$

Lemma 4.7. *All the rules of **GHOML** are height-preserving invertible.*

Proof. The lemma is proved by induction on the height h of the derivation of the conclusion.

Let us consider the rule $(x. \Rightarrow)$. The proof is obvious if $h = 0$, since the premise is an axiom in this case. Let $h > 0$ and let the last step in the derivation of the conclusion of $(x. \Rightarrow)$ be
$$\frac{x.\phi, \Gamma \Rightarrow \Delta, \psi_1 \quad \psi_2, x.\phi, \Gamma \Rightarrow \Delta}{\psi_1 \to \psi_2, x.\phi, \Gamma \Rightarrow \Delta} \; (\to\Rightarrow).$$

(The labels are omitted.) According to inductive hypothesis, $\vdash x.x \stackrel{\circ}{=} b, \phi[x := b], \Gamma \Rightarrow \Delta, \psi_1$, and $\vdash \psi_2, x.x \stackrel{\circ}{=} b, \phi[x := b], \Gamma \Rightarrow \Delta$. We apply ($\to\Rightarrow$) to these sequents and infer the required one
$$\psi_1 \to \psi_2, x.x \stackrel{\circ}{=} b, \phi[x := b], \Gamma \Rightarrow \Delta.$$

Let the last step in the derivation of the conclusion be
$$\frac{x.\phi, \Gamma, \psi_1 \Rightarrow \psi_2, \Delta}{x.\phi, \Gamma \Rightarrow \psi_1 \to \psi_2, \Delta} \; (\Rightarrow\to).$$

By inductive hypothesis, $\vdash x.x \stackrel{\circ}{=} b, \phi[x := b], \Gamma, \psi_1 \Rightarrow \psi_2, \Delta$. We apply ($\Rightarrow\to$) to this sequent and infer $x.x \stackrel{\circ}{=} b, \phi[x := b], \Gamma \Rightarrow \psi_1 \to \psi_2, \Delta$.

The remaining cases when the last step in the derivation of the conclusion is an application of another rule (r) of **GHOML** are dealt with in the same way: from the premise $x.\phi, \Gamma' \Rightarrow \Delta'$ of (r), we get $\vdash x.x \stackrel{\circ}{=} b, \phi[x := b], \Gamma' \Rightarrow \Delta'$, according to inductive hypothesis; the required sequent is attained by applying (r) to the latter sequent.

Invertibility of the remaining rules is proved in the same way, one can see also [7, 9]. □

Lemma 4.8. *The rules of contraction*

$$\frac{{}^i\phi, {}^i\phi, \Gamma \Rightarrow \Delta}{{}^i\phi, \Gamma \Rightarrow \Delta} \ (C \Rightarrow), \qquad \frac{\Gamma \Rightarrow {}^i\phi, {}^i\phi, \Delta}{\Gamma \Rightarrow {}^i\phi, \Delta} \ (\Rightarrow C)$$

are height-preserving admissible in **GHOML**.

Proof. The lemma is proved by induction on the derivation height h of the premise. If $h = 0$, then the proof is obtained, since the conclusion is an axiom. Let $h > 0$ and the derivation of the premise be concluded by application of a rule (r).

1. Let (r) be $(\Rightarrow \rightarrow)$ (the labels are omitted):

$$\frac{\phi, \phi, \Gamma, \psi_1 \Rightarrow \Delta, \psi_2}{\phi, \phi, \Gamma \Rightarrow \Delta, \psi_1 \rightarrow \psi_2} \ (\Rightarrow \rightarrow).$$

According to inductive hypothesis, $\vdash \phi, \Gamma, \psi_1 \Rightarrow \Delta, \psi_2$. We apply the rule $(\Rightarrow \rightarrow)$ to this sequent and infer $\phi, \Gamma \Rightarrow \Delta, \psi_1 \rightarrow \psi_2$.

All the cases when the principal formula of (r) is not the contraction formula ϕ are considered in the same way: from the premise of (r), we get $\vdash \phi, \Gamma' \Rightarrow \Delta'$ (or $\Gamma' \Rightarrow \Delta', \phi$, when the second contraction rule is considered), using inductive hypothesis; the required sequent is obtained by applying (r) to the latter sequent.

2. Let (r) be $(\Rightarrow x.)$ (the labels are omitted):

$$\frac{x.x \stackrel{\circ}{=} b, \Gamma \Rightarrow \Delta, x.\phi, \phi[x := b]}{\Gamma \Rightarrow \Delta, x.\phi, x.\phi} \ (\Rightarrow x.).$$

From the premise we get

$$\vdash x.x \stackrel{\circ}{=} c, x.x \stackrel{\circ}{=} b, \Gamma \Rightarrow \Delta, \phi[x := c], \phi[x := b],$$

based on Lemma 4.7. Applying Lemma 4.5 with $[c := b]$ to this sequent we obtain

$$\vdash x.x \stackrel{\circ}{=} b, x.x \stackrel{\circ}{=} b, \Gamma \Rightarrow \Delta, \phi[x := b], \phi[x := b].$$

Hence $\vdash x.x \stackrel{\circ}{=} b, \Gamma \Rightarrow \Delta, \phi[x := b]$, using inductive hypothesis twice. Applying the rule $(\Rightarrow x.)$ to this sequent, we infer the required one $\Gamma \Rightarrow \Delta, x.\phi$.

The admissibility of the rule $(C \Rightarrow)$ when the contraction formula $x.\phi$ is the principal formula of $(x. \Rightarrow)$ is considered in the same way.

3. Let (r) be $(\Rightarrow \Box)$:

$$\frac{i \mapsto j, \Gamma \Rightarrow \Delta, {}^i\phi, {}^j\Box\phi}{\Gamma \Rightarrow \Delta, {}^i\Box\phi, {}^i\Box\phi} \ (\Rightarrow \Box).$$

From the premise we get

$$\vdash i \mapsto j, i \mapsto k, \Gamma \Rightarrow \Delta, {}^j\phi, {}^k\phi,$$

based on Lemma 4.7. Substituting j for k in this sequent and using Lemma 4.4, we obtain

$$\vdash i \mapsto j, i \mapsto j, \Gamma \Rightarrow \Delta, {}^j\phi, {}^j\phi.$$

Hence $\vdash S \leftrightharpoons (i \mapsto j, \Gamma \Rightarrow \Delta, {}^j\phi)$, using inductive hypothesis twice. The required sequent $\Gamma \Rightarrow \Delta, {}^i\Box\phi$ is obtained by applying $(\Rightarrow \Box)$ to S.

4. Let (r) be $(\Box \Rightarrow)$:

$$\frac{i \mapsto j, {}^j\phi, {}^i\Box\phi, {}^i\Box\phi, \Gamma \Rightarrow \Delta}{i \mapsto j, {}^i\Box\phi, {}^i\Box\phi, \Gamma \Rightarrow \Delta} \ (\Box \Rightarrow).$$

According to inductive hypothesis, $\vdash i \mapsto j, {}^j\phi, {}^i\Box\phi, \Gamma \Rightarrow \Delta$. We apply the rule $(\Box \Rightarrow)$ to this sequent and infer the required one $i \mapsto j, {}^i\Box\phi, \Gamma \Rightarrow \Delta$.

The cases when (r) is an equality rule are considered in the same way: from the premise $\phi', \phi', \Gamma' \Rightarrow \Delta'$ (or $\Gamma' \Rightarrow \Delta', \phi', \phi'$) of (r) we get $\vdash S \leftrightharpoons (\phi', \Gamma' \Rightarrow \Delta')$ (or $\vdash S \leftrightharpoons (\Gamma' \Rightarrow \Delta', \phi')$). The required sequent is obtained by applying (r) to S. As a special case, see item 1 of the proof of Lemma 4.3.

5. Let (r) be $(\rightarrow \Rightarrow)$ (the labels are omitted):

$$\frac{\phi \rightarrow \psi, \Gamma \Rightarrow \Delta, \phi \quad \psi, \phi \rightarrow \psi, \Gamma \Rightarrow \Delta}{\phi \rightarrow \psi, \phi \rightarrow \psi, \Gamma \Rightarrow \Delta} \ (\rightarrow \Rightarrow).$$

We get $\vdash \Gamma \Rightarrow \Delta, \phi, \phi$, from the left premise and $\vdash \psi, \psi, \Gamma \Rightarrow \Delta$ from the right one, using the fact that the rule $(\rightarrow \Rightarrow)$ is invertible (Lemma 4.7). Hence $\vdash S_1 \leftrightharpoons (\Gamma \Rightarrow \Delta, \phi)$ and $\vdash S_2 \leftrightharpoons (\psi, \Gamma \Rightarrow \Delta)$, based on inductive hypothesis.

The required sequent $\phi \to \psi, \Gamma \Rightarrow \Delta$ is obtained by applying $(\to\Rightarrow)$ to S_1 and S_2.

The cases when (r) is one of the remaining propositional rules are considered in the same way, using Lemma 4.7 and inductive hypothesis. □

Proposition 4.9. *The rule*

$$\frac{S \leftrightharpoons (^{i}a = b, \Gamma \Rightarrow \Delta)}{\Gamma \Rightarrow \Delta},$$

where b does not occur in $\Gamma \Rightarrow \Delta$, *is height-preserving admissible in* **GHOML**.

Proof. If $\vdash S$, then $\vdash S[b := a] \leftrightharpoons (^{i}a = a, \Gamma \Rightarrow \Delta)$, based on Lemma 4.5; hence $\vdash \Gamma \Rightarrow \Delta$, according to Proposition 4.2. □

Proposition 4.10. *The rule*

$$\frac{^{i}x.x \stackrel{\circ}{=} b, \Gamma \Rightarrow \Delta}{\Gamma \Rightarrow \Delta},$$

where b does not occur in $\Gamma \Rightarrow \Delta$, *is height-preserving admissible in* **GHOML**.

Proof. The proposition is proved by induction on the height h of derivation of the premise. If $h = 0$, then the proof is obtained, since the conclusion is an axiom in this case. Let $h > 0$ and the derivation of the premise be concluded by:

1.
$$\frac{x.x \stackrel{\circ}{=} b, \Gamma, \phi \Rightarrow \psi, \Delta}{x.x \stackrel{\circ}{=} b, \Gamma \Rightarrow \phi \to \psi, \Delta} \; (\Rightarrow\to)$$

(the labels are omitted). It is true that $\vdash \Gamma, \phi \Rightarrow \psi, \Delta$, according to the inductive hypothesis. We apply $(\Rightarrow\to)$ to this sequent and infer $\Gamma \Rightarrow \phi \to \psi, \Delta$.

2. The remaining cases are considered in the same way as the previous one, using the inductive hypothesis. (Note that he derivation of the premise cannot be concluded by the application of $(b = c \Rightarrow)$, since the side condition of this rule requires that both b and c occur in the context formulas.)

□

Proposition 4.11. *The rule*

$$\frac{(^i x.\sigma_1 \, \theta \, \sigma_2), (^i y.\sigma_3 \, \xi \, \sigma_4), \Gamma \Rightarrow \Delta}{(^i x.\sigma_1 \, \theta \, \sigma_2), \Gamma \Rightarrow \Delta}$$

is height-preserving admissible in **GHOML**. *Here: 1)* $\theta, \xi \in \{=, \stackrel{\circ}{=}\}$, *2)* $\{\sigma_1, \sigma_2\} \leftrightharpoons \{x, b\}$, *and 3)* $\{\sigma_3, \sigma_4\} \leftrightharpoons \{y, b\}$.

Proof. The proposition is proved by induction on the height h of the derivation of the premise. If $h = 0$, then the proof is obtained, since the premise is an axiom in this case. Let $h > 0$:

$$\frac{\cdots}{(^i x.\sigma_1 \, \theta \, \sigma_2), (^i y.\sigma_3 \, \xi \, \sigma_4), \Gamma \Rightarrow \Delta} \, (r).$$

1. Let (r) be $(y. \Rightarrow)$ (the labels are omitted):

$$\frac{S \leftrightharpoons ((x.x \stackrel{\circ}{=} b), (y.y \stackrel{\circ}{=} b), d = b, \Gamma \Rightarrow \Delta)}{(x.x \stackrel{\circ}{=} b), (y.y = b), \Gamma \Rightarrow \Delta} \, (y. \Rightarrow).$$

We have

$$\vdash S[d := b] \leftrightharpoons (x.x \stackrel{\circ}{=} b), (y.y \stackrel{\circ}{=} b), b = b, \Gamma \Rightarrow \Delta,$$

according to Lemma 4.5. Hence

$$\vdash (x.x \stackrel{\circ}{=} b), (y.y \stackrel{\circ}{=} b), \Gamma \Rightarrow \Delta,$$

based on Proposition 4.2. The latter fact yields $\vdash (x.x \stackrel{\circ}{=} b), \Gamma \Rightarrow \Delta$, using inductive hypothesis.

2. The remaining cases are similar or considered using only the inductive hypothesis.

□

Proposition 4.12. *The rule*

$$\frac{\Gamma \Rightarrow \Delta, {}^i b = a,}{\Gamma \Rightarrow \Delta, {}^i a = b,}$$

is admissible in **GHOML**.

Proof. The proposition is proved by induction on the height h of derivation of the premise. Let $h = 0$ and the premise be the axiom $b = a, \Gamma \Rightarrow \Delta, b = a$ (the labels are omitted). The required sequent is derived as follows:

$$\frac{\Gamma[b := a] \Rightarrow \Delta[b := a], a = b[b := a]}{b = a, \Gamma \Rightarrow \Delta, a = b} \ (b := a).$$

In the remaining cases, the fact that the premise is an axiom implies that the conclusion is an axiom as well.

Let $h > 0$ (the labels are omitted):

$$\frac{\ldots}{\Gamma \Rightarrow \Delta, b = a} \ (r).$$

1. Let (r) be $(b := d)$:

$$\frac{\Gamma[b := d] \Rightarrow \Delta[b := d], b = a[b := d]}{b = d, \Gamma \Rightarrow \Delta, b = a} \ (b := d).$$

According to inductive hypothesis, $\vdash S \leftrightharpoons (\Gamma[b := d] \Rightarrow \Delta[b := d], a = b[b := d])$. The required sequent $b = d, \Gamma \Rightarrow \Delta, a = b$ is obtained by applying the rule $(b := d)$ to S.

2. The cases when (r) is $(\pi_1 := \pi_2)$, where π_1 and π_2 are some free variables and $\{\pi_1, \pi_2\} \cap \{a, b\} \neq \emptyset$, are considered similarly as case 1.

3. Let (r) be $(\Rightarrow \rightarrow)$:

$$\frac{\Gamma, \phi \Rightarrow \Delta, b = a, \psi}{\Gamma \Rightarrow \Delta, b = a, \phi \rightarrow \psi} \ (\Rightarrow \rightarrow).$$

According to inductive hypothesis, $\vdash S \leftrightharpoons (\Gamma, \phi \Rightarrow \Delta, a = b, \psi)$. The required sequent $\Gamma \Rightarrow \Delta, a = b, \phi \rightarrow \psi$ is obtained by applying the rule $(\Rightarrow \rightarrow)$ to S.

4. The cases where:

   (a) (r) is not $(\pi_1 := \pi_2)$ (where π_1 and π_2 are some free variables) or

   (b) (r) is $(\pi_1 := \pi_2)$, where $\{\pi_1, \pi_2\} \cap \{a, b\} \leftrightharpoons \emptyset$,

are considered similarly as the previous one: form the premise $\Gamma' \Rightarrow \Delta', b = a$ of (r), we get $S \leftrightharpoons (\Gamma' \Rightarrow \Delta', a = b)$, using inductive hypothesis; the required sequent is obtained by applying (r) to S.

□

Proposition 4.13. *The rule*

$$\frac{\Gamma \Rightarrow \Delta, \bot}{\Gamma \Rightarrow \Delta}$$

is height-preserving admissible in **GHOML**.

Proof. The proposition is proved by induction on the height of the premise. □

Theorem 4.14. *The rule of cut*

$$\frac{\Gamma \Rightarrow \Delta, {}^i\phi \quad {}^i\phi, \Pi \Rightarrow \Lambda}{\Gamma, \Pi \Rightarrow \Delta, \Lambda} \ (cut)$$

is admissible in **GHOML**.

Proof. The theorem is proved by induction on the ordered pair $\langle g, h \rangle$, where g is the complexity of the formula ϕ and h is the sum of the derivation heights of the cut premises.

If $h = 0$, then the (cut) premises are axioms irrespective of g. One can see that the (cut) conclusion is an axiom in this case as well, and the proof is obtained. Let $h > 0$ and $g \geq 0$. We use the expression '(cut)-g' or '(cut)-h' to denote that the rule of cut is admissible by induction on g or h, correspondingly. First we consider the cases when one of the (cut) premises is an axiom.

I.1. The left premise of (cut) be an axiom:

$$\frac{\Gamma \Rightarrow \Delta, \phi \quad \dfrac{\cdots}{\phi, \Pi \Rightarrow \Lambda}\ (r)}{\Gamma, \Pi \Rightarrow \Delta, \Lambda}\ (\text{cut})$$

If ϕ is an atomic formula and $\Gamma \leftrightharpoons (\phi, \Gamma')$, then the conclusion of (cut) is obtained from the right premise by weakening and Lemma 4.6.

If ϕ is $b = b$, then we get $\vdash \Pi \Rightarrow \Lambda$ from the right premise, according to Lemma 4.2. The conclusion of (cut) is obtained from this sequent by weakening and Lemma 4.6.

Otherwise, the sequent $\Gamma \Rightarrow \Delta$ is an axiom which implies that the conclusion of (cut) is an axiom, as well.

I.2. Let the right premise of (cut) be an axiom:

$$\frac{\dfrac{\cdots}{\Gamma \Rightarrow \Delta, \phi}\ (r) \quad \phi, \Pi \Rightarrow \Lambda}{\Gamma, \Pi \Rightarrow \Delta, \Lambda}\ (\text{cut})$$

If ϕ is an atomic formula and $\Lambda \leftrightharpoons (\phi, \Lambda')$, then the conclusion of (cut) is obtained from the left premise by weakening and Lemma 4.6.

If $\phi \rightleftharpoons \bot$, then we get $\vdash \Gamma \Rightarrow \Delta$ from the left premise, using Proposition 4.13. The required sequent is obtained from this sequent by the rule of weakening, using Lemma 4.6.

Otherwise, the sequent $\Pi \Rightarrow \Lambda$ is an axiom which implies that the conclusion of (cut) is an axiom, as well.

From now on, we assume that neither of the (cut) premises is an axiom.

II.1. Let the derivation of the left (cut) premise be concluded by the application of rule $(a := b)$:

$$\dfrac{\dfrac{(\Gamma \Rightarrow \Delta, \phi)[a := b]}{a = b, \Gamma \Rightarrow \Delta, \phi}\,(a := b) \quad \dfrac{\ldots}{\phi, \Pi \Rightarrow \Lambda}}{a = b, \Gamma, \Pi \Rightarrow \Delta, \Lambda}\,(\text{cut})$$

This derivation is transformed into

$$\dfrac{\dfrac{(\Gamma \Rightarrow \Delta, \phi)[a := b] \quad \dfrac{\dfrac{\phi, \Pi \Rightarrow \Lambda}{(\phi, \Pi \Rightarrow \Lambda)[a := b]}\,Sub[a := b]}{}}{\dfrac{(\Pi, \Gamma \Rightarrow \Lambda, \Delta)[a := b]}{a = b, \Pi, \Gamma \Rightarrow \Lambda, \Delta}\,(a := b)}\,(\text{cut})\text{-h}}{}$$

The rule $Sub[a := b]$ is height-preserving admissible, according to Lemma 4.5.

II.2. Let the derivation of the left (cut) premise be concluded by a rule application where the cut formula is not principal:

$$\dfrac{\dfrac{\Gamma' \Rightarrow \Delta', \phi}{\Gamma \Rightarrow \Delta, \phi}\,(r)_1 \quad \dfrac{\ldots}{\phi, \Pi \Rightarrow \Lambda}\,(r)_2}{\Gamma, \Pi \Rightarrow \Delta, \Lambda}\,(\text{cut})$$

(here $(r)_1$ is not $(a := b)$; this case has been dealt with in case II.1 of the present proof). These cases are considered as follows:

$$\dfrac{\dfrac{\Gamma' \Rightarrow \Delta', \phi \quad \phi, \Pi \Rightarrow \Lambda}{\Gamma', \Pi \Rightarrow \Delta', \Lambda}\,(\text{cut})\text{-h}}{\Gamma, \Pi \Rightarrow \Delta, \Lambda}\,(r)_1$$

II.3 Let the derivation of the right (cut) premise be concluded by the application of the rule $(a := b)$. The derivation:

A Labelled Sequent Calculus for HOML

$$\frac{\begin{array}{c}\cdots\\ \Gamma \Rightarrow \Delta, \phi\end{array} \quad \dfrac{(\phi, \Pi \Rightarrow \Lambda)[a := b]}{\phi, a = b, \Pi \Rightarrow \Lambda}\,(a := b)}{a = b, \Gamma, \Pi \Rightarrow \Delta, \Lambda}\,(\text{cut})$$

is transformed into

$$\dfrac{\dfrac{\dfrac{\Gamma \Rightarrow \Delta, \phi}{(\Gamma \Rightarrow \Delta, \phi)[a := b]}\,Sub[a := b] \quad (\phi, \Pi \Rightarrow \Lambda)[a := b]}{(\Gamma, \Pi \Rightarrow \Delta, \Lambda)[a := b]}\,(\text{cut})\text{-h}}{a = b, \Gamma, \Pi \Rightarrow \Delta, \Lambda}\,(a := b)$$

The rule $Sub[a := b]$ is height-preserving admissible, according to Lemma 4.5. The consideration of derivations of the shape

$$\dfrac{\begin{array}{c}\cdots\\ \Gamma \Rightarrow \Delta, a = b\end{array}\,(r)_1 \quad \dfrac{(\Pi \Rightarrow \Lambda)[a := b]}{a = b, \Pi \Rightarrow \Lambda}\,(a := b)}{\Gamma, \Pi \Rightarrow \Delta, \Lambda}\,(\text{cut})$$

is covered by cases II.1 and II.2 of the present proof.

II.4 Let the derivation of the right (cut) premise be concluded by a rule application where the cut formula is not principal:

$$\dfrac{\begin{array}{c}\cdots\\ \Gamma \Rightarrow \Delta, \phi\end{array}\,(r)_1 \quad \dfrac{\phi, \Pi' \Rightarrow \Lambda'}{\phi, \Pi \Rightarrow \Lambda}\,(r)_2}{\Gamma, \Pi \Rightarrow \Delta, \Lambda}\,(\text{cut})$$

(here $(r)_2$ is not $(a := b)$; this case has been dealt with in case II.3 of the present proof). These cases are considered as follows:

$$\dfrac{\dfrac{\Gamma \Rightarrow \Delta, \phi \quad \phi, \Pi' \Rightarrow \Lambda'}{\Gamma, \Pi' \Rightarrow \Delta, \Lambda'}\,(\text{cut})\text{-h}}{\Gamma, \Pi \Rightarrow \Delta, \Lambda}\,(r)_2$$

We have considered the cases where the derivation of the left or the right (cut) premise is concluded by an application of the rule $(a := b)$ and the cases where the cut formula is not principal in the derivation of the left or the right (cut) premise. From now on, we deal with the remaining cases where the derivation of neither (cut) premise is concluded by the rule $(a := b)$ and where the cut formula is principal in the last step of the derivations of the left and the right (cut) premises.

III.1. If the (cut) formula is $^i x.\phi$, where x occurs in ϕ, then:

(i) the derivation

$$\cfrac{\cfrac{^i x.x \stackrel{\circ}{=} b, \Gamma \Rightarrow \Delta, {}^i\phi[x := b]}{\Gamma \Rightarrow \Delta, {}^i x.\phi}(\Rightarrow x.) \quad \cfrac{^i x.x \stackrel{\circ}{=} c, {}^i\phi[x := c], \Pi \Rightarrow \Lambda}{^i x.\phi, \Pi \Rightarrow \Lambda}(x. \Rightarrow)}{\Gamma, \Pi \Rightarrow \Delta, \Lambda}(cut)$$

is transformed into

$$\cfrac{\cfrac{\cfrac{\cfrac{x.x \stackrel{\circ}{=} b, \Gamma \Rightarrow \Delta, \phi[x := b]}{x.x \stackrel{\circ}{=} d, \Gamma \Rightarrow \Delta, \phi[x := d]}Sub[b:=d] \quad \cfrac{x.x \stackrel{\circ}{=} c, \phi[x := c], \Pi \Rightarrow \Lambda}{x.x \stackrel{\circ}{=} d, \phi[x := d], \Pi \Rightarrow \Lambda}Sub[c:=d]}{x.x \stackrel{\circ}{=} d, x.x \stackrel{\circ}{=} d, \Gamma, \Pi \Rightarrow \Delta, \Lambda}(cut)\text{-h}}{x.x \stackrel{\circ}{=} d, \Gamma, \Pi \Rightarrow \Delta, \Lambda}(C \Rightarrow)}{\Gamma, \Pi \Rightarrow \Delta, \Lambda}\text{Proposition 4.10}$$

(The labels are omitted.) Here d does not occur in $\Gamma, \Pi \Rightarrow \Delta, \Lambda$. The rules $Sub[b := d]$, $Sub[c := d]$, and $(C \Rightarrow)$ are height-preserving admissible, according to Lemmas 4.5 and 4.8, respectively.

(ii) the derivation

$$\cfrac{\cfrac{x.x \stackrel{\circ}{=} d, \Gamma \Rightarrow \Delta, d = a}{\Gamma \Rightarrow \Delta, x.x = a}(\Rightarrow x.) \quad \cfrac{a = b, x.x = a, y.y = b, \Pi \Rightarrow \Lambda}{x.x = a, y.y = b, \Pi \Rightarrow \Lambda}(a = b \Rightarrow)}{y.y = b, \Gamma, \Pi \Rightarrow \Delta, \Lambda}(cut)$$

is transformed into

$$\cfrac{\cfrac{\cfrac{\cfrac{x.x \stackrel{\circ}{=} d, \Gamma \Rightarrow \Delta, d = a}{x.x \stackrel{\circ}{=} b, \Gamma \Rightarrow \Delta, b = a}(r)}{x.x \stackrel{\circ}{=} b, \Gamma \Rightarrow \Delta, a = b}(r_1) \quad \cfrac{\Sigma_1 \quad a = b, x.x = a, y.y = b, \Pi \Rightarrow \Lambda}{a = b, y.y = b, \Pi, \Gamma \Rightarrow \Delta, \Lambda}(cut)\text{-h}}{\cfrac{\cfrac{\cfrac{x.x \stackrel{\circ}{=} b, y.y = b, \Gamma, \Gamma, \Pi \Rightarrow \Delta, \Delta, \Lambda}{y.y = b, \Gamma, \Gamma, \Pi \Rightarrow \Delta, \Delta, \Lambda}\text{Proposition 4.11}}{y.y = b, \Gamma, \Pi \Rightarrow \Delta, \Delta, \Lambda}(C \Rightarrow)}{y.y = b, \Gamma, \Pi \Rightarrow \Delta, \Lambda}(\Rightarrow C)}(cut)\text{-g}$$

Here $\Sigma_1 \leftrightharpoons (\Gamma \Rightarrow \Delta, x.x = a)$. The rules $(r) \leftrightharpoons Sub[d := b]$ and (r_1) are admissible, according to Lemma 4.5 and Proposition 4.12, respectively; the rules $(\Rightarrow C)$ and $(C \Rightarrow)$ are admissible, according to Lemma 4.8.

III.2 The derivation

$$\dfrac{\dfrac{i\mapsto k, \Gamma \Rightarrow \Delta, {}^k\phi}{\Gamma \Rightarrow \Delta, {}^i\Box\phi}\ (\Rightarrow \Box) \qquad \dfrac{i\mapsto j, {}^j\phi, {}^i\Box\phi, \Pi \Rightarrow \Lambda}{i\mapsto j, {}^i\Box\phi, \Pi \Rightarrow \Lambda}\ (\Box\Rightarrow)}{i\mapsto j, \Gamma, \Pi \Rightarrow \Delta, \Lambda}\ (\text{cut})$$

is transformed into

$$\dfrac{\dfrac{\dfrac{i\mapsto k, \Gamma \Rightarrow \Delta, {}^k\phi}{i\mapsto j, \Gamma \Rightarrow \Delta, {}^j\phi}\ (\text{j/k}) \qquad \dfrac{\Gamma \Rightarrow \Delta, {}^i\Box\phi \qquad i\mapsto j, {}^j\phi, {}^i\Box\phi, \Pi \Rightarrow \Lambda}{i\mapsto j, {}^j\phi, \Gamma, \Pi \Rightarrow \Delta, \Lambda}\ (\text{cut})\text{-h}}{\dfrac{i\mapsto j, i\mapsto j, \Gamma, \Gamma, \Pi \Rightarrow \Delta, \Delta, \Lambda}{i\mapsto j, \Gamma, \Pi \Rightarrow \Delta, \Lambda}\ \text{Lemma 4.8}}\ (\text{cut})\text{-g}}$$

The rule (j/k) is admissible, according to Lemma 4.4.

III.3 The derivation

$$\dfrac{\dfrac{\Gamma, \phi \Rightarrow \Delta, \psi}{\Gamma \Rightarrow \Delta, \phi \to \psi}\ (\Rightarrow\to) \qquad \dfrac{\Pi \Rightarrow \Lambda, \phi \qquad \psi, \Pi \Rightarrow \Lambda}{\phi \to \psi, \Pi \Rightarrow \Lambda}\ (\to\Rightarrow)}{\Gamma, \Pi \Rightarrow \Delta, \Lambda}\ (\text{cut})$$

is transformed into

$$\dfrac{\dfrac{\dfrac{\Pi \Rightarrow \Lambda, \phi \qquad \Gamma, \phi \Rightarrow \Delta, \psi}{\Gamma, \Pi \Rightarrow \Delta, \Lambda, \psi}\ (\text{cut})\text{-h} \qquad \psi, \Pi \Rightarrow \Lambda}{\Gamma, \Pi, \Pi \Rightarrow \Delta, \Lambda, \Lambda}\ (\text{cut})\text{-g}}{\Gamma, \Pi \Rightarrow \Delta, \Lambda}\ \text{Lemma 4.8}$$

□

Proposition 4.15. *The rule*

$$\dfrac{\Rightarrow {}^i\phi}{\Rightarrow {}^i\Box\phi}\ (\Box)$$

is admissible in **GHOML**.

Proof. The proposition is proved as follows:

$$\frac{\dfrac{\Rightarrow {}^i\phi}{\Rightarrow {}^j\phi} \; (j/i), \text{ Lemma 4.4}}{\dfrac{i\mapsto j \Rightarrow {}^j\phi}{\Rightarrow {}^i\Box\phi} \; (\Rightarrow \Box)} \; (W), \text{ Lemma 4.6}$$

□

Proposition 4.16. *The sequent*

$$^i\Box(\phi_1 \to \phi_2) \Rightarrow {}^i\Box\phi_1 \to \Box\phi_2$$

is derivable in **GHOML**.

Proof. The proposition is proved as follows:

$$\frac{\dfrac{\Sigma_1 \quad i\mapsto j, {}^j\phi_1, {}^j\phi_2, {}^i\Box(\phi_1 \to \phi_2), {}^i\Box\phi_1 \Rightarrow {}^j\phi_2}{\dfrac{i\mapsto j, {}^j\phi_1, {}^j\phi_1 \to \phi_2, {}^i\Box(\phi_1 \to \phi_2), {}^i\Box\phi_1 \Rightarrow {}^j\phi_2}{\dfrac{i\mapsto j, {}^j\phi_1, {}^i\Box(\phi_1 \to \phi_2), {}^i\Box\phi_1 \Rightarrow {}^j\phi_2}{\dfrac{i\mapsto j, {}^i\Box(\phi_1 \to \phi_2), {}^i\Box\phi_1 \Rightarrow {}^j\phi_2}{\dfrac{{}^i\Box(\phi_1 \to \phi_2), {}^i\Box\phi_1 \Rightarrow \Box^i\phi_2}{{}^i\Box(\phi_1 \to \phi_2) \Rightarrow {}^i\Box\phi_1 \to \Box\phi_2} \; (\Rightarrow \to)} \; (\Rightarrow \Box)} \; (\Box \Rightarrow)} \; (\Box \Rightarrow)} \; (\to \Rightarrow)}$$

Here $\Sigma_1 \leftrightharpoons (i\mapsto j, {}^j\phi_1, {}^i\Box(\phi_1 \to \phi_2), {}^i\Box\phi_1 \Rightarrow {}^j\phi_2, {}^j\phi_1)$. If ϕ_1 and ϕ_2 are non-atomic formulas, then we make use of Lemma 4.1. □

Proposition 4.17. *The rule*

$$\frac{{}^i\phi \Rightarrow {}^i\psi}{{}^i\Box\phi \Rightarrow {}^i\Box\psi} \; (\Box)_1$$

is admissible in **GHOML**.

Proof. The proposition is proved as follows:

$$\frac{\dfrac{\dfrac{{}^i\phi \Rightarrow {}^i\psi}{\Rightarrow {}^i\phi \to \psi} \; (\Rightarrow \to)}{\Rightarrow {}^i\Box(\phi \to \psi)} \; (\Box) \quad {}^i\Box(\phi_1 \to \phi_2) \Rightarrow {}^i\Box\phi_1 \to \Box\phi_2}{{}^i\Box\phi \Rightarrow {}^i\Box\psi} \; (\text{cut})$$

The rules (□) and (cut) are admissible, based on Proposition 4.15 and Theorem 4.14, respectively. The right premise of (cut) is derivable, according to Proposition 4.16.

□

Proposition 4.18. *The sequents*

1. $^i x.(\phi_1 \to \phi_2) \Rightarrow {}^i\phi_1 \to x.\phi_2$, *where x does not occur in ϕ_1, and*

2. $^i\phi_1 \to x.\Box\phi_2 \Rightarrow {}^i\phi_1 \to \Box x.\phi_2$,

are derivable in **GHOML**.

Proof. The first sequent is considered by generating the following backward proof-search tree:

$$\cfrac{\cfrac{\cfrac{\cfrac{\cfrac{x.x \stackrel{\circ}{=} a, {}^i x.x \stackrel{\circ}{=} a, {}^i\phi_1 \Rightarrow {}^i\phi_2[x := a], {}^i\phi_1[x := a] \qquad S}{{}^i\phi_1[x := a] \to \phi_2[x := a], x.x \stackrel{\circ}{=} a, {}^i x.x \stackrel{\circ}{=} a, {}^i\phi_1 \Rightarrow {}^i\phi_2[x := a]}(\to\Rightarrow)}{\cfrac{b = a, {}^i\phi_1[x := b] \to \phi_2[x := b], x.x \stackrel{\circ}{=} b, {}^i x.x \stackrel{\circ}{=} a, {}^i\phi_1 \Rightarrow {}^i\phi_2[x := a]}{{}^i\phi_1[x := b] \to \phi_2[x := b], x.x \stackrel{\circ}{=} b, {}^i x.x \stackrel{\circ}{=} a, {}^i\phi_1 \Rightarrow {}^i\phi_2[x := a]}(b = a \Rightarrow)}(b := a)}{\cfrac{{}^i x.x \stackrel{\circ}{=} a, {}^i x.(\phi_1 \to \phi_2), {}^i\phi_1 \Rightarrow {}^i\phi_2[x := a]}{{}^i x.(\phi_1 \to \phi_2), {}^i\phi_1 \Rightarrow {}^i x.\phi_2}(\Rightarrow x.)}(x.\Rightarrow)}{{}^i x.(\phi_1 \to \phi_2) \Rightarrow {}^i\phi_1 \to x.\phi_2}(\Rightarrow\to)$$

Here it is true that $\phi_1[x := a]$ is the same formula as ϕ_1, since x does not occur in ϕ_1; hence the left leaf is derivable, based on Lemma 4.1; the right leaf

$$S \leftrightharpoons ({}^i\phi_2[x := a], x.x \stackrel{\circ}{=} a, {}^i x.x \stackrel{\circ}{=} a, {}^i\phi_1 \Rightarrow {}^i\phi_2[x := a])$$

is derivable, according to Lemma 4.1. □

The consideration of the second sequent is left to the reader.

Proposition 4.19. *The sequent*

$$^i x.(\phi_1 \hookrightarrow \cdots \hookrightarrow \phi_n \hookrightarrow \psi) \Rightarrow {}^i\phi_1 \hookrightarrow \cdots \hookrightarrow \phi_n \hookrightarrow x.\psi\,,$$

where the notation is the same as in the quantifier rule **Q2**, *is derivable in* **GHOML**.

Proof. The proposition is proved by induction on the number m of '$\hookrightarrow$'.

Let $m = 1$. We have:

$$\cfrac{^i x.(\phi \to \Box\psi) \Rightarrow {}^i\phi \to x.\Box\psi \qquad {}^i\phi \to x.\Box\psi \Rightarrow {}^i\phi \to \Box x.\psi}{{}^i x.(\phi \to \Box\psi) \Rightarrow {}^i\phi \to \Box x.\psi}(\text{cut}).$$

The (cut) premises are derivable (Proposition 4.18) and the rule (cut) is admissible in **GHOML** (Theorem 4.14).

Let $m > 1$. We denote $\phi_k \hookrightarrow \cdots \hookrightarrow \phi_n \hookrightarrow \psi$ by $F(k,n)$, and $\phi_k \hookrightarrow \cdots \hookrightarrow \phi_n \hookrightarrow x.\psi$ by $G(k,n)$. According to this notation, the required sequent becomes $x.F(1,n) \Rightarrow G(1,n)$. We have:

$$\frac{x.F(1,n) \Rightarrow \phi_1 \to x.\Box F(2,n) \qquad \phi_1 \to x.\Box F(2,n) \Rightarrow \phi_1 \to \Box G(2,n)}{x.F(1,n) \Rightarrow G(1,n)} \text{(cut)}$$

(the labels are omitted). The left premise is derivable, according to item 1 of Proposition 4.18. The right premise is considered as follows:

$$\frac{\phi_1 \to x.\Box F(2,n) \Rightarrow \phi_1 \to \Box x.F(2,n) \qquad \phi_1 \to \Box x.F(2,n) \Rightarrow \phi_1 \to \Box G(2,n)}{\phi_1 \to x.\Box F(2,n) \Rightarrow \phi_1 \to \Box G(2,n)} \text{(cut)}$$

The left premise is derivable, according to item 2 of Proposition 4.18. The right premise is considered as follows:

$$\frac{\phi_1 \Rightarrow \phi_1, \Box G(2,n) \qquad \dfrac{\dfrac{\dfrac{x.F(2,n) \Rightarrow G(2,n)}{\Box x.F(2,n) \Rightarrow \Box G(2,n)} (\Box)_1}{\Box x.F(2,n), \phi_1 \Rightarrow \Box G(2,n)} (W)}{\phi_1 \to \Box x.F(2,n), \phi_1 \Rightarrow \Box G(2,n)} (\to\Rightarrow)}{\phi_1 \to \Box x.F(2,n) \Rightarrow \phi_1 \to \Box G(2,n)} (\Rightarrow\to)$$

The left leaf is derivable, according to Proposition 4.1. The rules (W) and $(\Box)_1$ are admissible, based on Lemma 4.6 and Proposition 4.17, respectively. The right leaf is derivable, by the inductive hypothesis. □

We say that the formula ϕ is derivable in **GHOML**, iff the sequent $\Rightarrow {}^i\phi$ is derivable in **GHOML**.

Proposition 4.20. *The rule* **Q2*** *of* **HHOML** *is admissible in* **GHOML**.

Proof. The proposition is proved as follows:

$$\frac{\dfrac{\Rightarrow {}^i\phi[x := b]}{{}^ix.x \stackrel{\circ}{=} b \Rightarrow {}^i\phi[x := b]} (W)}{\Rightarrow {}^ix.\phi} (\Rightarrow x.)$$

The rule (W) is admissible, based on Lemma 4.6. □

Proposition 4.21. *The rule* **Q2** *of* **HHOML** *is admissible in* **GHOML**.

Proof. The proposition is proved as follows (the labels are omitted):

$$\frac{\dfrac{\Rightarrow \phi_1 \hookrightarrow \cdots \hookrightarrow \phi_n \hookrightarrow \psi[x := b]}{\Rightarrow x.(\phi_1 \hookrightarrow \cdots \hookrightarrow \phi_n \hookrightarrow \psi)} Q2^* \qquad \Sigma_1}{\Rightarrow \phi_1 \hookrightarrow \cdots \hookrightarrow \phi_n \hookrightarrow x.\psi} \text{(cut)}$$

Here $\Sigma_1 \leftrightharpoons (x.(\phi_1 \hookrightarrow \cdots \hookrightarrow \phi_n \hookrightarrow \psi) \Rightarrow \phi_1 \hookrightarrow \cdots \hookrightarrow \phi_n \hookrightarrow x.\psi)$. The right premise of (cut) is derivable, according to Proposition 4.19. The rules (cut) and $Q2^*$ are admissible, based on Theorem 4.14 and Proposition 4.20, respectively. □

Lemma 4.22. *All the axioms of **HHOML** are derivable in **GHOML** and all the rules of **HHOML** are admissible in **GHOML**.*

Proof. Propositional tautologies are derivable in **GHOML**, since the sub-calculus of **GHOML** consisting of the axiom schema $\Gamma, {}^i\alpha \Rightarrow \Delta, {}^i\alpha$ and the propositional rules is complete for classical propositional logic, [7, 9].

The derivability of axiom schema **K1** in **GHOML** follows from Proposition 4.16.
The rule **K2** is admissible in **GHOML**, according to Proposition 4.15.
To prove that the axiom schema **Q1** is derivable in **GHOML**, we have to show that the formulas

$$x.(\phi_1 \to \phi_2) \to (x.\phi_1 \to x.\phi_2) \quad \text{and} \quad (x.\phi_1 \to x.\phi_2) \to x.(\phi_1 \to \phi_2)$$

are derivable (the labels are omitted). Let us consider the first formula:

$$
\cfrac{
\cfrac{
\cfrac{
\cfrac{
\cfrac{
\cfrac{
\cfrac{
\cfrac{
\cfrac{x.x \stackrel{\circ}{=} a, x.x \stackrel{\circ}{=} c, x.\stackrel{\circ}{=} c, \phi_1[x := c] \Rightarrow \phi_2[x := a], \phi_1[x := c]}{b = c, x.x \stackrel{\circ}{=} a, x.x \stackrel{\circ}{=} b, x.\stackrel{\circ}{=} c, \phi_1[x := b] \Rightarrow \phi_2[x := a], \phi_1[x := c]} (b := c)}{x.x \stackrel{\circ}{=} a, x.x \stackrel{\circ}{=} b, x.\stackrel{\circ}{=} c, \phi_1[x := b] \Rightarrow \phi_2[x := a], \phi_1[x := c]} (b = c \Rightarrow) \quad S}{x.x \stackrel{\circ}{=} a, x.x \stackrel{\circ}{=} b, x.\stackrel{\circ}{=} c, (\phi_1 \to \phi_2)[x := c], \phi_1[x := b] \Rightarrow \phi_2[x := a]} (\to \Rightarrow)}{x.x \stackrel{\circ}{=} a, x.x \stackrel{\circ}{=} b, x.(\phi_1 \to \phi_2), \phi_1[x := b] \Rightarrow \phi_2[x := a]} (x. \Rightarrow)}{x.x \stackrel{\circ}{=} a, x.(\phi_1 \to \phi_2), x.\phi_1 \Rightarrow \phi_2[x := a]} (x. \Rightarrow)}{x.(\phi_1 \to \phi_2), x.\phi_1 \Rightarrow x.\phi_2} (\Rightarrow x.)}{x.(\phi_1 \to \phi_2) \Rightarrow x.\phi_1 \to x.\phi_2} (\Rightarrow \to)}{\Rightarrow x.(\phi_1 \to \phi_2) \to (x.\phi_1 \to x.\phi_2)} (\Rightarrow \to)
$$

Here the left leaf is derivable, according to Lemma 4.1; the right leaf

$$S \leftrightharpoons (x.x \stackrel{\circ}{=} a, x.x \stackrel{\circ}{=} b, x.x \stackrel{\circ}{=} c, \phi_1[x := b], \phi_2[x := a] \Rightarrow \phi_2[x := c])$$

is considered as follows:

$$
\cfrac{
\cfrac{x.x \stackrel{\circ}{=} c, x.x \stackrel{\circ}{=} b, x.x \stackrel{\circ}{=} c, \phi_1[x := b][a := c], \phi_2[x := a][a := c] \Rightarrow \phi_2[x := c][a := c]}{a = c, x.x \stackrel{\circ}{=} a, x.x \stackrel{\circ}{=} b, x.x \stackrel{\circ}{=} c, \phi_1[x := b], \phi_2[x := a] \Rightarrow \phi_2[x := c]} (a := c)}{S} (a = c \Rightarrow)
$$

Note that $\phi_2[x := a][a := c]$ and $\phi_2[x := c][a := c]$ are the same formula $\phi_2[x := c]$, because of the requirement that the variable introduced in the premise of the rule

($x. \Rightarrow$) or ($\Rightarrow x.$) does not occur in the conclusion, i. e., the variable a does not occur in $\phi_2[x := c]$ nor in $x.\phi_2$. Hence the topmost sequent is derivable, according to Lemma 4.1.

We leave to the reader to prove that the second formula and the remaining axiom schemata are derivable in **GHOML**.

The rules **Q2** and **Q2*** are admissible in **GHOML**, based on Propositions 4.21 and 4.20, respectively.

If the sequents $\Rightarrow {}^i\phi_1$ and $\Rightarrow {}^i\phi_1 \to \phi_2$ are derivable in **GHOML**, then the sequent ${}^i\phi_1 \Rightarrow {}^i\phi_2$ is derivable in **GHOML** as well, based on Lemma 4.7. Hence $GHOML \vdash {}^i\phi_2$, using Theorem 4.14. We get that rule **PROP2** is admissible in **GHOML**.

□

Theorem 4.23. *The calculus **GHOML** is complete for **HOML**$^{-f}$: if $\models \phi$, then the formula ϕ is derivable in **GHOML**.*

Proof. The proof follows from the fact that **HHOML** is complete for **HOML**$^{-f}$ and Lemma 4.22. □

To prove that **GHOML** is sound for **HOML**, we extend the definition of $\mathcal{M} \models$.

$$\mathcal{M} \models x.x \stackrel{\circ}{=} b \quad \text{iff} \quad \mathcal{M} \models x.x = b.$$

If some state in $\mathcal{S}$ of $\mathcal{M}$ is labelled with i and some state with j, then

$$\mathcal{M} \models {}^i\phi \quad \text{iff} \quad \mathcal{M}[s_0 := s_i] \models \phi,$$
$$\mathcal{M} \models i \mapsto j \quad \text{iff} \quad (s_i, s_j) \in \to_\Box,$$

where s_i and s_j are the states labelled with i and j, respectively.

Let

$$Sq \leftrightharpoons (i_1 \mapsto j_1, \ldots, i_k \mapsto j_k, {}^{l_1}\phi_1, \ldots, {}^{l_m}\phi_m \Rightarrow {}^{l_{m+1}}\phi_{m+1}, \ldots, {}^{l_{m+n}}\phi_{m+n})$$

be a sequent, and $\mathcal{M}$ be an interpretation. The members of $\mathcal{S}$ in $\mathcal{M}$ are labelled with the elements of the set $I \leftrightharpoons \{i_1, j_1, \ldots, i_k, j_k, l_1, \ldots, l_{m+n}\}$ according to the function

$$f_I : I \ni i \mapsto s_i \in \mathcal{S}' \subseteq \mathcal{S}.$$

We say:

1. $(\mathcal{M}, f_I) \models Sq$ iff:

(a) there is $\iota \in \{1, 2, \ldots, k\}$ such that $\mathcal{M} \not\models i_\iota \mapsto j_\iota$, or
(b) there is $\iota \in \{1, 2, \ldots, m\}$ such that $\mathcal{M} \not\models {}^{l_\iota}\phi_\iota$, or
(c) there is $\iota \in \{m+1, m+2, \ldots, m+n\}$ such that and $\mathcal{M} \models {}^{l_\iota}\phi_\iota$.

2. $\mathcal{M} \models Sq$ iff $(\mathcal{M}, f_I) \models Sq$ for each function f_I.

3. $\models Sq$ iff $\mathcal{M} \models Sq$ for each $\mathcal{M}$.

If $\Gamma \leftrightharpoons (\theta_1, \ldots, \theta_m)$, where each θ_i $(1 \leq i \leq m)$ is a labelled formula or a relation atom, then $\lambda\Gamma \leftrightharpoons (\theta_1\lambda \cdots \lambda\theta_m)$, where $\lambda \in \{\vee, \wedge\}$. We say $\mathcal{M} \models \vee\Gamma$ ($\mathcal{M} \models \wedge\Gamma$), iff there is $\iota \in \{1, \ldots, m\}$ such that (for each $1 \leq \iota \leq m$ it is true that) $\mathcal{M} \models \theta_\iota$.

Lemma 4.24. *If $GHOML \vdash S$, then $\models S$.*

Proof. The lemma is proved by induction on the derivation height h of S. The proof is obvious if $h = 0$. Let $h > 0$ and the derivation of S be concluded by

$$\frac{{}^ix.x \doteq b, {}^i\phi[x := b], \Gamma \Rightarrow \Delta}{{}^ix.\phi, \Gamma \Rightarrow \Delta} \ (x. \Rightarrow).$$

Let $\mathcal{M} \models \wedge({}^ix.\phi, \Gamma)$. If the value u_i associated with the state i is assigned to b, then $\mathcal{M} \models \wedge({}^ix.x \doteq b, {}^i\phi[x := b], \Gamma)$. Hence $\mathcal{M} \models \vee(\Delta)$, by inductive hypothesis. Assume that u_i is not assigned to b in $\mathcal{M}$. Let $\mathcal{M}'$ be obtained from $\mathcal{M}$ by assigning u_i to b. Since $\mathcal{M} \models \wedge({}^ix.\phi, \Gamma)$ and b does not occur in $(x.\phi, \Gamma)$, it is true that $\mathcal{M}' \models \wedge({}^ix.x \doteq b, {}^i\phi[x := b], \Gamma)$. We get $\mathcal{M}' \models \vee(\Delta)$, according to inductive hypothesis. Hence $\mathcal{M} \models \vee(\Delta)$, based on the fact that b does not occur in Δ.

Let the derivation of S be concluded by

$$\frac{x.x \doteq b, \Gamma \Rightarrow \Delta, {}^i\phi[x := b]}{\Gamma \Rightarrow \Delta, {}^ix.\phi} \ (\Rightarrow x.).$$

Let $\mathcal{M} \models \wedge(x.x \doteq b, \Gamma)$. This condition implies that: 1) $\mathcal{M} \models \vee(\Delta, {}^i\phi[x := b])$, according to inductive hypothesis, and 2) $\mathcal{M} \models x.x \doteq b$. From 2), we have that the value u_i associated with the state i is assigned to b. This fact and 1) yield $\mathcal{M} \models \vee(\Delta, {}^ix.\phi)$, based on the definition of the freeze quantifier.

Let the derivation of S be concluded by

$$\frac{i \mapsto j, \Gamma \Rightarrow {}^j\phi, \Delta}{\Gamma \Rightarrow {}^i\Box\phi, \Delta} \ (\Rightarrow \Box).$$

Let $\mathcal{M} \models (\wedge\Gamma)$. If $\mathcal{M} \models (\vee\Delta)$, then $\mathcal{M} \models S$. Assume that $\mathcal{M} \not\models (\vee\Delta)$. If there is no k such that $(s_i, s_k) \in \to_\Box$, then $\mathcal{M} \models {}^i\Box\phi$ and the proof is obtained. Otherwise, let us take any k such that $(s_i, s_k) \in \to_\Box$. According to Lemma 4.5,

$$\vdash S' \leftrightharpoons (i \mapsto k, \Gamma \Rightarrow {}^k\phi, \Delta),$$

where $S' \leftrightharpoons (i \mapsto j, \Gamma \Rightarrow {}^j\phi, \Delta)[j := k]$. We apply inductive hypothesis to this sequent and obtain $\mathcal{M} \models {}^k\phi$. Hence, $\mathcal{M} \models {}^i\Box\phi$ and the proof is obtained.

Let the derivation of S be concluded by

$$\frac{{}^ia = b, ({}^ix.x = a), ({}^iy.y = b), \Gamma \Rightarrow \Delta}{({}^ix.x = a), ({}^iy.y = b), \Gamma \Rightarrow \Delta} \ (a = b \Rightarrow).$$

If $\mathcal{M} \models \wedge(({}^ix.x = a), ({}^iy.y = b), \Gamma)$, then $\mathcal{M} \models a = b$, since if the free variables a and b are equal to the value associated with the same state i, then it is true that the values of a and b are equal between themselves. This yields $\mathcal{M} \models \wedge(a = b, ({}^ix.x = a), ({}^iy.y = b), \Gamma)$. Hence, $\mathcal{M} \models \vee(\Delta)$, according to inductive hypothesis and the proof is obtained.

The remaining cases are considered using the inductive hypothesis. □

Theorem 4.25. *The calculus* **GHOML** *is sound for* **HOML**$^{-f}$: *if an arbitrary formula ϕ is derivable in* **GHOML**, *then* $\models \phi$.

Proof. If $GHOML \vdash (\Rightarrow {}^i\phi)$, then $\models (\Rightarrow {}^i\phi)$, according to Lemma 4.24. Hence $\models \phi$, based on the definition of $\models$. □

5 Decision

Lemma 5.1. *The equality rules permute up with respect to each* **GHOML** *rule.*

Proof. The lemma is proved by transforming derivations, e. g., the derivation

$$\frac{\dfrac{x.x \stackrel{\circ}{=} d,\ b = c, y.y = b, z.z = c, \Gamma \Rightarrow \phi[x := d]}{b = c, y.y = b, z.z = c, \Gamma \Rightarrow x.\phi} (\Rightarrow x.)}{y.y = b, z.z = c, \Gamma \Rightarrow x.\phi} (b = c \Rightarrow)$$

is transformed into

$$\frac{\dfrac{b = c, x.x \stackrel{\circ}{=} d, y.y = b, z.z = c, \Gamma \Rightarrow \phi[x := d]}{x.x \stackrel{\circ}{=} d,\ y.y = b, z.z = c, \Gamma \Rightarrow \phi[x := d]} (b = c \Rightarrow)}{y.y = b, z.z = c, \Gamma \Rightarrow x.\phi} (\Rightarrow x.)$$

The remaining cases are considered similarly. □

An application of a non-equality rule in proof-search is called *irregular* if it is above an application of an equality rule on some path of the proof-search.

A derivation of a sequent is called *regular* if it has no irregular applications.

The *range* of a derivation is the number of irregular applications in it.

Lemma 5.2. *If a sequent is derivable in* **GHOML**, *then the sequent has a regular derivation.*

Proof. The lemma is proved by induction on the range r of derivation. If $r = 0$, then the proof is obtained. If $r > 0$, then we choose some uppermost irregular application ρ on some branch and permute up the corresponding applications of equality rules, using Lemma 5.1, so that ρ becomes regular. The induction parameter is reduced, and we apply the inductive hypothesis. □

According to Lemma 5.2, application of equality rules can be postponed till the moment when the propositional, modal, and quantifier rules are no more applied on the corresponding branch of the backward proof-search tree. This fact allows us to separate the variable unification, performed by applying the equality rules, from the rest of the proof-search in each branch.

Lemma 5.3. *The rule* $(\Box \Rightarrow)$ *permutes down with respect to each* **GHOML** *rule except* $(\Rightarrow \Box)$. *It permutes down with* $(\Rightarrow \Box)$ *if the principal relation atom of* $(\Box \Rightarrow)$ *is not active in* $(\Rightarrow \Box)$.

Proof. The lemma is proved in the same way as Lemma 6.3 in [7]. □

The application ν of $(\Box \Rightarrow)$ in a proof-search tree is called superfluous if there is another application of $(\Box \Rightarrow)$ with the same pair of principal formulas below ν on the same branch.

The sum of heights of all the branches of a derivation tree is called the absolute height of the tree.

Lemma 5.4. *If a sequent S is derivable in* **GHOML**, *then S has a derivation free of superfluous applications.*

Proof. The lemma is proved by induction on the absolute height h of derivation of S. If $h = 0$ or there are no superfluous applications in the derivation, then the proof is obtained. Otherwise, using Lemma 5.3, we eliminate one superfluous application by permuting it down and diminish the induction parameter in the same way as in the proof of Corollary 6.5 in [7]. □

Lemma 5.5. *Each backward* **GHOML** *proof-search free of superfluous applications terminates.*

Proof. Each premise of each **GHOML** rule, except $(\Box \Rightarrow)$, $(x. \Rightarrow)$, $(\Rightarrow x.)$, and $(a = b \Rightarrow)$ is simpler than the conclusion. The number of backward $(\Box \Rightarrow)$ applications is finite, based on the facts that there are no superfluous applications and the number of relation atoms is finite. The number of backward applications of $(x. \Rightarrow)$ and $(\Rightarrow x.)$ is finite, since each such application diminishes the number of formulas that have the shape of the principal formulas of these rules. One can see that the number of backward $(a = b \Rightarrow)$ applications is finite because of the side conditions introduced for this rule. $\square$

Let $Proc(S)$ be the following procedure: using calculus **GHOML**, perform backward, free of superfluous applications proof-search of the sequent $\Rightarrow {}^i\phi$, where ϕ is any formula. It follows from Lemma 4.7, Theorems 4.23, 4.25, and Lemmas 5.4, 5.5 that: 1) the proof-search terminates and 2) the formula ϕ is valid in **HOML**$^{-f}$, iff the sequent $\Rightarrow {}^i\phi$ is derivable in **GHOML**; that is to say $Proc(S)$ is a decision procedure for **HOML**$^{-f}$.

Let us consider some examples. Given the formula

$$\phi \rightleftharpoons (\alpha_1 \to \Box \alpha_2(b)) \to \Big(\big((\alpha_1 \to \Box \alpha_3(b)) \to \bot\big) \to \alpha_4\Big),$$

where α_i ($1 \leq i \leq 4$) are unequal in pairs. We want to determine if it is valid in **HOML**$^{-f}$. Using the calculus **GHOML**, we generate the following bottom-up proof-search tree with the sequent $\Rightarrow {}^i\phi$ at the root:

$$\cfrac{\cfrac{{}^i\alpha_1 \Rightarrow {}^i\alpha_4, {}^i\Box\alpha_3(b), {}^i\alpha_1 \quad \cfrac{\cfrac{\cfrac{{}^j\alpha_2(b), i\mapsto j, {}^i\Box\alpha_2(b), {}^i\alpha_1 \Rightarrow {}^i\alpha_4, {}^j\alpha_3(b)}{i\mapsto j, {}^i\Box\alpha_2(b), {}^i\alpha_1 \Rightarrow {}^i\alpha_4, {}^j\alpha_3(b)}(\Box\Rightarrow)}{{}^i\Box\alpha_2(b), {}^i\alpha_1 \Rightarrow {}^i\alpha_4, {}^i\Box\alpha_3(b)}(\Rightarrow\Box)}{{}^i\alpha_1 \to \Box\alpha_2(b), {}^i\alpha_1 \Rightarrow {}^i\alpha_4, {}^i\Box\alpha_3(b)}(\Rightarrow\to)}{{}^i\alpha_1 \to \Box\alpha_2(b) \Rightarrow {}^i\alpha_4, {}^i\alpha_1 \to \Box\alpha_3(b)}(\Rightarrow\to) \quad S}{\cfrac{{}^i\alpha_1 \to \Box\alpha_2(b), {}^i\big((\alpha_1 \to \Box\alpha_3(b)) \to \bot\big) \Rightarrow {}^i\alpha_4}{\cfrac{{}^i\alpha_1 \to \Box\alpha_2(b) \Rightarrow {}^i\big((\alpha_1 \to \Box\alpha_3(b)) \to \bot\big) \to \alpha_4}{\Rightarrow {}^i(\alpha_1 \to \Box\alpha_2(b)) \to \Big(\big((\alpha_1 \to \Box\alpha_3(b)) \to \bot\big) \to \alpha_4\Big)}(\Rightarrow\to)}(\Rightarrow\to)}}(\Rightarrow\to)$$

Here $S \rightleftharpoons ({}^i\alpha_1 \to \Box\alpha_2(b), {}^i\bot \Rightarrow {}^i\alpha_4)$. We see that the left leaf is an axiom, since it has the same atomic formula α_1 both on the left and on the right sides of '$\Rightarrow$'. The sequent S is an axiom as well, since it is of the shape $\Gamma, {}^i\bot \Rightarrow \Delta$. The middle leaf is not an axiom and no other rule can be backward applied to it, since the application of ($\Box \Rightarrow$) with the principal pair $(i \mapsto j, {}^i\Box\alpha_2(b))$ would be superfluous and is needless, according to Lemma 5.4. The fact that all rules of **GHOML** are invertible (Lemma 4.7) implies that the order of backward rule application have no impact on derivability of sequents. We conclude that the root sequent is not derivable in **GHOML**, and the considered formula is not valid, according to Theorem 4.23.

Let us consider another formula

$$\phi_1 \leftrightharpoons (\psi(a) \to y.\alpha(y)) \to \Big(a = b \to x.(\psi(b) \to \alpha(x))\Big),$$

where ψ is any formula. As in the previous case, we generate the following bottom-up proof-search tree with the sequent $\Rightarrow {}^i\phi_1$ at the root:

$$
\cfrac{
\cfrac{
\cfrac{
\cfrac{
\cfrac{
\cfrac{
\cfrac{
\cfrac{D \quad \cfrac{
\cfrac{
\cfrac{
\cfrac{
\cfrac{{}^iy.y \stackrel{\circ}{=} d, {}^ix.x \stackrel{\circ}{=} d, {}^i\alpha(d), {}^i\psi(b) \Rightarrow {}^i\alpha(d)}{{}^ie = d, {}^iy.y \stackrel{\circ}{=} e, {}^ix.x \stackrel{\circ}{=} d, {}^i\alpha(e), {}^i\psi(b) \Rightarrow {}^i\alpha(d)}(e := d)
}{{}^iy.y \stackrel{\circ}{=} e, {}^ix.x \stackrel{\circ}{=} d, {}^i\alpha(e), {}^i\psi(b) \Rightarrow {}^i\alpha(d)}(e = d \Rightarrow)
}{{}^ix.x \stackrel{\circ}{=} d, {}^iy.\alpha(y), {}^i\psi(b) \Rightarrow {}^i\alpha(d)}(y. \Rightarrow)
}{{}^ix.x \stackrel{\circ}{=} d, {}^iy.\alpha(y) \Rightarrow {}^i\psi(b) \to \alpha(d)}(\Rightarrow \to)
}{{}^iy.\alpha(y) \Rightarrow {}^ix.(\psi(b) \to \alpha(x))}(\Rightarrow x.)
}
}{{}^i\psi(b) \to y.\alpha(y) \Rightarrow {}^ix.(\psi(b) \to \alpha(x))}(\to \Rightarrow)
}{{}^i\psi(a) \to y.\alpha(y), {}^ia = b \Rightarrow {}^ix.(\psi(b) \to \alpha(x))}(a := b)
}{{}^i\psi(a) \to y.\alpha(y) \Rightarrow {}^ia = b \to x.(\psi(b) \to \alpha(x))}(\Rightarrow \to)
}{\Rightarrow {}^i(\psi(a) \to y.\alpha(y)) \to \Big(a = b \to x.(\psi(b) \to \alpha(x))\Big)}(\Rightarrow \to)
$$

Here D stands for

$$
\cfrac{
\cfrac{
\cfrac{{}^ix.x \stackrel{\circ}{=} d, {}^i\psi(b) \Rightarrow {}^i\alpha(d), {}^i\psi(b)}{{}^ix.x \stackrel{\circ}{=} d \Rightarrow {}^i\psi(b) \to \alpha(d), {}^i\psi(b)}(\Rightarrow \to)
}{\Rightarrow {}^ix.(\psi(b) \to \alpha(x)), {}^i\psi(b)}(\to x.)
$$

The left leaf of this tree is derivable, according to Lemma 4.1, while the right leaf is an axiom. We conclude that the root sequent is derivable in **GHOML** and the formula ϕ_1 is valid, based on Theorem 4.25.

6 Concluding remarks

In the present paper, we have introduced the sequent calculus **GHOML** and proved admissibility of the structural and cut rules in the calculus, invertibility of all the rules, soundness and completeness of **GHOML** with respect to the half-order modal logic without function symbols. We have showed that the considered half-order modal logic without function symbols is decidable by describing the decision procedure.

Similar results can be obtained for the other half-order modal logics such as the logic **HOML**$_{S5}$ based on modal logic **S5** and its sub-logics, e. g., **HOML**$_T$ based on modal logic **T**. The sequent calculi for these logics are obtained by adding additional rules for relation atoms so that the properties of the accessibility relation are captured, see [7].

The sequent calculus **GHOML**$_T$ is obtained from **GHOML** by adding the rule

$$\frac{i \mapsto i, \Gamma \Rightarrow \Delta}{\Gamma \Rightarrow \Delta} \ (Ref),$$

where $i \mapsto i$ does not occur in the conclusion, and the label i occurs in the conclusion.

The sequent calculus **GHOML**$_{S4}$ is obtained from **GHOML**$_T$ by adding the rule

$$\frac{i \mapsto l, \ i \mapsto j, \ j \mapsto l, \Gamma \Rightarrow \Delta}{i \mapsto j, \ j \mapsto l, \Gamma \Rightarrow \Delta} \ (Trans),$$

where $i \mapsto l$ does not occur in the conclusion.

The sequent calculus **GHOML**$_{S5}$ is obtained from **GHOML**$_{S4}$ by adding the rule

$$\frac{j \mapsto i, \ i \mapsto j, \Gamma \Rightarrow \Delta}{i \mapsto j, \Gamma \Rightarrow \Delta} \ (Sym),$$

where $j \mapsto i$ does not occur in the conclusion.

Since **GHOML** and the above calculi differ only in the rules for relation atoms, it is not difficult to adapt the proofs in the present paper to these new calculi. For decidability proof of **HOML**$_\theta$ ($\theta \in \{S4, S5\}$), Proposition 6.9 given in [7] is needed. The proof of the proposition can be adapted to the half-order logics. As far as the eigenvariables of rules $(x. \Rightarrow)$ and $(\Rightarrow x.)$ are concerned, we apply Lemma 4.5 along with Lemma 4.4 instead of Lemma 4.3 in the proof in [7], so that the corresponding formulas could be contracted, if the variables occur in them. We assume that all derivations are regular, which implies that the eigenvariables occur only in the active formulas of the quantifier rules and in their offspring at the moment of the substitution.

The calculus **GHOML** is not complete for **HOML** if the function symbols are presented, e. g., the sequent $b = f(b), \alpha(f(b)) \Rightarrow \alpha(b)$ is not derivable in **GHOML** (the labels are omitted):

$$\frac{\alpha(ff(b)) \Rightarrow \alpha(f(b))}{b = f(b), \alpha(f(b)) \Rightarrow \alpha(b)} (b := f(b)),$$

where f is a unary function symbol and $ff(b)$ stands for $f(f(b))$. Another equality rule, e. g.,

$$\frac{{}^i b = \pi, (\Gamma \Rightarrow \Delta)\langle b := \pi \rangle}{{}^i b = \pi, \Gamma \Rightarrow \Delta} (b := \pi),$$

where π is a first-order term, is needed.

References

[1] R. Alonderis, Proof-Theoretical Investigation of Temporal Logic with Time Gaps. *Lit. Math. Journal*, **40**(3), 255–276, (2000).

[2] T. A. Henzinger. Half-order Modal Logic: How To Prove Real-time Properties. *Proceedings of the 9th ACM Symposium on Principles of Distributed Computing*, pp. 43–56, 1990.

[3] Halpern, J. Y., and Moses, Y. A guide to completeness and complexity for modal logics of knowledge and belief. *Artificial Intelligence*, **54**, pp. 319–379 (1992)

[4] J. Hudelmaier. A Contraction-Free Sequent Calculus for S4. *Proof Theory of Modal Logic, Kluwer* pp. 3–15 (1996).

[5] N. Nide and S. Takata, Deduction systems for BDI logic using sequent calculus. In *Proc. AAMASâĂŹ02* 928–935 (2002).

[6] R. Pliuškevičius, Aida Pliuškevičienė. A New Method to Obtain Termination in Backward Proof Search For Modal Logic S4. *J. Log. Comput.* **20**(1) 353-379 (2010).

[7] S. Negri. Proof analysis in modal logic, *Journal of Philosophical Logic* **34** pp. 507–544, (2005).

[8] G. Takeuti. *Proof Theory*, North-Holland, Amsterdam (1975).

[9] A. S. Troelstra and H. Schwichtenberg. Basic Proof Theory, *Cambridge University Press second edition* (2000).

[10] H. Wansing. Sequent Calculi for Normal Modal Propositional Logics, *J. Logic Comput.* **4**(2) 125–142 (1994).

On epicomplete MV-algebras

Anatolij Dvurečenskij
Mathematical Institute, Slovak Academy of Sciences, Bratislava, Slovakia and Palacký University, Olomouc, Czech Republic.
dvurecen@mat.savba.sk

Omid Zahiri
University of Applied Science and Technology, Tehran, Iran.
zahiri@protonmail.com

Abstract

The aim of the paper is to study epicomplete objects in the category of MV-algebras. A relation between injective MV-algebras and epicomplete MV-algebras is found, an equivalent condition for an MV-algebra to be epicomplete is obtained, and it is shown that the class of divisible MV-algebras and the class of epicomplete MV-algebras coincide. Finally, the concept of epicompletion of an MV-algebra is introduced, and the conditions under which an MV-algebra has an epicompletion are obtained. As a result we show that each MV-algebra has an epicompletion.

AMS Mathematics Subject Classification (2010): 06D35, 06F15, 06F20

Keywords: MV-algebra, Epicomplete MV-algebra, Divisible MV-algebra, Injective MV-algebra, Epicompletion, a-closed MV-algebra.

1 Introduction

Epicomplete objects are interesting objects in each category. Many researches studied these objects in the category of lattice ordered groups (ℓ-group). Pedersen [28] defined the concept of an a-epimorphism in this category. It is an ℓ-homomorphism which is also an epimorphism in the category of all torsion free Abelian groups. Anderson and Conrad [1] proved that each epimorphism in the category of Abelian ℓ-groups is an a-epimorphism. They showed that an Abelian ℓ-group G is epicomplete if and only if it is divisible. They also studied epicomplete objects in some

---

This work was supported by grant VEGA No. 2/0069/16 SAV and GAČR 15-15286S.

subcategories of Abelian ℓ-groups. In particular, they proved that epicomplete objects in the category of Abelian o-groups (linearly ordered groups) with complete o-homomorphisms are the Hahn groups. Darnel [10] continued to study these objects and showed that any completely distributive epicomplete object in the category $\mathcal{C}$ of Abelian ℓ-groups with complete ℓ-homomorphisms is $V(\Gamma, \mathbb{R})$ for some root system Γ, where $V(\Gamma, \mathbb{R})$ is the set of functions $v : \Gamma \to \mathbb{R}$ whose support satisfies the ascending chain condition with a special order (see [11, Prop 51.2]). Also, he studied a new subcategory of $\mathcal{C}$ containing completely-distributive Abelian ℓ-groups with complete ℓ-homomorphisms. That is, the only epicomplete objects in this category are of the form $V(\Gamma, \mathbb{R})$. Ton [29] studied epicomplete archimedean ℓ-groups and proved that epicomplete objects in this category are ℓ-isomorphic to a semicomplete subdirect sum of real groups. Many other references can be found in [4, 3]. Recently, Hager [23] posed a question on the category of Archimedean ℓ-groups "Does the epicompleteness imply the existence of a compatible reduced f-ring multiplication? ". His answer to this question was "No" and he tried to find a partial positive answer for it.

There is an important class of structures called MV-algebras introduced by Chang [6] as an algebraic counterpart of many-valued reasoning. The principal result of the theory of MV-algebras is a representation theorem by Mundici [26] saying that there is a categorical equivalence between the category of MV-algebras and the category of unital Abelian ℓ-groups. Today the theory of MV-algebras is very deep and has many interesting connections with other parts of mathematics with many important applications to different areas. For more details on MV-algebras, we recommend the monographs [7, 27].

In the present paper, epicomplete objects in $\mathcal{MV}$, the category of MV-algebras, are studied. The concept of an a-extension in $\mathcal{MV}$ is introduced to obtain a condition on minimal prime ideals of an MV-algebra M under which M is epicomplete. Some relations between injective, divisible and epicomplete MV-algebras are found. In the final section, we introduce a completion for an MV-algebra which is epicomplete and has the universal mapping property. We called it the epicompletion and we show that any MV-algebra has an epicompletion.

2 Preliminaries

In the section, we gather some basic notions relevant to MV-algebras and ℓ-groups which will be needed in the next sections. For more details, we recommend to consult the books [2, 11] for the theory of ℓ-groups and [12, 7, 27] for MV-algebras.

We say that an *MV-algebra* is an algebra $(M; \oplus,', 0, 1)$ (and we will write simply

$M = (M; \oplus, ', 0, 1))$ of type $(2, 1, 0, 0)$, where $(M; \oplus, 0)$ is a commutative monoid with the neutral element 0 and, for all $x, y \in M$, we have:

(i) $x'' = x$;

(ii) $x \oplus 1 = 1$;

(iii) $x \oplus (x \oplus y')' = y \oplus (y \oplus x')'$.

In any MV-algebra $(M; \oplus, ', 0, 1)$, we can define the following further operations:

$$x \odot y := (x' \oplus y')', \quad x \ominus y := (x' \oplus y)'.$$

In addition, let $x \in M$. For any integer $n \geq 0$, we set

$$0.x = 0, \quad 1.x = x, \quad n.x = (n-1).x \oplus x, \ n \geq 2,$$

and

$$x^0 = 1, \quad x^1 = 1, \quad x^n = x^{n-1} \odot x, \ n \geq 2.$$

Moreover, the relation $x \leq y \Leftrightarrow x' \oplus y = 1$ is a partial order on M and $(M; \leq)$ is a lattice, where $x \vee y = (x \ominus y) \oplus y$ and $x \wedge y = x \odot (x' \oplus y)$. Let $(M, \oplus, ', 0, 1)$ and $(N, \oplus, ', 0, 1)$ be MV-algebras. A map $f : M \to N$ is called an MV-*homomorphism* if f preserves the operations $\oplus$, $'$, 0 and 1. We use $\mathcal{MV}$ to denote the category of MV-algebras whose objects are MV-algebras and morphisms are MV-homomorphisms. A non-empty subset I of an MV-algebra $(M; \oplus, ', 0, 1)$ is called an *ideal* of M if I is a down set which is closed under $\oplus$. The set of all ideals of M is denoted by $\mathcal{I}(M)$. For each ideal I of M, the relation θ_I on M defined by $(x, y) \in \theta_I$ if and only if $x \ominus y, y \ominus x \in I$ is a congruence relation on M, and x/I and M/I will denote $\{y \in M \mid (x, y) \in \theta_I\}$ and $\{x/I \mid x \in M\}$, respectively. A *prime* ideal is a proper ideal I of M such that M/I is a linearly ordered MV-algebra, or equivalently, for all $x, y \in M$, $x \ominus y \in I$ or $y \ominus x \in I$. The set of all minimal prime ideals of M is denoted by $Min(M)$. If M_1 is a subalgebra of an MV-algebra M_2, we write $M_1 \leq M_2$.

Remark 2.1. Let M_1 be a subalgebra of an MV-algebra M_2. For any ideal I of M_2, the set $\bigcup_{x \in M_1} x/I$ is a subalgebra of M_2 containing I which is denoted by $M_1 + I$ for simplicity.

An element a of an MV-algebra $(M; \oplus, ', 0, 1)$ is called *boolean* if $a \oplus a = a$. The set of all boolean elements of M is denoted by $B(M)$. An ideal I of M is called a *stonean* ideal if there is a subset $S \subseteq B(M)$ such that $I = \downarrow S$, where $\downarrow S = \{x \in M \mid x \leq a \text{ for some } a \in S\}$. An element $x \in M$ is called *archimedean* if there is an integer $n \in \mathbb{N}$ such that $n.x$ is boolean. An MV-algebra M is said to be *hyperarchimedean* if all elements of M are archimedean. For more details about hyperarchimedean MV-algebras see [7, Chap 6]

A group $(G; +, 0)$ is said to be *partially ordered* if it is equipped with a partial order relation $\leq$ that is compatible with $+$, that is, $a \leq b$ implies $x+a+y \leq x+b+y$ for all $x, y \in G$. An element $x \in G$ is called *positive* if $0 \leq x$. A partially ordered group $(G; +, 0)$ is called a *lattice ordered group* or simply an *ℓ-group* if G with its partially order relation is a lattice. The *lexicographic product* of two po-groups $(G_1; +, 0)$ and $(G_2; +, 0)$ is the direct product $G_1 \times G_2$ endowed with the lexicographic ordering $\leq$ such that $(g_1, h_1) \leq (g_2, h_2)$ iff $g_1 < g_2$ or $g_1 = g_2$ and $h_1 \leq h_2$ for $(g_1, h_1), (g_2, h_2) \in G_1 \times G_2$. The lexicographic product of po-groups G_1 and G_2 is denoted by $G_1 \overrightarrow{\times} G_2$.

An element u of an ℓ-group $(G; +, 0)$ is called a *strong unit* if, for each $g \in G$, there exists $n \in \mathbb{N}$ such that $g \leq nu$. A couple (G, u), where G is an ℓ-group and u is a fixed strong unit for G, is said to be a *unital ℓ-group*.

If $(G; +, 0)$ is an Abelian ℓ-group with strong unit u, then the interval $[0, u] := \{g \in G \mid 0 \leq g \leq u\}$ with the operations $x \oplus y := (x + y) \wedge u$ and $x' := u - x$ forms an MV-algebra, which is denoted by $\Gamma(G, u) = ([0, u]; \oplus, ', 0, u)$. Moreover, if $(M; \oplus, 0, 1)$ is an MV-algebra, then by Mundici's categorical equivalence, [26], there exists a unique (up to isomorphism) unital Abelian ℓ-group (G, u) with strong u such that $\Gamma(G, u)$ and $(M; \oplus, 0, 1)$ are isomorphic (as MV-algebras). Let $\mathcal{A}$ be the category of unital Abelian ℓ-groups whose objects are unital Abelian ℓ-groups and morphisms are unital ℓ-group morphisms (i.e. homomorphisms of ℓ-groups preserving fixed strong units). It is important to note that $\mathcal{MV}$ is a variety whereas $\mathcal{A}$ is not because it is not closed under infinite products. Then $\Gamma : \mathcal{A} \to \mathcal{MV}$ is a functor between these categories. Moreover, there is another functor from the category of MV-algebras to $\mathcal{A}$ sending M to a Chang ℓ-group induced by good sequences of the MV-algebra M, which is denoted by $\Xi : \mathcal{MV} \to \mathcal{A}$. For more details relevant to these functors, please see [7, Chaps 2 and 7].

Theorem 2.2. [7, Thms 7.1.2, 7.1.7] *The composite functors $\Gamma\Xi$ and $\Xi\Gamma$ are naturally equivalent to the identity functors of $\mathcal{MV}$ and $\mathcal{A}$, respectively. Therefore, the categories $\mathcal{A}$ and $\mathcal{MV}$ are categorically equivalent.*

Next theorem states that $\mathcal{MV}$ satisfies the amalgamation property.

Theorem 2.3. [27, Thm 2.20] *Given one-to-one homomorphisms $A \xleftarrow{\alpha} Z \xrightarrow{\beta} B$ of MV-algebras, there is an MV-algebra D together with one-to-one homomorphisms $A \xrightarrow{\mu} D \xleftarrow{\nu} B$ such that $\mu \circ \alpha = \nu \circ \beta$.*

An MV-algebra $(M; \oplus, ', 0, 1)$ is called *divisible* if, for all $a \in M$ and all $n \in \mathbb{N}$, there exists $x \in M$ such that

- $n.x = a$.
- $a' \oplus ((n-1).x) = x'$.

Let $(M; \oplus, ', 0, 1)$ be an MV-algebra and (G, u) be the unital Abelian ℓ-group corresponding to M, that is $M = \Gamma(G, u)$. It can be easily seen that M is divisible if and only if, for all $a \in M$ and for all $n \in \mathbb{N}$, there exists $x \in M$ such that the group element nx is defined in M and $nx = a$. Moreover, M is divisible if and only if G is divisible (see [12, Lem. 2.3] or [19, Prop 2.13]). It is possible to show that if $nx = a = ny$, then $x = y$ (see [14]). If $(G(M), u)$ is the unital Abelian ℓ-group corresponding to an MV-algebra M and $G(M)^d$ is the divisible hull of the ℓ-group $G(M)$, then $G(M)^d$ is an ℓ-group with strong unit u and we use M^d to denote the MV-algebra $\Gamma(G(M)^d, u)$. By [14], M^d is a divisible MV-algebra containing M; we call M^d the *divisible hull* of M. For more details about divisible MV-algebras we recommend to see [13, 14, 12, 25].

Definition 2.4. [19] An MV-algebra A is *injective* if for each MV-algebra B and each MV-homomorphism $h : C \to A$, where C is an MV-subalgebra of B, h can be extended to an MV-homomorphism from B into A.

Definition 2.5. [15] An ideal I of an MV-algebra M is called a *summand-ideal* if there exists an ideal J of M such that $\langle I \cup J \rangle = M$ and $I \cap J = \{0\}$, where $\langle I \cup J \rangle$ is the ideal of M generated by $I \cup J$. In this case, we write $M = I \boxplus J$. The set of all summand-ideals of M is denoted by $\mathfrak{Sum}(M)$. Evidently, $\{0\}, M \in \mathfrak{Sum}(M)$.

3 Epimorphisms on class of MV-algebras

In this section, epicomplete objects and an epimorphism in the category of MV-algebras are defined and their properties are studied. Some relations between epicomplete MV-algebras, a-extensions of MV-algebras and divisible MV-algebras are obtained. We show that any injective MV-algebra is epicomplete. Finally, we prove that an MV-algebra is epicomplete if and only if it is divisible.

Recall that a morphism $f : M_1 \to M_2$ of $\mathcal{MV}$ is called an *epimorphism* if, for each MV-algebra M_3 and all MV-homomorphisms $\alpha : M_2 \to M_3$ and $\beta : M_2 \to M_3$, the condition $\alpha \circ f = \beta \circ f$ implies $\alpha = \beta$. An object M of $\mathcal{MV}$ is called *epicomplete* if, for each MV-algebra A and for each one-to-one (note that monics coincide with one-to-one homomorphisms in $\mathcal{MV}$) epimorphism $\alpha : M \to A$ in $\mathcal{MV}$, we get that α is a surjection (see [23, p. 1969]).

Definition 3.1. Let M_1 be a subalgebra of an MV-algebra M_2. Then M_2 is an *a-extension* of M_1 if the map $f : \mathcal{I}(M_2) \to \mathcal{I}(M_1)$ defined by $f(J) = J \cap M_1$,

$J \in \mathcal{I}(M_2)$, is a lattice isomorphism. An MV-algebra is called *a-closed* if it has no proper *a*-extension.

It can be easily seen that M_2 is an *a*-extension for M_1 if and only if for all $0 < y \in M_2$ there are $n \in \mathbb{N}$ and $0 < x \in M_1$ such that $y < n.x$ and $x < n.y$.

Proposition 3.2. *If $f : M_1 \to M_2$ is an epimorphism, then M_2 is an a-extension for $f(M_1)$.*

Proof. Let I and J be two ideals of M_2 such that $I \cap f(M_1) = J \cap f(M_1)$. Then by the Third Isomorphism Theorem [5, Thm 6.18], we get that

$$\frac{M_2}{J} \supseteq \frac{f(M_1) + J}{J} \cong \frac{f(M_1)}{J \cap f(M_1)} = \frac{f(M_1)}{I \cap f(M_1)} \cong \frac{f(M_1) + I}{I} \subseteq \frac{M_2}{I}. \quad (3.1)$$

Let $\alpha_I : \frac{f(M_1)}{I \cap f(M_1)} \to \frac{M_2}{I}$ and $\alpha_J : \frac{f(M_1)}{J \cap f(M_1)} \to \frac{M_2}{J}$ be the canonical morphisms induced from (3.1). Then by the amalgamation property (Theorem 2.3), there exist an MV-algebra A and homomorphisms $\beta_I : \frac{M_2}{I} \to A$ and $\beta_J : \frac{M_2}{J} \to A$ such that $\beta_I \circ \alpha_I = \beta_J \circ \alpha_J$. Consider the following maps

$$\mu_I : M_2 \xrightarrow{\pi_I} \frac{M_2}{I} \xrightarrow{\beta_I} A, \quad \mu_J : M_2 \xrightarrow{\pi_J} \frac{M_2}{J} \xrightarrow{\beta_J} A,$$

where π_I and π_J are the natural projection homomorphisms. For all $x \in M_1$,

$$\mu_I(f(x)) = \beta_I(\frac{f(x)}{I}) = \beta_I(\alpha_I(\frac{f(x)}{I \cap f(M_1)})) = \beta_J(\alpha_J(\frac{f(x)}{J \cap f(M_1)})) =$$

$$= \beta_J(\frac{f(x)}{J}) = \mu_J(f(x)).$$

It follows that $\mu_I \circ f = \mu_J \circ f$ and so by the assumption $\mu_I = \mu_J$, which implies that $I = J$. Therefore, M_2 is an *a*-extension for $f(M_1)$. $\square$

The next theorem helps us to prove Corollaries 3.4 and 3.5.

Theorem 3.3. *Let M_1 be a subalgebra of an MV-algebra $(M_2; \oplus,',0,1)$ such that M_2 is an a-extension of M_1 and $M_1 + I = M_2$ for all $I \in Min(M_2)$. Then $M_1 = M_2$.*

Proof. Choose $b \in M_2 \setminus M_1$ and set $S := \{x \ominus b \mid x \in M_1, \ x \vee b \in M_1 \text{ and } x \ominus b > 0\}$. Clearly, $S \neq \emptyset$ and $0 \notin S$. First we show that S is closed under $\wedge$. Let $x, y \in M_1$ be such that $x \ominus b, y \ominus b \in S$. Then $x \vee b, y \vee b \in M_1$ and $x \ominus b, y \ominus b > 0$. We claim that $(x \wedge y) \ominus b > 0$. From [16, Props 1.15, 1.16, 1.21, 1.22] it follows that $(x \wedge y) \ominus b = (x \ominus b) \wedge (y \ominus b)$.

If $(x \wedge y) \ominus b = 0$, then
$$x \wedge y \leq b \Rightarrow (x \wedge y) \vee b = b \Rightarrow (x \vee b) \wedge (y \vee b) = b$$
but $(x \vee b) \wedge (y \vee b) \in M_1$ (since $x \vee b, y \vee b \in M_1$), which is a contradiction. So $0 < (x \wedge y) \ominus b$. Similarly, we can show that $(x \wedge y) \vee b \in M_1$. Hence $(x \ominus b) \wedge (y \ominus b) \in S$. It follows that there is a proper lattice filter of M_1 containing S which implies that there exists a maximal lattice filter of M_1 containing S, say $\overline{S}$, whence $\overline{S} = M_1 \setminus P$ for some minimal prime lattice ideal P of M_1. By [7, Cor 6.1.4], P is a minimal prime filter of M_1, and so there exists $Q \in Min(M_2)$ such that $P = Q \cap M_1$. By the assumption and by the Third Isomorphism Theorem,
$$\frac{M_1}{Q \cap M_1} \cong \frac{M_1 + Q}{Q} \cong \frac{M_2}{Q}.$$
Then there exists $a \in M_1$ such that $a/Q = b/Q$, so $b \ominus a, a \ominus b \in Q$. Clearly, $(b \ominus a) \vee (a \ominus b) \neq 0$ (otherwise, $b = a \in M_1$ which is a contradiction).
(i) If $a \ominus b = 0$, then $b \ominus a > 0$. Let $0 < b \ominus a = t \in Q$. Then there are $n \in \mathbb{N}$ and $z \in M_1$ such that $t < n.z$ and $z < n.t$, so $z, n.z \in Q$ which implies that $n.z \in Q \cap M_1$. From $b \ominus a \leq n.z$, we have $b \leq a \oplus n.z$. Clearly, $b < a \oplus n.z$ (since $a \oplus n.z \in M_1$). Thus $(a \oplus n.z) \ominus b > 0$ and $(a \oplus n.z) \vee b = a \oplus n.z \in M_1$ and hence by definition
$$(a \oplus n.z) \ominus b \in S. \tag{3.2}$$
On the other hand, in view of
$$\frac{(a \oplus n.z) \ominus b}{Q} = (\frac{a}{Q} \oplus \frac{n.z}{Q}) \ominus \frac{b}{Q} = \frac{a}{Q} \ominus \frac{b}{Q} = \frac{0}{Q},$$
we get
$$(a \oplus n.z) \ominus b \in Q. \tag{3.3}$$
From relations (3.2) and (3.3) it follows that $(a \oplus n.z) \ominus b \in S \cap Q$ which is a contradiction.
(ii) If $b \ominus a = 0$, then $b \leq a$ and $a \ominus b > 0$, so $a \ominus b \in S \cap Q$ (note that $b \vee a = a \in M_1$) which is a contradiction.
(iii) If $b \ominus a > 0$ and $a \ominus b > 0$, then $a \ominus b = t \in Q$, so similarly to (i) there are $n \in \mathbb{N}$ and $z \in M_1$ such that $a \ominus b < n.z \in Q \cap M_1$. It follows that $a \ominus n.z \leq b$. Since $a \ominus n.z \in M_1$, we have $a \ominus n.z < b$. Hence, $(a \ominus n.z) \ominus b = 0$, $\frac{a \ominus n.z}{Q \cap M_1} = \frac{a}{Q \cap M_1}$ and $\frac{a \ominus n.z}{P} = \frac{b}{P}$. Now, we return to (i) and replace a with $a \ominus n.z$. Then we get another contradiction. Therefore, the assumption was incorrect and there is no $b \in M_2 \setminus M_1$. That is, $M_2 = M_1$. $\square$

Corollary 3.4. *An MV-algebra $(A; \oplus,', 0, 1)$ is epicomplete if and only if for each epimorphism $f : A \to B$, we have $\mathrm{Im}(f) + I = B$ for all $I \in Min(B)$.*

Proof. The proof is straightforward by Proposition 3.2 and Theorem 3.3. □

Corollary 3.5. *An MV-algebra $(M; \oplus,', 0, 1)$ is divisible if and only if M/P is divisible for each $P \in Min(M)$.*

Proof. Let M^d be the divisible hull of the MV-algebra M. First, we claim that M^d is an a-extension of M. It suffices to show that, for each $y \in M^d$, there exists $x \in M$ and $n \in \mathbb{N}$ such that $y \leq n.x$ and $x \leq n.y$. Put $y \in M^d$. Consider the unital Abelian ℓ-groups $\Xi(M)$ and $\Xi(M)^d$ with a strong unit u, in Theorem 2.2. Then $y \in \Xi(M)^d$ and $y \leq u$. Since $\Xi(M)^d$ is an a-extension of $\Xi(M)$, see [1], then there is a positive element $x \in \Xi(M)$ and $n \in \mathbb{N}$ such that and $x \leq ny$ and $y \leq nx$. It follows from [11, Thm. 3.12] that $y = y \wedge u \leq ((nx) \wedge u) \wedge u \leq (n(x \wedge u)) \wedge u = n.(x \wedge u)$ and $x \wedge u \leq x \leq (ny) \wedge u = n.y$. Since $x \wedge u \in M$, the claim is true. So, M^d is an a-extension for M. It follows that $Min(M) = \{P \cap M \mid P \in Min(M^d)\}$. Moreover, for each $P \in Min(M^d)$, we have $\frac{M}{P \cap M} = \frac{P+M}{P} \subseteq \frac{M^d}{P}$ and so by the assumption $\frac{P+M}{P}$ is divisible. It follows that $\frac{P+M}{P} = \frac{M^d}{P}$ (since $\frac{M^d}{P}$ is a divisible extension of $\frac{P+M}{P}$), hence $P + M = M^d$. Now, by Theorem 3.3, we conclude that $M = M^d$. Therefore, M is divisible. The proof of the converse is straightforward. □

We recall that in Definition 3.1 an MV-algebra was called a-closed if has no proper a-extension. In Theorem 3.8, we show a condition under which an MV-algebra is a-closed.

Remark 3.6. If M_2 is an a-extension for an MV-algebra M_1, then for all $I \in \mathcal{I}(M_2)$, the MV-algebra $\frac{M_2}{I}$ is an a-extension for the MV-algebra $\frac{M_1+I}{I}$. Indeed, clearly, $M_1 \leq M_1 + I \leq M_2$. Let K_1 and K_2 be ideals of $\frac{M_2}{I}$. Then there exist two ideals H_1 and H_2 of M_2 containing I such that $\frac{H_1}{I} = K_1$ and $\frac{H_2}{I} = K_2$. If $K_1 \cap \frac{M_1+I}{I} = K_2 \cap \frac{M_1+I}{I}$, then

$$\frac{H_1 \cap (M_1+I)}{I} = \frac{H_1}{I} \cap \frac{M_1+I}{I} = \frac{H_2}{I} \cap \frac{M_1+I}{I} = \frac{H_2 \cap (M_1+I)}{I}.$$

Since $H_1 \cap (M_1 + I)$ and $H_2 \cap (M_1 + I)$ are ideals of $M_1 + I$ containing I, then we have

$$H_1 \cap (M_1+I) = \cup\{x \in M_1 + I \mid \frac{x}{I} \in \frac{H_1 \cap (M_1+I)}{I}\} =$$

$$\cup\{x \in M_1 + I \mid \frac{x}{I} \in \frac{H_2 \cap (M_1+I)}{I}\} = H_2 \cap (M_1+I).$$

It follows that $H_1 \cap M_1 = H_1 \cap (M_1+I) \cap M_1 = H_2 \cap (M_1+I) \cap M_1 = H_2 \cap M_1$, which implies that $H_1 = H_2$ and so $K_1 = K_2$. Clearly, the map $f : \mathcal{I}(\frac{M_2}{I}) \to \mathcal{I}(\frac{M_1+I}{I})$ sending K to $K \cap \frac{M_1+I}{I}$ is onto and a lattice homomorphism. Therefore, $\frac{M_2}{I}$ is an a-extension for $\frac{M_1+I}{I}$.

Definition 3.7. An ideal I of an MV-algebra $(M; \oplus,', 0, 1)$ is called an a-ideal if $\frac{M}{I}$ is an a-closed MV-algebra. Clearly, M is an a-closed ideal of M. Moreover, M is a-closed if and only if $\{0\}$ is an a-closed ideal.

Theorem 3.8. *If every minimal prime ideal of an MV-algebra $(M; \oplus,', 0, 1)$ is a-closed, then M is a-closed.*

Proof. Let A be an a-extension for M. For each $P \in Min(A)$, we have $\frac{M}{P \cap M} \cong \frac{M+P}{P} \subseteq \frac{A}{P}$. By the above remark, $\frac{A}{P}$ is an a-extension for $\frac{M+P}{P}$. Since $\frac{M+P}{P}$ is a-closed, then $\frac{M+P}{P} = \frac{A}{P}$. Now, from Theorem 3.3, it follows that $M = A$. $\square$

Clearly, the converse of Theorem 3.8 is true, when M is linearly ordered. Indeed, if M is a chain, $\{0\}$ is the only minimal prime ideal of M and so $M \cong \frac{M}{\{0\}}$ is a-closed. In the following proposition and corollary, we try to find a better condition under which the converse of Theorem 3.8 is true.

Proposition 3.9. *If I is a summand ideal of an a-closed MV-algebra $(M; \oplus,', 0, 1)$, then $\frac{M}{I}$ is a-closed.*

Proof. Let M be an a-closed MV-algebra and I be a summand ideal of M. By [15, Cor 3.5], there exists $a \in B(M)$ such that $I = \downarrow a$, $I^\perp = \downarrow a'$ and $M = \downarrow a \oplus \downarrow a' := \{x \oplus y \mid x \in \downarrow a,\ y \in \downarrow a'\}$. Moreover, for each $x \in M$, there are $x_1 \leq a$ and $x_2 \leq a'$ such that $x = x_1 \oplus x_2$ and so $x/I = x_2/I$. Hence for each $x, y \in M$,

$$\begin{aligned} x/I = y/I &\Leftrightarrow x_2/I = y_2/I \Leftrightarrow x_2 \ominus y_2, y_2 \ominus x_2 \in I \\ &\Rightarrow x_2 \ominus y_2 \leq x_2 \in I^\perp,\ y_2 \ominus x_2 \leq y_2 \in I^\perp \\ &\Rightarrow x_2 \ominus y_2, y_2 \ominus x_2 \in I \cap I^\perp = \{0\} \Rightarrow x_2 = y_2. \end{aligned}$$

That is, $\frac{M}{I} = \{x/I \mid x \in I^\perp\}$. Now, we define the operations $\boxplus$ and $*$ on $\downarrow a'$ by $x \boxplus y = x \oplus y$ and $x^* = t$, where t is the second component of x' in $\downarrow a \oplus \downarrow a'$. It can be easily seen that $I^\perp$ with these operations and 0 and $0'$ as the least and greatest elements, respectively, is an MV-algebra. Moreover, $\frac{M}{I} \cong I^\perp$. Similarly, I is an MV-algebra and $I = \downarrow a \cong \frac{M}{I^\perp}$. Now, let A be an a-extension for the MV-algebra $\frac{M}{I}$. Then

$$\phi : M \xrightarrow{x \mapsto x_1 \oplus x_2} \downarrow a \oplus \downarrow a' \xrightarrow{x \oplus y \mapsto (x/I, y/I^\perp)} \frac{M}{I} \times \frac{M}{I^\perp} \subseteq A \times \frac{M}{I^\perp}.$$

(1) Since $M \cong \frac{M}{I} \times \frac{M}{I^\perp}$, then $\frac{M}{I} \times \frac{M}{I^\perp}$ is a-closed.
(2) $A \times \frac{M}{I^\perp}$ is an a-extension for $\frac{M}{I} \times \frac{M}{I^\perp}$.

It follows that $\frac{M}{I} \times \frac{M}{I^\perp} = A \times \frac{M}{I^\perp}$ and so $A = \frac{M}{I^\perp}$. In a similar way, we can show that M/I is a-closed. $\square$

Corollary 3.10. *Let $(M; \oplus,' , 0, 1)$ be a closed hyperarchimedean MV-algebra. Then each principal ideal of M is an a-ideal.*

Proof. By [7, Thm 6.3.2] every principal ideal of M is a stonean ideal. Hence by [15, Cor 3.5(iii)], we get that every principal ideal of M is a summand ideal of M and so $\frac{M}{I}$ is a-closed for each principal ideal I of M. That is, each principal ideal of M is an a-ideal. $\square$

We note that an MV-algebra M is *simple* if $\mathcal{I}(M) = \{\{0\}, M\}$.

Example 3.11. (1) Consider the standard MV-algebra defined on the real unit interval $A = [0, 1]$. Let B be an a-extension for it. Then B is a simple MV-algebra that contains $[0, 1]$ (since $[0, 1]$ is simple and $\mathcal{I}(A) \cong \mathcal{I}(B)$). By [7, Thm 3.5.1], B is isomorphic to a subalgebra of $[0, 1]$; let $f : B \to [0, 1]$ be a one-to-one MV-algebra homomorphism. Then A and $f(A)$ are also isomorphic. Due to [7, Cor 7.2.6], two subalgebras of $[0, 1]$ are isomorphic if and only if they coincide. Therefore, $[0, 1] = A = f(A) \subseteq f(B) \subseteq [0, 1]$. Thus $f(A) = f(B)$ and $A = B$ which proves A is a-closed.

(2) Let A be a subalgebra of the real interval MV-algebra $[0, 1] = \Gamma(\mathbb{R}, 1)$. Then A is a-closed if and only if $A = [0, 1]$. Indeed, one direction was proved in the forgoing case (1). Now let A be a proper subalgebra of $[0, 1]$. Then the MV-algebra $[0, 1]$ is an a-extension of A such that $A \neq [0, 1]$.

(3) A simple MV-algebra A is a-closed if and only if A is isomorphic to the MV-algebra $[0, 1]$.

Theorem 3.12. *Every injective MV-algebra is epicomplete.*

Proof. The proof is straightforward by Figure 1. $\square$

It is well known that divisible and complete MV-algebras coincide with injective MV-algebras (see [24, Thm 1] and [19, Thm 2.14]). So we have the following result.

Corollary 3.13. *Every complete and divisible MV-algebra is epicomplete.*

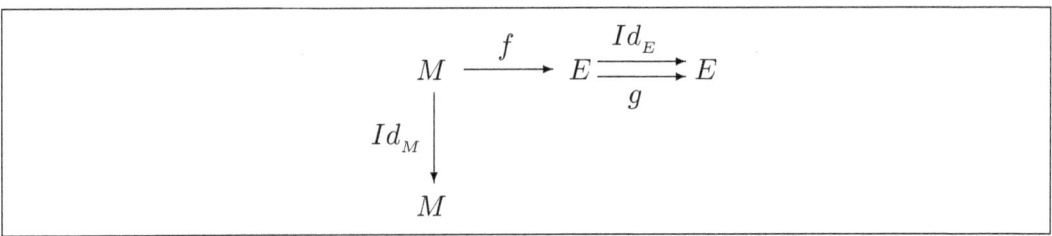

Figure 1: Injective MV-algebra is epicomplete

Let $(M; \oplus,', 0, 1)$ be an MV-algebra and (G, u) be a unital Abelian ℓ-group such that $M = \Gamma(G, u)$. Set $M^d = \Gamma(G^d, u)$, where G^d is the divisible hull of G. Let $i : M \to M^d$ be the inclusion map. Then i is an epimorphism. Indeed, if A is another MV-algebra and $\alpha, \beta : M^d \to A$ be MV-homomorphisms such that $\alpha \circ i = \beta \circ i$, then by Theorem 2.2, we have the following homomorphisms in $\mathcal{A}$

$$\Xi(i) : (G, u) \mapsto (G^d, u), \quad \Xi(\alpha), \Xi(\beta) : \Xi(M^d, u) \mapsto (\Xi(A), v),$$

where v is a strong unit of $\Xi(A)$ such that $\Gamma(\Xi(A), v) = A$. Since Ξ is a functor from $\mathcal{MV}$ to $\mathcal{A}$, then we have $\Xi(\alpha) \circ \Xi(i) = \Xi(\alpha \circ i) = \Xi(\beta \circ i) = \Xi(\beta) \circ \Xi(i)$. By [1, Sec 2], we know that the inclusion map $\Xi(i) : (G, u) \to (G^d, u)$ is an epimorphism, so $\Xi(\alpha) = \Xi(\beta)$, which implies that $\alpha = \beta$. That is, $i : M \to M^d$ is an epimorphism in $\mathcal{MV}$. Thus if M is epicomplete, i is onto. As i is always one-to-one, it is a bijection and so $M \cong M^d$, and we have the following result.

Theorem 3.14. *Epicomplete MV-algebras are divisible.*

Theorem 3.15. *Let $(M; \oplus,', 0, 1)$ be an MV-algebra. Then M is epicomplete if and only if each epimorphism of M into a linearly ordered MV-algebra is onto.*

Proof. Suppose that each epimorphism of M into a linearly ordered MV-algebra H is onto. If $f : M \to H$ is an epimorphism, then for each $P \in Min(H)$, the map $M \xrightarrow{f} H \xrightarrow{\pi_P} \frac{H}{P}$ is an epimorphism, where π_P is the natural homomorphism. Since $\frac{H}{P}$ is a linearly ordered MV-algebra, then by the assumption, $\pi_P \circ f$ is onto and so $\frac{f(M)}{P} = \frac{H}{P}$ or equivalently, $\bigcup_{x \in M} f(x)/P = H$. Hence $f(M) + P = H$ for all $P \in Min(H)$. By Proposition 3.2, we know that H is an a-extension for $f(M)$ and so by Theorem 3.3, $f(M) = H$. Therefore, M is epicomplete. The proof of the other direction is clear. $\square$

Definition 3.16. Let $(M_1; \oplus,', 0, 1)$, $(M_2; \oplus,', 0, 1)$ and $(M_3; \oplus,', 0, 1)$ be MV-algebras such that $M_1 \leq M_2$ and $M_1 \leq M_3$. An element $b \in M_2$ is *equivalent* to an element $c \in M_3$ if there exists an isomorphism f between $\langle M_1 \cup \{b\} \rangle_{M_2}$ and

$\langle M_1 \cup \{c\}\rangle_{M_3}$ such that $f(b) = c$ and $f|_{M_1} = Id_{M_1}$, where $\langle M_1 \cup \{b\}\rangle_{M_2}$ is the MV-subalgebra of M_2 generated by $M_1 \cup \{b\}$. An element $b \in M_2$ is *algebraic over* M_1 if no extension of M_1 contains two elements equivalent to b. Moreover, M_2 is an *algebraic extension* of M_1 if every element of M_2 is algebraic over M_1.

Proposition 3.17. *Let $(M_1; \oplus,',0,1)$ and $(M_2; \oplus,',0,1)$ be two MV-algebras such that $M_1 \leq M_2$. Then $y \in M_2$ is algebraic over M_1 if and only if the inclusion map $i: M_1 \to \langle M_1 \cup \{y\}\rangle_{M_2}$ is an epimorphism.*

Proof. Let $y \in M_2$ be algebraic over M_1. If $i: M_1 \to \langle M_1 \cup \{y\}\rangle$ is not an epimorphism, then there exist an MV-algebra M_3 and two homomorphisms $\alpha, \beta: \langle M_1 \cup \{y\}\rangle \to M_3$ such that $\alpha \circ i = \beta \circ i$ and $\alpha \neq \beta$. Then $\alpha(y) \neq \beta(y)$ (otherwise, $\alpha = \beta$). Consider the maps $\lambda, \mu: \langle M_1 \cup \{y\}\rangle_{M_2} \to M_3 \times \langle M_1 \cup \{y\}\rangle_{M_2}$ defined by $\lambda(x) = (\alpha(x), x)$ and $\mu(x) = (\beta(x), x)$ for all $x \in \langle M_1 \cup \{y\}\rangle_{M_2}$. Clearly, λ and μ are one-to-one homomorphisms. We have $\lambda(M_1) = \{(\alpha(x), x) \mid x \in M_1\} = \mu(M_1)$ and $M_1 \cong \lambda(M_1) \cong \mu(M_1)$. Set $M = M_3 \times \langle M_1 \cup \{y\}\rangle_{M_2}$. We identify M_1 with its image in M under λ. Then M is an extension for M_1. Since $\lambda: \langle M_1 \cup \{y\}\rangle_{M_2} \to \langle M_1 \cup \{(\alpha(y), y)\}\rangle_M$ and $\mu: \langle M_1 \cup \{y\}\rangle_{M_2} \to \langle M_1 \cup \{(\beta(y), y)\}\rangle_M$ are isomorphisms, then y is equivalent to $(\alpha(y), y)$ and $(\beta(y), y)$, which is a contradiction. Therefore, $i: M_1 \to \langle M_1 \cup \{y\}\rangle$ is an epimorphism.

Conversely, let $i: M_1 \to \langle M_1 \cup \{y\}\rangle$ be an epimorphism. We claim that y is algebraic over M_1. Otherwise, there are an extension E of M_1 and $e_1, e_2 \in E$ such that y is equivalent to e_1 and e_2. So, there exist two isomorphisms $f_1: \langle M_1 \cup \{y\}\rangle_{M_2} \to \langle M_1 \cup \{e_1\}\rangle_E$ and $f_2: \langle M_1 \cup \{y\}\rangle_{M_2} \to \langle M_1 \cup \{e_2\}\rangle_E$ such that $f_1|_{M_1} = Id_{M_1} = f_2|_{M_1}$ and $f_1(y) = e_1$ and $f_2(y) = e_2$. It follows that $f_1 \circ i = f_2 \circ i$ but $f_1 \neq f_2$, which is a contradiction. □

Corollary 3.13 showed that complete and divisible MV-algebras are epicomplete. In the sequel, we will use the same argument as in the proof of [1, Thm 2.1] with a little modification to show that every divisible MV-algebra is epicomplete.

Theorem 3.18. *Let $(M; \oplus,',0,1)$ be an MV-algebra.*

(i) *If $(L; \oplus,',0,1)$ is a linearly ordered MV-algebra and $f: M \to L$ is an epimorphism, then $L \subseteq f(M)^d$.*

(ii) *If M is divisible, then it is epicomplete.*

Proof. (i) Let (G, u) and (H, v) be the unital Abelian ℓ-groups such that $\Gamma(G, u) = M$ and $\Gamma(H, v) = L$. By Theorem 2.2, $\Xi(f): G \to H$ is a unital ℓ-group homomorphism. Let $K = B/Im(\Xi(f))$ be the torsion subgroup of $H/Im(\Xi(f))$ (clearly

B is a subgroup of H and $x \in B \Leftrightarrow nx \in Im(\Xi(f))$ for some $n \in \mathbb{N}$). Since $H/B \cong \frac{H/Im(\Xi(f))}{B/Im(\Xi(f))}$ is torsion free, by [2, Prop 1.1.7], H/B admits a linearly ordered group structure. By [17, Exm 3], $(H/B) \overrightarrow{\times} H$ is an ℓ-group. Since v is a strong unit of an ℓ-group H, for each $(x+B, y) \in (H/B) \overrightarrow{\times} H$, there exists $n \in \mathbb{N}$ such that $x, y < nv$ and so $(x+B, y) < n(v+B, v)$. It follows that $w := (v+B, v)$ is a strong unit for the ℓ-group $(H/B) \overrightarrow{\times} H$. Let $\alpha, \beta : L \to \Gamma((H/B) \overrightarrow{\times} H, w)$ be defined by $\alpha(x) = (B, x)$ and $\beta(x) = (x+B, x)$ for all $x \in L$ (both of them are MV-homomorphisms). We have $\alpha \circ f(x) = (B, f(x)) = \beta \circ f(x)$ for all $x \in L$. Since f is an epimorphism, then $\alpha = \beta$, hence for all $x \in L = \Gamma(H, v)$, we have $x \in B$ and so there is $n \in \mathbb{N}$ such that $nx \in Im(\Xi(f))$. That is, x belongs to the divisible hull of $Im(\Xi(f))$. Since $x \leq v$, then $x \in \Gamma((Im(\Xi(f)))^d, v) = f(M)^d$. Therefore, $L \subseteq f(M)^d$.

(ii) Let M be divisible. We use Theorem 3.15 to show that M is epicomplete. Let $(L; \oplus, ', 0, 1)$ be a linearly ordered MV-algebra and $f : M \to L$ be an epimorphism into L. By (i), $f(M) \subseteq L \subseteq (f(M))^d$. Clearly, $f(M)$ is divisible, so $f(M) = L = (f(M))^d$. It follows from Theorem 3.15 that M is epicomplete. □

Concerning the proof of (i) in the latter theorem, we note that since $f : M \to f(M)$ is onto, by [26, Lem. 7.2.1], $\Xi(f) : \Xi(M) \to \Xi(f(M))$ is onto ($\Xi(M) = G$), so $Im(\Xi(f)) = \Xi(f(M))$. It follows that $f(M)^d = \Gamma((\Xi(f(M)))^d, v) = \Gamma((Im(\Xi(f)))^d, v)$.

4 Epicompletion of MV-algebras

The main purpose of the section is to introduce an epicompletion for an MV-algebra and to discuss the conditions under which an MV-algebra has an epicompletion. First we introduce an epicompletion in $\mathcal{MV}$. An epicompletion for an MV-algebra A is an MV-algebra M epically containing A with the universal property. Then we use some results of the second section and prove that any MV-algebra has an epicompletion. Indeed, the epicompletion of A is A^d.

Definition 4.1. Let $(A; \oplus, ', 0, 1)$ be an MV-algebra.

(i) A pair $(\overline{A}, \alpha)$, where $\overline{A}$ is an MV-algebra and $\alpha : A \to \overline{A}$ is a one-to-one epimorphism (epiembedding for short), is called an *e-extension* for A. For simplicity, we called it $\overline{A}$ containing A epically.

(ii) An e-extension (E, α) for A is called an *epicompletion* for A if, for each epimorphism $f : A \to B$, there is an e-extension $(\overline{B}, \beta)$ for B and a surjective homomorphism $\mathbf{f} : E \to \overline{B}$ such that $\beta \circ f = \mathbf{f} \circ \alpha$, or equivalently, the diagram given by Figure 2 commutes.

$$\begin{array}{ccc} A & \xrightarrow{f} & B \\ \alpha \downarrow & & \downarrow \beta \\ E & \xrightarrow{\mathbf{f}} & \overline{B} \end{array}$$

Figure 2: Epicompletion property

Proposition 4.2. *Let* $(A; \oplus,', 0, 1)$ *be an MV-algebra.*

(i) *Each epicompletion of* A *is epicomplete.*

(ii) *If* A *has an epicompletion, then it is unique up to isomorphism.*

Proof. (i) Let $(\overline{A}, \alpha)$ be an epicompletion for A and $f : \overline{A} \to B$ be a one-to-one epimorphism. Then $f \circ \alpha : A \to B$ is an epimorphism and so there exists an e-extension $(\overline{B}, \beta)$ for B and an onto morphism $h : \overline{A} \to \overline{B}$ such that the diagram in Figure 3 commutes. From $h \circ \alpha = \beta \circ f \circ \alpha$ it follows that $h = \beta \circ f$, whence $\beta \circ f$

$$\begin{array}{ccc} A & \xrightarrow{f \circ \alpha} & B \\ \alpha \downarrow & & \downarrow \beta \\ \overline{A} & \xrightarrow{h} & \overline{B} \end{array}$$

Figure 3: Figure of Proposition 4.2(i)

is onto. Hence $\beta(f(\overline{A})) = \overline{B}$. Also, $\beta(f(\overline{A})) \subseteq \beta(B) \subseteq \overline{B}$, so $\beta(f(\overline{A})) = \beta(B) = \overline{B}$. Thus, B is an isomorphism, which implies that $f = \beta^{-1} \circ h$ must be onto. Therefore, $\overline{A}$ is epicomplete.

(ii) Let $(\overline{A}, \alpha)$ and (A', β) be two epicompletions for A. Since $\alpha : A \to \overline{A}$ is an epiembedding, then by Proposition 3.2, $\overline{A}$ is an a-extension for $f(A) \cong A$. By (i) and Theorem 3.14, $\overline{A}$ is divisible. Consider the functor Ξ from Theorem 2.2. Let $(G(A), u)$ and $(G(\overline{A}), v)$ be the unital Abelian ℓ-groups induced from MV-algebras A and $\overline{A}$, respectively. Since there is an epiembedding $A \hookrightarrow \overline{A}$, we have $u = v$. We have an embedding $\Xi(\alpha) : G(A) \to G(\overline{A})$, and $G(\overline{A})$ is divisible (see [14]). Also, $G(\overline{A})$

is an a-extension for the ℓ-group $G(A)$ (since there is a one-to-one correspondence between the lattice of ideals of A and the lattice of convex ℓ-subgroups of $G(A)$, see [8, Thm 1.2]), so by [22, Chap 1, Thm 20], $G(\overline{A}) \cong (G(A))^d$. In a similar way, $G(A') \cong (G(A))^d$. Therefore, $G(\overline{A}) \cong G(A')$. We note that the final isomorphism is an extension for the identity map on G, so it preserves the strong units of $G(\overline{A})$ and $G(A')$, which is a strong unit of $G(A)$, too. Using the functor Γ, it follows that $\overline{A} \cong A'$. □

Let $(A; \oplus, ', 0, 1)$ be an MV-algebra. By the last proposition if A has an epicompletion, then it is unique up to isomorphic image; this epicompletion is denoted by (A^e, α).

Corollary 4.3. *Let (A^e, α) be an epicompletion for an MV-algebra $(A; \oplus, ', 0, 1)$. Then the epicompletion of A^e is equal to A^e.*

Proof. It follows from Proposition 4.2(i). □

Now, we try to answer to a question "whether does an MV-algebra have an epicompletion". First we simply use Theorem 3.18(i) to show that each linearly ordered MV-algebra has an epicompletion. Then we prove it for any MV-algebra. For this purpose we try to extend the result of [28, Cor 1]. We show that each unital ℓ-group has an epicompletion. Then we use this result and we show that any MV-algebra has an epicompletion.

Proposition 4.4. *Let $(A; \oplus, ', 0, 1)$ be a linearly ordered MV-algebra. Then A has an epicompletion.*

Proof. Let $f : A \to B$ be an epimorphism. Since A is a chain, $f(A)$ is also a chain and so, $\mathcal{I}(f(M))$ is a chain. It follows from Proposition 3.2 that $\mathcal{I}(B)$ is a chain and so B is a linearly ordered MV-algebra. Hence by Theorem 3.18(i), $B \subseteq (f(A))^d$ (thus $B^d = (f(A))^d$). By [28, Prop 5] and Theorem 2.2, there is a homomorphism $g : A^d \to B^d$ such that the diagram in Figure 4 commutes. Then $g(A^d) \subseteq B^d$ is a

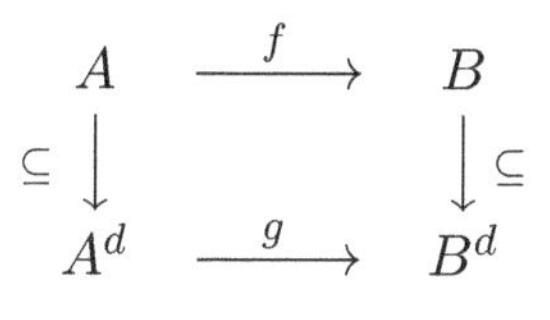

$$\begin{array}{ccc} A & \xrightarrow{f} & B \\ \subseteq \downarrow & & \downarrow \subseteq \\ A^d & \xrightarrow{g} & B^d \end{array}$$

Figure 4: Figure of Proposition 4.4

divisible MV-algebra containing B, so g is onto. Therefore, A^d is an epicompletion for the MV-algebra A. □

Remark 4.5. Let G and H be two ℓ-groups and $f : G \to H$ be an epimorphism. Let G^d and H^d be the divisible hull of G and H, respectively. By [1, p. 230], there is a unique extension of f to an epimorphism $\overline{f} : G^d \to H^d$. Clearly, if G and H are unital Abelian ℓ-groups and f is a unital ℓ-group morphism, then so is $\overline{f}$ (for more details see [1] the paragraph after Theorem 2.1 and [28, Prop 5]). We know that the inclusion maps $i : G \to G^d$ and $j : H \to H^d$ are epimorphisms (by the corollary of [1, Thm 2.1]), hence we have the following commutative diagram (Figure 5).

$$\begin{array}{ccc} G & \xrightarrow{i} & G^d \\ f \downarrow & & \downarrow \overline{f} \\ H & \xrightarrow{j} & H^d \end{array}$$

Figure 5: G^d is an epicompletion for G

Since $\overline{f} : G^d \to H^d$ is an epimorphism and G^d is epicomplete (by [1, Thm 2.1]), then $\overline{f}$ is onto and so G^d is an epicompletion for G.

Theorem 4.6. *Any MV-algebra has an epicompletion.*

Proof. Let $(A; \oplus,', 0, 1)$ be an MV-algebra and $f : A \to B$ be an epimorphism. Then $\Xi(f) : \Xi(A) \to \Xi(B)$ is a homomorphism of unital ℓ-groups. By Remark 4.5, we have the commutative diagram in Figure 6, where $\overline{\Xi(f)}$ is the unique extension of $\Xi(f)$. Applying the functor Γ to the diagram in Figure 6, we get the commutative

$$\begin{array}{ccc} \Xi(A) & \xrightarrow{\Xi(f)} & \Xi(B) \\ \subseteq \downarrow & & \downarrow \subseteq \\ (\Xi(A))^d & \xrightarrow{\overline{\Xi(f)}} & (\Xi(B))^d \end{array}$$

Figure 6: Applying the functor Ξ.

diagram in Figure 7 on $\mathcal{MV}$. Set $F := \Gamma(\overline{\Xi(f)})$. We claim that $F : A^d \to B^d$ is an epimorphism. Let $\alpha, \beta : B^d \to C$ be two homomorphisms of MV-algebras such that $\alpha \circ F = \beta \circ F$. Then clearly, $\alpha|_B \circ F = \beta|_B \circ F$, so by the assumption

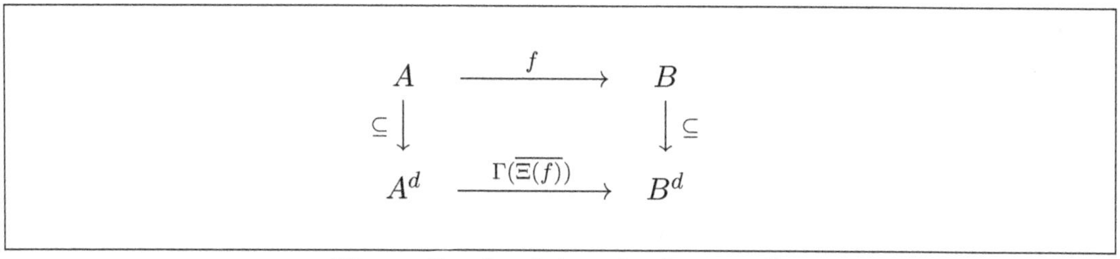

$$
\begin{array}{ccc}
A & \xrightarrow{f} & B \\
\subseteq \downarrow & & \downarrow \subseteq \\
A^d & \xrightarrow{\Gamma(\Xi(f))} & B^d
\end{array}
$$

Figure 7: Applying the functor Γ.

$\alpha|_B = \beta|_B$, which implies that $\Xi(\alpha|_B) = \Xi(\beta|_B)$. Thus by [1], $\overline{\Xi(\alpha|_B)} = \overline{\Xi(\beta|_B)}$, where $\overline{\Xi(\alpha|_B)}, \overline{\Xi(\beta|_B)} : (\Xi(B))^d \to (\Xi(C))^d$ are the unique extensions of $\Xi(\alpha|_B)$ and $\Xi(\beta|_B)$, respectively. It can be easily seen that the diagrams in Figure 8 are commutative. So by the uniqueness of the extension of $\Xi(\alpha|_B) : \Xi(B) \to \Xi(C)$

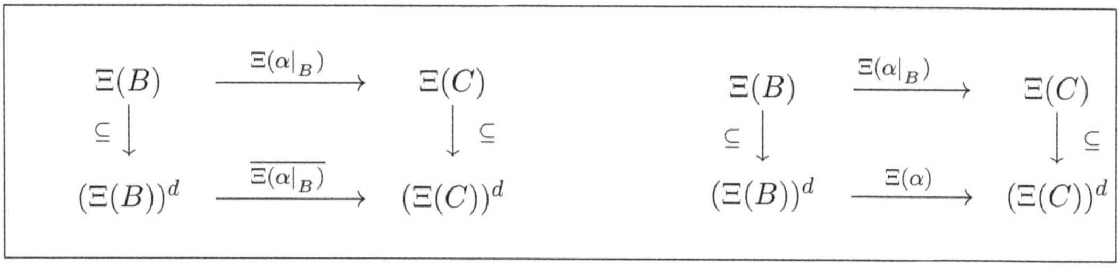

$$
\begin{array}{ccc}
\Xi(B) & \xrightarrow{\Xi(\alpha|_B)} & \Xi(C) \\
\subseteq \downarrow & & \downarrow \subseteq \\
(\Xi(B))^d & \xrightarrow{\overline{\Xi(\alpha|_B)}} & (\Xi(C))^d
\end{array}
\qquad
\begin{array}{ccc}
\Xi(B) & \xrightarrow{\Xi(\alpha|_B)} & \Xi(C) \\
\subseteq \downarrow & & \downarrow \subseteq \\
(\Xi(B))^d & \xrightarrow{\Xi(\alpha)} & (\Xi(C))^d
\end{array}
$$

Figure 8: Final step.

to a map $(\Xi(B))^d \to (\Xi(C))^d$, we get that $\Xi(\alpha) = \overline{\Xi(\alpha|_B)}$. In a similar way, $\Xi(\beta) = \overline{\Xi(\beta|_B)}$ and so $\Xi(\alpha) = \Xi(\beta)$. It follows that $\alpha = \Gamma(\Xi(\alpha)) = \Gamma(\Xi(\beta)) = \beta$. Therefore, F is an epimorphism. Since A^d is divisible, by Corollary 3.13, it is epicomplete and so F is onto. That is, A^d is an epicompletion for A. Therefore, any MV-algebra has an epicompletion. $\square$

Corollary 4.7. *Let $(A; \oplus,',0,1)$ be an MV-algebra. Then E is an epicompletion of A if and only if E is an epicomplete MV-algebra containing A epically.*

Proof. Let E be an epicomplete MV-algebra containing A epically. Then there is a one-to-one epimorphism $\alpha : A \to E$. By the proof of Theorem 4.6, we have an epimorphism $\alpha^d : A^d \to E^d$ which is one-to-one (so as α). Since A^d is epicomplete, then α^d is an isomorphism. On the other hand, by Theorem 3.14, $E \cong E^d$ and so $A^d \cong E$. Therefore, by the proof of Theorem 4.6, E is an epicompletion of A. The proof of the converse follows from definition and Proposition 4.2(i). $\square$

Acknowledgement: The authors are very indebted to anonymous referees for their careful reading and suggestions which helped us to improve the readability of the paper.

References

[1] M. Anderson, P. Conrad, Epicomplete ℓ-groups, *Algebra Universalis* **12** (1981), 224–241.

[2] M. Anderson and T. Feil, Lattice-Ordered Groups: An Introduction, *Springer Science and Business Media*, USA, 1988.

[3] R.N. Ball, A.W. Hager, Epicomplete archimedean ℓ-groups and vector lattices, *Transactions of the American Mathematical Society* **322** (1990), 459–478.

[4] R.N. Ball, A.W. Hager, Epicompletetion of archimedean ℓ-groups and vector lattices with weak unit, *Journal of the Australian Mathematical Society* (Series A) **48** (1990), 25–56.

[5] S. Burris, H.P. Sankappanavar, A Course in Universal Algebra, *Springer-Verlag*, New York, 1981.

[6] C.C. Chang, Algebraic analysis of many valued logics, *Transaction of the American Mathemetacal Society* **88** (1958), 467–490.

[7] R. Cignoli, I.M.L. D'Ottaviano and D. Mundici, Algebraic Foundations of Many-Valued Reasoning, *Springer Science and Business Media*, Dordrecht, 2000.

[8] R. Cignoli, A. Torrens, The poset of prime ℓ-ideals of an Abelian ℓ-group with a strong unit, *Journal of Algebra* **184** (1906), 604–612.

[9] P. Conrad, D. McAlister, The completion of a lattice ordered group, *Journal of the Australian Mathematical Society* **9** (1869), 182–208.

[10] M.R. Darnel, Epicomplete completely-distributive ℓ-groups, *Algebra Universalis* **21** (1985), 123–132.

[11] M.R. Darnel, Theory of Lattice-Ordered Groups, *Marcel Dekker, Inc.*, New York, Basel, Hong Kong, 1995.

[12] A. Di Nola, S. Sessa, On MV-algebras of continuous functions, In: Non-classical Logics and Their Applications to Fuzzy Subsets. A Handbook of the Mathematical Foundations of Fuzzy Set Theory, U. Höhle et al. (eds), *Kluwer Academic Publishers*, Dordrecht, 1995, pp. 23–32.

[13] D. Diaconescu, I. Leuştean, The Riesz hull of a semisimple MV-algebra, *Mathematica Slovaca* **65** (2015), 801–816.

[14] A. Dvurečenskij, B. Riečan, Weakly divisible MV-algebras and product, *Journal of Mathematical Analysis and Applications* **234** (1999), 208–222.

[15] A. Dvurečenskij, O. Zahiri, Orthocomplete pseudo MV-algebras, *International Journal of General Systems* **45** (2016), 889–909. DOI: 10.1080/03081079.2016.1220008

[16] G. Georgescu and A. Iorgulescu, Pseudo MV-algebras, *Multiple-Valued Logics* **6** (2001), 193–215.

[17] A.M.W. Glass, W. Holland, Lattice-Ordered Groups: Advances and Techniques, **48**, Kluwer Academic Publishers, Dordrecht, 1989.

[18] A.M.W. Glass, J. Rachůnek, R. Winkler, Functional representations and universals for MV- and GMV-algebras, *Tatra Mountains Mathematical Publications* **27** (2003), 91–110.

[19] D. Gluschankof, Prime deductive systems and injective objects in the algebras of Łukasiewicz infinite-valued calculi, *Algebra Universalis* **29** (1992), 354–377.

[20] K.R. Goodearl, Partially Ordered Abelian Groups with Interpolation, Mathematical Surveys and Monographs No. 20, *American Mathematical Society*, Providence, Rhode Island, 1986.

[21] S. Gottwald, Many-valued logic and fuzzy set theory, In: Mathematics of Fuzzy Sets Logic, Topology and Measure Theory, The Handbooks of Fuzzy Sets Series, 3, U. Höhle, S.E. Rodabough (eds), *Kluwer Academic Publishers*, Dordrecht, 1999, pp. 5–90.

[22] P.A. Griffith, Infinite Abelian Group Theory, *University of Chicago Press*, Chigaco, London, 1970.

[23] A.W. Hager, Some unusual epicomplete Archimedean lattice-ordered groups, *Proceedings of the American Mathematical Society*, **143** (2015), 1969–1980.

[24] F. Lacava, Sulle L-algebre iniettive, *Bolletino della Unione Matemàtica Italiana* **3-A**(3) (1989), 319–324.

[25] S. Lapenta, I. Leuştean, Notes on divisible MV-algebras, *Soft Computing*, (2016), To appear. doi: 10.1007/s00500-016-2339-z

[26] D. Mundici, Interpretation of AF C^*-algebras in Łukasiewicz sentential calculus, *Journal of Functional Analysis* **65** (1986), 15–63.

[27] D. Mundici, Advanced Łukasiewicz calculus and MV-algebras, *Springer, Dordrecht, Heidelberg, London, New York*, 2011.

[28] F.D. Pedersen, Epimorphisms in the category of abelian ℓ-groups, *Proceedings of the American Mathematical Society* **53** (1975), 311–317.

[29] D.R. Ton, Epicomplete archimedean lattice-ordered groups, *Bulletin of the Australian Mathematical Society* **39** (1989), 277–286. doi: 10.1017/S0004972700002768.

Paraconsistent Rule-Based Reasoning with Graded Truth Values

Francesco Luca De Angelis, Giovanna Di Marzo Serugendo
Institute of Services Science, University of Geneva, Switzerland.
{francesco.deangelis,giovanna.dimarzo@unige.ch}@unige.ch

Andrzej Szałas
Institute of Informatics, University of Warsaw, Poland and Department of Computer and Information Science, Linköping University, Sweden.
andrzej.szalas@{mimuw.edu.pl, liu.se}

Abstract

Modern artificial systems, such as cooperative traffic systems or swarm robotics, are made of multiple autonomous agents, each handling uncertain, partial and potentially inconsistent information, used in their reasoning and decision making. Graded reasoning, being a suitable tool for addressing phenomena related to such circumstances, is investigated in the literature in many contexts – from graded modal logics to various forms of approximate reasoning. In this paper we first introduce a family of many-valued paraconsistent logics parametrised by a number of truth/falsity/inconsistency grades allowing one to handle multiple truth-values at the desired level of accuracy. Second, we define a corresponding family of rule-based languages with graded truth-values as first-class citizens, enjoying tractable query evaluation. In addition, we introduce introspection operators allowing one to resolve inconsistencies and/or lack of information in a non-monotonic manner. We illustrate and discuss the use of the framework in an autonomous robot scenario.

1 Introduction and Motivations

Modern artificial systems exhibit characteristics such as autonomy, collectiveness, situatedness and uncertain and changing environment. Examples of such systems include autonomous cars, intelligent cooperative traffic systems, smart systems, swarm robotics, systems exploiting edge computing [41], spatial computing [9, 71] and spatial services [23], or more generally collective adaptive systems [3]. Entities or agents, constituting these systems, are autonomous, spatially-distributed, geographically dispersed, interconnected and interacting through a communication network. Every individual agent holds its own perceived local and therefore partial, incomplete and potentially inconsistent information about the system, and uses it to identify its current situation, and subsequently take adaptation actions.

Due to technological limitations of sensors, the analysis of multiple factors, and dynamically changing environments, information perception is often affected by a certain grade of uncertainty or associated with several levels of quality. Such phenomena are, for instance, exhibited by knowledge bases of intelligent context-aware systems: information sources may be equipped with different sensors or classifiers, providing better or worse approximations of the perceived reality. Also, the transition from an absolute to a graded (multiple) partial perception of reality is implicitly prone to the emergence of contradictory information. Latency of information dissemination also contributes to propagating slightly outdated or contradictory information.

Therefore, in such circumstances, several factors have to be addressed to perform adequate formal reasoning, like: modelling the uncertainty and quality level of information; aggregating coherently multiple graded information, arising from distinct partial views of the system; resolving inconsistencies generated by contradictory knowledge, obtaining representative information useful to underpin decision making analysis and further reasoning processes.

One way to model such uncertainty is to resort to many-valued or paraconsistent logics [2, 12, 16, 17]. To avoid triviality, paraconsistent approaches handle inconsistent information by resorting to non-explosive consequence relations, limiting the set of conclusions inferred by contradictory premises [12, 17, 18, 50, 58].

Contemporary many-valued logics, underpinning query languages that accommodate positive and negative literals in premises and conclusions of rules, are based on predefined sets of truth-degrees. Also, such languages do not fully develop logical mechanisms to reason about truth-degrees of inferred information.

The current paper first defines a family of paraconsistent many-valued logic, parametrised by a set of finitely many truth-degrees. Second, it embeds this logic into a tractable rule-based language usable by teams of heterogeneous agents with different perception and reasoning capabilities. Third, it introduces a generalised logical mechanism to reason on truth-degrees of inferred literals.

We develop paraconsistency through a family of many-valued logic that supports arbitrarily large finite sets of logical values of truth, falseness and inconsistency; this aspect empowers knowledge with desired grades of accuracy. For example, a single agent may use rules of different quality. When heterogeneous techniques are involved in reasoning, their relative strengths may vary: conclusions based on sure facts and certain rules are stronger than those ones obtained from heuristic non-monotonic rules. However, in common sense reasoning, non-monotonic conclusions typically have the same status as monotonic ones which may lead to wrong decisions, especially when rules provide conflicting conclusions. Of course, there are approaches where the strength of arguments and conclusions is one among many important factors (see, e.g. [26, 31, 60]).

Our language allows agents to distinguish among conclusions, to compare their relative strengths, as well as to react on potential conflicts and lack of knowledge. More precisely, given an integer $N \geq 1$, allowing one to fine-tune the accuracy of reasoning, we define a logic where we consider the following truth-values τ_N:

- representing degrees of truth: $\boldsymbol{t}_1, \ldots, \boldsymbol{t}_N$, where $\boldsymbol{t}_1$ is the weakest and $\boldsymbol{t}_N$ the strongest truth;

- representing degrees of falsity: $\boldsymbol{f}_1, \ldots, \boldsymbol{f}_N$, where $\boldsymbol{f}_1$ is the weakest and $\boldsymbol{f}_N$ the strongest falsity;

- representing degrees of inconsistency: $\boldsymbol{i}_{1,1} \ldots, \boldsymbol{i}_{N,N}$, where $\boldsymbol{i}_{i,j}$ is the inconsistency level involving $\boldsymbol{t}_i$ and $\boldsymbol{f}_j$;

- representing unknown: $\boldsymbol{u}$.

In addition, we develop a rule-based language, RL^N, involving these graded truth-values and allowing agents to compute queries over finite domains in deterministic polynomial time. The language accommodates positive and negative literals both in premises and conclusions of rules. It is based on the Open World Assumption and we also introduce introspection operators, a logical machinery used to close the world locally and globally and to apply other forms of non-monotonic reasoning.

The RL^N language turns out to be a natural candidate for reasoning with contextual information in multi-agent systems. Also, we notice that the set of truth-degrees allows one to easily model information obtained through approximate reasoning techniques [22]. For example, a natural way to understand τ_N in the context of fuzzy reasoning [69, 70, 46, 59], where inconsistencies are not explicitly present, is to select $2*N$ pairwise disjoint subintervals $\iota_N^f, \ldots, \iota_1^f, \iota_1^t, \ldots \iota_N^t$ from the interval $[0, 1]$ and then define a mapping $\delta_F : [0, 1] \longrightarrow \{\boldsymbol{f}_N, \ldots, \boldsymbol{f}_1, \boldsymbol{u}, \boldsymbol{t}_1, \ldots, \boldsymbol{t}_N\}$, deriving truth-degrees from values belonging to the subintervals. Such a methodology can be extended to many other approximate techniques, such as intuitionistic fuzzy sets [7], rough sets [25, 56, 57], graded rough sets [68], etc. For details, see [22], where we also present other scenarios motivating the use of RL^N. By tuning the underlying family of logics, we obtain a rule-based language used to reason on graded paraconsistent information, gathered by combining heterogeneous approximate reasoning techniques. The language can also be adopted as the main internal language in Logic Fragments [20, 19] to handle coordination in context-aware self-organizing systems, extend paraconsistent approaches to knowledge bases [18, 27, 48, 50, 63], belief structures [29, 30], argumentation [28], or defeasible reasoning [37, 54].

Paper Structure

The paper is structured as follows. Section 2 introduces our family of logics, in particular knowledge, monotonicity-preservering ordering and connectives. Section 3 defines syntax and model-theoretic semantics of the proposed family of rule languages. Section 4 defines fixpoint semantics providing the basis for computational engine for the language. It also discusses relationships among monotonicity and the monotonicity-preservering ordering. To provide tools for resolving inconsistencies, lack of knowledge and introducing arbitrary truth orderings, Section 5 extends the language with introspection operators. In Section 6 we show some examples of the use of our language. In Section 7 we discuss related work. Finally, Section 8 concludes the paper.

2 The Family of Logics

2.1 Knowledge-ordering and Monotonicity-preservering ordering

Definition 2.1: (*Truth-degrees*) Given $N \geq 1$, we define the set of *truth-degrees* (also called *truth-values*) $\tau_N \stackrel{\text{def}}{=} \{0, 1, ..., N\} \times \{0, 1, ..., N\}$. For the sake of clarity, we adopt the following notation, where $1 \leq i, j \leq N$:

$$\boldsymbol{u} \stackrel{\text{def}}{=} (0,0) \quad \boldsymbol{t}_i \stackrel{\text{def}}{=} (i,0) \quad \boldsymbol{f}_i \stackrel{\text{def}}{=} (0,i) \quad \boldsymbol{i}_{i,j} \stackrel{\text{def}}{=} (i,j). \tag{1}$$

For $(p, q) \in \tau_N$, we call p and q respectively the *positive* and *negative component* of the truth-degree. Truth-values $(i, 0)$ and $(0, i)$ are called *positive* and *negative* truth-degrees, respectively. ◁

Let us now define two partial orders over τ_N (see also Figure 1):

- the knowledge partial order, to manage information at the level of multiple information sources;

- the truth-partial order, to perform computations on the information of a single source.

Definition 2.2: (*Knowledge-ordering*) We define the *knowledge ordering* $\leq_k$ as the transitive closure of the binary relation $\leq^k$, defined by:

$$(i,j) \leq^k (p,q) \quad \text{iff} \quad i \leq p \text{ and } j \leq q. \tag{2}$$

◁

The monotonicity-preservering ordering defined below is useful in many contexts (see, e.g. Section 6 and examples of applications in [22]). It appears that it is the only ordering making program operators monotonic (see Lemmas 4.1 and 4.2). Further details about this ordering and its use in RL^N are explained Section 4.2. It is also worth emphasizing that arbitrary truth-orderings, required in other application domains, can be introduced using introspection operators and then applied in a stratified manner (see Section 5.4).

Definition 2.3: (*Monotonicity-preservering ordering*) We define the *monotonicity-preservering ordering* $\leq_m$ as the reflexive and transitive closure of the binary relation $\leq^m$, defined by:

$$\begin{aligned}(0,i) \leq^m (0,j) \leq^m (0,0) \leq^m (k,l) \leq^m (m,0) \leq^m (n,0),\\ \text{for } 1 \leq j \leq i \leq N, 1 \leq k,l \leq N, 1 \leq m \leq n \leq N,\\ (i,j) \leq^m (p,q) \quad \text{if } (i = p+1 \text{ and } j = q) \text{ or } (i = p \text{ and } j = q+1).\end{aligned} \tag{3}$$

◁

Observe that in terms of notation introduced in (1), with respect to knowledge ordering we obtain: (i) $\boldsymbol{u} \leq_k \tau$ for all $\tau \in \tau_N$ as it represents the absence of information; (ii) $\boldsymbol{t}_i$ and $\boldsymbol{f}_j$ are never comparable in terms of ammount of information; (iii) $\boldsymbol{t}_i \leq_k \boldsymbol{i}_{p,q}$ and $\boldsymbol{f}_j \leq_k \boldsymbol{i}_{p,q}$ for all $i \leq p$ and $j \leq q$, given that those specific inconsistent literals contain more information. For what concerns the definition of monotonicity-preservering ordering, we have:

$$\begin{aligned}\boldsymbol{f}_i \leq_m \boldsymbol{f}_j \leq_m \boldsymbol{u} \leq_m \boldsymbol{i}_{p,q} \leq_m \boldsymbol{t}_m \leq_m \boldsymbol{t}_n\\ \text{for } 1 \leq j \leq i \leq N, 1 \leq p,q \leq N, 1 \leq m \leq n \leq N,\\ \boldsymbol{i}_{i,j} \leq_m \boldsymbol{i}_{p,q} \text{ iff } \boldsymbol{i}_{p,q} \leq_k \boldsymbol{i}_{i,j}, \text{ for } 1 \leq p \leq i \leq N \text{ and } 1 \leq q \leq j \leq N.\end{aligned} \tag{4}$$

That is (see Figure 1): (i) $\boldsymbol{f}_N$ is less true than $\boldsymbol{f}_{N-1}, \ldots, \boldsymbol{f}_2$, being less true than $\boldsymbol{f}_1$; (ii) *unknown* ($\boldsymbol{u}$) is less false than all $\boldsymbol{f}_i$ and less true than all $\boldsymbol{i}_{i,j}, \boldsymbol{t}_k$; (iii) all $\boldsymbol{i}_{i,j}$ are more true than $\boldsymbol{u}$ and less true than $\boldsymbol{t}_k$ (they have a negative component greater than zero); (iv) $\boldsymbol{t}_1$ is less true than $\boldsymbol{t}_2, \ldots, \boldsymbol{t}_{N-1}$ being less true than $\boldsymbol{t}_N$ The ordering among inconsistencies is the reversed knowledge-ordering. Such a choice allows us to evaluate truth-degrees of conjunctions of inconsistent literals w.r.t. the monotonicity-preservering ordering, used in bodies of rules, treating the degree of truth and falsity symmetrically, without favoring one over the other; the conjunction keeps track of truth and falsity levels of its operands. Such properties are better discussed in Section 4.2.

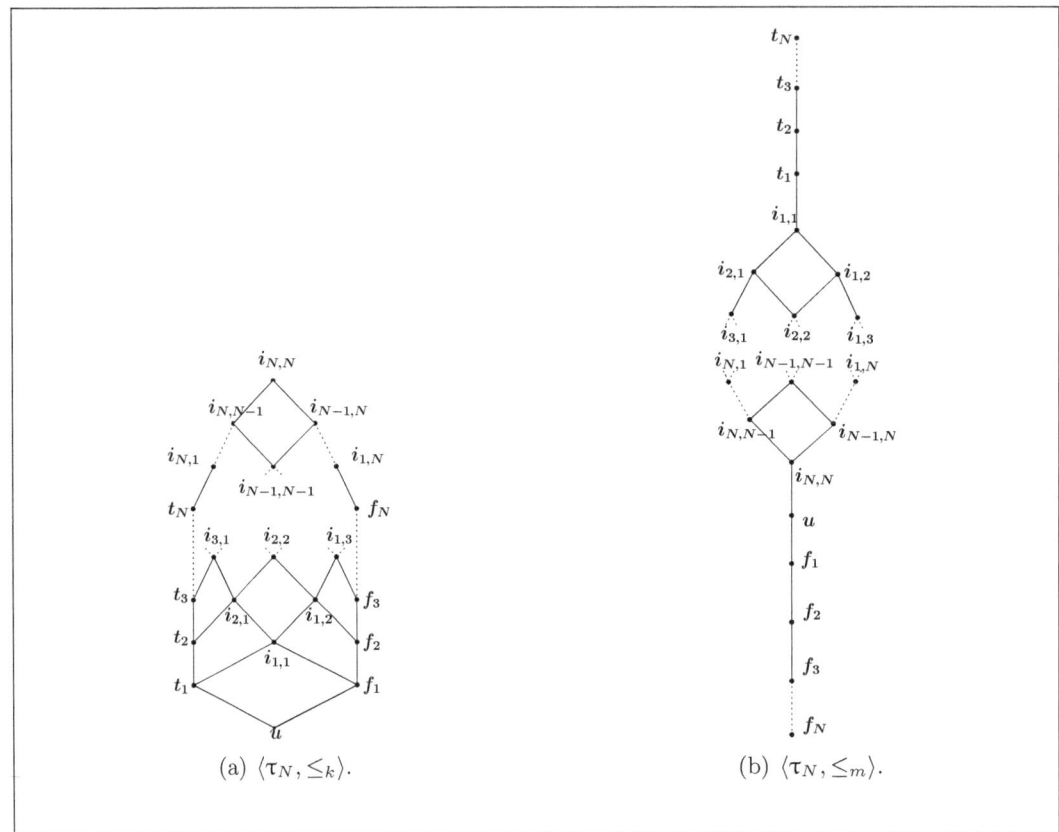

Figure 1: (a) Knowledge and (b) monotonicity-preservering ordering over τ_N.

Notice that $\langle \tau_N, \leq_k \rangle$ and $\langle \tau_N, \leq_m \rangle$ are complete lattices.[1]

Definition 2.4: (*Infimum and supremum*) Given a subset $S \subseteq \tau_N$ and a partial order over τ_N, we define $\mathrm{glb}_p S$ and $\mathrm{lub}_p S$ respectively as the *greatest lower bound* and *least upper bound* of S w.r.t. the ordering $\leq_p$. ◁

We notice that:

$$\mathrm{glb}_k\{(i,j),(p,q)\} = (min\{i,p\}, min\{j,q\}), \\ \mathrm{lub}_k\{(i,j),(p,q)\} = (max\{i,p\}, max\{j,q\}). \tag{5}$$

---

[1] A complete lattice is a partially ordered set $(L, \leq)$ in which every subset of L has both a *greatest lower bound* and a *least upper bound* in $(L, \leq)$.

Paraconsistent Rule-Based Reasoning with Graded Truth Values

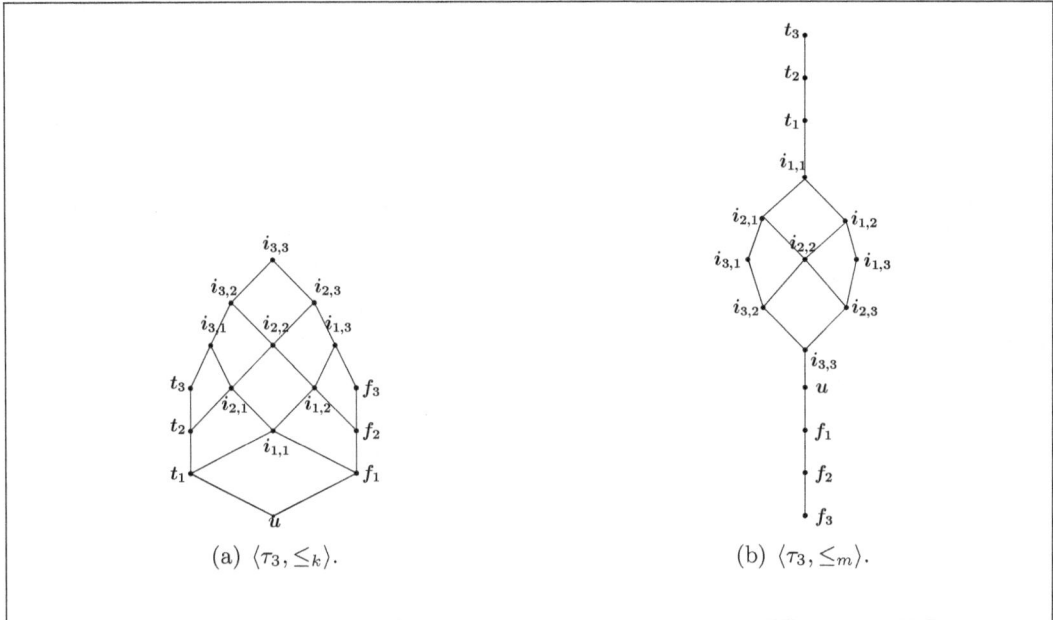

Figure 2: (a) Knowledge and (b) monotonicity-preservering ordering over τ_3.

According to Definitions 2.1, 2.2 and 2.3, the set of truth-degrees and the orderings are parametrised w.r.t. N. This means that one can derive specific instances of truth-degrees and orderings; for example, Figure 2 shows the orderings for the instance τ_3.

2.2 Logical Connectives

Let us now define logical connectives: conjunctions $\wedge_k, \wedge_m$, disjunctions $\vee_k, \vee_m$, implication $\Rightarrow_k$ and negation $\neg$.

Definition 2.5: Given $\{a_1,...,a_n\} \subseteq \tau_N$ (with $n \geq 2$), we define:

$$a \wedge_k b \stackrel{\text{def}}{=} \text{glb}_k\{a,b\}, \quad a \vee_k b \stackrel{\text{def}}{=} \text{lub}_k\{a,b\},$$
$$a \wedge_m b \stackrel{\text{def}}{=} \text{glb}_m\{a,b\}, \quad a \vee_m b \stackrel{\text{def}}{=} \text{lub}_m\{a,b\}.$$

For $a, b \in \tau_N$, implication is defined as follows:

$$a \Rightarrow_k b \stackrel{\text{def}}{=} \begin{cases} \boldsymbol{t}_N & \text{if } a \leq_k b \\ & \text{or } a = \boldsymbol{f}_i \text{ for some } 1 \leq i \leq N; \\ \boldsymbol{f}_N & \text{otherwise.} \end{cases}$$

For $(i,j) \in \tau_N$, $\neg(i,j) \stackrel{\text{def}}{=} (j,i)$. ◁

Example 2.1: In τ_3: $\boldsymbol{i}_{3,1} \wedge_k \boldsymbol{f}_3 = \boldsymbol{f}_1$, $\boldsymbol{i}_{3,1} \vee_m \boldsymbol{t}_2 = \boldsymbol{t}_2$ and $\boldsymbol{t}_2 \Rightarrow_k \boldsymbol{i}_{2,2} = \boldsymbol{t}_3$. ◁

The implication $\Rightarrow_k$ is an extension of classical implication. In rule languages one attempts to derive conclusions in order to satisfy all rules understood as implications (typically, but not necessarily, in a minimal manner). Implications should be already satisfied when conclusion need not be derived (program rules interpreted w.r.t. $\Rightarrow_k$ are like *true assertions* concerning a given scenario). In the case of RL^N we derive conclusions only on the basis of premises evaluating to truth values involving some truth (that is, to $\boldsymbol{t}_i$, $\boldsymbol{i}_{i,j}$ with $1 \leq i,j \leq N$). Given an implication $B \Rightarrow_k H$, when the truth value of B is smaller or equal (w.r.t. $\leq_k$) than the truth value of H,[2] or B evaluates to $\boldsymbol{f}_i$ for some $1 \leq i \leq N$, the implication is true $(\boldsymbol{t}_N)$ and there is no need to "correct" its value.

Given $a, b \in \tau_N$ and $1 \leq i \leq N$, we have the following inference rule, extending the Modus Ponens rule of classical logic:

$$\frac{a \geq_k \boldsymbol{t}_i \quad (a \Rightarrow_k b) = \boldsymbol{t}_N}{b \geq_k \boldsymbol{t}_i}. \tag{6}$$

Rule (6) passes the degree of truth of premises into conclusions (notice that this is true also w.r.t. positive components of inconsistent truth-degrees associated with conclusions). As detailed in the subsequent sections, in particular Section 3.2, the semantics of RL^N programs is founded on (6), with the final goal of minimising the value of b with respect to a and the knowledge ordering; in particular, no inconsistent conclusions are inferred when no contradictory information is expressed through rules of programs.

3 The Family of Rule Languages

3.1 Syntax

In the rest of the paper, by *Pred* we denote a set of predicate symbols, by *Var* a set of variables and by *Cons* a finite set of constants. We assume that these sets are pairwise disjoint.

---

[2]In particular, when B evaluates to $\boldsymbol{u}$.

Definition 3.1: (*Literals*) Given a set of truth-degrees τ_N, we define:

$$\mathcal{L}_+ \stackrel{\text{def}}{=} \{P(t_1,...,t_n) \mid P \in Pred \text{ and for } 1 \leq i \leq n, t_i \in Cons \cup Var\};$$
$$\mathcal{L}_- \stackrel{\text{def}}{=} \{\neg P(t_1,...,t_n) \mid P \in Pred \text{ and for } 1 \leq i \leq n, t_i \in Cons \cup Var\};$$
$$\mathcal{L} \stackrel{\text{def}}{=} \mathcal{L}_+ \cup \mathcal{L}_- \cup \bigcup_{i=1}^{N} \{true_i, false_i\} \cup \bigcup_{1 \leq i,j \leq N} \{inc_{i,j}\},$$

where each n, called the arity of a predicate, is a nonnegative integer ($n \geq 0$). Every element of $\mathcal{L}$ is called a *literal*. $\mathcal{L}_+$ (respectively, $\mathcal{L}_-$) is the set of *positive* (respectively, *negative*) literals. A literal without variables is called *ground literal*.
◁

Semantically, logical constants $true_i, false_i, inc_{i,j}$ are interpreted as $t_i, f_i, i_{i,j}$, respectively. We simplify expressions of the form $\neg\neg l$ to l.

Definition 3.2: (*Rules*) A *rule* R is an expression of the form:[3]

$$H \leftarrow B_1, ..., B_n \tag{7}$$

where $n \geq 0$, $H \in \mathcal{L}_+ \cup \mathcal{L}_-$ and $B_1, ..., B_n \in \mathcal{L}$.

H is called the *head* or a *conclusion* of the rule and B is called its *body*. If $n = 0$ and H is ground then (7) is called a *fact* and is understood as an abbreviation for the rule $H \leftarrow true_N$.
◁

We sometimes abbreviate rules as $H \leftarrow D$, assuming $B - B_1, \ldots, B_n$.

Definition 3.3: (*Logic programs*) A (*base* RL^N) *logic program* P is a finite set of rules. By P' we denote the *ground version of* P, i.e., the program with all ground instances of rules from P. By $\mathcal{G}_P$ we denote the set of all ground literals of P'. By the set of *positive ground literals* of P' we understand the set $\mathcal{G}_P^+ \stackrel{\text{def}}{=} \mathcal{G}_P \cap \mathcal{L}_+$ and by the set of *negative ground literals* of P' we understand $\mathcal{G}_P^- \stackrel{\text{def}}{=} \mathcal{G}_P \cap \mathcal{L}_-$.
◁

Example 3.1: Given the logic program $P \stackrel{\text{def}}{=} \{Q(X) \leftarrow P(X), P(a) \leftarrow\}$, its ground version is $P' = \{Q(a) \leftarrow P(a), P(a) \leftarrow\}$. Also, $\mathcal{G}_P^+ = \{Q(a), P(a)\}$, $\mathcal{G}_P^- = \{\neg Q(a), \neg P(a)\}$.
◁

3.2 Model-Theoretic Semantics

We define the interpretation of logic programs in terms of *many-valued Herbrand models*, which are many-valued extensions of traditional Herbrand models [24], as

---
[3] As usual, we assume that all the rules are (implicitly) universally quantified.

explained in Section 3.2.1. In Section 4 we show how to define such semantics in terms of fixpoints of program operators providing a basic engine for computing least Herbrand models.

3.2.1 Many-valued Herbrand Interpretations

Definition 3.4: (*Many-valued Herbrand interpretations*) Let P be a program. A *many-valued Herbrand interpretation* for P is a set:

$$I \subseteq \mathcal{G}_P^+ \times (\tau_N \setminus \{u\})$$

such that each positive literal of $\mathcal{G}_P^+$ appears in I in at most one pair. By $\mathcal{V}$ we denote the set of all many-valued Herbrand interpretations. ◁

We treat many-valued Herbrand interpretations as canonical interpretations of symbols, in which every constant is interpreted as itself and the interpretation of predicate symbols is defined by truth-values assigned to literals. Note that Definition 3.4 allows one to obtain compact representations for interpretations, dropping all the elements that are unknown (u) and simplifying further definitions in the sections that follow.

Definition 3.5: (*Interpretation of literals*) Given a many-valued Herbrand interpretation I and a ground literal $l \in \mathcal{G}_P^+$, the *truth-degree associated* to l is defined by:

$$I(l) \stackrel{\text{def}}{=} \begin{cases} \tau & \text{if } (l, \tau) \in I, \text{ for (a unique) } \tau \in \tau_N; \\ u & \text{otherwise.} \end{cases}$$

We extend the above definition to truth constants, negative literals, conjunctions of literals and rules by setting:

$$I(true_i) \stackrel{\text{def}}{=} t_i, I(false_i) \stackrel{\text{def}}{=} f_i, I(inc_{ij}) \stackrel{\text{def}}{=} i_{ij};$$
$$I(\neg l) \stackrel{\text{def}}{=} (q, p) \text{ iff } I(l) = (p, q);$$
$$I(l_1 \wedge_m \ldots \wedge_m l_k) \stackrel{\text{def}}{=} I(l_1) \wedge_m \ldots \wedge_m I(l_k);$$
$$I(H \leftarrow B) \stackrel{\text{def}}{=} I(B) \Rightarrow_k I(H).$$

If $B = B_1, ..., B_n$ is a body of a rule and $n \geq 1$ then $I(B) \stackrel{\text{def}}{=} I(B_1) \wedge_m \ldots \wedge_m I(B_n)$. If $n = 0$ then $I(B) \stackrel{\text{def}}{=} t_N$. ◁

Note that negation transforms a true literal (t_i) into a false literal (f_i) and vice versa, whereas it swaps the components of inconsistent literals (from $i_{p,q}$ to $i_{q,p}$). Negation of u remains u.

Example 3.2: Given $I \stackrel{\text{def}}{=} \{(P(a), t_2)\}$, $I(\neg P(a)) = f_2$, $I(P(c)) = u$. ◁

We extend the partial order $\leq_k$ to many-valued Herbrand interpretations as follows.

Definition 3.6: (*Knowledge-ordering over interpretations*) Given a program P and many-valued Herbrand interpretations I_1 and I_2, we define:

$$I_1 \leq_k I_2 \text{ iff for every literal } l \text{ in } \mathcal{G}_P, I_1(l) \leq_k I_2(l).$$ ◁

By resorting to standard results of domain theory, it can be shown that $\langle \mathcal{V}, \leq_k \rangle$ is a complete lattice (the reader can find an *ad hoc* proof in [21]). We also have the following lemma (again a proof is reported in [21]).

Lemma 3.1: Let I_1 and I_2 be two many-valued Herbrand interpretations for a program P and let $H \leftarrow B$ a ground rule in P' with $B = B_1, ..., B_n$. If $I_1 \leq_k I_2$ and $t_1 \leq_k I_1(B)$ then $I_1(B) \leq_k I_2(B)$. ◁

Definition 3.7: (*Many-valued Herbrand models*) Let P be a program and P' its ground version. A many-valued Herbrand interpretation I is a *model* of P if, for every ground rule $H \leftarrow B \in P'$, $I(H \leftarrow B) = t_N$. ◁

Example 3.3: Let P be the following program over τ_3:

$$T(a) \leftarrow true_1$$
$$T(b) \leftarrow true_2$$

For many-valued Herbrand interpretation $I = \{(T(a), t_1), (T(b), t_2)\}$, we have:

$$I(T(a) \leftarrow true_1) = (t_1 \Rightarrow_k t_1) = t_3$$
$$I(T(b) \leftarrow true_1) = (t_1 \Rightarrow_k t_2) = t_3$$
$$I(T(a) \leftarrow T(a)) = (t_1 \Rightarrow_k t_2) = t_3$$

Note that I satisfies all the ground rules of P', so I is also a model of P. ◁

We are now ready to define the model-theoretic semantics of RL^N.

Definition 3.8: (*Entailment*) Let τ_N be a set of truth values, P be an RL^N program, A be a formula and $1 \leq n \leq N$. Then we say that P *entails* A *to the degree* n, denoted by $P \models_n A$, iff $M(A) \geq_k t_n$ for every model M of P.

We say that P *entails* A ($P \models A$) iff for every many-valued Herbrand model M of P we have $M(A) \geq_k t_1$. ◁

We have the following theorem.

Theorem 3.1: (*Soundness of RL^N semantics*) Given a program P, its least many-valued Herbrand model I_P and a ground literal A, if $I_P(A) \geq_k t_n$ then $P \models_n A$. In particular, if $I_P(A) \geq_k t_1$ then $P \models A$.

Proof Straightforward as I_P is the least Herbrand model of P. ◁

Note that the least model I_P, referred to in Theorem 3.1, exists for any RL^N program P, as shown in Theorem 4.3(b).

4 Fixpoint Semantics

In this section we introduce the semantics of logic programs in terms of minimal many-valued Herbrand models. Like for other model-theoretic semantics [24, 34], the strategy used to find minimal models consists in a recursive construction of a fixpoint for a specific operator defined over the set of many-valued Herbrand interpretations.

4.1 Program Operators

We define the operator used to build minimal many-valued Herbrand models as follows.

Definition 4.1: (*Many-valued program operator*) Let P be a program, P' its ground version and let I be a many-valued Herbrand interpretation. We define the operator $T_P : \mathcal{V} \to \mathcal{V}$ as follows:

$$T_P(I) \stackrel{\text{def}}{=} \{(l, \tau) \mid l \in \mathcal{G}_P^+ \text{ and}$$
$$\tau = \text{lub}_k(\{(p,q) \mid (l \leftarrow B) \in P' \text{ and } I(B) = (p,q) \geq_k t_1\} \cup$$
$$\{(q,p) \mid (\neg l \leftarrow B) \in P' \text{ and } I(B) = (p,q) \geq_k t_1\})$$
$$\}$$

◁

T_P is a many-valued generalisation of the *immediate consequence operator* for definite programs [24]. According to the definition, $T_P(I)$ is a many-valued Herbrand interpretation that "minimally" satisfies the rules of P' whose bodies are evaluated by I to a value greater than or equal to t_1. Given that $\Rightarrow_k$ is satisfied when the antecedent is *unknown* or *false*, we only consider bodies with $I(B) \geq_k t_1$ that refer to the same head literal (perhaps negated).

Example 4.1: Let the set of truth values be τ_2 and P consist of rules:

$$Q(X) \leftarrow P(X)$$
$$\neg Q(a) \leftarrow$$

Then, for $I = \{(P(a), t_2)\}$ we have $T_P(I) = \{(P(a), t_2), (Q(a), i_{1,2})\}$. ◁

Definition 4.2: (*Semantics of programs*) The semantics of a logic program P is defined as the least many-valued Herbrand model w.r.t. the knowledge-ordering. ◁

As shown in Theorem 4.3(b), the least many-valued Herbrand model is the least fixpoint of operator T_P. In terms of RL^N logic programs and many-valued Herbrand interpretations, given a program P and a rule $H \leftarrow B \in P'$, we can rewrite Equation (6) as follows:

$$\frac{I(B) \geq_k t_i \quad I(H \leftarrow B) = t_N}{I(H) \geq_k t_i}.$$

In particular, if I is the least many-valued Herbrand model of P and $H \leftarrow B$ is the unique rule with head H or $\neg H$ in P' then we obtain a many-valued extension of *modus ponens*:

$$\frac{I(B) = t_i \quad I(H \leftarrow B) = t_N}{I(H) = t_i}.$$

If $I(H) = t_j > t_i$ then we conclude that there is a rule $H \leftarrow \hat{B} \in P'$ such that $I(\hat{B}) = t_j$. $I(H)$ is inconsistent if (i) there is a rule in $H \leftarrow \hat{B} \in P'$ with $I(\hat{B})$ inconsistent or (ii) there are two rules $H \leftarrow \hat{B}_1 \in P'$ and $H \leftarrow \hat{B}_2 \in P'$ such that $I(B_1) \geq_k t_1$ and $I(B_2) \geq_k t_1$. In all cases, the semantics of a program preserves the grade of truth of premises into conclusions (eventually in the positive components of inconsistent truth-degrees).

In the next theorems we analyse some fundamental results used to prove that the many-valued semantics of a program P is the least fixpoint of the operator T_P.

Theorem 4.1: Let P be a program, I_1 and I_2 be many-valued Herbrand interpretations. If $I_1 \leq_k I_2$ then $T_P(I_1) \leq_k T_P(I_2)$.

Proof The truth-degrees considered in the least upper bound of Definition 4.1 are associated with interpretations of bodies of rules that are greater than t_1 w.r.t. $\leq_k$. Thus, from Lemma 3.1, for every ground body B whose evaluation appears in the set of truth-degrees, we have $t_1 \leq_k I_1(B) \leq_k I_2(B)$. Given that in the operator T_P we consider least upper bounds of such (possibly negatd) evaluations, we obtain $T_P(I_1) \leq_k T_P(I_2)$. ◁

Theorem 4.1 states that $T_P : \mathcal{V} \to \mathcal{V}$ is monotone w.r.t. $\leq_k$. This is an important property: being $\langle \mathcal{V}, \leq_k \rangle$ a complete lattice, it paves the way for applying the Knaster-Tarski theorem [45, 65] to the operator $T_P : \mathcal{V} \to \mathcal{V}$, as shown in the following theorem.

Theorem 4.2: The operator $T_P : \mathcal{V} \to \mathcal{V}$ has a unique least fixpoint w.r.t. the knowledge-ordering.[4]

Proof By Theorem 4.1, T_P is monotone w.r.t. $\leq_k$ and $\langle \mathcal{V}, \leq_k \rangle$ is a complete lattice. By the Knaster-Tarski theorem we then conclude that T_P has a unique least fixpoint lfp_P. ◁

A proof of the following theorem is reported in [21].

Theorem 4.3: Let P be a program. Then:

(a) Every fixpoint of $T_P : \mathcal{V} \to \mathcal{V}$ is a model of P.

(b) The least fixpoint of $T_P : \mathcal{V} \to \mathcal{V}$ is the least model of P w.r.t. $\leq_k$. ◁

The fixpoint characterisation gives rise to the following important theorem.

Theorem 4.4: Over finite domains, computing the least many-valued Herbrand model of P can be done in deterministic polynomial time w.r.t. the size of the domain.

Proof Given that we assume logic programs to have a fixed number of rules, this is similar to the computation of the *data complexity* of Datalog [55]. Let $|U|$ be the size of the universe ($|U| = |Cons|$) and c_v the maximum number of variables appearing in the bodies of rules (c_v is a constant). Then $|P'| = O(|U|^{c_v})$. In the worst case, the evaluation of a new rule implies a change in the evaluation of bodies of the rules already evaluated. Thus, the least many-valued Herbrand model can be computed in $O(|U|^{2c_v})$ steps (where a step depends on computing the truth-degree associated with the whole body of a rule). ◁

4.2 Monotonicity-preservering ordering and Monotonicity of consequence operators

We observe that the result of Theorem 4.3 is achieved by resorting to the property expressed in Lemma 3.1. Such a property is supported by the current definitions of the orderings (see Definition 2.2). The rationale behind the definition of $\leq_m$ is the following one.

I. In conjunctions of consistent truth-degrees (*true*, *unknown* or *false*), $\leq_m$ intuitively preserves the truth-degree associated with the minimum level of truth (see Figure 1 and Definition 2.5).

---

[4] $F \in \mathcal{V}$ is a fixpoint of T_P iff $T_P(F) = F$. F is the *least* fixpoint of T_P w.r.t $\leq_k$ iff F is a fixpoint of T_P and $F \leq_k F'$ for every F' such that $T_P(F') = F'$.

II. In conjunctions of inconsistent truth-degrees, $\leq_m$ preserves the maximum values of the positive and negative components of the truth-degrees (i.e. $\boldsymbol{i}_{i,j} \wedge_m \boldsymbol{i}_{p,q} = \boldsymbol{i}_{max(i,p),max(j,q)}$). This assures that a truth-value obtained from a conjunction of inconsistent literals is in turn inconsistent and it has the maximum positive and negative components of the inconsistent truth-degrees.

Example 4.2: Consider the rule $A \leftarrow B, C, D$ when B is evaluated to $\boldsymbol{i}_{2,3}$, C to $\boldsymbol{i}_{4,1}$ and D to $\boldsymbol{t}_3$. When interpreted according to the semantics of Section 4, we expect A to be associated with the truth-degree $\boldsymbol{i}_{4,3}$ (or with a greater one w.r.t. $\leq_k$ in presence of further rules with head A or $\neg A$); rephrased, we say that: (i) there is evidence supporting the thesis that A is inconsistent and (ii) they attest that the level of inconsistency is at most 4 for the positive component and 3 for the negative one. ◁

Point (ii) in the previous example is useful when using introspection operators (Section 5) to reason on inconsistency bounded above by some given values. Moreover, point (i) refers to a quite general and appealing property for truth-degrees. This means that:

$$\begin{aligned} \boldsymbol{i}_{p,q} <_m \boldsymbol{t}_i & \quad \text{for every } p \geq 0, q \geq 0, i > 1; \\ \boldsymbol{i}_{p,q} >_m \boldsymbol{f}_i & \quad \text{for every } p \geq 0, q \geq 0, i > 1; \\ \boldsymbol{t}_i <_m \boldsymbol{t}_j & \quad \text{for every } 0 < i < j; \\ \boldsymbol{f}_i <_m \boldsymbol{f}_j & \quad \text{for every } 0 < j < i. \end{aligned} \quad (8)$$

Thus, the definition of further truth-orderings satisfying (8) involves the definition of relations among inconsistent truth-degrees. Technically, these relations can affect the monotonicity of T_P, as stated in the following lemmas.

Lemma 4.1: For every linear order $\langle \tau_N, \leq_{nt} \rangle$ satisfying (8) with $N \geq 2$ there exists a program P such that T_P of Definition 4.1 is not monotonic.
Proof Consider the program:

$$P \stackrel{\text{def}}{=} \{H \leftarrow B_1, B_2\}. \quad (9)$$

We define:

$$I_1 \stackrel{\text{def}}{=} \{(B_1, \boldsymbol{i}_{p,q}), (B_2, \boldsymbol{t}_1)\} \text{ and } I_2 \stackrel{\text{def}}{=} \{(B_1, \boldsymbol{i}_{p,q}), (B_2, \boldsymbol{i}_{r,s})\}$$

such that $1 \leq p, q, r, s \leq N$, $\boldsymbol{i}_{r,s} \leq_{nt} \boldsymbol{i}_{p,q}$ and $\boldsymbol{i}_{r,s}$ and $\boldsymbol{i}_{p,q}$ are not comparable w.r.t. $\leq_k$ (notice that such conditions can be always satisfied for $N \geq 2$). Then $I_1 \leq_k I_2$ but $T_P(I_1)(H) = \boldsymbol{i}_{p,q}$ and $T_P(I_2)(H) = \boldsymbol{i}_{r,s}$ are not comparable w.r.t. $\leq_k$. ◁

Lemma 4.2: Let $\langle \tau_N, \leq_{nt} \rangle$ be a complete lattice satisfying (8) with $N \geq 2$. If there exist two elements $\boldsymbol{i}_{p,q} \in \tau_N$ and $\boldsymbol{i}_{r,s} \in \tau_N$ such that $1 \leq p, q, r, s \leq N$, $\boldsymbol{i}_{r,s} <_k \boldsymbol{i}_{p,q}$ and $\boldsymbol{i}_{r,s} <_{nt} \boldsymbol{i}_{p,q}$ then there exists a program P such that T_P of Definition 4.1 is not monotone w.r.t. $\leq_k$.

Proof Let P be the program defined by (9) and let $I_1 \stackrel{\text{def}}{=} \{(B_1, \boldsymbol{i}_{p,q}), (B_2, \boldsymbol{t}_1)\}$ and $I_2 \stackrel{\text{def}}{=} \{(B_1, \boldsymbol{i}_{p,q}), (B_2, \boldsymbol{i}_{r,s})\}$. Then $I_1 \leq_k I_2$ but:

$$T_P(I_2)(H) = \boldsymbol{i}_{r,s} <_k T_P(I_1)(H) = \boldsymbol{i}_{p,q}. \qquad \triangleleft$$

Thus, Lemmas 4.1–4.2 state that to preserve the monotonicity of T_P, truth-orderings satisfying point (i) (i.e., (8)) must not be linear and its inconsistent truth values have to satisfy $\boldsymbol{i}_{p,q} <_{nt} \boldsymbol{i}_{r,s}$ when $\boldsymbol{i}_{r,s} <_k \boldsymbol{i}_{p,q}$, for every $\boldsymbol{i}_{p,q} \in \tau_N$, $\boldsymbol{i}_{r,s} \in \tau_N$. We observe that such conditions are satisfied by $\langle \tau_N, \leq_m \rangle$.

5 Introspection Operators

In this section we extend the language defined so far by tools to express non-monotonic/defeasible rules. Non-monotonicity can appear when inconsistencies or lack of knowledge is resolved with rules drawing tentative conclusions (e.g., reflecting some heuristics) that are assumed to be defeasible when gathering more information. The mechanism that we introduce makes it possible to compare truth-values of (sets of) literals, handling inconsistent information and lack of knowledge. Moreover, it provides support to enrich the language with further orderings that would break the monotonicity of the T_P operator of Definition 4.1. In this way, ad-hoc orderings accommodating different interpretations of conjunctions and disjunctions of literals can be employed.

5.1 Definition of Introspection Operators

Definition 5.1: (*Introspection operators*) An *introspection operator* is an expression of the form $\mathcal{O}(S, T)$, where $\mathcal{O}$ is the operator's name, $S \subseteq \mathcal{L}$ and $T \subseteq \tau_N$.[5] From the semantic point of view, introspections operators map sets of literals and truth-values into τ_N. $\triangleleft$

In the rest of the paper we assume that the considered introspection operators are computable in deterministic polynomial time in the size of the domain. This assumption is needed to retain tractability of query evaluation.

---
[5] For the sake of readability, we use here truth-degrees rather than logical constants. Of course, an equivalent definitions can be given defining T as a subset of logical constant.

Let I be a many-valued Herbrand interpretation. We define the following sample introspection operators, where we also provide more convenient notation for these operators.

- Operator $\mathcal{O}_\in(\{l\}, \{\tau_1, \ldots, \tau_N\})$, denoted by $l \in \{\tau_1, \ldots, \tau_N\}$:

$$I(l \in \{\tau_1, \ldots, \tau_N\}) \stackrel{\text{def}}{=} \begin{cases} \boldsymbol{t}_N & \text{when } I(l) \in \{\tau_1, \ldots, \tau_N\}; \\ \boldsymbol{f}_N & \text{otherwise.} \end{cases} \quad (10)$$

- Operator $\mathcal{O}_{\leq_k, \emptyset}(\{l_1, l_2\}, \emptyset)$, denoted by $l_1 \leq_k l_2$:

$$I(l_1 \leq_k l_2) \stackrel{\text{def}}{=} \begin{cases} \boldsymbol{t}_N & \text{when } I(l_1) \leq_k I(l_2); \\ \boldsymbol{f}_N & \text{otherwise.} \end{cases} \quad (11)$$

- Operator $\mathcal{O}_{\leq_k}(\{l\}, \{\tau\})$, denoted by $l \leq_k \tau$:

$$I(l \leq_k \tau) \stackrel{\text{def}}{=} \begin{cases} \boldsymbol{t}_N & \text{when } I(l) \leq_k \tau; \\ \boldsymbol{f}_N & \text{otherwise;} \end{cases} \quad (12)$$

- Operator $\mathcal{O}_{\leq_m, \emptyset}(\{l_1, l_2\}, \emptyset)$, denoted by $l_1 \leq_m l_2$:

$$I(l_1 \leq_m l_2) \stackrel{\text{def}}{=} \begin{cases} \boldsymbol{t}_N & \text{when } I(l_1) \leq_m I(l_2); \\ \boldsymbol{f}_N & \text{otherwise;} \end{cases} \quad (13)$$

- Operator $\mathcal{O}_{\leq_m}(\{l\}, \{\tau\})$, denoted by $l \leq_m \tau$:

$$I(l \leq_m \tau) \stackrel{\text{def}}{=} \begin{cases} \boldsymbol{t}_N & \text{when } I(l) \leq_m \tau; \\ \boldsymbol{f}_N & \text{otherwise;} \end{cases} \quad (14)$$

- Operator $\mathcal{O}_{\Delta_n^N}(\{l\}, \emptyset)$, denoted by $l \leq_\Delta n$:

$$I(l \leq_\Delta n) \stackrel{\text{def}}{=} \begin{cases} \boldsymbol{t}_N & \text{when } I(l) \in \Delta_n^N; \\ \boldsymbol{f}_N & \text{otherwise,} \end{cases} \quad (15)$$

where $\Delta_n^N \stackrel{\text{def}}{=} \{(p, q) \mid (p, q) \in \tau_N \text{ and } p - q \leq n\}$.

- Other useful operators can be defined using the strict partial order $<_{truth}$ defined by:

$$(p_1, p_2) <_{truth} (q_1, q_2) \text{ when } \begin{array}{l} (p_1 - p_2) < (q_1 - q_2) \text{ or} \\ (p_1 - p_2) = (q_1 - q_2) \text{ and } p_1 < q_1. \end{array} \quad (16)$$

In particular, $\mathcal{O}_{\leq_{truth}}(\{l_1, l_2\}, \emptyset)$, denoted by $l_1 \leq_{truth} l_2$:

$$I(l_1 \leq_{truth} l_2) \stackrel{\text{def}}{=} \begin{cases} \boldsymbol{t}_N & \text{when } I(l_1) \leq_{truth} I(l_2); \\ \boldsymbol{f}_N & \text{otherwise.} \end{cases} \tag{17}$$

For all such operators we consider also their versions corresponding to respective strict partial orders $l_1 <_k l_2$, $l_1 <_m l_2$, $l_1 <_k \tau$, $l_1 <_m \tau$, $l_1 <_{truth} l_2$ and $l_1 <_\Delta n$.

5.2 Extending Programs with Introspection Operators

Let us now define extended rules allowing for introspection operators.

Definition 5.2: (*Extended rules*) Let be O the set of introspection operators. An *extended rule* is an expression of the form:

$$H \leftarrow B_1, ..., B_n \tag{18}$$

with $n \geq 0$, $H \in \mathcal{L}_+ \cup \mathcal{L}_-$ and $B_1, ..., B_n \in \mathcal{L} \cup O$. ◁

Introspection operators introduce non-monotonicity of reasoning. Therefore, to keep our solutions tractable, we have to structure rules in layers in a way similar to stratification used in logic programming [1, 4].

Definition 5.3: (*Stratification and extended programs*) Let S be a finite set of extended rules. Then S is an *extended program* (or RL^N *program*) iff there is a mapping $\kappa_S : \mathcal{L}_+ \cup \mathcal{L}_- \longrightarrow \mathbb{N}$ such that for every literal $l \in \mathcal{L}_+$, $\kappa_S(l) = \kappa_S(\neg l)$, and for every rule $H \leftarrow B_1, \ldots, B_n \in S$ and every $0 \leq i \leq n$,

1. if B_i is a literal then $\kappa_S(H) \geq \kappa_S(B_i)$;

2. if B_i is an introspection expression and l is a literal occurring in B_i then $\kappa_S(H) > \kappa_S(l)$. ◁

One can easily observe that κ_S defines a partition of rules such that a rule $H \leftarrow B_1, \ldots, B_n \in S$ belongs to a component $i \in \mathbb{N}$ when $\kappa(H) = i$. Moreover, when a given literal occurs in a body of a rule in the scope of an introspection operator then it is fully defined by rules "smaller" (w.r.t. κ_S) than the current rule. This allows us to compute rules component by component:

1. first rules with heads having the smallest κ_s are computed (observe that no introspection operators occur in bodies of such rules);

2. next rules with heads in the next (w.r.t. κ_S) component are interpreted; truth-degrees of literals appearing in introspection operators (if any) are already computed in the previous component;

3. *iteration step*: rules with heads in the component $i+1$ are computed after all component $j \leq i$; once again, truth-degrees of literals belonging to introspection operators (if any) are obtained from previous iterations.

Such a procedure allows one to compute extended rules incrementally, without increasing too much the complexity of fixpoint semantics.

Example 5.1: The sets of rules:
$$P_1 = \{Q(X) \leftarrow P(X) \in \{t_1\}\} \text{ and } P_2 = \{P(a) \leftarrow, \neg P(a) \leftarrow\}$$
represent a partition of the extended program P:

$$\begin{aligned} Q(X) &\leftarrow P(X) \in \{t_1\} \\ P(a) &\leftarrow \\ \neg P(a) &\leftarrow \end{aligned}$$
◁

In the following definition we provide semantics of extended programs.

Definition 5.4: (*Semantics of extended programs*) Let $P = P_1 \cup ... \cup P_m$ be an extended program with a partition $P_1, ..., P_m$ and let $I_\emptyset \stackrel{\text{def}}{=} \emptyset$. We define a sequence of programs $P_1^t, ..., P_m^t$ related to $P_1, ..., P_m$ as follows:

- $P_0^t = \emptyset$ (no rules and no facts).

- Let T stand for an introspection operator in P_i or a literal appearing in the head of a rule of a program P_j with $j < i$. Then P_i^t is obtained from P_i by replacing, in its rules, every such T by the logical constant representing the truth value $(\bigcup_{j=0}^{i-1} M_j)(T)$, where M_j is the least Herbrand model of P_j^t ($M_0 = \emptyset$ for $P_0^t = \emptyset$).

The semantics of P is $\bigcup_{j=0}^{m} M_j$ and we call it the *least Herbrand model of the extended program P*.
◁

We call $\bigcup_{j=0}^{m} M_j$ the "least model" of P because it is the least Herbrand model of a program consisting of $P_1^t, ..., P_m^t$ (see Lemma 3.1 in [21]).

Example 5.2: Let P be the following program, evaluated w.r.t $N = 2$:

$$\begin{aligned} b &\leftarrow a \geq_m \boldsymbol{u} \\ a &\leftarrow \end{aligned}$$

We have $P = P_1 \cup P_2$, where $P_1 = \{a \leftarrow\}$ and $P_2 = \{b \leftarrow a \geq_m \boldsymbol{u}\}$. Thus we obtain $M_0 = \emptyset$, $P_1^t = P_1$ with $M_1 = \{(a, t_2)\}$. It follows that $P_2^t = \{b \leftarrow true_2\}$, thus $M_2 = \{(b, t_2)\}$. The semantics of P is then $M_0 \cup M_1 \cup M_2 = \{(a, t_2), (b, t_2)\}$.
◁

5.3 Fixpoint Semantics of Extended Programs

The semantics of extended programs is defined by an iterated fixpoint construction analogous to the one for stratified programs [1, 4].

Definition 5.5: (*Consequence operator for extended programs*)
Let be $P = P_1 \cup ... \cup P_m$ an extended program with a partition $P_1, ..., P_m$. We define the "progressive" version of *immediate consequence operator*:

$$T'_{P_i}(I) \stackrel{def}{=} \{(l,\tau) \mid l \in \mathcal{G}_P^+ \text{ and} \\ \tau = \text{lub}_k(\{(p,q) \mid (l \leftarrow B) \in P'_i \text{ and } I(B) = (p,q) \geq_k t_1\} \cup \\ \{(q,p) \mid (\neg l \leftarrow B) \in P'_i \text{ and } I(B) = (p,q) \geq_k t_1\} \cup \\ \{I(l)\})\}. \qquad (19)$$

◁

For each $i \geq 1$, the least fixpoint of T'_{P_i} always exists over finite domains and is defined by $T'_{P_i} \uparrow n_i$, for a natural number n_i such that $T'_{P_i} \uparrow n_i = T'_{P_i} \uparrow (n_i + 1)$, where $T'_{P_0} \uparrow n_0 \stackrel{def}{=} \emptyset$, and:

$$T'_{P_i} \uparrow 0 \stackrel{def}{=} T'_{P_{i-1}} \uparrow n_{i-1}, \\ T'_{P_i} \uparrow (k+1) \stackrel{def}{=} T'_{P_i}(T'_{P_i} \uparrow k). \qquad (20)$$

We notice that this version of the consequence operator is very close to the one for definite programs. In this case, the set $\{I(l)\}$ in (19) keeps track of the truth-degrees for literals appearing in the heads of components associated with lower indexes. Such a definition assures that the interpretation contains at most one truth-degree for a positive literal (Definition 3.4).

We have the following theorem (for a proof see [21]).

Theorem 5.1: Let be $P = P_1 \cup ... \cup P_m$ an extended program with a partition $P_1, ..., P_m$. Then $T'_{P_m} \uparrow n_m$ is the *least Herbrand model of P* (Figure 3).   ◁

We observe that the way of partitioning the extended program does not affect its interpretation. Concerning the complexity of the semantics computation, we have the following theorem.

Theorem 5.2: Let P be any extended program. Computing the least Herbrand model of P can be done in deterministic polynomial time in the size of the number of constants occurring in P.   ◁

It is important to note that one can verify whether a finite set S of rules is an extended program in deterministic time polynomial in the number of the size of

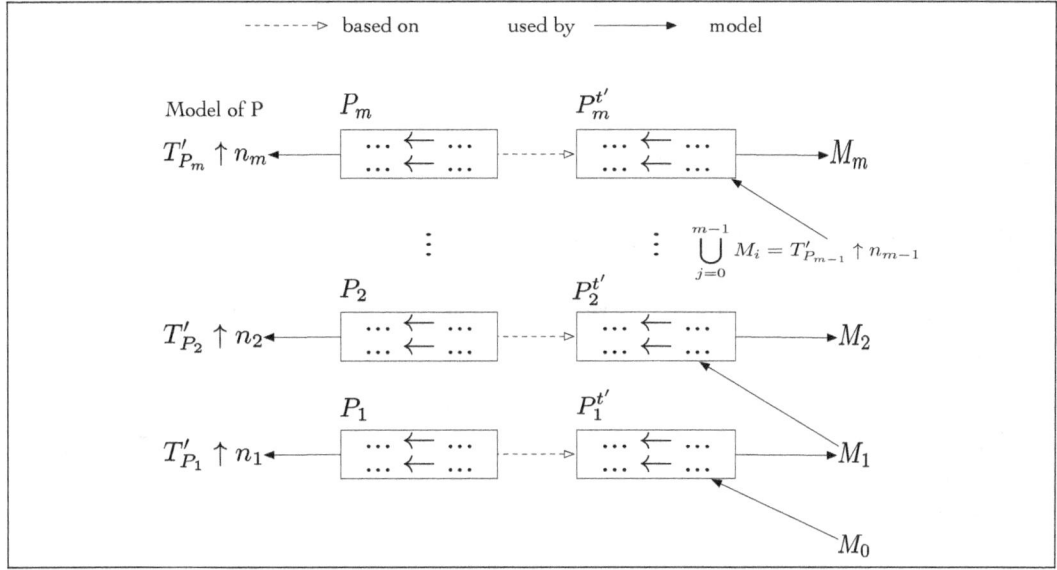

Figure 3: Fixpoint construction for extended programs.

the domain (or, equivalently, literals occurring in the program). For that purpose one can adjust the construction for stratified logic programs (see, e.g. [1]). By a *positive part* of a literal l we understand l when l is positive, and l' when $l = \neg l'$. We construct a graph G_S with nodes labeled by positive literals. For each rule $H \leftarrow B_1, \ldots, B_n \in S$:

- if B_i is a literal then there is an edge from the positive part of B_i to the positive part of H, labeled with '+' (a "positive" edge);

- if B_i is an introspection expression and l is a literal occurring in B_i then there is an edge from the positive part of l to the positive part of H, labeled with '−' (a "negative" edge).

We have the following property.

Lemma 5.1: A finite set of rules S is an extended program if G_S has no cycle containing a negative edge. ◁

Of course, checking for existence of a cycle with negative cycle indicated in Lemma 5.1 takes deterministic polynomial time in the size of S.

5.4 Arbitrary Truth-orderings as Introspection Operators

Introspection operators can be used to introduce alternative orderings including those violating monotonicity of the operator T_P of Definition 4.1. This is useful especially when a different interpretation of conjunctions and disjunctions of literals is needed to evaluate a subset of rules, i.e., when a truth-ordering replacing the monotonicity-preservering ordering can capture some important semantic aspects in a given application domain.

Let $\langle \tau_N, \leq_o \rangle$ be a complete lattice. By $lub_o S$ and $glb_o S$ we denote the least upper bound of a set of truth values $S \subseteq \tau_N$ w.r.t. $\leq_o$ and the greatest lower bound of S w.r.t. $\leq_o$, respectively.

Definition 5.6: (*Orderings over introspection operators*) Let $L = \langle \tau_N, \leq_o \rangle$ be a complete lattice. We then define the interpretations for the following *introspection operators induced by* L:

$$I(\mathcal{O}_{\wedge_o}(\{l_1, ..., l_n\}, \emptyset)) \stackrel{def}{=} glb_o\{l_1, ..., l_n\}; \tag{21}$$

$$I(\mathcal{O}_{\vee_o}(\{l_1, ..., l_n\}, \emptyset)) \stackrel{def}{=} lub_o\{l_1, ..., l_n\} \tag{22}$$

◁

In what follows,

$\mathcal{O}_{\wedge_o}(\{l_1, ..., l_n\}, \emptyset)$ is abbreviated by $l_1 \wedge_o ... \wedge_o l_n$;
$\mathcal{O}_{\vee_o}(\{l_1, ..., l_n\}, \emptyset)$ is abbreviated by $l_1 \vee_o ... \vee_o l_n$.

These operators can be used to evaluate respectively conjunctions and disjunctions of literals w.r.t. $\leq_o$. Such a way of introducing generic orderings is restricted in the sense that the stratification requirement is to be met. Even though, in some cases, this aspect may represent a too restrictive constraint, such a limitation represents the trade-off for benefiting from the flexibility of easily introducing additional orderings exploiting, at the same time, the results of Theorem 5.1 and Theorem 5.2; indeed, such theorems assure that the extended program using the additional ordering still has a many-valued Herbrand model computable in polynomial time w.r.t. the size of the domain.

Example 5.3: In the following program we want the conjunction $u(X), r(X)$ in rule (23) to be interpreted according to the ordering $\leq_{truth}$ defined by (16) instead of the basic one $\leq_m$.

$$w(X) \leftarrow u(X), r(X) \qquad (23)$$
$$u(X) \leftarrow p(X), q(X) \qquad (24)$$
$$p(a) \leftarrow true_1 \qquad (25)$$
$$q(a) \leftarrow true_3 \qquad (26)$$
$$q(b) \leftarrow true_1 \qquad (27)$$
$$r(b) \leftarrow true_2 \qquad (28)$$

We then use $\mathcal{O}_{\wedge_{truth}}(\{l_1, ..., l_n\}, \emptyset)$, replacing (23) by:

$$w(X) \leftarrow u(X) \wedge_{truth} r(X). \qquad (29)$$

We notice that the new program is extended; one possible stratification is:
$\{(24), \ldots, (28)\} \cup \{(29)\}$. ◁

Note that conjunctions and disjunctions based on any other (polynomially computable) truth-orderings can be defined by introducing new relations and defining them in terms of rules. In such a case introspection operators are not needed so stratification is not required.

6 Examples of Applications

6.1 A Robotics Scenario

Example 6.1: Consider an exploratory robot moving in a hostile environment to search and collect ground samples for analysis of minerals. The robot is equipped with some sensors that analyse the surrounding space detecting:

- the presence of a good concentration of minerals in a given area (*min* predicate);

- rocky areas (*rock* predicate);

- potential holes in the ground (*holes* predicate).

The robot also receives information from other sources that may cause inconsistencies in its knowledge base.

A dangerous area for the robot (*dang* predicate) is an area that is either rocky or containing some holes. Every safe (not dangerous) area containing minerals has to be analysed (*move* predicate). The behavior of the robot is defined as follows:

$$\begin{aligned} move(X) &\leftarrow min(X), \neg dang(X) \\ \neg dang(X) &\leftarrow \neg rock(X), \neg holes(X) \\ dang(X) &\leftarrow rock(X) \\ dang(X) &\leftarrow holes(X) \end{aligned} \qquad (30)$$

We now consider an instance of the problem in which sensor measurements for areas a_1 and a_2 are associated with grades, reflecting the perception quality. The following set of facts represents sample sensed contextual-information:

$$\begin{aligned} rock(a_1) &\leftarrow true_2 & \neg rock(a_2) &\leftarrow true_2 \\ \neg rock(a_1) &\leftarrow true_1 & \neg holes(a_2) &\leftarrow true_3 \\ \neg holes(a_1) &\leftarrow true_3 & min(a_2) &\leftarrow true_3 \\ min(a_1) &\leftarrow true_3 \end{aligned} \qquad (31)$$

We notice that two distinct sensors have generated contradictory information about the geomorphology of area a_1, so we expect to obtain some inconsistency regarding the action to take in that location. We now consider the program P given by the union of the rules of equations (30) and (31); computing its fixpoint semantics we obtain:

$$\begin{aligned} T_P\uparrow 1 = T_P(I_\emptyset) &= \{(rock(a_1), \boldsymbol{i}_{2,1}), ((rock(a_2), \boldsymbol{f}_2), \\ &\quad (holes(a_1), \boldsymbol{f}_3), (holes(a_2), \boldsymbol{f}_3), \\ &\quad (min(a_1), \boldsymbol{t}_3), (min(a_2), \boldsymbol{t}_3)\}; \\ T_P\uparrow 2 = T_P(T_P\uparrow 1) &= T_P\uparrow 1 \cup \{(dang(a_1), \boldsymbol{i}_{2,1}), (dang(a_2), \boldsymbol{f}_2)\}; \\ T_P\uparrow 3 = T_P(T_P\uparrow 2) &= T_P\uparrow 2 \cup \{(move(a_1), \boldsymbol{i}_{1,2}), (move(a_2), \boldsymbol{t}_2)\}; \\ T_P\uparrow 4 = T_P(T_P\uparrow 3) &= T_P\uparrow 3. \end{aligned} \qquad (32)$$

$T_P\uparrow 1$ is obtained from the rules of (31) and it has assigned $\boldsymbol{i}_{2,1}$ to $rock(a_1)$ due to the contradictory facts remarked above. In this case $T_P\uparrow 1(rock(a_1)) = \boldsymbol{i}_{2,1}$ is founded on $rock(a_1) \leftarrow true_2$ (the first component) and $\neg rock(a_1) \leftarrow true_1$ (the second component).

$T_P\uparrow 2$ and $T_P\uparrow 3$ are obtained from (30); the inconsistency of $rock(a_1)$ propagates through the rules of (30) and is reflected in $dang(a_1)$ and $move(a_1)$. For the area a_2 there are no contradictory facts and $T_P\uparrow 2$ assigns the degree $\boldsymbol{t}_2$ to $\neg dang(a_2)$. $\neg holes(a_2)$ is associated with $\boldsymbol{t}_3$ and the truth-level assigned by sensors to $rock(a_2)$ is $\boldsymbol{f}_2$; thus, the maximum degree of certainty that can be associated with the conjunction $\neg rock(a_2), \neg holes(a_2)$ is $\boldsymbol{t}_2$ (i.e., there is enough information to assign $\boldsymbol{t}_2$

to $\neg dang(a_2)$, but it is not sufficient to ensure the highest value t_3). An analogous reasoning explains the value t_2 entailed for $move(a_2)$. ◁

Example 6.2: We now consider program P given by rules (30) and facts (31) with the additional fact $\neg min(a_1) \leftarrow true_1$. In this case there is contradictory information also about $min(a_1)$. Computing the fixpoint semantics of the program we obtain:

$$\begin{aligned}
T_P{\uparrow}1 = T_P(I_\emptyset) &= \{(rock(a_1), i_{2,1}), ((rock(a_2), f_2), (holes(a_1), f_3), \\
&\quad (holes(a_2), f_3), (min(a_1), i_{3,1}), (min(a_2), t_3)\}; \\
T_P{\uparrow}2 = T_P(T_P{\uparrow}1) &= T_P{\uparrow}1 \cup \{(dang(a_1), i_{2,1}), (dang(a_2), f_2)\}; \\
T_P{\uparrow}3 = T_P(T_P{\uparrow}2) &= T_P{\uparrow}2 \cup \{(move(a_1), i_{3,2}), (move(a_2), t_2)\}; \\
T_P{\uparrow}4 = T_P(T_P{\uparrow}3) &= T_P{\uparrow}3.
\end{aligned}$$

The facts concerning area a_2 are the same as in the case of facts (31), that is the reason why $T_P{\uparrow}3$, for the literals referring to a_2, infers the same truth-degrees of (32).

Due to the contradictions in the facts for a_1, $min(a_1)$ and $rock(a_1)$ are both inconsistent ($T_P \uparrow 2(min(a_1)) = i_{3,1}$, $T_P \uparrow 2(dang(a_1)) = i_{2,1}$) and $move(a_1)$ is evaluated to $i_{3,1} \wedge_m i_{1,2} = i_{max(3,1),max(1,2)} = i_{3,2}$ in $T_P{\uparrow}3$. Thus, when inconsistent truth degrees are involved in conjunctions, the evaluation of bodies preserves the maximum values of the literal components being evaluated. This mechanism makes it possible, during the computation, to keep track of the global amount of collected information supporting both the head of a rule and its negation; such knowledge can be exploited to manage contradictory information, as shown in the next section. ◁

6.2 Resolving Inconsistencies

In this section we show how to use the introspection operators defined in Section 5 to resolve inconsistent information. Graded values allow for fine-grained modelling of predicate truth-degrees: the intuition behind inconsistency resolution is then grounded on the comparison of truth-degree components, as a form of analysis and categorization of inferred information.

Example 6.3: We consider an improved version of the exploratory robot of Section 6.1: the robot moves in an area affected by inconsistent information if the global knowledge indicates that the location is "sufficiently attractive" to encourage the movement. Area attractiveness is defined by analysing literals concerning the existence of dangers and minerals; more specifically, in the case of contradictory knowledge the robot moves when two primary conditions are met:

(i) There may be evidence confirming that the location is dangerous but globally there are also stronger facts supporting the opposite (i.e. the positive component of $dang(X)$ is greater then the negative one).

(ii) The certitude of finding minerals is higher than a given threshold (i.e. the difference between the positive and negative component of $min(X)$ is greater than a specific value).

We introduce the concept of explorable area (explorable predicate - $expl$) to designate non-dangerous areas containing minerals. In the following rules we use introspection operators $\leq_m$, $\leq_{truth}$ and $<_\Delta$ defined by (14), (17) and (15), respectively.

$$\begin{aligned}
move(X) &\leftarrow expl(X) \geq_m t_1 \\
move(X) &\leftarrow expl(X) >_m \boldsymbol{u}, expl(X) <_m t_1, \\
& \qquad \neg dang(X) >_{truth} dang(X), min(X) \geq_\Delta 2 \\
\neg move(X) &\leftarrow expl(X) >_m \boldsymbol{u}, expl(X) <_m t_1, \\
& \qquad \neg dang(X) \leq_{truth} dang(X) \\
\neg move(X) &\leftarrow expl(X) >_m \boldsymbol{u}, expl(X) <_m t_1, min(X) <_\Delta 2 \\
expl(X) &\leftarrow min(X), \neg dang(X) \\
\neg dang(X) &\leftarrow \neg rock(X), \neg holes(X) \\
dang(X) &\leftarrow rock(X) \\
dang(X) &\leftarrow holes(X)
\end{aligned} \qquad (33)$$

Consistent explorable areas ($expl(X) \geq_m t_1$) can be directly analysed by the robot. In case of contradictory information about $expl(X)$, the second, third and fourth rule are concerned. $\neg dang(X) >_{truth} dang(X)$ expresses condition (i): it is satisfied when the first component of $\neg dang(X)$ is greater than the second one, i.e., when $\neg dang(X)$ is more strongly supported than $dang(X)$. $min(X) \geq_\Delta 2$ defines condition (ii): it is satisfied when the difference between the components supporting $min(X)$ and $\neg min(X)$ is greater or equal to 2; this value implies that $min(X)$ must have a truth-degree greater or equal to t_2 w.r.t. the knowledge-ordering. The condition $expl(X) >_m \boldsymbol{u}, expl(X) <_m t_1$ ensures that these rules are applied only to resolve inconsistent explorable areas.

We consider the following set of facts in which we add information also about

Paraconsistent Rule-Based Reasoning with Graded Truth Values

area a_3:

$$\begin{aligned}
rock(a_1) &\leftarrow true_2 & rock(a_3) &\leftarrow true_1 \\
\neg rock(a_1) &\leftarrow true_1 & \neg rock(a_3) &\leftarrow true_2 \\
\neg holes(a_1) &\leftarrow true_3 & \neg holes(a_3) &\leftarrow true_3 \\
min(a_1) &\leftarrow true_3 & min(a_3) &\leftarrow true_4 \\
\neg min(a_1) &\leftarrow true_1 & \neg min(a_3) &\leftarrow true_2 \\
\neg rock(a_2) &\leftarrow true_2 \\
\neg holes(a_2) &\leftarrow true_3 \\
min(a_2) &\leftarrow true_3
\end{aligned} \qquad (34)$$

P given by (33) and (34) is an extended program (Definition 5.3). One possible partition is given by $P = P_1 \cup P_2$, with P_1 containing all the facts and:

$$\begin{aligned}
expl(X) &\leftarrow min(X), \neg dang(X) \\
\neg dang(X) &\leftarrow \neg rock(X), \neg holes(X) \\
dang(X) &\leftarrow rock(X) \\
dang(X) &\leftarrow holes(X)
\end{aligned} \qquad (35)$$

and P_2 containing:

$$\begin{aligned}
move(X) &\leftarrow expl(X) \geq_m t_1 \\
move(X) &\leftarrow expl(X) >_m u, expl(X) <_m t_1, \\
 & \quad dang(X) \succ_{truth} dang(X), min(X) \geq_\Delta 2 \\
\neg move(X) &\leftarrow expl(X) >_m u, expl(X) <_m t_1, \\
 & \quad \neg dang(X) \leq_{truth} dang(X) \\
\neg move(X) &\leftarrow expl(X) >_m u, expl(X) <_m t_1, min(X) <_\Delta 2
\end{aligned} \qquad (36)$$

Computing the fixpoint semantics we obtain:

$$\begin{aligned}
T'_{P_1}\uparrow 1 = T'_{P_1}(I_\emptyset) &= \{(rock(a_1), i_{2,1}), ((rock(a_2), f_2), (rock(a_3), i_{1,2}), \\
 & \quad (holes(a_1), f_3), (holes(a_2), f_3), (holes(a_1), f_3), \\
 & \quad (min(a_1), i_{3,1}), (min(a_2), t_3), (min(a_3), i_{4,2})\}; \\
T'_{P_1}\uparrow 2 = T'_{P_1}(T'_{P_1}\uparrow 1) &= T'_{P_1}\uparrow 1 \cup \{(dang(a_1), i_{2,1}), (dang(a_2), f_2), \\
 & \quad (dang(a_3), i_{1,2})\}; \\
T'_{P_1}\uparrow 3 = T'_{P_1}(T'_{P_1}\uparrow 2) &= T'_{P_1}\uparrow 2 \cup \{(ex(a_1), i_{3,2}), (ex(a_2), t_2), (ex(a_3), i_{4,2})\}; \\
T'_{P_1}\uparrow 4 = T'_{P_1}(T'_{P_1}\uparrow 3) &= T'_{P_1}\uparrow 3 = T'_{P_1}\uparrow n_1; \\
T'_{P_2}\uparrow 1 = T'_{P_2}(T'_{P_1}\uparrow n_1) &= T'_{P_1}\uparrow n_1 \cup \{(move(a_1), f_3), (move(a_2), t_3), \\
 & \quad (move(a_3), t_3)\}; \\
T'_{P_2}\uparrow 2 = T'_{P_2}(T'_{P_2}\uparrow 1) &= T'_{P_2}\uparrow 1 = T'_{P_2}\uparrow n_2.
\end{aligned}$$

Area a_2 is directly explorable (since $T'_{P_1}\uparrow 3(ex(a_2)) = t_2$). Therefore we obtain $T'_{P_2}\uparrow 1(move(a_2)) = t_3$. Area a_1 is associated with inconsistent information as to $dang$ and $T'_{P_1}\uparrow n_1(\neg dang(a_1)) = i_{1,2} <_{truth} i_{2,1} = T'_{P_1}\uparrow n_1(dang(a_1))$, which entails $T'_{P_2}\uparrow 1(move(a_1)) = f_3$.

Also for area a_3 there is contradictory information as to $dang$ but:

$$T'_{P_1}\uparrow n_1(\neg dang(a_3)) = i_{2,1} >_{truth} i_{1,2} = T'_{P_1}\uparrow n_1(dang(a_3)), \text{ and}$$
$$T'_{P_1}\uparrow n_1(min(a_3)) = i_{4,2} \geq_\Delta 2.$$

Thus we obtain $T'_{P_2}\uparrow 1(move(a_3)) = f_3$.

The only difference between a_1 and a_3 concerns the information about dangers. Indeed, even though the evaluation of $min(a_1)\geq_\Delta 2$ and $min(a_3)\geq_\Delta 2$ is the same, facts contain stronger evidence supporting the positive component of $dang(a_1)$, violating condition (i).

By resorting to introspection all contradictions are solved, producing as output truth-values f_3, t_3 for the *move* predicate. The intuition applied to handle inconsistent information consisted in comparing the positive and negative components of the truth-degrees obtained for predicates $expl(X)$, $dang(X)$ and $min(X)$, figuring out a strategy to define some thresholds to consider an area directly explorable or at least sufficiently not dangerous and rich in minerals to be explored. ◁

6.3 Negation as Failure and Universal Quantification

Negation as failure [15] is a non-monotonic inference rule used to entail a negative sentence of the form **not** p when the predicate p cannot be derived from inference rules of the program (depending on the formal system used, the sentence **not** p and $\neg p$ can have different interpretations – see, e.g., [38]). In two-valued logic negation as failure can be used to easily model universal quantification; without such an inference rule, expressing universal quantification is more difficult because the whole set of facts must be known in advance. This restriction imposes some limitations when we consider programs as deductive databases (see, e.g., [1]), where logic rules represent reasoning components (called *intensional database - IDB*) using collections of facts (called *extensional database - EDB*) that may change over time; in this context, to define universal quantification, *EDB* should be known before defining *IDB*.

In our logic, the meaning of a sentence of type **not** p can be expressed using a sentence of type $p \in \{u\}$, i.e. we are able to reproduce negation as failure. For more details, we refer to [21].

7 Related Work

Graded reasoning is investigated in many contexts, including modal reasoning (for a survey see, e.g., [47]) and approximate reasoning (see, e.g., [26, 68]), to mention just a few of them. The language presented in this paper aims to support monotonic as well as non-monotonic graded reasoning. Non-monotonicity is particularly useful to reason about incomplete and/or inconsistent information arising from partial views of the system, making conclusions defeasible. Historically, many techniques have been proposed to realize non-monotonic reasoning. In particular, these ones include: drawing negative conclusions from a global/local limitation of the accessible information (e.g., *(Local) Closed World Assumption* [32]); extending theories with additional rules/operators handling non-conclusive information (e.g., *default logic* [61, 62], *circumscription* [52]), filling the lack of knowledge (*autoepistemic logic* [53]) or resolving inconsistencies [54] (*defeasible reasoning*).

Our language is based on a family of many-valued paraconsistent logics to tackle the principle of explosion (*ex contradictione quodlibet*) of the consequence relation; graded truth-values underpin both paraconsistency and paracompleteness. Observing Definition 2 we notice three important properties of our family of many-valued logics: (i) $x = \neg\neg x$ for every x in τ_N; (ii) $x \leq_k y \Rightarrow \neg x \leq_k \neg y$ for every x, y in τ_N; (iii) $\leq_m$ does not satisfy: $x \leq_m y \Rightarrow \neg y \leq_m \neg x$ for every x, y in τ_N. We conclude that $\langle \tau_N, \leq_k, \leq_m, \neg \rangle$ is not a bilattice [39, 40]. Bilattices are truth-value structures with interesting properties for handling inconsistent and incomplete information. The simplest example of a bilattice is the one used in Belnap's four-valued logic [10], based on Kleene's strong three-valued logic [44]. Bilattices have been widely used during the last three decades to define several types of many-valued logics having general unified properties. Among other important results, Ginsberg defined bilattices for truth maintenance systems, first order, default and prioritised default logic [39, 40], allowing for a given theorem prover to reason, at the same time, on these three different domains. Also, Fitting [33] proved that the semantics of a wide family of logic languages, based on interlaced bilattices and devoid of negations in the heads of rules, can be expressed in terms of fixpoint semantics of a specific monotone operator. In our case, even though the current definition of the monotonicity-preservering ordering prevents $\langle \tau_N, \leq_k, \leq_m, \neg \rangle$ from being a bilattice, introspection operators can help introducing the same orderings employed in logics defined on bilattices (e.g., the ones in [39, 40]), as shown in Section 5.4.

The first modern approach to address paraconsistent reasoning is provided in [42] (for English version see [43]). The need for addressing inconsistencies is discussed in many sources, including [35, 36]. For many other references see, e.g. [12]. Among other important applications, many-valued logics represent a natural framework to model inconsistent information using paraconsistent logic programs [17]. One of the first approaches towards this direction dates back to Belnap's four-valued logic [10].

Paraconsistent rough sets [51, 66] are different, in spirit, from our approach and are usually restricted to four truth-values. 4QL [48, 49, 50] is a rule-based language supporting negation both in bodies and heads of rules. It extends proposals included in [51, 66] by allowing disjunction in bodies of rules and a specific form of introspection operators, called external literals. Compared to Belnap's lattices, 4QL uses a different (linear) truth-ordering entailing more intuitive results in practical reasoning. It is important to notice that 4QL can be instantiated from our language by considering τ_1, i.e., it represents a particular instance with only one level for truth and falseness.

Another approach, based on quasi-possibilistic logic [26] also uses pairs of degrees. However, rather than representing degrees of truth and falseness, these values reflect the possibility and the necessity of a property expressed by a formula. Also, unlike our approach, quasi-possibilistic logic addresses paraconsistent reasoning by considering consistent fragments of knowledge bases as well as via consequence relation allowing one to isolate formulas with degree not smaller than a given one. The complexity of the inference problem is co-NP complete. In our approach we concentrate on a family of rule languages enjoying tractability. Moreover, introspection operators we introduced allow to treat inconsistency and uncertainty in a more flexible way. This is important when such disambiguation is dependent on a particular application domain.

A well-known approach based on stable models, Answer Set Programming [13, 38], also allows for negation in bodies and heads of rules. Its default negation can be considered as a particular introspection operator. However, this approach is basically three-valued and does not address inconsistent information. Also, it is not tractable.

Another direction of research on inconsistent knowledge bases depends on repairing inconsistencies and computing consistent answers to queries [5, 6, 11]. However, we apply a different methodology: rather than compute consistent answers to queries by (locally) repairing databases, we provide introspection operators as a tool too disambiguate inconsistencies in a nonmonotonic and highly contextual manner. Grading truth values allows one to compute meaningful answers also when they are inconsistent. Indeed, when an answer's truth value is $\boldsymbol{i}_{ij}$, we know what is the support of its truth and its falsity and comparing i with j provides additional information not present in consistent answers. Also, repair checking is sometimes Π_1^P or even Π_2^P-complete [14].

Fuzzy set-based reasoning [70, 69] is frequently used as a basis for decision making. It belongs to a larger area of quantitative approaches to reasoning, like those concentrated around models involving probability, credibility and plausibility, possibility and necessity, degrees of belief and disbelief (mass distributions), fuzzy truth-degrees (see [46, 59]). *Intuitionistic fuzzy sets* [7, 8] serve to model incomplete information and provide separate grades for truth and falsity. This idea is further developed to *paraconsistent intuitionistic fuzzy sets* [67], applied to model uncertainty, lack of knowledge as well as inconsistency. Understanding of paraconsistent fuzziness [64] is closer to the approach developed in the current paper; however, specific examples are restricted to four truth-values. Also, orderings considered are different and no rule language is developed.

8 Conclusions

In this paper we have presented a family of rule-based languages, RL^N, grounded on paraconsistent many-valued logics with graded truth-values. Truth-degrees are selected from an arbitrarily large finite set of logical values; this aspect empowers modelling to shape information with a desired grade of accuracy, endowing expressiveness with an arbitrarily large set of values for defining truth, falseness and inconsistency. Such truth values appear natural in many real-world scenarios, as indicated in the current paper as well as in [22].

Every RL^N language is founded on a twofold basis: a core language based on the Open World Assumption and a logical machinery, called introspection, is used to confine and analyse inferred information, generating a local/global world closure. The rules of the core language are suitable to enforce non-monotonic reasoning on graded paraconsistent information, allowing for the presence of negative and positive literals both in conclusions and premises.

For a wider scope, we introduce introspection; its flexibility permits, for example, to realize Negation As Failure through instantiation of a particular operator, whereas further defined operators can be used to resolve inconsistent information arising during the computation or to introduce new truth-orderings.

The semantics of programs has been defined in terms of many-valued models and over finite domains. It enjoys deterministic polynomial data complexity for any instantiation of the generic family of paraconsistent many-valued logics.

Acknowledgments

The third author has been supported by the Polish National Science Centre grant 2015/19/B/ST6/02589.

References

[1] S. Abiteboul, R. Hull, and V. Vianu. *Foundations of Databases*. Addison-Wesley, 1995.

[2] J. Alcântara, C.V. Damásio, and L.M. Pereira. An encompassing framework for paraconsistent logic programs. *J. Applied Logic*, 3(1):67–95, 2005.

[3] S. Anderson, N. Bredeche, A. E. Eiben, G. Kampis, and M. van Steen. *Adaptive Collective Systems Herding black sheep*. VU University Amsterdam, 2013.

[4] K.R. Apt, H. A. Blair, and A. Walker. Towards a theory of declarative knowledge. In J. Minker, editor, *Foundations of Deductive Databases and Logic Programming*, pages 89–148. Morgan Kaufmann Publishers Inc., 1988.

[5] M. Arenas, L.E. Bertossi, and J. Chomicki. Consistent query answers in inconsistent databases. In V. Vianu and C.H. Papadimitriou, editors, *Proc. of the 18th ACM SIGACT-SIGMOD-SIGART Symp. on Principles of Database Systems*, pages 68–79. ACM Press, 1999.

[6] M. Arenas, L.E. Bertossi, and J. Chomicki. Answer sets for consistent query answering in inconsistent databases. *TPLP*, 3(4-5):393–424, 2003.

[7] K.T. Atanassov. Intuitionistic fuzzy sets. *Fuzzy Sets and Systems*, 20:87–96, 1986.

[8] K.T. Atanassov. *On Intuitionistic Fuzzy Sets Theory*, volume 283 of *Studies in Fuzziness and Soft Computing*. Springer, 2012.

[9] J. Beal, S. Dulman, K. Usbeck, M. Viroli, and N. Correll. *Organizing the Aggregate: Formal and Practical Aspects of Domain-Specific Languages: Recent Developments*, pages 436–501. Hershey: IGI Global, 2013.

[10] N.D. Belnap. A useful four-valued logic. In G. Epstein and J. M. Dunn, editors, *Modern Uses of Multiple-Valued Logic*, pages 7–37. Reidel Publishing Company, Boston, 1977.

[11] L.E. Bertossi and J. Chomicki. Query answering in inconsistent databases. In J. Chomicki, R. van der Meyden, and G. Saake, editors, *Logics for Emerging Applications of Databases*, pages 43–83. Springer, 2003.

[12] J-J. Bézieau, W. Carnielli, and D.M. Gabbay, editors. *Handbook of Paraconsistency*. College Publications, 2007.

[13] G. Brewka, T. Eiter, and M. Truszczynski. Answer set programming at a glance. *Commun. ACM*, 54(12):92–103, 2011.

[14] J. Chomicki. Consistent query answering: Five easy pieces. In T. Schwentick and D. Suciu, editors, *Database Theory - ICDT 2007, 11th Int. Conf.*, volume 4353 of *LNCS*, pages 1–17. Springer, 2007.

[15] K.L. Clark. Negation as failure. In J. Minker, editor, *Logic and Data Bases*, volume 1, pages 293–322. Plenum Press, New York, London, 1978.

[16] N.C.A. da Costa and E.H Alves. Relations between paraconsistent logic and many-valued logic. *Bulletin of the Section of Logic*, 10(4):185–190, 1981.

[17] C.V. Damásio and L.M. Pereira. A survey of paraconsistent semantics for logic programs. In Ph. Besnard and A. Hunter, editors, *Reasoning with Actual and Potential Contradictions*, pages 241–320. Springer, 1998.

[18] S. de Amo and M.S. Pais. A paraconsistent logic approach for querying inconsistent databases. *International Journal of Approximate Reasoning*, 46:366–386, 2007.

[19] F.L. De Angelis and G. Di Marzo Serugendo. Logic fragments: A coordination model based on logic inference. In *Coordination Models and Languages: 17th IFIP WG 6.1 International Conference, COORDINATION 2015, DisCoTec 2015, Grenoble, France*, pages 35–48, 2015.

[20] F.L. De Angelis and G. Di Marzo Serugendo. *Logic Fragments: Coordinating Entities with Logic Programs*, pages 589–604. Springer International Publishing, Cham, 2016.

[21] F.L. De Angelis, G. Di Marzo Serugendo, and A. Szałas. Foundation of paraconsistent rule-based reasoning with graded truth values. *Technical Report Archive Ouverte University of Geneva*, 2017. https://archive-ouverte.unige.ch/unige:94264.

[22] F.L. De Angelis, B. Dunin-Kęplicz, G. Di Marzo Serugendo, and A. Szałas. Heterogeneous approximate reasoning with graded truth values. In *Proc. of International Joint Conference on Rough Sets*. Springer, 2017. To appear.

[23] G. Di Marzo Serugendo, J.L. Fernandez-Marquez, and F.L. De Angelis. Engineering spatial services: Concepts, architecture, and execution models. In R. Ramanathan and K. Raja, editors, *Handbook of Research on Architectural Trends in Service-Driven Computing*, pages 136–159. IGI Global, 2014.

[24] K. Doets. *From logic to logic programming*. Foundations of computing. MIT Press, Cambridge (Mass.), 1994.

[25] P. Doherty, W. Łukaszewicz, A. Skowron, and A. Szałas. *Knowledge Representation Techniques. A Rough Set Approach*, volume 202 of *Studies in Fuziness and Soft Computing*. Springer-Verlag, 2006.

[26] D. Dubois, S. Konieczny, and H. Prade. Quasi-possibilistic logic and its measures of information and conflict. *Fundamenta Informaticae*, 57(2-4):101–125, 2003.

[27] B. Dunin-Kęplicz, A.L. Nguyen, and A. Szałas. A framework for graded beliefs, goals and intentions. *Fundamenta Informaticae*, 100(1-4):53–76, 2010.

[28] B. Dunin-Kęplicz and A. Strachocka. Paraconsistent argumentation schemes. *Web Intelligence*, 14(1):43–65, 2016.

[29] B. Dunin-Kęplicz and A. Szałas. Taming complex beliefs. *Transactions on Computational Collective Intelligence XI*, LNCS 8065:1–21, 2013.

[30] B. Dunin-Kęplicz and A. Szałas. Indeterministic belief structures. In *Proc. KES-AMSTA 2014: Agents and Multi-agent Systems: Technologies and Applications*, volume 296 of *Advances in Intelligent and Soft Computing*, pages 57–66. Springer, 2014.

[31] E.P. Dunne, A. Hunter, P. McBurney, S. Parsons, and M. Wooldridge. Weighted argument systems: Basic definitions, algorithms, and complexity results. *Artificial Intelligence*, 175(2):457 – 486, 2011.

[32] O. Etzioni, K. Golden, and D.S. Weld. Tractable closed world reasoning with updates. In J. Doyle, E. Sandewall, and P. Torasso, editors, *Proc. KR'94*, pages 178–189. Morgan Kaufmann, 1994.

[33] M. Fitting. Bilattices and the semantics of logic programming. *The Journal of Logic Programming*, 11(2):91 – 116, 1991.

[34] M. Fitting. Fixpoint semantics for logic programming a survey. *Theoretical Computer Science*, 278(1–2):25 – 51, 2002. Mathematical Foundations of Programming Semantics 1996.

[35] D.M. Gabbay and A. Hunter. Making inconsistency respectable: A logical framework for inconsistency in reasoning, part I — a position paper. In Ph. Jorrand and J. Kelemen, editors, *Fundamentals of Artificial Intelligence Research: Int. Workshop FAIR'91*, pages 19–32. Springer, 1991.

[36] D.M. Gabbay and A. Hunter. Making inconsistency respectable: Part 2 - meta-level handling of inconsistency. In M. Clarke, R. Kruse, and S. Moral, editors, *Proc. ECSQARU'93*, volume 747 of *LNCS*, pages 129–136. Springer, 1993.

[37] D.M. Gabbay, P. Smets, and J. Kohlas. *Handbook of Defeasible Reasoning and Uncertainty Management Systems: Volume 5: Algorithms for Uncertainty and Defeasible Reasoning*. Springer, 2000.

[38] M. Gelfond and Y. Kahl. *Knowledge Representation, Reasoning, and the Design of Intelligent Agents - The Answer-Set Programming Approach*. Cambridge University Press, 2014.

[39] M.L. Ginsberg. Multi-valued logics. In *Proc. of AAAI-86*, pages 243–247, 1986.

[40] M.L. Ginsberg. Multivalued Logics: A Uniform Approach to Inference in Artificial Intelligence. *Computational Intelligence*, 4:256–316, 1988.

[41] M. Hajibaba and S. Gorgin. A review on modern distributed computing paradigms: Cloud computing, jungle computing and fog computing. *CIT - Journal of Computing and Information Technology*, 22(2), 2014.

[42] S. Jaśkowski. Rachunek zdań dla systemów dedukcyjnych sprzecznych. *Studia Soc. Sci. Torunensis*, 5:55–77, 1948.

[43] S. Jaśkowski. Propositional calculus for contradictory deductive systems. *Studia Logica*, 24:143–157, 1969.

[44] S. Kleene. Introduction to Metamathematics, 1952.

[45] B. Knaster. Un théorème sur les fonctions d'ensembles. *Annales de la Société Polonaise de Mathématique*, (6):133–134, 1928.

[46] R. Kruse, E. Schwecke, and J. Heinsohn. *Uncertainty and Vagueness in Knowledge Based Systems. Numerical Methods*. Springer-Verlag, 1991.

[47] D. Lassiter. *Graded Modality: Qualitative and Quantitative Perspectives*. Oxford University Press, 2016. to appear.

[48] J. Małuszyński and A. Szałas. Living with inconsistency and taming nonmonotonicity. In O. de Moor et al., editor, *Datalog Reloaded*, volume 6702 of *LNCS*, pages 384–398. Springer, 2011.

[49] J. Małuszyński and A. Szałas. Logical Foundations and Complexity of 4QL, a Query Language with Unrestricted Negation. *Journal of Applied Non-Classical Logics*, 21(2):211–232, 2011.

[50] J. Małuszyński and A. Szałas. Partiality and inconsistency in agents' belief bases. In D. Barbucha et al., editor, *Proc. KES-AMSTA*, volume 252 of *Frontiers of AI and Applications*, pages 3–17. IOS Press, 2011.

[51] J. Małuszyński, A. Szałas, and A. Vitória. Paraconsistent logic programs with four-valued rough sets. In C-C. Chan, J.W. Grzymala-Busse, and W.P. Ziarko, editors, *Proc. RSCTC 2008*, volume 5306 of *LNCS*, pages 41–51. Springer, 2008.

[52] J. McCarthy. Circumscription a form of non-monotonic reasoning. *Artificial Intelligence*, 13(1–2):27 – 39, 1980. Special Issue on Non-Monotonic Logic.

[53] R.C. Moore. Semantical considerations on nonmonotonic logic. *Artif. Intell.*, 25(1):75–94, 1985.

[54] D. Nute. Defeasible logic. In *Handbook of Logic in Artificial Intelligence and Logic Programming*, pages 353–395, 1994.

[55] Ch.H. Papadimitriou and M. Yannakakis. On the complexity of database queries. *Journal of Computer and System Sciences*, 58(3):407 – 427, 1999.

[56] Z. Pawlak. *Rough Sets. Theoretical Aspects of Reasoning about Data*. Kluwer Academic Publishers, Dordrecht, 1991.

[57] Z. Pawlak, L. Polkowski, and A. Skowron. Rough set theory. In B.W. Wah, editor, *Wiley Encyclopedia of Computer Science and Engineering*. John Wiley & Sons, Inc., 2008.

[58] S. G. Pimentel and W. L. Rodi. Belief revision and paraconsistency in a logic programming framework. In A. Nerode, W. Marek, and V. S. Subrahmanian, editors, *Proc. Logic Programming and Non-Monotonic Reasoning*, pages 228–242. MIT Press, Cambridge, MA, 1991.

[59] H. Prade. A quantitative approach to approximate reasoning in rule-based expert systems. In L. Bolc and M.J. Coombs, editors, *Expert System Applications*, pages 199–256. Springer-Verlag, 1988.

[60] G. Rainbolt and S. Dwyer. *Critical Thinking: The Art of Argument*. Cengage Learning, 2014.

[61] R. Reiter. A logic for default reasoning. *Artificial Intelligence*, 13(1–2):81 – 132, 1980. Special Issue on Non-Monotonic Logic.

[62] E. Sandewall. A functional approach to non-monotonic logic. In *Proc. of the 9th IJCAI - Volume 1*, IJCAI'85, pages 100–106. Morgan Kaufmann Publishers Inc., 1985.

[63] A. Szałas. How an agent might think. *Logic Journal of the IGPL*, 21(3):515–535, 2013.

[64] A. Szałas. Symbolic explanations of generalized fuzzy reasoning. In R. Neves-Silva, G.A. Tshirintzis, V. Uskov, R.J. Howlett, and L.C. Jain, editors, *Smart Digital Futures 2014*, page 7–16. IOS PRESS, 2014.

[65] A. Tarski. A lattice-theoretical fixpoint theorem and its applications. *Pacific J. Math.*, 5(2):285–309, 1955.

[66] A. Vitória, J. Maluszyński, and A. Szałas. Modeling and reasoning in paraconsistent rough sets. *Fundamenta Informaticae*, 97(4):405–438, 2009.

[67] H. Wang and R. Sunderraman. A data model based on paraconsistent intuitionistic fuzzy relations. In M-S. Hacid, N.V. Murray, Z.W. Ras, and S. Tsumoto, editors, *ISMIS*, volume 3488 of *LNCS*, pages 669–677. Springer, 2005.

[68] Y.Y Yao and T.Y. Lin. Graded rough set approximations based on nested neighborhood systems. In *Proc. 5th European Congress on Intelligent Techniques and Soft Computing*, volume 1, pages 196–200, 1997.

[69] L. Zadeh. From computing with numbers to computing with words – from manipulation of measurements to manipulation of perceptions. *Int. J. Appl. Math. Comput. Sci.*, 12(3):307–324, 2002.

[70] L.A. Zadeh. Fuzzy sets. *Information and Control*, 8:333–353, 1965.

[71] F. Zambonelli and M. Mamei. Spatial computing: An emerging paradigm for autonomic computing and communication. In *Proc. of the 1st Int. IFIP Conf. on Autonomic Communication*, WAC'04, pages 44–57. Springer, 2005.

Paracomplete Logic Kl — Natural Deduction, its Automation, Complexity and Applications

Alexander Bolotov*
University of Westminster, London, UK.
a.bolotov@westminster.ac.uk

Daniil Kozhemiachenko
Lomonosov Moscow State University, Russian Federation.
kodaniil@yandex.ru

Vasilyi Shangin[†]
Lomonosov Moscow State University, Russian Federation.
shangin@philos.msu.ru

Abstract

In the development of many modern software solutions where the underlying systems are complex, dynamic and heterogeneous, the significance of specification-based verification is well accepted. However, often parts of the specification may not be known. Yet reasoning based on such incomplete specifications is very desirable. Here, paracomplete logics seem to be an appropriate formal setup: opposite to Tarski's theory of truth with its principle of bivalence, in these logics a statement and its negation may be both untrue. An immediate result is that the law of excluded middle becomes invalid. In this paper we show how to apply an automatic proof searching procedure for the natural deduction formulation of the paracomplete logic Kl to reason about incomplete information systems. We provide an original account of complexity of natural deduction systems, which leads us closer to the efficiency of the presented proof search algorithm. Moreover, we have turned the assumptions management into an advantage by showing the applicability of the proposed technique to assume-guarantee reasoning.

The authors are grateful to the referees for their fruitful advice which greatly improved the paper.

*The first author thanks the University of Westminster for supporting his Sabbaticals in January-June 2017.

[†]The third author is supported by Russian Foundation for Humanities, grant 16-03-00749 *Logical-epistemic problems of knowledge representation*.

1 Introduction

1.1 Problem Setup — Reasoning with Incomplete Information

The significance of formal specification with subsequent verification in Software Engineering is well accepted. It is quite standard to classify two types of verification — the explorative approach (with model checking as its typical representative) and the deductive one. In this paper, we are interested in specification-based deductive verification. Incorporating the notation of [22], we represent the task of deductive verification, DV, of a system Sys with its specification $Spec$ by the following signature:

$$DV :: Sys \times Spec \longrightarrow B \times [Proof]$$

where the Boolean result of deductive verification based on theorem proving is either a proof that a system satisfies a given property or a demonstration that no proof can be established — $B \times [Proof]$.

Traditionally, specifications follow two general classical principles: completeness and consistency. The former assumes that a statement — ϕ — about the specification $Spec$, or its negation — $\neg\phi$ — is true. Under the latter, a member of $Spec$ — ϕ — and its negation — $\neg\phi$ — cannot be both true. As a consequence, completeness and consistency govern reasoning applied to such formal specifications — regardless whether it is model checking, or deductive reasoning, classical or not (temporal, modal, etc). Inconsistent or incomplete specifications are results of rejecting one of (or both) the principles mentioned above. Here paraconsistent and paracomplete logics come into play [1]. We strongly believe such cases are of more interest when one considers the development of modern software solutions with their underlying complex, dynamic and heterogenous systems. This definitely applies to such areas as clouds or robotics, where software systems are defined to work in a complex, dynamic and heterogeneous environment. However, our thorough research of software engineering formal methods literature has not shown many works where authors tackle incomplete specifications. Perhaps one of the main reasons for this is the lack of deductive methods for such a non-standard setting. Among few of those that address this problem are [17, 18, 29, 28, 43]. However, none of the techniques proposed in these papers, gives any account of automation, and, to our believe, they are not open to an easy way of automation. Below we identify the following cases relevant to the account of incompleteness of specifications:

(a) the problem to simplify complex software requirements in incomplete specifications,

(b) a typical integration task of various resources, which could be the problem of forming of heterogeneous resources into networks or clouds, or component-based system engineering where components are not fully specified, or

(c) the problem of finding assumptions in assume-guarantee reasoning in the context of incomplete specifications.

We argue that reasoning following the classical principles is unsuitable when one deals with incomplete information as such reasoning validates the excluded middle (bivalence) principle[1]. Informally, it says that a truth-value of any statement is either true or false. One may also say that under the given specification, any statement is fully defined. In case this principle does not hold (i.e. some statement is not fully defined, or, in other words, we have here a truth-value gap) we are required to propose both specification languages of high level and corresponding deductive methods.

In the paper, we deal with the paracomplete logic where the law of excluded middle and some other classical laws are invalid. For example, one can not deduce $A \supset B$ from $\neg A \vee B$. We confine ourselves to the sentential reasoning, and at the moment, abstract from temporal or dynamic dimensions. We assume the language of the paracomplete logic **Kl** [1] (which is called PComp in [38] and [8]) is the language to write incomplete specifications. One must find efficient deductive techniques to deal with the reasoning which corresponds to clauses (a)–(c) above. When we choose among available formalisms and methods of deduction which use assumptions, we believe it is reasonable to take into account the following considerations.

(i) Efficient management of assumptions: tracking assumptions, making sure the assumptions occur in the proof with some reasons, not randomly, and to managing the way how assumptions occur in the proof.

(ii) Availability of automated proof searching that enables implementation.

(iii) Potential to reuse and adapt deductive techniques and proof searching for to the various kinds of formal specifications; for instance, an option to deal with incomplete or inconsistent specifications as well as with specifications which are both incomplete and inconsistent, or an option to extend our results to such richer formalisms as dynamic systems.

We argue now that natural deduction seems to be an appropriate framework if one wants to satisfy (i)–(iii).

---

[1] Although the principles of bivalence and excluded middle are different, in the paper we will use them as synonyms.

In the framework of automated reasoning, provers are usually based upon either resolution method, analytic tableaux or Fitch-style natural deduction (see, for example [31, 34, 27, 35, 30, 40] for provers based on classical natural deduction). Automated theorem proving in many-valued logic is usually conducted via the method of analytic tableaux, which provides a useful way of constructing counter-models to non-provable formulas. Our target is different. We are interested in a proof technique that explicitly constructs proofs. Considering automated natural deduction for the three-valued paracomplete logic, we use a Fitch-style calculus. Furthermore, in contrast to analytic tableaux we aim at developing a proof search algorithm which constructs explicit proofs for tautologies, not only counter-models for non-provable formulae.

In the rest of this introductory section we first provide some argumentation in favour of our choice of the underlying logic, **Kl**, to reason about incomplete specifications and then we will analyse possible approaches to build a desired natural deduction proof technique.

1.2 Choice of Logic — Paracomplete Logic Kl as a Many-valued Logic

The logic **Kl** was originally introduced by Avron [1]. In Avron's paper, **Kl** plays an important role in the definition of a family of paracomplete natural logics (though, Avron himself doesn't use the term 'paracomplete'; in his terminology, such logics are logics with the 'undefined' interpretation). This family includes strong Kleene's logic, logic of partial functions LPF and Łukasiewicz's 3-valued logic. In the following we highlight the importance of **Kl** in the context of these logics and explore some arguments in favour of our natural deduction presentation in comparison to Carnielli's approach to systematization of finite many-valued logics [14].

Considering strong Kleene's logic we note its famous property of not having theorems. As in our paper we want to tackle both derivations and proofs we find this logic inappropriate for our purposes. The logic of partial functions, LPF, has an additional unary connective (so to speak, another kind of negation), and for this reason we consider LPF being not in the scope of our research. Finally, Łukasiewicz's 3-valued logic lacks the deduction theorem which is crucial for our proof searching procedure, where the deduction theorem is incorporated in the form of the implication introduction rule. However, these arguments only justify our choice of logic and do not mean that proof searching procedures for these logics won't be a task for a future research that may be carried out. We note that these systems can be tackled, for example, in the spirit of [16].

It is also worth to analyse here Sette and Carnielli's weakly-intuitionistic logic I^1

[39]. Note that I^1 is both a counterpart of Sette's maximal paraconsistent logic P^1 and an extension of strong three-valued Kleene's logic $K3$. First, we observe that I^1 is different from our target logic, **Kl**, with respect to the validity of the formulae representing the law of excluded middle. In particular, $A \vee \neg A$ is invalid in **Kl** for an arbitrary A while in I^1 it is invalid for an atomic A only. Another difference lies within the matrix definitions of both implication and negation. The valuation of $A \supset B$ when $A = 1$ and $B = f$ is 'f' in the semantics of **Kl** but it is '0' in the semantics of I^1. The valuation of $\neg A$ when $A = f$ is 'f' in the semantics of **Kl** but it is '0' in the semantics of I^1. Last, not least, logics I^1 and **Kl** don't coincide in respect to their notions of theoremhood. For example, only a restricted version of $\neg\neg A \supset A$ is valid in I^1 while in **Kl** this law holds without restriction.

1.3 Choice of Deductive Approach — Natural Deduction for Kl

Considering the nature of our approach to build a natural deduction system, it is worth to compare it to [14, 13] and [1]. Carnielli's approach essentially uses the idea of signed formulae. Following this approach, a prefix of a formula used in a tableaux or natural deduction, would have been the corresponding matrix evaluation for this formula. For instance, given three values 1, T, 0 we would formulate in a sequent calculus (and with a slight adaptation, a natural deduction system) exactly three rules for each signed formula, $1 \supset$, $T \supset$, and $0 \supset$. A different approach was adapted by Avron, (see in particular [1], p. 277, footnote 2) and we follow this approach. Also note that some natural deduction system can be routinely extracted from Avron's paper, however, it would be considerably different from our natural deduction construction.

Natural deduction allows not only to establish that the proof one wants to achieve exists, but it also makes it very explicit. Both a natural deduction system for **Kl** and its proof searching (as presented in [10]) satisfy (i)–(ii). To the best of our knowledge, no other (direct) natural deduction system for **Kl** has been proposed. We believe this can be explained by the following. Both paraconsistent and paracomplete logics are likely to be analysed with some philosophical motivation and, therefore, in computer science framework the preferential methods have been Hilbert-style systems [24], analytic tableaux [12] or sequent-style calculi [20]. The only exception here is [4], where a kind of natural deduction system for a paracomplete setting is introduced. However, one can't consider such an approach as a direct method of deduction as it is based on the translation techniques to Isabelle [26]. We remind the reader that the system PCont, the dual of **Kl** (named as three-valued paraconsistent logic [1, p.278]), deals with inconsistent systems. Both a natural deduction system for PCont and its proof searching can be found in [7] and [33]. Consequently, the latter paper together with the results of this paper, imply that our choice of natural deduction satisfies (iii).

The novelty of our paper is in the following. First, we show the way an automated natural deduction for **Kl** in [10] is applicable to reason about incomplete information systems. We also provide proofs of some statements previously announced and presented without proof in [8], thus significantly improving and expanding the latter. We present substantial conceptual and methodological considerations, introduce new technical concepts, refine and polish proofs and provide several examples. Finally, we provide an account of complexity and efficiency.

The paper is organised as follows. To make reading self-contained, §2 reviews the formulation of the natural deduction system for classical propositional logic. Next, §3 introduces the underlying logic **Kl**, its axiomatics, and natural deduction calculus, it also contains sketches of results in [10]. In §4 we discuss the complexity account. This follows by an overview of the proof searching procedure and the core algorithm in §5. We also provide a detailed example of the algorithmic proof search. The next section, §6, classifies problems to which natural deduction is applicable as a tool for deductive verification. We also present a methodology for solving some of the problems of the type (a)–(c) mentioned above and consider typical scenarios of component-based system synthesis and assume-guarantee technique. Finally, §7 contains the conclusion and the roadmap to future work.

2 Natural Deduction System for Classical Propositional Logic — CPL$_{\text{ND}}$

We commence with the review of the natural deduction system for classical propositional logic, **CPL$_{\text{ND}}$**. The natural deduction system presented below is a standard Fitch-style natural deduction system. One of the specifics of this type of natural deduction systems is that a derivation is defined in a linear format, opposite to Gentzen-style, or tree-like format. The rules of derivation are traditionally divided into elimination and introduction rules — the former allow to decompose compound formulae while the latter allow to construct compound formulae. Recall that in constructing proofs in natural deduction systems, we introduce assumptions. In some cases we need to discard alive assumptions. To indicate that a natural deduction rule with the conclusion C discards the last alive assumption, A, and all formulae $A, \ldots, C^-$ (where C^- is the formula preceding C), we will use a standard abbreviation, $[A]C$.

The system **CPL$_{\text{ND}}$** has the following rules of derivation.

Elimination rules:

$$\wedge_{el_1} \frac{A \wedge B}{A}, \quad \wedge_{el_2} \frac{A \wedge B}{B}, \quad \neg_{el} \frac{\neg\neg A}{A}, \quad \vee_{el} \frac{\neg A, A \vee B}{B}, \quad \supset_{el} \frac{A \supset B, A}{B}$$

Introduction rules:

$$\wedge_{in} \frac{A, B}{A \wedge B}, \quad \vee_{in_1} \frac{A}{A \vee B}, \quad \vee_{in_2} \frac{B}{A \vee B}, \quad \supset_{in} \frac{[A]B}{A \supset B}, \quad \neg_{in} \frac{[A]B, [A]\neg B}{\neg A}$$

Definition 1 (**CPL**$_{ND}$-derivation). An **CPL**$_{ND}$-derivation of a formula A from a set of formulae Γ is a finite sequence of formulae, each of which is either a member of Γ (an assumption) or is derived from the previous formulae by one of the elimination or introduction rules. In case $\supset in$ or $\neg in$ are used, all formulae from the last alive assumption to the resulting formula should be discarded from the derivation.

Definition 2 (Proof). A *proof* in the system **CPL**$_{ND}$ is a derivation with the empty set of alive assumptions.

Note that this and the other definitions of a derivation in natural deduction systems in the paper are 'standard' textbook ones and are sufficient for the purposes of the paper. For a more accurate definition of proof see [42].

It has been shown that **CPL**$_{ND}$ is sound and complete [5]. The natural deduction system for paracomplete logic **Kl** given in §3 is a modification of the **CPL**$_{ND}$ which reflects its characteristic features.

3 Paracomplete Logic Kl and its natural deduction calculus Kl$_{ND}$

Here, to make the presentation self-contained, we define fully the logic **Kl**, its syntax and semantics, give a full set of rules of the natural deduction calculus and provide an account of its metatheoretical properties — the main results of [10].

3.1 Kl and Its Axiomatics

Kl is a propositional logic with the infinite number of propositional symbols $Prop = p, q, r, \ldots$ and the semantics assigning to each propositional symbol from $Prop$ one of the three truth-values 1 — 'true' (the designated one), 0 — 'false', and 1/2 — 'none' such that $A \vee B = \max(A, B)$ and $A \wedge B = \min(A, B)$ The matrices for connectives are defined as follows.

$\vee$	1	1/2	0		$\wedge$	1	1/2	0		$\supset$	1	1/2	0		p	$\neg p$
1	1	1	1		1	1	1/2	0		1	1	1/2	0		1	0
1/2	1	1/2	1/2		1/2	1/2	1/2	0		1/2	1	1	1		1/2	1/2
0	1	1/2	0		0	0	0	0		0	1	1	1		0	1

It is the presence of the third truth assignment, 1/2, that makes the calculus paracomplete allowing to identify the cases of incompleteness (uncertainty, etc.) and thus allowing to consider systems with incomplete information, (see §6 for details). Often the properties and the flavour of the logic become more transparent in the axiomatic construction. For these reasons we export the axiomatic of **Kl** from [1] which is a subset of the set of axioms of classical propositional logic.

1. $(A \supset B) \supset ((B \supset C) \supset (A \supset C))$
2. $A \supset (A \vee B)$
3. $A \supset (B \vee A)$
4. $(A \supset C) \supset ((B \supset C) \supset ((A \vee B) \supset C))$
5. $(A \wedge B) \supset A$
6. $(A \wedge B) \supset B$
7. $(C \supset A) \supset ((C \supset B) \supset (C \supset (A \wedge B)))$
8. $A \supset (B \supset A)$
9. $(A \supset (B \supset C)) \supset ((A \supset B) \supset (A \supset C))$
10. $((A \supset B) \supset A) \supset A$
11. $\neg(A \vee B) \supset (\neg A \wedge \neg B)$
12. $(\neg A \wedge \neg B) \supset \neg(A \vee B)$
13. $\neg(A \wedge B) \supset (\neg A \vee \neg B)$
14. $(\neg A \vee \neg B) \supset (\neg A \wedge \neg B)$
15. $\neg(A \supset B) \supset (A \wedge \neg B)$
16. $(A \wedge \neg B) \supset \neg(A \supset B)$
17. $\neg\neg A \supset A$
18. $A \supset \neg\neg A$
19. $\neg A \supset (A \supset B)$

The only rule of inference of **Kl** is modus ponens: from A and $A \supset B$ infer B.

Note that this axiomatics reflects the failure of the law of excluded middle so, for example, $p \vee \neg p$ is not provable in this system. We also observe that Axiom 19 is equivalent to $(B \supset \neg A) \supset ((B \supset A) \supset \neg B)$ [1, p.288].

3.2 Kl$_{ND}$ — Natural Deduction Calculus for Kl.

Definition 3 (Kl$_{ND}$-derivation). A *derivation* in the system Kl$_{ND}$ is a finite non-empty sequence of formulae where each formula is an alive assumption or is derived from the previous ones by one of the following Kl$_{ND}$-rules.

Elimination rules:

$$\wedge_{el_1} \frac{A \wedge B}{A}, \quad \wedge_{el_2} \frac{A \wedge B}{B}, \quad \neg\wedge_{el} \frac{\neg(A \wedge B)}{\neg A \vee \neg B}, \quad \neg_{el} \frac{\neg\neg A}{A},$$

$$\neg\vee_{el_1} \frac{\neg(A \vee B)}{\neg A}, \quad \neg\vee_{el_2} \frac{\neg(A \vee B)}{\neg B}, \quad \supset_{el} \frac{A, A \supset B}{B},$$

$$\neg\supset_{el_1} \frac{\neg(A \supset B)}{A}, \quad \neg\supset_{el_2} \frac{\neg(A \supset B)}{\neg B}, \quad \vee_{el} \frac{A \vee B, [A]C, [B]C}{C},$$

$$\vee\supset_{el_1} \frac{(A \vee B) \supset C}{A \supset C}, \quad \vee\supset_{el_2} \frac{(A \vee B) \supset C}{B \supset C}.$$

Introduction rules:

$$\wedge_{in} \frac{A, B}{A \wedge B}, \quad \neg\wedge_{in} \frac{\neg A \vee \neg B}{\neg(A \wedge B)}, \quad \vee_{in_1} \frac{A}{A \vee B}, \quad \vee_{in_2} \frac{B}{A \vee B},$$

$$\neg\vee_{in} \frac{\neg A, \neg B}{\neg(A \vee B)}, \quad \supset_{in} \frac{[A]B}{A \supset B}, \quad \neg\supset_{in} \frac{A, \neg B}{\neg(A \supset B)},$$

$$\neg_{in} \frac{B}{\neg\neg B}, \quad \supset_p \frac{[A \supset B]A}{A}, \quad Kl_{\neg in} \frac{A, \neg A}{B}.$$

Definition 4 (Proof). A *proof* in the system Kl$_{ND}$ is a derivation with the empty set of alive assumptions.

Let us give now a short, but indicative, example of proof for $((p \wedge q) \vee (p \wedge r)) \supset (p \wedge (q \vee r))$ in the described natural deduction calculus. Below we use the square brackets to indicate which formulae are discarded from the proof. Thus, the application of $\vee_{el}$ rule to $p \wedge (q \vee r)$ on step 12 requires to discard all formulae from the assumption $p \wedge q$ on step 2 up to $p \wedge (q \vee r)$ on step 6 and all formulae from the assumption $p \wedge r$ on step 7 up to formula $p \wedge (q \vee r)$ on step 11. Finally, applying $\supset_{in}$ rule to $p \wedge (q \vee r)$ on step 12, we obtain the desired derivation for $((p \wedge q) \vee (p \wedge r)) \supset (p \wedge (q \vee r))$ discarding all formulae from the last alive assumption $(p \wedge q) \vee (p \wedge r)$ on step 1, up to the conclusion of this rule.

$$\begin{array}{l}
\left[\begin{array}{l}
1.\ (p \wedge q) \vee (p \wedge r) \text{ --- assumption} \\
\left[\begin{array}{l}
2.\ p \wedge q \text{ --- assumption} \\
3.\ p \text{ --- } \wedge_{el_1}\colon 2 \\
4.\ q \text{ --- } \wedge_{el_2}\colon 2 \\
5.\ p \vee r \text{ --- } \vee_{in_1}\colon 4 \\
6.\ p \wedge (q \vee r) \text{ --- } \wedge_{in}\colon 3,\ 5
\end{array}\right. \\
\left[\begin{array}{l}
7.\ p \wedge r \text{ --- assumption} \\
8.\ p \text{ --- } \wedge_{el_1}\colon 7 \\
9.\ r \text{ --- } \wedge_{el_2}\colon 7 \\
10.\ q \vee r \text{ --- } \vee_{in_2}\colon 9 \\
11.\ p \wedge (q \vee r) \text{ --- } \wedge_{in}\colon 8,\ 10
\end{array}\right. \\
12.\ p \wedge (q \vee r) \text{ --- } \vee_{el}\colon 1,\ 6,\ 11 \\
\end{array}\right. \\
13.\ ((p \wedge q) \vee (p \wedge r)) \supset (p \wedge (q \vee r)) \text{ --- } \supset_{in}\colon 12
\end{array}$$

As the derivation does not have any alive assumptions it is also a proof for $((p \wedge q) \vee (p \wedge r)) \supset (p \wedge (q \vee r))$.

The presented natural deduction calculus is sound and complete, below $\models$ stands for **Kl$_{ND}$** logical consequence:

Theorem 1. $\Gamma \vdash_{\textbf{Kl}_{\textbf{ND}}} A \iff \Gamma \models A$ [10]

Theorem 1 semantically justifies applications of derivations based on natural deduction. We argue that the natural deduction style of a proof is a powerful technique to tackle formal specification/verication software engineering problems. It is particularly important when there is an obvious need to not only establish if a desired proof exists but to also explicitly show how the proof (for some desired property) is constructed. Let us give here an informal insight into the way how the proof in natural deduction is formed. Assume we have a specification S, and would like to investigate if some statement $B \in S$ holds under some set of assumptions Γ. In this introductory case, we have a task to derive B from the specification S, given the assumptions Γ. Following the specifics of natural deduction, now, we either simplify compound formulae in the proof by elimination rules, or synthesise formulae by introduction rules. In the subsequent sections we present a proof search algorithm which guides such applications of elemination/introduction rules in an efficient manner, and give an annotated example.

4 Complexity of Natural Deduction

Convention 1. We will, according to Reckhow [41] and Pelletier [32], say that a given calculus is *natural* if it allows to use arbitrary assumptions in the proofs of theorems and incorporates the deduction theorem as one of its rules.

It is evident then that systems $\mathbf{CPL_{ND}}$ and $\mathbf{Kl_{ND}}$ are "natural" systems.

Now we will consider three sound and complete classical propositional natural calculi, namely, $\mathbf{CPL_{ND}}$ described in §2, nested deduction Frege system and general deduction Frege system described in [11].

Definition 5 (Nested deduction Frege system — $nd\mathscr{F}$). The system $nd\mathscr{F}$ is characterised by the following constraints:

- it has two rules of derivation:

    1. mp_n — $\dfrac{A \quad A \supset B}{B}$

    2. dr_n — $\dfrac{[A]B}{A \supset B}$ (where A is the last alive assumption)

- it uses a finite number of axiom schemas.

An $nd\mathscr{F}$-derivation of a formula A from a set of formulae Γ is a finite sequence of formulae, each of which is

- either a member of Γ (an assumption), or

- an instance of an axiom schema or

- is derived from previous formulae by mp_n or dr_n. In case dr_n rule is used, all formulae from the last alive assumption up to (but not including) formula A should be discarded from derivation.

We write $\Gamma \vdash^{nd\mathscr{F}}_n A$ if there is an $nd\mathscr{F}$-derivation of A from Γ with the length of no more than n formulae. We use here and below, in the formulation of the rules, a lower index n to indicate that these are derivations and rules in Nested deduction Frege system.

Definition 6 (General deduction Frege system — $d\mathscr{F}$). Derivations in $d\mathscr{F}$ have steps presented as sequents of the form $\Gamma \mapsto A$ with Γ being a set of formulae and A being a formula. We use below, in the formulation of the rules, a lower index g to indicate that these are derivations and rules in General deduction Frege system. There are four rules of derivation in $d\mathscr{F}$:

1. $\mapsto A$, where A is an instance of an axiom schema of a consistent and complete set of axioms taken, for example, from [21].

2. $\{A\} \mapsto A$, where A is either a member of Γ or an assumption

3. $mp_g \quad \dfrac{\Gamma_1 \mapsto A \quad \Gamma_2 \mapsto A \supset B}{\Gamma_1 \cup \Gamma_2 \mapsto B}$

4. $dr_g \quad \dfrac{\Gamma \mapsto B}{\Gamma \setminus \{A\} \mapsto A \supset B}$

We define a $d\mathscr{F}$-derivation of a formula A from a set of formulae Γ as a finite sequence of sequents, each of which is obtained by one of the rules above, and the last sequent is $\Gamma \mapsto A$. We write $\Gamma \vdash^{d\mathscr{F}}_{n} A$ to indicate that there is a $d\mathscr{F}$-derivation of $\Gamma \mapsto A$ containing no more than n sequents.

We will now prove some theorems related to speedups (better performance) of these calculi.

Theorem 2. $\Gamma \vdash^{\mathbf{CPL_{ND}}}_{n} C \Rightarrow \Gamma \vdash^{nd\mathscr{F}}_{O(n)} C$

Proof. We prove the theorem by induction on the number of steps n of $\mathbf{CPL_{ND}}$-derivation.

The proof splits into two cases depending on how the last formula C in $\mathbf{CPL_{ND}}$-derivation was inferred.

Case 1 C is an assumption or a member of Γ. Then an $nd\mathscr{F}$-derivation consists of only one formula — C itself.

Case 2 C was derived by a rule of a derivation. We will now show that the conclusion of every $\mathbf{CPL_{ND}}$-rule can be derived from its premises in $nd\mathscr{F}$ in a constant number of steps. This is obvious in case of rules $\wedge_{in}$, $\wedge_{el_1}$, $\wedge_{el_2}$, $\vee_{in_1}$, $\vee_{in_2}$, $\supset_{el}$, and $\neg_{el}$. Next, we substitute each application of $\supset_{in}$ with dr_n and each application of $\supset_{el}$ with mp_n.

In case C was derived by $\neg_{in}$, let $C = \neg A$. We have a $\mathbf{CPL_{ND}}$-derivation of length n. We proceed as follows. We will also provide necessary comments explaining how steps of the proofs are derived.

$$\vdots$$
$$\left[\begin{array}{l} A \text{ --- the last alive assumption} \\ \vdots \\ B \\ \vdots \\ \neg B \end{array} \right.$$
$$\neg A \text{ --- } \neg_{in} \text{ applied to } B \text{ and } \neg B$$

The $nd\mathscr{F}$-derivation will be as follows:

$$
\begin{array}{l}
\vdots \\
\left[\begin{array}{l} A \text{ --- the last alive assumption} \\ \vdots \\ B \\ \vdots \\ \neg B \\ \vdots \\ B \wedge \neg B \text{ --- in a constant number of steps using } A \supset (B \supset (A \wedge B)) \end{array}\right. \\
A \supset (B \wedge \neg B) \text{ --- } dr_n \\
\left[\begin{array}{l} A \text{ --- assumption} \\ \vdots \\ B \text{ --- in a constant number of steps using } (B \wedge \neg B) \supset B \end{array}\right. \\
A \supset B \text{ --- } dr_n \\
\left[\begin{array}{l} A \text{ --- assumption} \\ \vdots \\ \neg B \text{ --- in a constant number of steps using } (B \wedge \neg B) \supset B \end{array}\right. \\
A \supset \neg B \text{ --- } dr_n \\
\vdots \\
\neg A \text{ --- in a constant number of steps using } (A \supset B) \supset ((A \supset \neg B) \supset \neg A)
\end{array}
$$

$\square$

Theorem 3. $\Gamma\vdash^{nd\mathscr{F}}_{n} C \Rightarrow \Gamma\vdash^{\mathbf{CPL_{ND}}}_{O(n)} C$

Proof. To prove this we simply note that assumptions and formulae of Γ in an $nd\mathscr{F}$-derivation become, respectively, assumptions and formulae of Γ in a **CPL$_{\mathbf{ND}}$**-derivation. Similarly, each application of mp_n becomes an application of $\supset_{el}$, and each application of dr_n becomes an application of $\supset_{in}$. We substitute all instances of axiom schemata with their proofs which are constructed in a constant number of steps. $\square$

Theorem 4. *Assume, there is a* **CPL$_{\mathbf{ND}}$**-*derivation of C from Γ in n steps. Then, there is a $d\mathscr{F}$-derivation of C from Γ in $O(n)$ steps.*

The proof is similar to the proof of Theorem 2.

Theorem 5. *Assume, there is a $d\mathscr{F}$-derivation of C from Γ in n steps. Then, there is a $\mathbf{CPL_{ND}}$-derivation of C from Γ in $O(n^2)$ steps.*

Proof. For this theorem, let $\bigwedge\limits_{i=1}^{m} A_i$ be a conjunction of m formulae A_i which are ordered arbitrarily. Also, if Γ is a finite set of formulae, then $\bigwedge(\Gamma_1 \cup \Gamma_2)$ is a conjunction of its members ordered and associated arbitrarily.

It suffices to prove that if $\{A_1, \ldots, A_m\} \mapsto C$ has a $d\mathscr{F}$-proof of the length n, then $\bigwedge\limits_{i=1}^{m} A_i \supset C$ has a $\mathbf{CPL_{ND}}$-proof of the length $O(n^2)$. The proof of this theorem is similar to the proof of THEOREM 4 in [11]. We substitute each sequent in a $d\mathscr{F}$-derivation with its relevant formula and then fill in the gaps. Now we show that all gaps can be filled in $O(n)$ steps. The proof splits into four cases depending on how the sequent in a $d\mathscr{F}$-derivation was inferred.

Case 1 The sequent has the form $\mapsto A$, where A is an instance of an axiom schema. Then we substitute it with the formula A which can be proved in a constant number of steps (since A is a tautology).

Case 2 The sequent has the form $A \mapsto A$, where A is an assumption. We substitute it with the formula $A \supset A$ which has a $\mathbf{CPL_{ND}}$-derivation of a constant number of steps.

Case 3 The sequent was inferred by mp_g. Then it has the form $\Gamma_1 \cup \Gamma_2 \mapsto B$ and there are also two sequents prior to it, namely, $\Gamma_1 \mapsto A \supset B$ and $\Gamma_2 \mapsto A$. It suffices to show that $\bigwedge(\Gamma_1 \cup \Gamma_2) \supset B$ can be inferred from $\bigwedge \Gamma_1 \supset (A \supset B)$ and $\bigwedge \Gamma_2 \supset A$. The derivation proceeds as follows.

$$\vdots$$
$$\bigwedge \Gamma_1 \supset (A \supset B)$$
$$\vdots$$
$$\bigwedge \Gamma_2 \supset A$$

$$\left[\begin{array}{l} \bigwedge(\Gamma_1 \cup \Gamma_2) \text{ — assumption} \\ \vdots \\ \bigwedge \Gamma_1 \text{ applying } \wedge_{el} \text{ to } \bigwedge(\Gamma_1 \cup \Gamma_2) \text{ and then } \wedge_{in} \\ \vdots \\ \bigwedge \Gamma_2 \text{ applying } \wedge_{el} \text{ to } \bigwedge(\Gamma_1 \cup \Gamma_2) \text{ and then } \wedge_{in} \\ A \supset B \text{ — applying } \supset_{el} \text{ to } \bigwedge \Gamma_1 \supset (A \supset B) \text{ and } \bigwedge \Gamma_1 \\ A \text{ — } \supset_{el} \text{ applying } \supset_{el} \text{ to } \bigwedge \Gamma_2 \supset A \text{ and } \bigwedge \Gamma_2 \\ B \text{ — } \supset_{el} \text{ applying } \supset_{el} \text{ to } A \supset B \text{ and } A \end{array}\right.$$
$\bigwedge(\Gamma_1 \cup \Gamma_2) \supset B$ — applying $\supset_{in}$ to B

If there are m formulae in $\bigwedge(\Gamma_1 \cup \Gamma_2)$, it can be shown by induction on m that $\bigwedge \Gamma_1$ and $\bigwedge \Gamma_2$ can be inferred from $\bigwedge(\Gamma_1 \cup \Gamma_2)$ in $O(m)$ steps via $\wedge_{el}$ and $\wedge_{in}$ rules. Since $m \leqslant n$, we can infer both $\bigwedge \Gamma_1$ and $\bigwedge \Gamma_2$ in $O(n)$ steps which proves the case.

Case 4 The sequent was inferred by dr_g. Then it has the form $\Gamma \mapsto A \supset B$ and there is also the sequent $\Gamma \setminus \{A\} \mapsto B$ prior to it. It suffices to show that $\bigwedge(\Gamma \setminus \{A\}) \supset (A \supset B)$ can be inferred from $\bigwedge \Gamma \supset B$ in $O(n)$ steps. We proceed as follows.

$$\begin{array}{l} \vdots \\ \bigwedge \Gamma \supset B \\ \left[\begin{array}{l} \bigwedge(\Gamma \setminus \{A\}) \text{ — assumption (if } A \notin \Gamma) \\ \left[\begin{array}{l} A \text{ — assumption} \\ \vdots \\ \bigwedge \Gamma \text{ — from } \bigwedge(\Gamma \setminus \{A\}) \text{ and } A \text{ using } \wedge_{el} \text{ and } \wedge_{in} \\ B \text{ — } \supset_{el} \end{array}\right. \\ A \supset B \text{ — } \supset_{in} \end{array}\right. \\ \bigwedge(\Gamma \setminus \{A\}) \supset (A \supset B) \text{ — } \supset_{in} \end{array}$$

If there are m formulae in Γ, then it can be shown by induction on m that $\bigwedge \Gamma$ can be derived in $O(m)$ steps from A and $\bigwedge(\Gamma \setminus \{A\})$. Since $m \leqslant n$, we infer $\bigwedge(\Gamma \setminus \{A\}) \supset (A \supset B)$ from $\bigwedge \Gamma \supset B$ in $O(n)$ steps which proves the case. □

We will prove theorems showing the speedup of **Kl$_{ND}$** over the axiomatic calculus for **Kl** presented above which we will further designate as **Kl$_{Ax}$**.

Definition 7 (proof simulation, speedup). A proof system S_1 simulates S_2 with an $f(n)$ increase in number of steps if for any S_2-proof of formula A in n steps there is a proof of A in S_1 in $O(f(n))$ steps. We say that S_2 provides at most $f(x)$ speedup w.r.t. S_1 if S_1 simulates S_2 with an increase of number of steps in $f(x)$.

Theorem 6. $\mathbf{Kl_{ND}}$ *linearly simulates* $\mathbf{Kl_{Ax}}$.

The proof of this theorem is straightforward since $\mathbf{Kl_{ND}}$ has modus ponens rule ($\supset_{el}$) and all axioms have $\mathbf{Kl_{ND}}$-proofs of a constant length. The details are left to the reader.

As it had been shown in [1], $\mathbf{Kl_{Ax}}$ is sound and complete (and so is $\mathbf{Kl_{ND}}$). This means that we can add $\supset_{in}$ rule to $\mathbf{Kl_{Ax}}$ thus transforming it into the natural calculus which we will further denote as $\mathbf{Kl_{Axn}}$. One can see that $\mathbf{Kl_{Axn}}$ is actually a nested deduction Frege system for $\mathbf{Kl}$ — hence our use of the index n for this system.

Theorem 7. $\mathbf{Kl_{ND}}$ *and* $\mathbf{Kl_{Axn}}$ *linearly simulate one another.*

Proof. It is obvious that $\mathbf{Kl_{ND}}$ linearly simulates $\mathbf{Kl_{Axn}}$ since all axioms can be proven in a constant number of steps while instances of modus ponens and $\supset_{in}$ as well as assumptions in a $\mathbf{Kl_{Axn}}$-derivation become, without loss of generality, instances of $\supset_{el}$, $\supset_{in}$ and assumptions in a $\mathbf{Kl_{ND}}$-derivation.

Next we show that $\mathbf{Kl_{Axn}}$ linearly simulates $\mathbf{Kl_{ND}}$. It suffices to show that we can obtain conclusions of all rules of derivation from their premises in a constant number of steps. We will prove the cases of $\vee_{el}$ and $\supset_p$ rules only.

$\vee_{el}$ $\mathbf{Kl_{ND}}$-proof has the following form:

$$
\begin{array}{l}
A \vee B \\
\left[\begin{array}{l} A \text{ — assumption} \\ \vdots \\ C \end{array}\right. \\
\left[\begin{array}{l} B \text{ — assumption} \\ \vdots \\ C \end{array}\right. \\
C - \vee_{el}
\end{array}
$$

We proceed here as follows.

$A \vee B$

$$\left[\begin{array}{l} A \text{ --- assumption} \\ \vdots \\ C \end{array}\right.$$

$A \supset C$ --- $\supset_{in}$ --- to C

$$\left[\begin{array}{l} B \text{ --- assumption} \\ \vdots \\ C \end{array}\right.$$

$B \supset C$ --- $\supset_{in}$ --- to C

$\vdots$

C --- in a constant number of steps using $(A \supset C) \supset ((B \supset C) \supset ((A \vee B) \supset C))$

$\supset_p$ **Kl$_{ND}$**-proof has the following form:

$$\left[\begin{array}{l} A \supset B \text{ --- assumption} \\ \vdots \\ A \end{array}\right.$$

A --- $\supset_p$

We proceed here as follows.

$$\left[\begin{array}{l} A \supset B \text{ --- assumption} \\ \vdots \\ A \end{array}\right.$$

$(A \supset B) \supset A$ --- $\supset_{in}$ to A

$\vdots$

A --- in a constant number of steps using $((A \supset B) \supset A) \supset A$

$\square$

Theorem 8. *If there is an **Kl$_{ND}$**-proof of A of length n, then there is a **Kl$_{Ax}$**-proof of A of length $O(n \cdot \alpha(n))$ with α being the inverse Ackermann function.*

Theorem 7 shows that **Kl$_{Axn}$** linearly simulates **Kl$_{ND}$**. The former is, by virtue of definition, a nested deduction Frege system. This means that we can apply the result of Buss and Bonet (Main Theorem 6 proved in [11]) which states that nested deduction Frege systems provide a near-linear speedup over Frege systems (and $\mathbf{Kl_{Ax}}$ is a Frege system).

The above observations at least give us an idea how some fragments of proof search technique perform from the point of view of complexity. It also gives us grounds to expect that similar developments can be applied to the case of non-classical logics.

Concluding this section, we note that Theorems 6-8 provide us with an important tool of checking whether or not our proof-search algorithm presented in the next section is optimal. We know that natural deduction for **Kl** gives at most a near-linear speedup over Frege system for **Kl**. This means that we can test our algorithm on known examples that are hard for proof systems like analytical tableaux but have Frege proofs in a polynomial number of steps. If an algorithmic proof happens to be near-linearly faster than Frege proof, the algorithm works optimally at least on these examples. On the other hand, if the algorithm proves these examples polynomially slower than Frege system does, we will learn that it is not optimal. Finally, if the algorithm proves these formulae in an exponential number of steps, we will find out that it is considerably less effective than Frege systems.

Concluding this section, we observe that these general theoretical discussions should be supported by the study of the implementation of the proof searching algorithm, which forms part of our future work.

5 Algorithmic Proof Searching for $\mathbf{Kl_{ND}}$

The potential of the application of a logical deductive method to some practical specification/verification problem depends on the existence of the proof search and its efficiency. Here, we review the proof search technique for the logic **Kl** originally defined in [10]. To keep the presentation self-contained, we describe the procedures behind this search and then present the searching algorithm referring an interested reader to [10] for full details.

The proof search strategy is *goal-directed*, which means that it runs over two sequences: list_proof and list_goals. The former is a list of formulae in the proof, while the latter is a list of goals to be reached. A specific goal, the last goal in list_goals, is called *current_goal*. We identify three types of goals in list_goals.

Definition 8 (Types of goals). A goal, $G_i, 0 \leq i \leq n$, occurring in list_goals = $\langle G_0, G_1, \ldots, G_n \rangle$, is one of the following

- G_i is a formula B, or

- G_i is of the form $[A]B$, i.e, it is a derivation of a formula B from an assumption A, or

- G_i is a contradiction, i.e. two contradictory **Kl** formulae, A and $\neg A$. In this case we will write $G_i = \bot$.

In our introductory case, we have a task to derive B from the specification S, given the assumptions Γ, or $S, \Gamma \Vdash B$. Note that here and below we distinguish the task of establishing that B is derivable from S, Γ (abbreviated by $S, \Gamma \Vdash B$) from the statement that such a derivation exists ($S, \Gamma \vdash B$). We will see that our searching procedures transform derivation tasks. Thus, list_proof = $\{A | A \in S \cup \Gamma\}$ and list_goals = B. Now, if our goal is not reachable, we either simplify compound formulae in list_proof invoking applicable elimination rules, or manage list_goals to generate new goals, applying introduction rules only when and if necessary. Each step of the algorithmic proof is associated with the *current_goal*. In our introductory case *current_goal* = B. Checking the reachability of the current goal, one of the core procedures, is introduced below and is based on Definition 8.

Definition 9 (Current goal reachability). Current goal, G_n, $0 \leq n$, occurring in list_goals = $\langle G_0, G_1, \ldots, G_n \rangle$, is reached if

- G_n is some formula B and there is a formula $A \in$ list_proof such that A is not discarded and $A = B$ or

- G_n is of the form $[A]B$ and there is a derivation of B from a non-discarded assumption A, or

- G_n is a contradiction and there are two contradictory formulae, $A \in$ list_proof and $\neg A \in$ list_proof.

5.1 Proof-Searching Algorithm Kl$_{\text{NDALG}}$

Now we are ready to introduce the notion of an *algo-derivation* and searching procedures involved.

Definition 10 (Algo-derivation Kl$_{\text{NDALG}}$). A **Kl** *algo-derivation*, abbreviated as Kl$_{\text{NDALG}}$, is a pair (list_proof, list_goals) whose construction is determined by the searching Procedures (1)–(4) outlined below.

5.1.1 Searching Procedures

Searching Procedures below update list_proof, list_goals or both of them.

Procedure (1) Here we follow one of the main ideas of natural deduction proof to simplify structures of obtained formulae: list_proof is updated due to an applicable elimination rule. If we find a formula, or two formulae, which can serve as premises of one of these rules, the rule is enforced and the sequence list_proof is updated by the relevant conclusion.

Procedure (2) We apply Procedure (2) when Procedure (1) terminates but the current goal is not reached. Here we distinguish two subroutines.

Procedure (2.1). This procedure applies when the current goal is not reached. Analysing the structure of the current goal we update list_proof and list_goals, respectively, by new goals or new assumptions. Let list_proof $= P_1, \ldots P_k$ and list_goals $= G_1, \ldots, G_n$, where G_n is the current goal. A new goal, G_{n+1}, is generated by applying the subroutines (2.1.1)–(2.1.9) below which depends on the possible structures of G_n:

$$G_n = A \wedge B | A \vee B | A \supset B | \neg(A \wedge B) | \neg(A \vee B) | \neg(A \supset B) | L | \neg\neg A | \bot | [C]A$$

where A, B are any formulae, $L \in Lit$ and $[C]A$ states for the derivation of A from assumption C. The rules below have structure $\Gamma \Vdash \alpha \longrightarrow \Gamma' \Vdash \alpha'$ indicating that the rule modifies some given derivation task $\Gamma \Vdash \alpha$ to a new derivation task $\Gamma' \Vdash \alpha'$. The procedures depend on the structure of the current goal: they tackle the cases when the current goal is a compound **Kl** formula. The last type of the goal — $\bot$ — is managed as follows.

(2.1.1) $\quad \Gamma \Vdash \Delta, A \wedge B \quad \longrightarrow \quad \Gamma \Vdash \Delta, A \wedge B, B, A$

In the above, Procedure (2.1.1) splits the current conjunctive goal into two conjuncts.

(2.1.2.1) $\quad \Gamma \Vdash \Delta, A \vee B \quad \longrightarrow \quad \Gamma \Vdash \Delta, A \vee B, A$
(2.1.2.2) $\quad \Gamma \Vdash \Delta, A \vee B \quad \longrightarrow \quad \Gamma \Vdash \Delta, A \vee B, B$

Procedure (2.1.2) tackles a disjunctive goal $A \vee B$ setting each disjunct as a separate goal. We need some clarifications for Procedures (2.1.2.1) and (2.1.2.2) to explain the way how we avoid infinite loops invoking a dedicated marking technique. For the former, when the current goal is disjunction, we try to reach the left disjunct (Procedure 2.1.2.1), and if we fail this subroutine is deleted and we apply Procedure (2.1.2.2). Similarly, if the latter fails we delete this subroutine and terminate the whole Procedure (2.1.2).

(2.1.3) $\quad \Gamma \Vdash \Delta, A \supset B \quad \longrightarrow \quad \Gamma, A \Vdash \Delta, A \supset B, B$

Procedure (2.1.3) tackles $A \supset B$ as a goal, requiring to update list_proof with A and list_goals with B.

$$
\begin{aligned}
(2.1.4) & \quad \Gamma \Vdash \Delta, \neg(A \supset B) & \longrightarrow & \quad \Gamma \Vdash \Delta, \neg(A \supset B), A, \neg B \\
(2.1.5) & \quad \Gamma \Vdash \Delta, \neg(A \vee B) & \longrightarrow & \quad \Gamma \Vdash \Delta, \neg(A \vee B), \neg A, \neg B \\
(2.1.6) & \quad \Gamma \Vdash \Delta, \neg(A \wedge B) & \longrightarrow & \quad \Gamma \Vdash \Delta, \neg(A \wedge B), \neg A \vee \neg B
\end{aligned}
$$

Procedures (2.1.4)–(2.1.6) transform negative compound goals $\neg(A \supset B)$, $\neg(A \vee B)$, $\neg(A \wedge B)$ into $A \wedge \neg B$, $\neg A \wedge \neg B$ and $\neg A \vee \neg B$, respectively.

$$
\begin{aligned}
(2.1.7.1) & \quad \Gamma \Vdash \Delta, F & \longrightarrow & \quad \Gamma \Vdash \Delta, F, \bot \\
(2.1.7.2) & \quad \Gamma \Vdash \Delta, F & \longrightarrow & \quad \Gamma, F \supset p \wedge \neg p \Vdash \Delta, [F \supset p \wedge \neg p] F
\end{aligned}
$$

Here F is a literal (a proposition or its negation) or $F = A \vee B$ and variable p should be fresh.

In the paracomplete setting, we also reason by refutation. When the current goal is not reached, and it is either a literal or disjunction (not reached by Procedure (2.1.2)) we first look for the contradictions in the proof — Procedure (2.1.7.1) which sets up a new goal, $\bot$.

If no contradictions are found, then we turn into the refutation style proof applying Procedure (2.1.7.2). The application of this procedure is linked to the rule $\supset_p$ which allows us to introduce to list_proof the derivation of F from $F \supset p \wedge \neg p$, the goal of Procedure (2.1.7.2), once this goal is achieved.

$$
\begin{aligned}
(2.1.8) & \quad \Gamma \Vdash \Delta, \neg\neg A & \longrightarrow & \quad \Gamma \Vdash \Delta, \neg\neg A, A \\
(2.1.9) & \quad \Gamma \Vdash \Delta, [A]B & \longrightarrow & \quad \Gamma, A \Vdash \Delta, B
\end{aligned}
$$

Procedure (2.1.9) corresponds to our interpretation of assumptions — the given goal $[A]B$ means to infer B from the assumption A, hence we update list_goals by B and list_proof by the assumption A.

Marking Various marking routines are applied to prevent infinite looping during the search. For example, applying Procedure (2.1) we mark literals and formulae of the type $A \vee B$. This mark serves proof by refutation — in reaching relevant goals we cannot any longer apply reasoning by refutation. Also, applying Procedure (2.1.7.2), we mark the assumption that this procedure defines, and these marks indicate that this assumption, and any formula which is derivable from it, cannot serve as source of a new goal, i.e. Procedure (2.2) described below, is not applicable (otherwise, the proof search will enter an infinite loop). Our example in §5.2 will further clarify how marking technique affects proof search.

Procedure (2.2). Here we analyse compound disjunctive and implicative formulae (but not of the type $A \supset \bot$, where $\bot$ is any contradiction, as explained above) contained in list_proof in order to find sources for new goals. If one of these formulae is found then its structure determines the generation of a new goal.

$$
\begin{array}{ll}
(2.2.1) & \Gamma, A \vee B \Vdash \Delta, C \longrightarrow \Gamma \Vdash \Delta, [A]C \quad \Gamma \Vdash \Delta, [B]C \\
(2.2.2) & \Gamma, A \supset B \Vdash \Delta, C \longrightarrow \Gamma \Vdash \Delta, C, A
\end{array}
$$

Procedure (3) Here we check the application of Definition 9. If the current goal G_n, $(n > 0)$ is reached, we delete G_n from the sequence list_goals and set G_{n-1} as the current goal. If the current goal G_0 is reached, we delete G_0 from the sequence list_goals.

Procedure (4) This is a search for an applicable introduction rule. It is based on the association of Procedures (2.1.1)–(2.1.8) with correspondent introduction rules presented below.

$$
\begin{array}{llll}
\text{Procedure (2.1.1)} & \longrightarrow & \wedge_{in} & \text{Procedure (2.1.5)} & \longrightarrow & \neg\vee_{in} \\
\text{Procedure (2.1.2.1)} & \longrightarrow & \vee_{in_1} & \text{Procedure (2.1.6)} & \longrightarrow & \neg\wedge_{in} \\
\text{Procedure (2.1.2.2)} & \longrightarrow & \vee_{in_2} & \text{Procedure (2.1.7)} & \longrightarrow & \supset_p \\
\text{Procedure (2.1.3)} & \longrightarrow & \supset_{in} & \text{Procedure (2.1.8)} & \longrightarrow & \neg_{in} \\
\text{Procedure (2.1.4)} & \longrightarrow & \neg\supset_{in} & & &
\end{array}
$$

Note that Procedure (4) represents the unique specifics of our searching technique — it makes the application of the introduction rules completely determined by the analysis of the structure of the current goal (reached) and its preceding goals.

5.1.2 Algorithm KI$_{ND}$ALG

Let us introduce the following abbreviations

- 'G_{cur}' abbreviates the current goal in list_goals
- 'last(list_goals)' returns the last element of list_goals, and
- list_goals — G_n deletes the last formula, G_n, from list_goals.

Now, based on the procedures (1)-(4) we introduce the proof search algorithm **KI$_{ND}$ALG** making comments to the steps of the algorithm within the '//'.

(0) list_proof(), list_goals(), *go to* (1) // initialisation of sequences list_proof and list_goals//

(1) Given a task $\Gamma \Vdash G_0$, $G_{cur} = G_0$ // initialisation of G_{cur} as G_0

$(\Gamma \neq \emptyset) \longrightarrow$ (list_proof = Γ, list_goals = G_0, $go\ to\ (2))$// when Γ is not empty update list_proof with formulae of Γ and list_goals with G_0//

ELSE

list_goals = G_0, $go\ to\ (2)$ // when there are no given assumptions in Γ only update list_goals with G_0//

(2) Procedure $(3)(G_{cur}) = \mathbf{true}$ //checks the reachability of the current goal//

(2a) IF Reached $(G_{cur}) = \mathbf{true}$, then list_goals = list_goals $- G_{cur}$ // when the current goal is reached it is deleted from list_goals, the new current goal is the previous goal in list_goals//

THEN

IF $(G_{cur} = G_0) \longrightarrow go\ to\ (6a)$ // If the initial goal is reached, go to the terminating step//

ELSE

$G_{cur} \neq G_0$, then $G_{cur} = last(\mathsf{list\_goals})\ go\ to\ (3)$//If the reached goal is not the initial goal determine a new G_{cur} as the last goal in $G_{cur} = last(\mathsf{list\_goals})$ and proceed with the relevant introduction rule//

(2b) IF Reached $(G_{cur}) = \mathbf{false}$, THEN $go\ to\ (4)$//If (G_{cur}) is not reached proceed further with elimination rules//

(3) Procedure $(4)(\langle \mathsf{list\_proof}, \mathsf{list\_goals}\rangle) = \mathbf{true}$ //apply a relevant introduction rule// $go\ to\ (2)$.

(4) Procedure $(1)(\langle \mathsf{list\_proof}\rangle) = \mathbf{true}$ //apply elimination rules//

(4a) Elimination rule is applicable, $go\ to\ (2)$ ELSE

(4b) if there are no compound formulae in list_proof to which an elimination rule can be applied, $go\ to\ (5)$.

(5) Procedure $(2)((\langle \mathsf{list\_proof}, \mathsf{list\_goals}\rangle) = \mathbf{true})$ // update list_proof and list_goals based on the structure of G_{cur}//

(5a) Procedure $(2.1)(\langle \mathsf{list\_proof}, \mathsf{list\_goals}\rangle) = \mathbf{true})$ //analysis of the structure of G_{cur}//

go to

(2) ELSE

(5b) Procedure (2.2)($\langle$list_proof, list_goals$\rangle$) = **true**) //searching for the sources of new goals in list_proof//

go to

(2) ELSE

(5c) if all compound formulae in list_proof are marked, i.e. have been considered as sources for new goals, *go to* (6b).

(6) Terminate **Kl$_{ND}$ALG**.

(6a) The desired ND proof has been found. EXIT.

(6b) No ND proof has been found, counterexample found. EXIT.

5.2 Algo-Proof Example

As an example of an algorithmic ND proof we apply **Kl$_{ND}$ALG** as an attempt to prove the following formula

$$(\natural) \quad (p \supset q) \supset (\neg p \vee q)$$

Note that this formula is valid in the classical setting and is not in the setting of paracomplete logic. Its validity would have led to the validity of $\neg p \vee p$ as shown in the following: if ($\natural$) is valid then so would be

$$(\sharp) \quad (p \supset p) \supset (\neg p \vee p),$$

now since $p \supset p$ is valid, by modus ponens, we would derive $\neg p \vee p$.

This is an indicative formula which contains a disjunctive constraint and as the reader will see in the proof attempt, all core procedures related to disjunctive formulae are invoked.

Let us introduce a useful concept of *algo-step* which will make the understanding of the application of proof search easier. Recall that an algo-proof is a pair (list_proof, list_goals). At each step of the application of the procedures described above we have the sequences list_proof and list_goals of specific lengths, say i and j. Let's abbreviate them by (list_proof$_i$, list_goals$_j$), respectively, and let list_proof = $B_1, \ldots, B_i$ and list_goals = $G_0, \ldots, G_j$, where G_j is the last goal, that is it is the current goal. So an algo-step is the task to find a derivation $B_1, \ldots, B_i \Vdash G_0, \ldots, G_j$. Thus, the algo-proof for some formula C (with no given assumptions) commences with the first algo-step $\Vdash G_0$, where $G_0 = C$.

Now, for the input $(p \supset q) \supset (\neg p \vee q)$, we commence the proof with the main goal, $(p \supset q) \supset (\neg p \vee q)$. According to the classical search Procedure (2.1.3), the antecedent of the main goal, $p \supset q$, becomes the new assumption, and its consequent, $\neg p \vee q$ — the new goal, $G_1 = \neg p \vee q$. So the next algo-step would be $p \supset q \Vdash \neg p \vee q$. In the representation of the algo-proof below we will have the following columns indicating, in order, a step of the algo proof (step), so the abbreviation $as0$ stands for the first algo-step, formulae in the proof (list_proof), an annotation explaining how a formula appears in list_proof, and finally, a list of the goals (list_goals).

step	list_proof	annotation	list_goals
$as0$			$G_0 = (p \supset q) \supset (\neg p \vee q)$
$as1$	1. $p \supset q$	assumption	$G_0, G_1 = \neg p \vee q$

The current goal $G_1 = \neg p \vee q$ cannot be reached so we apply Procedure (2.1.2.1) and set a new goal $G_2 = \neg p$, hence list_goals $= \neg p \vee q, \neg p$. Since $\neg p$ is not reachable, we delete it from list_goals, and applying Procedure (2.1.2.2) we set a new goal $G_2 = q$, hence list_goals $= \neg p \vee q, q$. Since q is not reachable, we delete it from list_goals. At this stage we have failed to reach both disjuncts of G_1. Hence we start the refutation, applying first Procedure (2.1.7.1). Thus, we set up a new goal $G_2 = \bot$. This new goal, in turn, is not derivable, so we delete G_2 from list_goals, and apply Procedure (2.1.7.2) adding (a) a new assumption, $(\neg p \vee q) \supset (r \wedge \neg r)$ and (b) a new goal, $[(\neg p \vee q) \supset (r \wedge \neg r)] \neg p \vee q$. Here we mark the assumption on step 'as3' indicating that it should not be subject to Procedure (2.2).

Note that, according to the definition of Procedure (2.1.7.1), in the $r \wedge \neg r$ constraint, the variable r should be fresh.

step	list_proof	annotation	list_goals
$as0$			$G_0 = (p \supset q) \supset (\neg p \vee q)$
$as1$	1. $p \supset q$	assumption	$G_0, G_1 = \neg p \vee q$
$as2$			$G_0, G_1, G_2 = \bot$
$as3$	2. $(\neg p \vee q) \supset (r \wedge \neg r)$	assumption	$G_0, G_1, G_2 = $ $[(\neg p \vee q) \supset (r \wedge \neg r)] \neg p \vee q$

Now, looking for the applicable elimination rule, we notice that $\vee \supset_{el_1}$ and $\vee \supset_{el_2}$ are applicable to formula 2, thus we derive steps 3 and 4.

step	list_proof	annotation	list_goals
$as0$			$G_0 = (p \supset q) \supset (\neg p \vee q)$
$as1$	1. $p \supset q$	assumption	$G_0, G_1 = \neg p \vee q$
$as2$			$G_0, G_1, G_2 = \bot$
$as3$	2. $(\neg p \vee q) \supset (r \wedge \neg r)$	assumption	$G_0, G_1, G_2 =$ $[(\neg p \vee q) \supset (r \wedge \neg r)]\neg p \vee q$
$as4$	3. $\neg p \supset (r \wedge \neg r)$	$\vee \supset_{el_1}$	G_0, G_1, G_2
$as5$	4. $q \supset (r \wedge \neg r)$	$\vee \supset_{el_2}$	G_0, G_1, G_2

At this stage, the current goal, G_2 is not reachable, so we look for the sources of new goals analysing compound formulae in the proof applying Procedure (2.2.2). Hence by analysing step 1, we set up a new goal $G_3 = p$. This is not reachable, so we again apply Procedure (2.1.7.1), setting a new goal, $G_4 = \bot$.

step	list_proof	annotation	list_goals
$as0$			$G_0 = (p \supset q) \supset (\neg p \vee q)$
$as1$	1. $p \supset q$	assumption	$G_0, G_1 = \neg p \vee q$
$as2$			$G_0, G_1, G_2 = \bot$
$as3$	2. $(\neg p \vee q) \supset (r \wedge \neg r)$	assumption	$G_0, G_1, G_2 =$ $[(\neg p \vee q) \supset (r \wedge \neg r)]\neg p \vee q$
$as4$	3. $\neg p \supset (r \wedge \neg r)$	$\vee \supset_{el_1}$	G_0, G_1, G_2
$as5$	4. $q \supset (r \wedge \neg r)$	$\vee \supset_{el_2}$	G_0, G_1, G_2
$as6$			$G_0, G_1, G_2, G_3 = p, G_4 = \bot$

The current goal, G_4 is not reachable, so we delete it and applying Procedure (2.1.7.2) we set up a new assumption, $p \supset (s \wedge \neg s)$ and a new goal, $G_4 = [p \supset (s \wedge \neg s)]p$. Note that s is a fresh variable.

step	list_proof	annotation	list_goals
$as0$			$G_0 = (p \supset q) \supset (\neg p \vee q)$
$as1$	1. $p \supset q$	assumption	$G_0, G_1 = \neg p \vee q$
$as2$			$G_0, G_1, G_2 = \bot$
$as3$	2. $(\neg p \vee q) \supset (r \wedge \neg r)$	assumption	$G_0, G_1, G_2 =$ $[(\neg p \vee q) \supset (r \wedge \neg r)]\neg p \vee q$
$as4$	3. $\neg p \supset (r \wedge \neg r)$	$\vee \supset_{el_1}$	G_0, G_1, G_2
$as5$	4. $q \supset (r \wedge \neg r)$	$\vee \supset_{el_2}$	G_0, G_1, G_2
$as6$			$G_0, G_1, G_2, G_3 = p, G_4 = \bot$
$as7$	5. $p \supset (s \wedge \neg s)$	assumption	$G_0, G_1, G_2, G_3, G_4 =$ $[p \supset (s \wedge \neg s)]p$

At this stage the searching algorithm terminates as there are no procedures to apply and all formulae in list_proof are marked: as a result, we still have goals to reach, however, no more elimination rules can be applied, we do not have any more formulae in list_proof that could give us new goals and, once again, introduction rules are only applied as a result of Procedure (4), which is now void. Note that although formula 2 in list_proof is compound, it was set up as an assumption due to Procedure (2.1.7.2), hence it is marked and is not considered as a source for new goals. These marks are carried on for the derivable formulae on steps 3 and 4.

Now, looking at the list_proof we can extract the counterexample as follows. Formula $p \supset (s \wedge \neg s)$ means that p has the value f while $q \supset (r \wedge \neg r)$ means q has the value f. Under these values for p and q, formula $(p \supset q) \supset (\neg p \vee q)$ also takes the value f.

5.3 Correctness

The following theorems reflect the metatheoretical properties of the above algorithm [10].

Theorem 9. $\mathbf{Kl_{ND}}_{ALG}$ *terminates for any input formula.*

Theorem 9 guarantees that for any input formula for the $\mathbf{Kl_{ND}ALG}$ the sequences list_proof and list_goals are finite.

Theorem 10. $\mathbf{Kl_{ND}ALG}$ *is sound.*

Theorem 10 ensures that every formula for which an ND proof is constructed according with $\mathbf{Kl_{ND}ALG}$ is valid.

Theorem 11. $\mathbf{Kl_{ND}ALG}$ *is complete.*

Theorem 11 establishes that for every valid formula, A, $\mathbf{Kl_{ND}ALG}$ finds a $\mathbf{Kl_{ND}}$ proof.

Altogether, theorems 9, 10 and 11 imply the following fundamental property of our algorithm:

Theorem 12. *For any input formula A, the $\mathbf{Kl_{ND}ALG}$ terminates either building up a $\mathbf{Kl_{ND}}$-proof for A or providing a counter-model.*

Let us now present some important observations on the proof search and on some of its core and important features.

As in the other ND calculi, in constructing an ND derivation, we are allowed to introduce arbitrary formulae as new assumptions. Note that for many researchers, this opportunity to introduce arbitrary formulae as assumptions has been a point of great scepticism regarding the very possibility of the automation of the proof search. It is true that without the proof search technique assumptions can be introduced arbitrarily. However, due to the goal-directed feature of the presented algorithm, any assumption that appears in the proof is well justified serving a specific target. Let us emphasise that we also turned the assumptions management into an advantage showing the applicability of the proposed technique to assume-guarantee reasoning as shown in §6.

We also note that, according to the algorithm, the order in which assumptions are discharged, is the reverse order to their introduction into the proof.

Finally, introduction rules that have been another point of scepticism concerning the automation of natural deduction, in our algorithm are completely determined. Namely, the reachability of the current goal and the type of the previous goal determine the relevant introduction rule. Also, though $\neg_{in}$ rule of our system $\mathbf{Kl_{ND}}$, in general, allows to derive any formula from the contradiction, the application of this rule is strictly determined by the searching procedures. Therefore, the formula that we derived from a contradiction is always the one mentioned in list_goals.

6 Applications in Specification-Based Verification

Our development of the automated reasoning technique tackles at this stage only the propositional basis. However, even at this more or less simple level, we argue that it can significantly contribute in specification-based verification.

6.1 Methodology of applying $\mathbf{Kl_{ND_{ALG}}}$ as Deductive Verification

Here we draw several routes of applying natural deduction enhanced with the proof search.

Below we list relevant problems and indicate the relevant methodology of their solution based on natural deduction.

1. To find if a system satisfies some desired property

    1.1. obtain the specification of the system, *Spec*, with some core properties, Γ and the specification of the desired property, say, B;

    1.2. find an ND derivation $\Gamma \Vdash B$.

2. To reason about requirements

    2.1. specify the requirements;

2.2. for a given requirement B, find if there is an ND proof of B;

2.3. drop such requirements since they are valid regardless of a system.

3. To check the consistency of a given system

   3.1. obtain the specification, *Spec*, of a system and run the searching technique to obtain the contradiction, i.e. setting up the goal $\bot$;

   3.2. if $\bot$ has been reached, the given system is inconsistent.
   We will show in the present section how this works in the framework of component-based system.

4. To look for non-explicit assumptions, apply the presented **Kl** proof search algorithm, and the procedures will automatically upgrade list_proof with new assumptions.
   We will show in this section how this works in finding assumptions in the framework of assume-guarantee reasoning.

In the following subsections we tackle problem setting 3–4 leaving the discussion of problems 1–2 for the conclusion.

6.2 Component-Based Systems

Here we justify the application of the natural deduction to component-based system assembly. Thus, we aim to apply the searching algorithm **Kl**$_{\text{NDALG}}$ as the deductive verification technique for a component system.

As an example, let us consider a simple component system interpreted in The Grid Component Model (GCM) based on Fractal [3].

Let our component system, Sys have the following specification *Spec*. Components interact together by being bound through interfaces. The system has four core components P, Q, R and S. Let p, q, r and s represent properties that core components, P, Q, R and S are bound to the system (one that should be always available and should not be "touched").

Consider as an example the following set of global requirements and their formalisation:

- whenever P is bound R should be bound: $p \supset r$

- whenever P is not bound S should be bound: $\neg p \supset s$.

- whenever Q is bound both R and S should not be bound: $q \supset (\neg r \land \neg s)$.

- Q should be bound to the system: q.

Consider now the verification task to establish if the above configuration of components is consistent. We commence the proof (see below) by the given conditions of the *Spec* and set up the goal of the procedure to derive the contradiction, abbreviated in the proof annotation below as $\bot$. If the contradiction is derivable, then we would have been able to see its sources tracing the proof backwards. Otherwise, the *Spec* would have been shown consistent.

We commence the proof by listing all four given formulae on steps 1-4. From 3 and 4 by eliminating implication we derive $\neg r \wedge \neg s$ and then eliminating conjunction from the latter, derive steps 6 and 7. We have not reached the goal $\bot$. By Procedure (2.2) we analyse compound formulae in the proof. Thus, analysing formula on step 1 we apply Procedure (2.2.2) and set up p, the antecedent of 1, as the new goal.

step	list_proof	annotation	goals
$as0$			$\bot$
$as1$	1. $p \supset r$	given	$\bot$
$as2$	2. $\neg p \supset s$	given	$\bot$
$as3$	3. $q \supset (\neg r \wedge \neg s)$	given	$\bot$
$as4$	4. q	given	$\bot$
$as5$	5. $\neg r \wedge \neg s$	$3, 4 \supset_{el}$	$\bot$
$as6$	6. $\neg r$	$5, \wedge_{el}$	$\bot$
$as7$	7. $\neg s$	$5, \wedge_{el}$	$\bot$
$as8$			$\bot, p$

The current goal, p has not been reached — we apply Procedure (2.1.7.1) setting up the new goal, $\bot$. If we derive $\bot$, then by $\mathbf{Kl}_{\neg in}$ we would be able to derive the desired p. However, $\bot$ is not reachable so we delete it and apply Procedure (2.1.7.2) so the new assumption is $p \supset (t \wedge \neg t)$ (where $t \wedge \neg t$ is the formula $\bot$ in the formulation of Procedure (2.1.7.2)) and our task is now to derive p. Since we cannot do it we apply Procedure (2.2.2) and analyse formula 2 putting its antecedent, $\neg p$, as the new goal.

Again, as it is reachable we apply Procedures (2.1.7.1) and (2.1.7.2) consequently. The latter procedure sets up the new assumption $\neg p \supset (u \wedge \neg u)$ on step 9 and the new goal $\neg p$, where $u \wedge \neg u$ is $\bot$ in Procedure (2.1.7.2).

step	list_proof	annotation	list_goals
$as0$			$\bot$
$as1$	1. $p \supset r$	given	$\bot$
$as2$	2. $\neg p \supset s$	given	$\bot$
$as3$	3. $q \supset (\neg r \wedge \neg s)$	given	$\bot$
$as4$	4. q	given	$\bot$
$as5$	5. $\neg r \wedge \neg s$	$3, 4 \supset_{el}$	$\bot$
$as6$	6. $\neg r$	$5, \wedge_{el}$	$\bot$
$as7$	7. $\neg s$	$5, \wedge_{el}$	$\bot$
$as8$			$\bot, p$
$as9$	8. $p \supset (t \wedge \neg t)$	assumption	$\bot, p, p$
$as10$			$\bot, p, p, \neg p$
$as11$	9. $\neg p \supset (u \wedge \neg u)$		$\bot, \neg p, \neg p$

At this moment, the proof search stops. A model is extractable as follows: p is assigned f because $p \supset (t \wedge \neg t)$ is in the list_proof or because $\neg p \supset (t \wedge \neg t)$ is in the list_proof. Note that p is assigned f if, and only if, $\neg p$ is assigned f. Next, r gets the value 0 because $\neg r$ is in the list_proof and s is assigned 0 because $\neg s$ is in the list_proof. Under this valuation, each formula $p \supset r, \neg p \supset s, q \supset (\neg r \wedge \neg s)$ and q is assigned 1. So, this set of formulae in $Spec$ is consistent.

This explicitly shows the nature of the applicability of paracomplete logic — the given $Spec$ does not have a precise information about p — if this component should be bound or not. So the reasoning stops.

Had we reasoned about this specification in the classical set up, we would have been able to use classically valid formula $p \vee \neg p$ (which is not valid in **Kl**) to derive the contradiction. We will give the corresponding proof a little later, after presenting a derivable rule which we will use in the proof:

$$\frac{A \supset B, \ C \supset D}{(A \vee C) \supset (B \vee D)}$$

Note that this rule is also derivable in logic **Kl$_{ND}$**, so we will construct the proof applying our algorithm **Kl$_{NDALG}$**. It will return the conclusion of this rule, $(A \vee C) \supset (B \vee D)$, given that the premises are constituted.

The **Kl$_{NDALG}$** (hence the classical algorithm [9] as well) would set up $(A \vee C) \supset (B \vee D)$ as the main goal G_0 to be derived from the given set $A \supset B, C \supset D$. Because the goal is implicative, by Procedure (2.1.3), its antecedent $A \vee C$ becomes the new assumption, and its consequent, $B \vee D$ — the new goal, G_1.

step	list_proof	annotation	list_goals
$as0$			$G_0 = ((A \vee C) \supset (B \vee D))$
$as1$	1. $A \supset B$	given	G_0
$as2$	2. $C \supset D$	given	G_0
$as3$	3. $A \vee C$	assumption	$G_0, G_1 = B \vee D$

The current goal, G_1, is disjunctive, therefore, by Procedure (2.1.2.1), the left disjunct of G_1 is set up as the new goal $G_2 = B$.

This goal cannot be reached so it is deleted from list_goals, and, by Procedure (2.1.2.2), the right disjunct is set up as the new goal $G_2 = D$.

This goal cannot be reached so it is deleted from list_goals. Therefore, we have a disjunctive goal G_1 which so far has not been reached.

Next, the Procedure (2.2.1) is fired. The algorithm finds a disjunctive formula $A \vee C$ in list_proof and it should take in turn two branches.

First, to derive G_1 adding A as the new assumption and then to derive G_1 adding C as the new assumption.

Solving the first derivation, A is the new assumption on step 4 as below. Now G_1 is a disjunctive goal and its antecedent becomes the new goal $G_2 = B$.

This can be reached by eliminating implication from 1 and 4 obtaining B on step 5 and then introducing disjunction to the latter obtaining $B \vee D$ on step 6.

step	list_proof	annotation	list_goals
$as0$			$G_0 = ((A \vee C) \supset (B \vee D))$
$as1$	1. $A \supset B$	given	G_0
$as2$	2. $C \supset D$	given	G_0
$as3$	3. $A \vee C$	assumption	$G_0, G_1 = B \vee D$
$as4$			$G_0, G_1, G_2 = B$
$as5$	4. A	assumption	G_0, G_1, G_2
$as6$	5. B	$1, 4 \supset_{el}$	G_0, G_1
$as7$	6. $B \vee D$	$5, \vee_{in}$	G_0, G_1

Although we have obtained $B \vee D$ on step 6, we have not reached the goal G_1 — to reach the latter we also need to achieve the second subderivation — from the set of formulae $1, 2, 3, C$ where C is the new assumption, to derive $B \vee D$. The application of the algorithm is similar to the above, so the proof continues as follows:

step	list_proof	annotation	list_goals
$as0$			$G_0 = ((A \vee C) \supset (B \vee D))$
$as1$	1. $A \supset B$	given	G_0
$as2$	2. $C \supset D$	given	G_0
$as3$	3. $A \vee C$	assumption	$G_0, G_1 = B \vee D$
$as4$			$G_0, G_1, G_2 = B$
$as5$	4. A	assumption	G_0, G_1, G_2
$as6$	5. B	$1, 4 \supset_{el}$	G_0, G_1
$as7$	6. $B \vee D$	$5, \vee_{in}$	G_0, G_1
$as8$			$G_0, G_1, G_2 = D$
$as9$	7. C	assumption	G_0, G_1, G_2
$as10$	8. D	$2, 7 \supset_{el}$	G_0, G_1
$as11$	9. $B \vee D$	$8, \vee_{in}$	G_0, G_1

Both subderivations tasks have been completed so the algorithm applies $\vee_{el}$ rule as we have the disjunctive formula $A \vee C$ in the proof and from either of its disjuncts we have derived $B \vee D$. The result of this rule is $B \vee D$ on step 10 with annotations as below.

Finally, introducing implication to the formula on step 10 we derive the desired goal G_0 from the formulae on steps 1 and 2.

step	list_proof	annotation	list_goals
$as0$			$G_0 = (A \vee C) \supset (B \vee D)$
$as1$	1. $A \supset B$	given	G_0
$as2$	2. $C \supset D$	given	G_0
$as3$	3. $A \vee C$	assumption	$G_0, G_1 = B \vee D$
$as4$			$G_0, G_1, G_2 = B$
$as5$	4. A	assumption	G_0, G_1, G_2
$as6$	5. B	$1, 4 \supset_{el}$	G_0, G_1
$as7$	6. $B \vee D$	$5, \vee_{in}$	G_0, G_1
$as8$			$G_0, G_1, G_2 = D$
$as9$	7. C	assumption	G_0, G_1, G_2
$as10$	8. D	$2, 7 \supset_{el}$	G_0, G_1
$as11$	9. $B \vee D$	$8, \vee_{in}$	G_0, G_1
$as12$	10. $B \vee D$	$3, 4, 7, [4-6], [7-9]$	G_0
$as13$	11. $(A \vee C) \supset (B \vee D)$	$10, \supset_{in}, [3-10]$	

Now we use this derivable rule returning to the task of showing that in the classical setting the given $SPEC$ is inconsistent. We will not show below the algo-steps as they would correspond to the steps of the proof.

list_proof	annotation
1. $p \supset r$	given
2. $\neg p \supset s$	given
3. $q \supset (\neg r \wedge \neg s)$	given
4. q	given
5. $(p \vee \neg p) \supset (r \vee s)$	1, 2, derived rule
6. $\neg r \wedge \neg s$	3, 4, $\supset_{el}$
7. $\neg r$	6, $\wedge_{el}$
8. $\neg s$	6, $\wedge_{el}$
9. $p \vee \neg p$	classical validity
10. $r \vee s$	5, 9, $\supset_{el}$
11. s	7, 10, $\vee_{el}$

Now steps 8 and 11 constitute the contradiction hence the classical reasoning would have detected the contradiction while, in fact, in our initial setup with incomplete knowledge on p we do not have any inconsistency due to this lack of the exact information about the truth conditions of p.

6.3 Assume-Guarantee Reasoning

We consider here how the reasoning based upon natural deduction can be applied to the automation of the assume-guarantee reasoning [19, 36] technique, the most used technique in the framework of compositional analysis.

In assume-guarantee reasoning, a verification problem is represented as a triple, $\langle A \rangle S \langle P \rangle$, where S is the subsystem being analyzed, P is the property to be verified, and A is an assumption about the environment in which S is used.

The standard interpretation of $\langle A \rangle S \langle P \rangle$ suggests that A is a constraint on S and if S as constrained by A satisfies P, then the formula $\langle A \rangle S \langle P \rangle$ is true.

Let us formulate the semantics of $\langle A \rangle S \langle P \rangle$ in the following way: $S/A \models P$ where S/A means the system S with the additional information A. Now, the typical example of the application of assume-guarantee reasoning is in the context of decomposing a given system S into two subsystems S_1 and S_2 that run in parallel. Suppose we need to verify that the property P is satisfied in S. Then we can apply the assume-guarantee rule † as follows.

$$(\dagger) \quad \frac{\langle A\rangle S_1\langle P\rangle \quad \langle \mathbf{true}\rangle S_2\langle A\rangle}{\langle \mathbf{true}\rangle S_1||S_2\langle P\rangle}$$

Here $\langle \mathbf{true}\rangle S_2\langle A\rangle$ and $\langle \mathbf{true}\rangle S_1||S_2\langle P\rangle$ mean, respectively, that A is verified in S_2 (without any constraints) and P is verified in $S_1||S_2$ (without any constraints).

In terms of natural deduction we can rewrite this rule as $\ddagger$ below.

$$(\ddagger) \quad \frac{S_1, A \vdash P \quad S_2 \vdash A}{S_1||S_2 \vdash P}$$

Now new tasks are to find the natural deduction derivations $S_1, A \Vdash P$ and $S_2 \Vdash A$ in order to conclude that $S_1||S_2 \vdash P$ and the application of the proof search technique is the next logical step here.

One of the major obstacles in the efficient application of assume-guarantee approach [15] is that once decomposition is selected, to manually find an assumption A to complete an assume-guarantee proof is difficult. Indeed, the assumption must be strong enough to sufficiently constrain the behavior of S_1 so that $S_1, A \vdash P$ holds, and must be weak enough so that $S_2 \vdash A$ holds. The problem of finding such as assumption A would become even more difficult if the systems in question are constrained with an incomplete information. The application of the proof search algorithm of paracomplete logic **Kl** described above would represent an efficient solution. (Of course we would need to introduce the rigorous reasoning here defining what are 'strong' and 'weak' conditions.)

Let us draw here some directions of the application of the presented proof search towards the automation of assume-guarantee technique.

In the reasoning below we rigorously follow the proof search algorithm for $\mathbf{Kl_{ND}}$.

When solving the problem $S_1||S_2 \Vdash P$ we look for the assumption A such that $S_1, A \Vdash P$ and $S_2 \Vdash A$. Assume that S_1 and S_2 are systems with the specifications containing statements $B_1, \ldots, B_m$ and $C_1, \ldots, C_n$, respectively. Our task is to find an assumption A, following rule ($\dagger$) above, such that $B_1, \ldots, B_m, A \Vdash P$ and $C_1, \ldots, C_n \Vdash A$. In the description of our reasoning, we will use the concept of the algo-step, introduced above. Now we commence $\mathbf{Kl_{ND}}$ proof setting list_proof $= B_1, \ldots, B_n$ and list_goals $= P$:

step	list_proof	annotation	list_goals
$as0$	1. B_1	given	P
$as1$	.	given	P
$as2$	.	given	P
$as\ m$	$m.\ B_m$	given	P
$as\ m+1$	$m+1.$		$P, \bot$

On algo-step m, since the goal P is not reachable, we update list_goals by $\bot$. If on algo-step $m+1$ list_proof contains contradictory elements, then the new goal $\bot$ would be reachable and we would have two contradictory statements within $B_1, \ldots, B_m$, say, C and $\neg C$ at the stages $1 \leq i < j \leq m$. Thus, our new goal would have been P again which we would reach by applying $\mathbf{Kl}_{\neg in}$ rule:

step	list_proof	annotation	list_goals
$as0$	1.	given	P
$as1$	.	given	P
$as3$	$i.\ C$	given	P
.		given	P
$as\ j.\ \neg C$		given	P
$as\ m$	$m.$	given	P
$as\ m+1.$			$P, \bot$
$as\ m+2.$			P
$as\ m+3.\ P$	$i, j, \mathbf{Kl}_{\neg in}$		

Now we found our first candidate for A — contradiction. Hence we set up the new task — $C_1, \ldots, C_n \Vdash \bot$ and thus check if we can establish the latter.

Alternatively, we consider the second case on step m above, when the goal $\bot$ on algo-step m is not reachable. In this case we would have the following continuation of the proof:

step	list_proof	annotation	list_goals
$as0$	1. B_1	given	P
.	.	given	P
.	.	given	P
$as\ m$	$m.\ B_m$	given	P
$as\ m+1$	$m+1.\ P \supset r \wedge \neg r$	assumption	$P, [P \supset r \wedge \neg r]\ P$

At this stage, since P was not reachable, it is not contained in list_proof hence no elimination rules are applicable and we search for new assumptions. Namely, we would be looking for disjunctive and implicative formulae in list_proof (but ignoring the formula $P \supset r \wedge \neg r$ on step $m+1$).

If successful, we would introduce into the proof the corresponding assumption and proceed further applying the searching algorithm until it terminates with either finding the desired proof for P or failing to do so.

In the former case, P would be the last formula of list_proof and we will be able to consider assumptions appearing in list_proof between algo-step $m+1$ to test them in the second task $C_1, \ldots, C_n \Vdash A$.

7 Conclusion and Roadmap to Future Work

The contribution of this paper is twofold. On one hand, we provided the complexity analysis of the classical natural deduction system and its modified version, for paracomplete logic **Kl**. This has led us closer to the important question on the efficiency of the presented proof search technique and enables us to speak about the second aspect of the contribution of the paper — application issues. We have shown how paracomplete logic **Kl** can be used in providing high level specifications for incomplete systems and how natural deduction system for this logic, supported by the algorithmic proof search, can be used to reason about obtained specifications. To the best of our knowledge, there is no other similar work on the automation of paracomplete natural deduction systems or on an application of natural deduction techniques in general to the reasoning about incomplete specifications.

We have shown how these developments can be integrated into the existing approaches dealing with component-based system assembly.

It is notable that for many researchers, one of the core features of natural deduction, the opportunity to introduce arbitrary formulae as assumptions, has been a point of great scepticism regarding the very possibility of the automation of the proof search. In this paper, not only we show the contrary, but we also turned the assumptions management into an advantage showing the applicability of the proposed technique to assume-guarantee reasoning.

The results presented in this paper have important methodological aspects forming the basis for the development of automated goal-directed techniques for more expressive formalisms, for example, temporal and normative extensions. The feasibility of these extensions is based on the systematic, generic nature of the natural deduction construction and algorithmic proof search. This will, in turn, enable the application of the powerful natural deduction based reasoning to tackle dynamic systems defined in heterogeneous environments, with such complicated cases as the combinations of time / paraconsitency / paracompleteness. Thus we envisage the extensions of the applicability of our methodology to the specification of complex dynamic systems, to the specification of normative systems (i.e. protocols) and to reasoning about systems that are both inconsistent and incomplete.

One specific area, where we have obtained some preliminary results, is *Requirements Engineering*. In a series of works authors indicate the importance of the specification of high-level requirements of a partial model such that these specifications are built incrementally from higher-level goal formulations in a way that guarantees their correctness by construction [23]. In [44] the approach to tackle the problem of reduction of complex software requirements to simpler ones and to reason about the requirements is given.

However, we are not aware of any approach which would tackle this task under the following constraints:

(i) considering this problem in the context of incomplete specifications;

(ii) using the advances of automated deduction.

We argue that the natural deduction searching technique, which enables us to trace the dependencies of the formulae in the proof, opens a very important prospect of finding solutions to the above (i) and (ii). The methodology here is as follows: set the formally specified requirements as the goals for the searching technique so the latter returns the set of assumptions upon which these goals depend.

This corresponds to the layer of 'global invariants' mentioned in [23], where the authors give a very reasonable taxonomy of goal patterns (see [23, P.26]).

Now, our solution looks as follows: setting the requirements Req as goals for the proof searching technique, we aim at finding such global invariants.

Thus, applying to each such requirement $r \in Req$ our proof searching algorithm, **Kl$_{\text{NDALG}}$**, we aim at finding the assumptions, $Depend(r)$, on which r depends in the proof. This set of formulae $Depend(r)$ represents the desired set of reduced requirements (global invariants).

Acknowledgements

The authors should acknowledge here Prof. Vladimir Popov (Lomonosov Moscow State University), who acquainted them with matrix definitions for **Kl** which we are utilised in this paper.

References

[1] A. Avron. Natural 3-valued Logics — Characterization and Proof Theory. The Journal of Symbolic Logic, Vol. 56(1): 276–294, 1991.

[2] A. Avron and I. Lev. A formula-preferential Base for Paraconsistent and Plausible Non-monotonic Reasoning. In Proceedings of the Workshop on Inconsistency in Data and Knowledge (KRR-4), Int. Joint Conf. on AI (IJCAI 2001), pages 60-70, 2001.

[3] Basic Features of the Grid Component Model Deliverable D.PM.04. CoreGRID, March 2007 (http://coregrid.ercim.eu/mambo/).

[4] D. Basin, S. Matthews, and L. Vigano. Natural deduction for non-classical Logics. Studia Logica, 60(1), (1998): 119–160.

[5] V. Bocharov and V. Markin, Introduction to Logic. Moscow, Higher Education, 2008 (in Russian).

[6] A. Bolotov, O. Grigoriev and V. Shangin: Automated Natural Deduction for Propositional Linear-Time Temporal Logic. Proceedings of TIME 2007: 47–58.

[7] A. Bolotov and V. Shangin. Natural Deduction System in Paraconsistent Setting: proof search for PCont. Journal of Intelligent Systems, Vol. 21(1), (2012): 1–24.

[8] A. Bolotov and V. Shangin. Tackling Incomplete System Specifications Using Natural Deduction in the Paracomplete Setting. Proceedings of COMPSAC 2014: 91–96.

[9] A. Bolotov, V. Bocharov, A. Gorchakov and V. Shangin. Automated First Order Natural Deduction. Proceedings of IICAI 2005: 1292–1311.

[10] A. Bolotov and V. Shangin. Natural Deduction in a Paracomplete Setting. Logical Investigations, Vol. 20, (2014): 224–247.

[11] M. Bonet and S. Buss. The Deduction Rule and Linear and Near-Linear Proof Simulations. Journal of Symbolic Logic, Vol. 58/2 (1993): 688-709.

[12] A. Buchsbaum and T. Tarcisio. A Reasoning Method for a Paraconsistent Logic. Studia Logica, 52(2), (1993): 281-290.

[13] W. Carnielli. Systematization of Finite Many-Valued Logics Through the Method of Tableaux. Journal of Symbolic Logic Volume 52, Issue 2 (1987), 473–493.

[14] W. Carnielli. On Sequents and Tableaux for Many-valued Logics. Journal of Non-Classical Logic 8(1), (1991): 59–76.

[15] J. Cobleigh and G. Avrunin and L. Clarke. Breaking Up Is Hard To Do: An Evaluation of Automated Assume-guarantee Reasoning. ACM Transactions on Software Engineering and Methodology 17(2), (2008): 104–155.

[16] V. Degauquier. Partial and Paraconsistent Three-valued Logics. Logic and Logical Philosophy, Volume 25, (2016): 143–171.

[17] A. Hunter and B. Nuseibeh. Managing Inconsistent Specifications: Reasoning, Analysis and Action. ACM Transactions on Software Engineering and Methodology, 7(4), (1998): 335–367.

[18] A. Hunter and S. Parsons. Introduction to Uncertainty Formalisms. Hunter, A and Parsons, S, (eds.) Applications of Uncertainty Formalisms. (pp. 1-7). Springer. Lecture notes in computer science; Vol. 1455, ISBN 3-540-65312-0, Springer, (2001): 1–7

[19] C. Jones. Specification and Design of (parallel) Programs. Proceedings of the IFIP 9th World Congress: IFIP: North Holland, (1983): 321–332.

[20] N. Kamide. Natural Deduction Systems for Nelson's Paraconsistent Logic and its Neigh-

bors. Journal of Applied Non-Classical Logics. Vol. 15 (4), (2005): 405–435.

[21] S. Kleene. Introduction to Metamathematics. Wolters-Noordhoff Publishing and North Holland Publishing Company, Amsterdam, 1971, 7th Edition.

[22] J. Kreiker, A. Tarlecki, M. Vardi, and R. Wilhelm. Modeling, Analysis, and Verification — The Formal Methods Manifesto 2010 (Dagstuhl Perspectives Workshop 10482). Dagstuhl Manifestos, 1(1), (2011): 21–40.

[23] E. Letier and A. van Lamsweerde. Deriving Operational Software Specifications from System Goals. SIGSOFT 2002/FSE-10, Charleston, SC, USA, (2002): 18–22.

[24] C. Middelburg. A Survey of Paraconsistent Logics. The Computing Research Repository (CoRR), vol. 1103/4324, 2011.

[25] A. Naddeo. Axiomatic Framework Applied to Industrial Design Problem Formulated by Paracomplete Logics Approach: the Power of Decoupling on Optimization-Problem solving. Proceedings of Fourth International Conference on Axiomatic Design, (2006): 1-8.

[26] T. Nipkow, L. Paulson and M. Wenzel. Isabelle/HOL — A Proof Assistant for Higher-Order Logic. Springer, 2002.

[27] D. Li. Using the Prover ANDP to Simplify Orthogonality. Annals of Pure and Applied Logic, 124-1, (2003): 49–70.

[28] J. Noppen, P. van den Broek and M. Aksit. Software Development with Imperfect Information. Soft Computing: 12, (2008): 3–28.

[29] B. Nuseibeh, S. Easterbrook, and A. Russo. Making Inconsistency Respectable in Software Development. Journal of Systems and Software, Vol. 58, No. 2, (2001): 171–180.

[30] D. Pastre. Muscadet2.3 : A Knowledge-based Theorem Prover based on Natural Deduction. International Joint Conference on Automated Reasoning - Conference on Automated Deduction (2001): 685—689.

[31] F. J. Pelletier. Natural Deduction Theorem Proving in THINKER. Studia Logica, 60 (1998): 3–43.

[32] F. J. Pelletier. A Brief History of Natural Deduction. History and Philosophy of Logic, 20 (1999): 1–31.

[33] Y. Petrukhin and V. Shangin Automated correspondence analysis for the binary extensions of the logic of paradox. The Review of Symbolic Logic, 1–26 (doi:10.1017/S1755020317000156).

[34] John L. Pollock. Rational Cognition in OSCAR. Agent Theories, Architectures, and Languages (1999): 71—90.

[35] F. Portoraro. Strategic Construction of Fitch-style Proofs. Studia Logica, 60 (1998): 45–66.

[36] A. Pnueli. In Transition from Global to Modular Temporal Reasoning about Programs. Logics and Models of Concurrent Systems, K. R. Apt, Ed. NATO ASI: vol. 13. Springer-Verlag, (1984): 123–144.

[37] V. Popov. Between the logic Par and the set of all formulae in 'The Proceeding of the

6th Smirnov Readings in logic', Contemporary notebooks, Moscow, 93–95 (In Russian).

[38] V. Popov. Between Int$<\omega,\omega>$ and Intuitionistic Propositional Logic. Logical Investigations, Issue 19, 2013, 197–199 (in Russian).

[39] A. Sette and W. Carnielli. Maximal Weakly-intuitionistic Logics. Studia Logica, 55, 1 (1995): 181—203.

[40] W. Quine. On Natural Deduction. The Journal of Symbolic Logic, 2-15 (1950): 93–102.

[41] R. A. Reckhow. On the Lengths of Proofs in the Propositional Calculus. Ph. D. thesis / Reckhow R.A. ; University of Toronto, 1976.

[42] Shangin V.O. A Precise Definition of an Inference (by the example of natural deduction systems for logics $I<\alpha,\beta>$) Logical investigation 23(1), (2017): 83–104

[43] L. V. Tien, Q. T. Tho, and L. D. Anh. Specification-based Verification of Incomplete Programs. ACEEE Int. Journal on Information Technology, Vol. 02, No. 02, (2012): 56–61.

[44] B. Wei, Z. Jin, D. Zowghi, and B. Yin. Automated Reasoning with Goal Tree Models for Software Quality Requirements. Proceedings of COMPSAC 2012 Workshops, (2012): 373–378.

Suzumura Consistency, an Alternative Approach

Peter Schuster
Università degli Studi di Verona, Verona, Italy.
peter.schuster@univr.it

Daniel Wessel
Università degli Studi di Trento, Trento, Italy.
daniel.wessel@unitn.it

Abstract

Suzumura consistency is known as a sufficient and necessary condition for a binary relation to have an order extension. We advocate the use of equivalent but negation-free forms of Suzumura consistency and of the related notion of compatible extension. From a methodological perspective, our proposals make possible to work more abstractly, in the algebra of relations, and to give more direct proofs. To illustrate this we reconsider various forms and proofs of the order extension principle. As a complement we adopt to quasi-orders J.L. Bell's argument that Gödel–Dummett logic is necessary for order extension.

1 Introduction

Order extension principles are originally due to Szpilrajn [47] for strict partial orders; for quasi-orders they were phrased by Arrow [3] and proved by Hansson [34]. They

---

The present paper has emerged from the project "Abstract Mathematics for Actual Computation: Hilbert's Program in the 21st Century" funded by the John Templeton Foundation. The opinions expressed in this publication are those of the authors and do not necessarily reflect the views of the John Templeton Foundation. Initial studies were undertaken when the first author was visiting the Munich Center for Mathematical Philosophy upon kind invitation by Hannes Leitgeb and with a research fellowship "Erneuter Aufenthalt" by the Alexander-von-Humboldt Foundation. Both authors would like to thank Sara Negri for the enlightening discussions during her "Cooperint Azione 3" senior visiting fellowship provided by the University of Verona in 2016; and Thomas Streicher for the helpful advice he gave during his INdAM-GNSAGA visiting professorship at the University of Padua in 2016. Both authors would further like to thank the anonymous referees for carefully reading the manuscript and giving fair critique and helpful suggestions.

play a seminal role in mathematical economics, game theory, and in the theory of social choice, preferences, and utility (see, e.g., [2, 18] for an exhaustive overview, and [16] for the rising field of computational social choice). Suzumura [45] specified a notion of consistency which is sufficient and necessary for a binary relation to have an order extension. With the customary definition of consistency, however, proofs often require indirect reasoning and arguments on pairs, i.e., arguments involving specific elements of the underlying set. We now show how Suzumura consistency can be put in a logically equivalent and negation-free form, which allows for a somewhat slicker treatment. In fact, this has an interesting methodological effect: we can keep to a minimum arguments on pairs, and largely avoid proofs by contradiction—instead we argue abstractly within the algebra of relations.

Yet one cannot do with constructive means only, as J.L. Bell made clear. While Zorn's Lemma, the key tool for order extension, allegedly is "constructively neutral" [5], order extension is not: it results in Gödel–Dummett logic not only for partial orders [6] but also, as we show below (Section 8.4), for quasi-orders. Negri et al. [38] proved practicable a proof–theoretic study of order extension; see also [43].

This paper is organised as follows. We first list the most necessary preliminaries in Section 2. In Section 3 we discuss the notion of consistency, and in Section 4 we make precise a notion of (compatible) extension intimately related with consistency. Then, in Section 5, we concentrate on the extendability of consistent relations to complete quasi-orders, while in Section 6 we rephrase a classic result of Dushnik and Miller [28] in terms of consistent relations. In Section 7 we present another proof of Arrow's generalization, along the lines of [3, 36, 46, 18]. In the complementary Section 8 we explain an alternative proof of the order extension principle by way of Open Induction [40] rather than Zorn's Lemma; carry over from partial orders to quasi-orders J.L. Bell's argument [6] that Gödel–Dummett logic is necessary for the order extension principle; and revisit Richter's theorem [41] on rationalizability of choice functions.

2 Preliminaries

For the purposes of this paper, a certain amount of fairly standard terminology needs to be fixed. In the following, let R and S denote binary relations on a set X, i.e., subsets of the cartesian product $X \times X$. By "relation" we shall always mean "binary relation", and henceforth we skip "binary". The *opposite* (or *reciprocation*) of R is

$$R^\circ = \{\,(x,y) \in X \times X : (y,x) \in R\,\}.$$

Note that $(R^\circ)^\circ = R$, and if $R \subseteq S$, then $R^\circ \subseteq S^\circ$. Furthermore, $(R \cup S)^\circ = R^\circ \cup S^\circ$. The *asymmetric part* of R is $P(R) = R - R^\circ$, which is to say that

$$P(R) = \{\, (x,y) : (x,y) \in R \wedge (y,x) \notin R \,\}.$$

The *transitive closure* of R is

$$\mathrm{tc}(R) = \bigcup_{i \geqslant 1} R^i,$$

with $R^1 = R$ and $R^{i+1} = R^i \circ R$, where

$$R \circ S = \{\, (x,z) : \exists\, y \in X\ (x,y) \in R \wedge (y,z) \in S \,\}$$

denotes the *relational composition*.[1] A relation R is *transitive* if and only if $R \circ R \subseteq R$, which in turn holds if and only if $\mathrm{tc}(R) = R$. If R, S are relations, then $R \subseteq \mathrm{tc}(S)$ if and only if $\mathrm{tc}(R) \subseteq \mathrm{tc}(S)$; that is to say that tc is a closure operator on the powerset $\mathcal{P}(X \times X)$. As composition obeys $(R \circ S)^\circ = S^\circ \circ R^\circ$ and distributes over unions, we see that the transitive closure commutes with reciprocation, $\mathrm{tc}(R^\circ) = \mathrm{tc}(R)^\circ$.

We say that R is *complete*[2] if $R \cup R^\circ = X \times X$, which is also known as R being *linear* or *total*. Mind that a complete relation is *reflexive*, i.e., $\Delta \subseteq R$, where

$$\Delta = \{\, (x,y) \in X \times X : x = y \,\}$$

is the *diagonal*, and notice that $\Delta \cap R = \Delta \cap R^\circ$. The diagonal is neutral for composition, i.e., $\Delta \circ R = R = R \circ \Delta$. The *reflexive closure* of R is $R \cup \Delta$.

A *quasi-order* (or *preorder*) is a reflexive transitive relation. The *hull* $\overline{R}$ of a relation R, viz.

$$\overline{R} = \bigcup_{i \geqslant 0} R^i$$

where $R^0 = \Delta$, is the least quasi-order which contains R. Note that $\overline{R} = \mathrm{tc}(R) \cup \Delta = \mathrm{tc}(R \cup \Delta)$.

If Y is a subset of X, then

$$R|_Y = R \cap (Y \times Y)$$

is the *restriction* of R on Y. The restricted diagonal is denoted by Δ_Y. Occasionally we write $R, (x,y)$ instead of $R \cup \{\,(x,y)\,\}$. If S is a set and $X, Y \subseteq S$, then $X \between Y$ is shorthand for $X \cap Y$ being inhabited.[3]

---

[1] We adhere to the traditional, Tarskian convention about composition which is customary in the context of preference relations [14], and even in certain abstract categorical settings [31].

[2] The notion of "completeness" is prevalent in the context of logical theories and Lindenbaum's Lemma.

[3] We have adopted this notation from Giovanni Sambin.

An antisymmetric quasi-order R on X, i.e., one for which $R \cap R^\circ \subseteq \Delta$, is a *partial order*; the underlying set X in which case is called a *poset*. If R is complete, then X is said to be *linearly ordered*. By a *chain* in a poset X we understand an inhabited subset of X that is linearly ordered by the restricted relation. We say that X is *chain-complete*, if X is inhabited and every chain C has a least upper bound $\bigvee C$ in X. A *maximal element* x in X is such that

$$\forall y \in X \ (x \leqslant y \to x = y).$$

One of the standard forms of *Zorn's Lemma* (ZL) reads as follows:

ZL. *Every chain-complete poset has a maximal element.*

It is as such that the Axiom of Choice (AC) gets involved in proving the order extension principle in its full generality. In fact, a strictly weaker form of AC suffices [37, 29], but this shall not be of our concern.[4]

3 Consistency

Suzumura [45] gave a sufficient and necessary condition for a relation R to have a complete quasi-order extension which preserves the asymmetric part $P(R)$. A relation R is *Suzumura consistent* if

$$\forall x, y \in X \ [\ (x,y) \in \mathrm{tc}(R) \ \to \ (y,x) \notin P(R) \ ].$$

Unfolding the definition of $P(R)$, Suzumura consistency amounts to

$$\forall x, y \in X \ [\ (x,y) \in \mathrm{tc}(R) \ \to \ \neg(\ (y,x) \in R \land (x,y) \notin R \ ) \ ],$$

which (with classical logic) is equivalent to

$$\forall x, y \in X \ [\ (x,y) \in \mathrm{tc}(R) \land (y,x) \in R \ \to \ (x,y) \in R \ ].$$

This condition on R can now be written succinctly as set containment. We replace Suzumura consistency by this equivalent, and simply call it consistency, as follows.

Definition 1. A relation R is *consistent* if

$$\mathrm{tc}(R) \cap R^\circ \subseteq R.$$

---

[4]The Axiom of Choice is not entirely indispensable: syntactical conservation works for Horn sequents [38]; see also [43].

Remark 1. A relation is consistent if and only if $\overline{R} \cap R^\circ \subseteq R$. Moreover, a relation R is consistent if and only if every cycle in R is "reversible", by which we mean that every cycle in R forces its reciprocal to be in R as well. Really this concerns cycles of any length—here is another, equivalent way to put consistency:

$$\forall n \geqslant 0 \quad R^n \cap R^\circ \subseteq R.$$

In terms of preferences, consistency "rules out ... all cycles with at least one strict preference" [14, p. 36].

Remark 2. Every transitive relation is consistent. In particular, Δ, $\mathrm{tc}(R)$ and $\overline{R}$ are consistent. On the other hand, it is well-known that consistency is weaker than transitivity. For instance, $R = \{(x,y), (y,x)\}$ is consistent on $X = \{x, y\}$ but not transitive unless $x = y$, in fact $\mathrm{tc}(R) = R \cup \Delta$ and $R^\circ = R$.

But what is missing for a consistent relation to be transitive? Compositions need to be comparable.

Proposition 1. *For a relation R, each of the following items implies the next.*
(i) R is transitive,
(ii) $\mathrm{tc}(R) \subseteq R \cup R^\circ$,
(iii) $R \circ R \subseteq R \cup R^\circ$.
If R is consistent, then the above assertions are equivalent. In particular, transitivity is equivalent to consistency together with any of (ii) and (iii) above.

Proof. Of course, if R is transitive, then $\mathrm{tc}(R) = R$, whence (ii) follows from (i). Furthermore, from $R \circ R \subseteq \mathrm{tc}(R)$, we know that (ii) implies (iii). Next, if R is consistent and $R^2 = R \circ R \subseteq R \cup R^\circ$, then

$$R^2 = R^2 \cap (R \cup R^\circ) = (R^2 \cap R) \cup (R^2 \cap R^\circ) \subseteq R \cup (\mathrm{tc}(R) \cap R^\circ) \subseteq R$$

Therefore, transitivity is implied by (iii), given that R is consistent. $\square$

Remember that we have defined a relation R on X to be complete if $R \cup R^\circ = X \times X$. The following corollary is a direct consequence of Proposition 1. This observation has also been made in [14].

Corollary 1. *A complete consistent relation is transitive.*[5] *In particular, a relation is complete and consistent if and only if it is a complete quasi-order.* $\square$

---

[5]This is readily proved element-wise too. Here is another direct argument: if R is complete and consistent, then

$$\mathrm{tc}(R) = \mathrm{tc}(R) \cap (X \times X) = \mathrm{tc}(R) \cap (R \cup R^\circ) = \mathrm{cc}(R) = R,$$

whence R is transitive—see below for the *consistent closure* $\mathrm{cc}(R)$ of R.

Consider again conditions (ii) and (iii) in Proposition 1 above. Neither of them follows from consistency, just because a consistent relation need not be transitive. In turn, neither (ii) nor (iii) implies consistency. For example, if $R = \{(x,y),(y,z),(z,x)\}$ on a set $\{x,y,z\}$ with pairwise distinct elements x, y, z, then the reflexive closure $R \cup \Delta$ is not a consistent relation, yet it satisfies (ii). Therefore, consistency is independent of each of these assertions (ii) and (iii). Furthermore, as the example we have just given also shows (ii) and (iii) to be strictly weaker than transitivity, by Proposition 1 we have at hand a proper decomposition of transitivity.

It has been observed [13] that consistency—just as transitivity—can be expressed by means of a closure condition. Here this takes the following form.

Definition 2. The *consistent closure* of R is

$$\mathrm{cc}(R) = \mathrm{tc}(R) \cap (R \cup R^\circ).$$

Note that $R \subseteq \mathrm{cc}(R) \subseteq \mathrm{tc}(R)$. The consistent closure reverses cycles and thus "eliminates" strict preference from any such cycle. Consider for example once more a set $X = \{x, y, z\}$ with three pairwise distinct elements x, y, z, together with the "cyclic" relation $R = \{(x,y),(y,z),(z,x)\}$. This relation is not consistent, and the transitive closure of R is universal, i.e., $\mathrm{tc}(R) = X \times X$. The consistent closure, on the other hand, adds the opposite, but neither is reflexive nor transitive.

Lemma 1. *Let R and S be relations.*
(i) R is consistent if and only if $\mathrm{cc}(R) = R$.
(ii) $R \subseteq \mathrm{cc}(S)$ if and only if $\mathrm{cc}(R) \subseteq \mathrm{cc}(S)$.

Proof.
(i) We have $\mathrm{cc}(R) = (\mathrm{tc}(R) \cap R) \cup (\mathrm{tc}(R) \cap R^\circ) = R \cup (\mathrm{tc}(R) \cap R^\circ)$. Therefore, $\mathrm{cc}(R) = R$ if and only if $\mathrm{tc}(R) \cap R^\circ \subseteq R$.
(ii) Notice that the consistent closure is defined as intersection of transitive closure and symmetrization $R \mapsto R \cup R^\circ$. Therefore, it suffices to show that the latter satisfies the corresponding equivalence, which is immediate from the properties of reciprocation: if $R \subseteq S \cup S^\circ$, then $R^\circ \subseteq (S \cup S^\circ)^\circ = S^\circ \cup S^{\circ\circ} = S^\circ \cup S$, whence $R \cup R^\circ \subseteq S \cup S^\circ$. The converse implication is trivial. $\square$

In other words, the assignment $R \mapsto \mathrm{cc}(R)$ defines a closure operator the fixed points of which are precisely the consistent relations. Furthermore, $\mathrm{cc}(R)$ is the least consistent relation which contains R.

4 Compatible Extensions

The following definition is equivalent to the one employed in the context of preference relations [26].

Definition 3. Let R, S be relations. We say that S is a *compatible extension* of R if
$$R \subseteq S \quad \text{and} \quad S \cap R^\circ \subseteq R.$$

In fact, if $R \subseteq S$, then $S \cap R^\circ \subseteq R$ precisely when $P(R) \subseteq P(S)$ holds for the asymmetric parts.

Remark 3. If S is a compatible extension of $R \cup \{(y,x)\}$, then $(x,y) \in S$ implies $(x,y) \in R$, provided that either R is reflexive or $x \neq y$. For compatibility of S over $R \cup \{(y,x)\}$ means that
$$R \cup \{(y,x)\} \supseteq S \cap (R \cup \{(y,x)\})^\circ = S \cap (R^\circ \cup \{(x,y)\}) = (S \cap R^\circ) \cup (S \cap \{(x,y)\}).$$

Compatible extension can thus be regarded as "reflecting opposite elements". On the other hand, Remark 3 also has the following reading: if S is a compatible extension of $R \cup \{(y,x)\}$, and if $(x,y) \notin R$, then $(x,y) \notin S$.

The containment $R \subseteq X \times X$ is a compatible extension if and only if $R^\circ \subseteq R$, which is to say that R is symmetric. The reason why compatibility needs to be involved, is to avoid the universal relation to be an extension of every R [34, p. 453], and hence to be a solution of the problem of extending a relation to a complete quasi-order in the absence of any further restrictive assumption as, say, antisymmetry.[6] Mind that every relation R is a compatible extension of itself; whence a compatible extension need not necessarily be consistent. Not even a compatible extension of a consistent relation needs to be consistent.[7]

Caveat. *For brevity's sake, following a certain tradition [45, 25, 18], whenever referring to an extension we will henceforth always mean a compatible extension.*

Remark 4. Suppose that $R \subseteq S \subseteq T$. If T extends R, then so does S, because $S \cap R^\circ \subseteq T \cap R^\circ \subseteq R$.

Remark 5. Suppose that S is an extension of R. Then $S = R$ already if $S \subseteq R \cup R^\circ$, for in that case $S = S \cap (R \cup R^\circ) = (S \cap R) \cup (S \cap R^\circ) \subseteq R$. In particular, every

---

[6]On the other hand, notice that if R is reflexive, then every extension of R by an antisymmetric relation S automatically is compatible, for in that case $S \cap R^\circ \subseteq S \cap S^\circ = \Delta \subseteq R$.

[7]In fact, the empty relation $\emptyset$ is consistent, and is compatibly extended by any—possibly non-consistent—relation whatsoever.

complete relation R is maximal for extension, i.e., if S extends R and R is complete, then $S = R$. An extension might thus be very close; in fact, the consistent closure $\mathrm{cc}(R)$ of a relation R cannot extend R unless R itself is consistent, simply because $\mathrm{cc}(R) \subseteq R \cup R^\circ$.

In any case, the reflexive closure always gives an extension, which is the special case $R = S$ of the following.

Remark 6. Extensions carry over to reflexive closures. In fact, $R \subseteq S$ is compatible if and only if $R \subseteq S \cup \Delta$ is compatible, because

$$(S \cup \Delta) \cap R^\circ = (S \cap R^\circ) \cup (\Delta \cap R^\circ) = (S \cap R^\circ) \cup (\Delta \cap R).$$

Lemma 2. *Extension defines a partial order on relations.*

Proof. Extension clearly is reflexive, and inherits antisymmetry from inclusion. It remains to verify transitivity, i.e., if S extends R and T extends S, then T extends R. To this end, we calculate, using that $R^\circ \subseteq S^\circ$ whenever $R \subseteq S$,

$$T \cap R^\circ = (T \cap R^\circ) \cap R^\circ \subseteq (T \cap S^\circ) \cap R^\circ \subseteq S \cap R^\circ \subseteq R. \qquad \square$$

With the following proposition we adapt and extend an interesting result from [18]. The proof is straightforward in terms of our notion of consistency.

Proposition 2. *The following are equivalent for every relation R.*
(i) R is consistent.
(ii) $\mathrm{cc}(R)$ extends R.
(iii) $\mathrm{tc}(R)$ extends R.
(iv) $\overline{R}$ extends R.
(v) R has a consistent extension.
(vi) R has a transitive extension.
(vii) R has a quasi-order extension.

Proof. Notice first that whenever a quasi-order S contains R, we actually have

$$R \subseteq \mathrm{cc}(R) \subseteq \mathrm{tc}(R) \subseteq \overline{R} \subseteq S.$$

Hence, if S extends R, then (Remark 4) so do $\overline{R}$, $\mathrm{tc}(R)$, and $\mathrm{cc}(R)$; the latter extension is tantamount to R being consistent, by way of Remark 5. On the other hand, if R is consistent, then $\mathrm{cc}(R) = R$, and $\mathrm{tc}(R)$ extends R, simply by definition. Adding the diagonal does not do any harm, so $\overline{R}$ extends R, if $\mathrm{tc}(R)$ does, in which case R has a quasi-order extension. $\qquad \square$

As noticed in [18], because of $R \subseteq \mathrm{cc}(R) \subseteq \mathrm{tc}(R)$, it follows that $\mathrm{tc}(R) = \mathrm{tc}(\mathrm{cc}(R))$. In view of Lemma 1 and Proposition 2, then $\mathrm{tc}(R)$ is an extension of $\mathrm{cc}(R)$. Moreover, notice that $\overline{R} = \overline{\mathrm{cc}(R)}$.

5 Extension Principles

Suzumura singled out that for a relation R to have a *complete* quasi-order extension, it suffices for R to be consistent. But as long as there are no further assumptions made on the underlying set X, some form of the Axiom of Choice has to be involved.

We still need some preparation on our way to Suzumura's variant of order extension: first we have to make sure that the consistent extensions of a relation form a chain-complete poset (Lemma 3 below). Then we need to verify, typical indeed for many an application of Zorn's Lemma [5], that a consistent relation can be "step-wise" extended. Once all this has been done, Zorn's Lemma may be invoked.

Lemma 3.
   (i) *Every union of a chain $(R_i)_{i \in I}$ of consistent relations is consistent.*
   (ii) *If $(R_i)_{i \in I}$ is a chain with respect to extension, then $\bigcup_{i \in I} R_i$ is the least upper bound of $(R_i)_{i \in I}$ also with respect to extension.*

Proof.
   (i) This is a standard argument. Suppose that $(R_i)_{i \in I}$ is a chain of consistent relations, and let
$$(x,y) \in \text{tc}(\bigcup_{i \in I} R_i) \cap (\bigcup_{i \in I} R_i)^\circ.$$
   By the definition of tc, finitely many R_i suffice. As we have a chain, there in fact is $i_0 \in I$ with $(x,y) \subset \text{tc}(R_{i_0}) \cap R_{i_0}^\circ$. Then $(x,y) \in R_{i_0}$ since R_{i_0} is consistent, and therefore $(x,y) \in \bigcup_{i \in I} R_i$.
   (ii) Of course $R_{i_0} \subseteq \bigcup_{i \in I} R_i$ for every $i_0 \in I$. Furthermore, as we have a chain of extensions, for every $i \in I$ either R_i extends R_{i_0} or vice versa, and in each case we have $R_i \cap R_{i_0}^\circ \subseteq R_{i_0}$. From this we get
$$(\bigcup_{i \in I} R_i) \cap R_{i_0}^\circ = \bigcup_{i \in I}(R_i \cap R_{i_0}^\circ) \subseteq R_{i_0}.$$

Next, if S is a relation such that S extends R_i for every $i \in I$, then of course $\bigcup_{i \in I} R_i \subseteq S$, and
$$S \cap (\bigcup_{i \in I} R_i)^\circ = S \cap (\bigcup_{i \in I} R_i^\circ) = \bigcup_{i \in I}(S \cap R_i^\circ) \subseteq \bigcup_{i \in I} R_i,$$
which is to say that S extends the union $\bigcup_{i \in I} R_i$. □

Simply adding some pair to a consistent relation need not in general result in a consistent relation. If x, y, z are pairwise distinct, then $R = \{(x,y), (y,z)\}$ is consistent, yet $R' = R \cup \{(z,x)\}$ is not. Now one might be tempted to work with $cc(R')$ instead, which, as we have seen, is the least consistent relation to contain R'. But still there is a problem, since $R \subseteq cc(R')$ is not compatible, for $cc(R') \cap R^\circ = R^\circ \not\subseteq R$.

The following, somewhat technical lemmas are crucial in this regard, and when it comes to proving the order extension principle. We need to provide means for extending consistent relations by suitable pairs of elements.

Lemma 4. *If R is a quasi-order, then for all $x, y \in X$*
(i) if $(R \circ \{(x,y)\} \circ R) \between R^\circ$, then $(y,x) \in R$,
(ii) $\mathrm{tc}(R,(x,y)) = R \cup (R \circ \{(x,y)\} \circ R)$,
(iii) if $(y,x) \in \mathrm{tc}(R,(x,y))$, then $(y,x) \in R$, and
(iv) if $\mathrm{tc}(R,(x,y))$ extends R, then $R,(x,y)$ is consistent.

Proof.
(i) Suppose that $(a,b) \in (R \circ \{(x,y)\} \circ R) \cap R^\circ$. This means $(a,x), (y,b) \in R$ and $(b,a) \in R$. By transitivity of R we get $(y,x) \in R$.[8]

(ii) One inclusion is easily verified; as for the converse inclusion we show
$$(R,(x,y))^n \subseteq R \cup (R \circ \{(x,y)\} \circ R)$$
for every $n \geqslant 1$. We have
$$R \cup \{(x,y)\} = R \cup \Delta \circ \{(x,y)\} \circ \Delta \subseteq R \cup (R \circ \{(x,y)\} \circ R)$$
which takes care of $n = 1$ (mind that R needs to be reflexive in order for this to go through). Next we argue by induction, which gives
$$(R,(x,y))^{n+1} = (R,(x,y))^n \circ (R,(x,y))$$
$$\subseteq [R \cup (R \circ \{(x,y)\} \circ R)] \circ (R,(x,y)).$$
This is left to the reader; take into account the transitivity of R, and
$$\{(x,y)\} \circ R \circ \{(x,y)\} \subseteq \{(x,y)\}.$$

(iii) By (ii), if $(y,x) \in \mathrm{tc}(R,(x,y))$, then $(y,x) \in R$ or $(y,x) \in R \circ \{(x,y)\} \circ R$. In case of the latter, due to the definition of relational composition, we get again $(y,x) \in R$.

(iv) From (iii) we know that $\mathrm{tc}(R,(x,y)) \cap \{(y,x)\} \subseteq R$. Therefore, if $\mathrm{tc}(R,(x,y))$ is an extension of R, then
$$\mathrm{tc}(R,(x,y)) \cap (R,(x,y))^\circ = [\mathrm{tc}(R,(x,y)) \cap R^\circ] \cup [\mathrm{tc}(R,(x,y)) \cap \{(y,x)\}] \subseteq R.$$

---
[8]Notice that reflexivity of R is irrelevant for this argument.

Lemma 5. *If R is a quasi-order and $(y,x) \notin R$, then $R,(x,y)$ is a consistent extension of R.*

Proof. By Lemma 4(i), if $(y,x) \notin R$, then $(R \circ \{(x,y)\} \circ R) \cap R^\circ = \emptyset$. Now Lemma 4(ii) implies $\text{tc}(R,(x,y)) \cap R^\circ \subseteq R$, which is to say that $\text{tc}(R,(x,y))$ extends R; whence $R,(x,y)$ is consistent, according to Lemma 4(iv). Finally, $R,(x,y)$ is an extension of R, simply because $(y,x) \notin R$. □

In terms of [26], Lemma 5 establishes that the set of quasi-orders on X is *arc-receptive*. We will now see that, with a maximal extension at hand, proving completeness boils down to just one application of Lemma 5, as we will end up with a quasi-order, anyway.

Lemma 6. *For consistent relations, ordered by extension, maximality implies completeness.*

Proof. Let R be maximal among consistent relations, ordered by extension. According to Proposition 2, $\overline{R}$ extends R, hence $\overline{R} = R$ by maximality, i.e., R is a quasi-order. This R cannot fail to compare any two elements from X, hence must be complete. To be precise, we need to verify $X \times X = R \cup R^\circ$, and to this end consider $x, y \in X$ such that $(x,y) \notin R^\circ$, i.e., $(y,x) \notin R$. Then $R,(x,y)$ is consistent and extends R by Lemma 5. Therefore, again by way of maximality, $R,(x,y) = R$, which is to say that $(x,y) \in R$. □

We are now ready to put everything together. Recall that a complete relation is consistent if and only if it is a quasi-order (Corollary 1). We have been working towards the following extension principle for consistent relations [45, Theorem 3]:

Consistent Extension Principle (CEP). *Every consistent relation can be extended to a complete quasi-order.*

Proof. If R is consistent, then the set $\mathcal{E}$ of consistent extensions of R is inhabited, and it is chain-complete by Lemma 3. As we have seen in Lemma 6, every maximal element of $\mathcal{E}$ is a complete consistent extension of R, hence a complete quasi-order. The existence of at least one such maximal extension is ensured by Zorn's Lemma. □

It is in order to list also the following two slight variants and immediate consequences of CEP:

Transitive Extension Principle (TEP). *Every transitive relation can be extended to a complete quasi-order.*

Quasi-Order Extension Principle (QEP). *Every quasi-order can be extended to a complete quasi-order.*

Proof. QEP is a special case of TEP, and TEP follows from CEP as transitivity implies consistency. □

6 Intersection Principles

We also want to adapt a well-known observation due to Dushnik and Miller [28], which has been phrased for quasi-orders by Donaldson and Weymark [25], and Bossert [11], and put into general terms by Duggan [26]. We present a slight variation (Proposition 3) from which some immediate consequences can be drawn.

If R is consistent, then it has at least one complete consistent extension, according to CEP. Hence we can reasonably talk about the intersection S of all such extensions of R. Since every complete consistent extension of R is a quasi-order, S too is a quasi-order. This observation sets S apart from R whenever R happens to lack either reflexivity or transitivity. However, we will now see that S coincides with the hull $\overline{R}$, which is the intersection of all quasi-orders containing R. In fact, every pair of elements which is comparable under every complete extension of a consistent relation R must already be comparable by way of its hull.

Lemma 7. *For a quasi-order R, the intersection S of all complete consistent extensions of R compares the same elements as R, i.e., $R \cup R^\circ = S \cup S^\circ$.*

Proof. Of course $R \cup R^\circ \subseteq S \cup S^\circ$. In order to show the reverse inclusion, suppose $(x,y) \notin R \cup R^\circ$. i.e., $(x,y) \notin R$ and $(y,x) \notin R$. Then, according to Lemma 5, both $R, (x,y)$ and $R, (y,x)$ are consistent extensions of R, and both have complete consistent extensions by CEP, say $S_{(x,y)}$ and $S_{(y,x)}$, respectively, the former of which avoids (y,x), the latter (x,y) (Remark 3). Hence neither (x,y) nor (y,x) is common to all complete consistent extensions of R, which is to say that $(x,y) \notin S \cup S^\circ$. □

The key observation is that any pair of elements $x, y \in X$ which a quasi-order R fails to compare provides a choice: either adjoin (x,y) or go with (y,x), arbitrarily. In general there is no hope for a unique complete extension.

We can now state and prove the following *Intersection Principle* which in fact is equivalent to CEP.

Proposition 3. *The hull of a relation R is the intersection of all complete consistent extensions of $\mathrm{cc}(R)$.*

Proof. Recall that $\overline{R} = \overline{\mathrm{cc}(R)}$ extends $\mathrm{cc}(R)$, hence every extension of $\overline{R}$ is an extension of $\mathrm{cc}(R)$. Therefore

$$\bigcap\{\,T : T \supseteq \mathrm{cc}(R) \text{ is compatible}\,\} \subseteq \bigcap\{\,T : T \supseteq \overline{R} \text{ is compatible}\,\},$$

where T ranges over complete consistent relations, i.e., complete quasi-orders. As an intersection of quasi-orders all of which contain R, the left-hand side contains $\overline{R}$. On the other hand, since any intersection of extensions still is an extension, we know that the right-hand side S extends $\overline{R}$. The assertion now follows from Lemma 7 and Remark 5. □

While CEP results in Proposition 3, it is clear that Proposition 3 in turn implies CEP. In fact, if R is consistent, and $\overline{R}$ is not yet complete, then every pair (x, y) avoided by $\overline{R}$ yields a complete quasi-order which extends $\mathrm{cc}(R) = R$, and which avoids (x, y), as well.

Here is an equivalent way to put the Intersection Principle; recall that $\mathrm{cc}(R)$ is consistent even if R is not, and that $\overline{R} = \overline{\mathrm{cc}(R)}$.

Corollary 2. *The hull of a consistent relation R is the intersection of all complete consistent extensions of R.* □

The following is an immediate consequence. It is implicit already in Lemma 7.

Corollary 3. *Every quasi-order is the intersection of its complete consistent extensions.* □

We have been very careful in distinguishing compatible extension from simple containment: the former is a special case of the latter, so, given a quasi-order R, the intersection of all complete consistent relations containing R cannot exceed the intersection of all complete consistent *compatible* extensions of R; and R is contained in the former. Since a complete consistent relation is the same as a complete quasi-order, we thus have

Corollary 4. *Every quasi-order is the intersection of all complete quasi-orders containing it.* □

7 Relative Extensions

Arrow [3] gave a slightly more general form of the extension principle; Inada [36] provided a brief and detailed proof of this variant. A further variation was phrased by Suzumura [46] for consistent relations; recently Cato [18] suggested another generalization. We want to give a short account, focusing on Cato's result. We are not going to go into painstaking detail, and leave out a few details which can be easily verified by "chasing elements". The point we wish to make is that the algebraic method suffices at large.

Lemma 8. *If Q is a quasi-order on $Y \subseteq X$, and R is a quasi-order with $R|_Y = \Delta_Y$, then*

(i) $Q \circ R \circ Q = Q$,
(ii) $(Q \cup R)^n \subseteq Q \cup (Q \circ R) \cup (R \circ Q \circ R) \cup (R \circ Q) \cup R$ for every $n \geqslant 1$, and
(iii) $Q \cup R$ is consistent.

Proof. We omit the symbol $\circ$ for relational composition, writing RS for $R \circ S$, etc.
 (i) With $R|_Y = \Delta$ we have $QRQ = QR|_Y Q = Q\Delta_Y Q = QQ = Q$.
 (ii) This is a simple argument by induction; apply (i) and take into account transitivity of R and Q.
 (iii) From (ii) we get

$$\mathrm{tc}(Q \cup R) = Q \cup QR \cup RQR \cup RQ \cup R.$$

This can be used to show that $Q \cup R$ is consistent, which [36, 14] demonstrate in detail, but element-wise. However, an algebraic proof is possible too, the key to which is given by the *law of modularity* [31]:

$$RS \cap T \subseteq (R \cap TS^\circ)S, \tag{$\dagger$}$$

which can be put equivalently as

$$RS \cap T \subseteq R(S \cap R^\circ T). \tag{$\ddagger$}$$

E.g., we calculate

$$RQ \cap R^\circ \stackrel{(\ddagger)}{\subseteq} R(Q \cap R^\circ R^\circ) \subseteq R(Q \cap R^\circ) \subseteq R\Delta_Y \subseteq R, \tag{$*$}$$

and then

$$RQR \cap R^\circ \stackrel{(\dagger)}{\subseteq} (RQ \cap R^\circ R^\circ)R = (RQ \cap R^\circ)R \stackrel{(*)}{\subseteq} RR = R.$$

The remaining inclusions can be shown similarly. □

The first additional principle we consider reads as follows [3, 36, 46, 18]:

Relative Extension Principle (REP). *Let Q be a relation on $Y \subseteq X$, and let P be a relation on X such that $\overline{P}|_Y = \Delta_Y$. If both P and Q are consistent, then P has a complete consistent extension which restricts to a complete consistent extension on Y of Q,*

Notice that the assumption is put in positive form: instead of assuming [18] that P satisfies $(x,y) \notin \text{tc}(P)$ for every pair of distinct elements $x, y \in Y$, we stipulate $\text{tc}(P)|_Y \subseteq \Delta_Y$. This makes possible a more perspicuous proof.

Proof of REP. Suppose that P and Q are consistent as in CEP. The hull $\overline{P}$ is a quasi-order on X which extends P. By means of CEP, there is a complete consistent extension Q^* of Q on Y. According to Lemma 8(iii), the union $Q^* \cup \overline{P}$ is consistent, and it extends $\overline{P}$ on X, because of

$$Q^* \cap \overline{P}^\circ \subseteq \overline{P}^\circ|_Y \subseteq (\overline{P}|_Y)^\circ = \Delta_Y \subseteq \overline{P}.$$

Another invocation of CEP gives rise to a complete consistent extension S of $\overline{P} \cup Q^*$ on X. This S is an extension of $\overline{P}$, and the restriction of S on Y coincides with Q^*. In fact, since Q^* is complete on Y, it suffices to show that $S|_Y$ extends Q^* (remember Remark 5):

$$S|_Y \cap Q^{*\circ} = (S \cap Q^{*\circ})|_Y \subseteq [S \cap (\overline{P} \cup Q^*)^\circ]|_Y \subseteq (\overline{P} \cup Q^*)|_Y = Q^*. \quad \square$$

This REP is the special case $n = 1$ of the following principle:

Nested Extension Principle (NEP). *If $Y_0 \subseteq Y_1 \subseteq \cdots \subseteq Y_n$ is a chain of sets, each of which is equipped with a consistent relation P_i in such a way that $\overline{P_{i+1}}|_{Y_i} = \Delta_{Y_i}$ for every $i < n$, then P_n has a complete consistent extension, which restricts for every $i < n$ to a complete consistent extension on Y_i of P_i.* $\quad \square$

Proof. By a straightforward inductive argument NEP follows from REP. $\quad \square$

8 Complements

8.1 Complementing Consistency

In Proposition 1 we have seen that if a consistent relation R ranks composed pairs, i.e., if R is consistent and such that $R \circ R \subseteq R \cup R^\circ$, then R is transitive. If R

is even reflexive, then transitivity follows already in case R ranks (endpoints of) compositions for *some* length $n \geqslant 2$, which is to say that

$$R^n \subseteq R \cup R^\circ.$$

For if $\Delta \subseteq R$, then $R^2 = R^2 \circ \Delta^{n-2} \subseteq R^n$, and one can proceed with a similar argument as in the proof of Proposition 1.

Another condition on consistent relations, which brings about transitivity, has been given by Bossert and Suzumura [15]. Their *SC-complementarity* can be put as

$$\forall x, y, z \in X \, [\, xRy \wedge yRz \, \rightarrow \, (\, xRz \, \vee \, (\, z\mathrm{tc}(R)x \, \wedge \, \neg(zRy \wedge yRx) \,) \,) \,],$$

writing xRy for $(x, y) \in R$. Along with Proposition 1, it then follows that a consistent relation R satisfies SC-complementarity if and only if $R \circ R \subseteq R \cup R^\circ$.

We should point out that $R \circ R \subseteq R \cup R^\circ$ occurs in [15] equivalently as *TSC-complementarity*

$$\forall x, y, z \in X [\, xRy \wedge yRz \, \rightarrow \, \neg xNz \,],$$

where $N = \{ \, (x, y) : (x, y) \notin R \wedge (y, x) \notin R \, \}$ is the *non-comparable factor* of R. We have preferred to put it positively.

8.2 Equivalent Principles

While to prove the order-extension principles CEP, TEP, QEP, REP and NEP we have tacitly worked in customary Zermelo–Fraenkel Set Theory with the Axiom of Choice (**ZFC**), to establish their equivalence requires to drop the Axiom of Choice and move to Zermelo–Fraenkel Set Theory (**ZF**) without the Axiom of Choice. Most likely even weaker set theories would suffice, but this shall not be our concern here.

Proposition 4. *In* **ZF** *the following principles are equivalent:* CEP, TEP, QEP, REP *and* NEP.

Proof. We already know the following implications:

$$\mathrm{QEP} \leftarrow \mathrm{TEP} \leftarrow \mathrm{CEP} \rightarrow \mathrm{REP} \leftrightarrow \mathrm{NEP}$$

In view of this we only have to verify that each of QEP and REP implies CEP.

As for QEP implies CEP, let R be consistent. Now $\overline{R}$ is a quasi-order which, by Proposition 2, extends R. By QEP, this hull can be extended to a complete quasi-order S, which is an extension of R too (Lemma 2). As for REP implies CEP, to prove the latter apply the former with $X = Y$, $Q = R$ and $P = \Delta$. □

8.3 Order Extension by Open Induction

Several theorems which commonly are proved by means of Zorn's Lemma have been reproved in a more direct way via the principle of *Open Induction* [40].[9] In this vein we now present an alternative proof of CEP which rests on Open Induction. First some terminology is required.

Let $(E, \leqslant)$ be a chain-complete poset, and let O be a predicate on E.[10] One says that O is *progressive* if

$$\forall x \, (\, \forall y > x \; O(y) \; \to \; O(x) \,),$$

where $y > x$ is understood as the conjunction of $x \leqslant y$ and $x \neq y$. Furthermore, O is said to be *open* if

$$O(\bigvee C) \; \to \; \exists x \in C \; O(x)$$

for every chain $C \subseteq E$; recall that $\bigvee C$ stands for the least upper bound of C. For example, a predicate O is open whenever it is *downward monotone*, i.e., satisfies

$$O(x) \wedge y \leqslant x \; \to \; O(y).$$

Indeed, for if O is downward monotone, and if C is a chain such that $O(\bigvee C)$, then even $\forall x \in C \; O(x)$; note that in this paper every chain is required to have an element.

Raoult [40] has coined the following principle:

Open Induction (OI). *If E is a chain-complete poset, and O is open and progressive, then $\forall x \; O(x)$.*

Moreover, Raoult [40] has deduced OI from ZL; in fact, both principles are equivalent by complementation and thus in **ZF**—see, e.g., [42]. Here is how to prove CEP by means of OI:

We have seen (Lemma 2 and Lemma 3) that the set $\mathcal{E}$ of consistent relations on X is partially ordered and chain-complete with respect to the order of (compatible) extension. On $\mathcal{E}$ we consider the predicate O of "being completely extendable", formally, for $R \in \mathcal{E}$:

$$O(R) \; \equiv \; \exists S \in \mathcal{E} \, (\, R \subseteq S \wedge S \cap R^\circ \subseteq R \wedge S \cup S^\circ = X \times X \,).$$

E.g., the universal relation $X \times X$ is completely extendable, for trivial reasons. This predicate O is downward monotone, hence open. As for O being progressive, suppose that $R \in \mathcal{E}$ is such that every *strict* extension of R is completely extendable. The

---
[9] For the use of Open Induction in diverse contexts see [8, 21, 44, 42, 20].
[10] This O may be identified with its extension $\{ x \in E : O(x) \}$ in E.

hull $\overline{R}$ is a consistent extension of R, which either is complete—by which R has itself as complete extension—or else fails to compare a certain pair of elements. In the latter case, say $x, y \in X$ are such that $(x, y) \notin \overline{R} \cup \overline{R}^\circ$. Then $\overline{R} \cup \{(x, y)\}$ is a consistent extension of $\overline{R}$ by Lemma 5, and strictly extends $\overline{R}$ because of $(x, y) \notin \overline{R}$. Now $\overline{R} \cup \{(x, y)\}$ is completely extendable, whence R is, as well. Then, by way of OI, we get $\forall R \in \mathcal{E} \ O(R)$, which is to say that every consistent relation R on X has a complete consistent extension.

Any concrete enough instance, i.e., one for which the underlying set X of alternatives is finite, should then allow to reduce the invocation of OI to one of what in [44] is called *Finite Induction*, which in turn can be proved by means of mathematical induction only.

8.4 From Order Extension to Gödel–Dummett Logic

As alluded to in the Introduction, we now briefly sketch how Gödel–Dummett logic [33, 27] is necessary for QEP, adapting to quasi-orders an argument given by Bell [6, p. 162] for partial orders. We recall that Gödel–Dummett logic,[11] which "naturally turns up in different fields in logic and computer science" [30, 4], is an intermediate logic between intuitionistic and classical logic. Roughly speaking, intuitionistic logic [35, 48] is classical logic without the law of excluded middle but with the principle *ex falso sequitur quodlibet*. Now Gödel–Dummett logic is intuitionistic logic plus

Gödel–Dummett Principle (GDP). $(\varphi \to \psi) \vee (\psi \to \varphi)$ *for all well-formed formulas φ and ψ.*

In order to adapt Bell's argument, we first make the following observation:

Lemma 9. *Let X be a set and $R \subseteq X \times X$ antisymmetric. Assume that X has top element 1, i.e., $xR1$ for every $x \in X$. If S is a compatible extension of R, then 1 is S-maximal.*

Proof. Since 1 is R-top we have $1R^\circ x$ for every $x \in X$. Therefore, if $1Sy$, we get $1Ry$ because of $S \cap R^\circ \subseteq R$. As R is antisymmetric, $y = 1$ follows. □

Note that if 1 is R-top and $R \subseteq S$, then of course 1 is S-top too, but this does not mean that 1 is S-maximal unless S is antisymmetric.

In the following, we work in Friedman's Intuitionistic Zermelo-Fraenkel set theory **IZF** [32, 1, 22, 7]. This **IZF** is as standard Zermelo-Fraenkel set theory (**ZF**) but with intuitionistic rather than classical logic; to make this move possible, the axiom

---

[11] As von Plato points out, Gödel–Dummett logic "was actually introduced by Skolem already in 1913" [49].

of foundation needs to be replaced by the schema of set induction, whereas the principles of power set and full separation are part of **IZF**.

In **IZF** one thus has the so-called set of intuitionistic truth values $\Omega = \mathcal{P}(1)$, i.e., the set of subsets of $1 = \{0\}$ partially ordered by inclusion $\subseteq$. Every formula φ in the first-order language of set theory gives rise to its truth value $V_\varphi \in \Omega$, viz.

$$V_\varphi = \{x \in 1 : \varphi\},$$

for which φ is equivalent to $0 \in V_\varphi$ and thus to $V_\varphi = 1$. Conversely, every $U \in \Omega$ is of the form V_φ, for φ being $U = 1$. Note that an implication $\varphi \to \psi$ between formulas φ and ψ is equivalent to $V_\varphi \subseteq V_\psi$ in Ω; and that $U \subseteq W$ in Ω amounts to $U = 1 \to W = 1$.

Bell [6, p. 162] deduced GDP from the principle that every partial order is contained in a complete one. To do so he needed that $\subseteq$ is maximal, with respect to containment, among antisymmetric relations on Ω. As quasi-orders lack antisymmetry, we have to adapt Bell's tool as follows.

Lemma 10. *Every compatible extension $\leqslant$ of $\subseteq$ on Ω coincides with $\subseteq$, i.e., $\subseteq$ is maximal with respect to compatible extension of relations on Ω.*

Proof. Now suppose that $\leqslant$ is a compatible extension of $\subseteq$ on Ω. By Lemma 9, and since 1 is $\subseteq$-top, we get

$$U \leqslant W \to (U = 1 \to W = 1)$$

or, equivalently, $U \leqslant W \to U \subseteq W$, for all $U, W \in \Omega$. $\square$

Proposition 5 (IZF). QEP *implies* GDP.

Proof. Applying QEP, and taking into account Lemma 10, we may consider $\subseteq$ on Ω to be complete, which is tantamount to $(\varphi \to \psi) \vee (\psi \to \varphi)$ for arbitrary formulas φ and ψ. $\square$

The same assertion holds true if QEP in Proposition 5 is replaced by any of its equivalents from Theorem 4, because proving these forms equivalent is possible already in **IZF**.

To get GDP we have used the same data as Bell [6, p. 162]: the relation $\subseteq$ on the set Ω. In particular, we have invoked the consequence of QEP that every partial order $\subseteq$ can be extended to a complete quasi-order. Any such extension of $\subseteq$ on Ω, however, *a fortiori* is a partial order anyway (Lemma 10).

8.5 Further directions

By now our focus has been on methodological advantage which our choice of positive notions for consistency and compatible extension entails. Now we sketch an important application of order extension in the theory of preference relations—that is, rationalizability.

We follow [14, 41]. Let again the set X denote our domain of discourse. A *choice function* is a mapping
$$\mathfrak{c}: \mathcal{T} \to \mathcal{P}(X)$$
which assigns to each inhabited member $Y \in \mathcal{T}$, where $\mathcal{T} \subseteq \mathcal{P}(X)$, an inhabited subset $\mathfrak{c}(Y) \subseteq Y$. A relation R on X *rationalizes* $\mathfrak{c}$ if
$$\mathfrak{c}(Y) = \{\, x \in Y : \forall y \in Y \; xRy \,\}$$
for every Y in the domain of $\mathfrak{c}$; notice that no further assumption on R is made. This version of rationalizability is known as *greatest-element rationalizability* [14] of the choice function $\mathfrak{c}$. The (indirect) *revealed preference relation* of a choice function $\mathfrak{c}$ is defined to be (the transitive closure of)
$$\{\, (x, y) : \exists Y \in \mathcal{T} \, (\, x \in \mathfrak{c}(Y) \wedge y \in Y\,)\,\}.$$

A choice function $\mathfrak{c}$ is said to satisfy the *congruence axiom* [41] if it is rationalized by its indirect revealed preference relation. This is the setting for a fundamental application of order extension in the theory of preference relations, viz.

Richter's Theorem ([41, Theorem 1]). *A choice function satisfies the congruence axiom if and only if it can be rationalized by a complete quasi-order.*

A key step to proving this is the following observation.

Lemma 11. *Let $\mathfrak{c}: \mathcal{T} \to \mathcal{P}(X)$ be a choice function. If R is transitive and rationalizes $\mathfrak{c}$, then so does every compatible extension of R.*

Proof. We redraft an argument laid out in [14, Theorem 3.2]. Suppose that S is a compatible extension of R. Given that R rationalizes $\mathfrak{c}$, and since $R \subseteq S$, it suffices to show
$$\{\, x \in Y : \forall y \in Y \; xSy \,\} \subseteq \{\, x \in Y : \forall y \in Y \; xRy \,\}$$
whenever $Y \in \mathcal{T}$. To this end, let $Y \in \mathcal{T}$, and let $x \in Y$ be such that $\forall y \in Y \; xSy$. Pick any $z \in \mathfrak{c}(Y)$. As R rationalizes $\mathfrak{c}$, we have $\forall y \in Y \; zRy$, so in particular zRx. But we also know xSz. Hence xRz by compatibility of S over R. Now $\forall y \in Y \; xRy$ is immediate, since R is supposed to be transitive. □

Therefore, if a choice function c satisfies the congruence axiom, then every compatible complete extension of its indirect revealed preference relation rationalizes c as well. This is how Richter's theorem rests on order extension.[12] Conversely, it is not hard to show that if c can be rationalized by means of a complete quasi-order, then it satisfies the congruence axiom. In his proof Richter applies Szpilrajn's theorem in its original reading that every *irreflexive* transitive relation, i.e., every strict partial order, is contained in one which compares every pair of distinct elements. While Cato has recently deemed this form of Szpilrajn's theorem "not useful for economic analyses because partial orders do not allow two alternatives to be indifferent" [19, p. 60], there is again a definite point to make from the methodological perspective. In order to have (strict) partial orders at hand, Richter performs a quotient construction—a move which has turned out avoidable by way of an appropriate extension principle for quasi-orders, as considered before.

Further applications of our method might be possible in the directions that research on order and extension principles has taken. For instance, the topological notion of continuity comes into play in [12, 10, 9]. In [24], transitive closure is replaced by several other closure operators, thus leading to further extension theorems. In [23] conditions for a collection of binary relations to have a common ordering extension are provided. The classic closure-complement problem has been revisited for consistent closure in [17]. Last but not least, extensions have been considered with regard to the existence of maximal elements in quasi-orders. In [39] it is shown that any maximal element of a quasi-order R is the greatest element for some complete extension of R.

References

[1] Peter Aczel and Michael Rathjen. Constructive set theory. Book draft, 2010.

[2] Athanasios Andrikopoulos. Szpilrajn-type theorems in economics. MPRA Paper No. 14345, 2009. URL: https://mpra.ub.uni-muenchen.de/14345/.

[3] Kenneth J. Arrow. *Social Choice and Individual Values*. Wiley, New York, 1951.

[4] Federico Aschieri, Agata Ciabattoni, and Francesco A. Genco. Gödel logic: from natural deduction to parallel computation. In *Proceedings of the Thirty-Second Annual ACM/IEEE Symposium on Logic in Computer Science (LICS), Reykjavik 2017*, pages 1–12.

---

[12]Incidentally, Richter mentions that "both representability and rationality have existential clauses in their definitions, so proofs of these properties are likely to involve tools like the axiom of choice and other nonconstructive techniques." [41, p. 637]

[5] John L. Bell. Zorn's lemma and complete Boolean algebras in intuitionistic type theories. *Journal of Symbolic Logic*, 62(4):1265–1279, 1997.

[6] John L. Bell. *Set Theory. Boolean-Valued Models and Independence Proofs*. Oxford Logic Guides. Oxford University Press, 2005.

[7] John L. Bell. *Intuitionistic Set Theory*, volume 50 of *Studies in Logic*. College Publications, 2014.

[8] Ulrich Berger. A computational interpretation of open induction. In F. Titsworth, editor, *Proceedings of the Nineteenth Annual IEEE Symposium on Logic in Computer Science*, pages 326–334. IEEE Computer Society, 2004.

[9] Gianni Bosi and Gerhard Herden. On a strong continuous analogue of the Szpilrajn theorem and its strengthening by Dushnik and Miller. *Order*, 22(4):329–342, 2005.

[10] Gianni Bosi and Gerhard Herden. On a possible continuous analogue of the Szpilrajn theorem and its strengthening by Dushnik and Miller. *Order*, 23(4):271–296, 2006.

[11] Walter Bossert. Intersection quasi-orderings: An alternative proof. *Order*, 16(3):221–225, 1999.

[12] Walter Bossert, Yves Sprumont, and Kotaro Suzumura. Upper semicontinuous extensions of binary relations. *Journal of Mathematical Economics*, 37(3):231–246, 2002.

[13] Walter Bossert, Yves Sprumont, and Kotaro Suzumura. Consistent rationalizability. *Economica*, 72(286):185–200, 2005.

[14] Walter Bossert and Kotaro Suzumura. *Consistency, Choice, and Rationality*. Harvard University Press, Cambridge, 2010.

[15] Walter Bossert and Kotaro Suzumura. Quasi-transitive and Suzumura consistent relations. *Social Choice and Welfare*, 39(2):323–334, 2012.

[16] Felix Brandt, Vincent Conitzer, Ulle Endriss, Jérôme Lang, and Ariel D. Procaccia, editors. *Handbook of Computational Social Choice*. Cambridge University Press, New York, 2016.

[17] Susumu Cato. Complements and consistent closures. *Discrete Mathematics*, 312(6):1218–1221, 2012.

[18] Susumu Cato. Szpilrajn, Arrow and Suzumura: concise proofs of extension theorems and an extension. *Metroeconomica*, 63(2):235–249, 2012.

[19] Susumu Cato. *Rationality and Operators. The Formal Structure of Preferences*. Springer, Singapore, 2016.

[20] Francesco Ciraulo, Davide Rinaldi, and Peter Schuster. Lindenbaum's lemma via open induction. In R. Kahle, T. Strahm, and T. Studer, editors, *Advances in Proof Theory*, volume 28 of *Progress in Computer Science and Applied Logic*, pages 65–77. Springer International Publishing Switzerland, Cham, 2016.

[21] Thierry Coquand. A note on the open induction principle. Technical report, Göteborg University, 1997. URL: www.cse.chalmers.se/~coquand/open.ps.

[22] Laura Crosilla. Set Theory: Constructive and Intuitionistic ZF. In Edward N. Zalta, editor, *The Stanford Encyclopedia of Philosophy*. Metaphysics Research Lab, Stanford University, summer 2015 edition, 2015. URL: https://plato.stanford.edu/

archives/sum2015/entries/set-theory-constructive/.

[23] Thomas Demuynck. Common ordering extensions. Working paper, Ghent University, Faculty of Economics and Business Administration, 2009. URL: http://econpapers.repec.org/paper/rugrugwps/09_2f593.htm.

[24] Thomas Demuynck. A general extension result with applications to convexity, homotheticity and monotonicity. *Mathematical Social Sciences*, 57(1):96–109, 2009.

[25] David Donaldson and John A. Weymark. A quasiordering is the intersection of orderings. *Journal of Economic Theory*, 78(2):382–387, 1998.

[26] John Duggan. A general extension theorem for binary relations. *Journal of Economic Theory*, 86(1):1–16, 1999.

[27] Michael Dummett. A propositional calculus with denumerable matrix. *Journal of Symbolic Logic*, 24(2):97–106, 1959.

[28] Ben Dushnik and E. W. Miller. Partially ordered sets. *American Journal of Mathematics*, 63(3):600–610, 1941.

[29] Ulrich Felgner and John K. Truss. The independence of the prime ideal theorem from the order-extension principle. *Journal of Symbolic Logic*, 64(1):199–215, 1999.

[30] Christian G. Fermüller and Agata Ciabattoni. From intuitionistic logic to Gödel-Dummett logic via parallel dialogue games. In *Proceedings of IEEE International Symposium on Multiple-Valued Logic (ISMVL 2003)*, pages 188–193, 2003.

[31] Peter J. Freyd and Andre Scedrov. *Categories, Allegories*. North-Holland, Amsterdam, 1990.

[32] Harvey Friedman. The consistency of classical set theory relative to a set theory with intuitionistic logic. *Journal of Symbolic Logic*, 38(2):315–319, 1973.

[33] Kurt Gödel. Zum intuitionistischen Aussagenkalkül. In Solomon Feferman, editor, *Kurt Gödel: Collected Works*, volume 1, pages 222–225. Oxford University Press, 1986.

[34] Bengt Hansson. Choice structures and preference relations. *Synthese*, 18(4):443–458, 1968.

[35] Arend Heyting. *Intuitionism. An Introduction*. North-Holland, Amsterdam, 1956.

[36] Ken-ichi Inada. Elementary proofs of some theorems about the social welfare function. *Annals of the Institute of Statistical Mathematics*, 6(1):115–122, 1954.

[37] Thomas Jech. *The Axiom of Choice*. North-Holland, Amsterdam, 1973.

[38] Sara Negri, Jan von Plato, and Thierry Coquand. Proof-theoretical analysis of order relations. *Archive for Mathematical Logic*, 43:297–309, 2004.

[39] Vladislav V. Podinovski. Non-dominance and potential optimality for partial preference relations. *European Journal of Operational Research*, 229(2):482–486, 2013.

[40] Jean-Claude Raoult. Proving open properties by induction. *Information Processing Letters*, 29(1):19–23, 1988.

[41] Marcel K. Richter. Revealed preference theory. *Econometrica*, 34(3):635–645, 1966.

[42] Davide Rinaldi and Peter Schuster. A universal Krull-Lindenbaum theorem. *Journal of Pure and Applied Algebra*, 220(9):3207–3232, 2016.

[43] Davide Rinaldi, Peter Schuster, and Daniel Wessel. Eliminating disjunctions by disjunction elimination. *Bulletin of Symbolic Logic*, 23(2):181–200, 2017.

[44] Peter Schuster. Induction in algebra: a first case study. *Logical Methods in Computer Science*, 9(3:20), 2013.

[45] Kotaro Suzumura. Remarks on the theory of collective choice. *Economica*, 43(172):381–390, 1976.

[46] Kotaro Suzumura. An extension of Arrow's lemma with economic applications. Technical report, Institute of Economic Research, Hitotsubashi University, 2004.

[47] Edward Szpilrajn. Sur l'extension de l'ordre partiel. *Fundamenta Mathematicae*, 16(1):386–389, 1930.

[48] Dirk van Dalen. *Logic and Structure*. Springer-Verlag, Berlin, fourth edition, 2004.

[49] Jan von Plato. Skolem's discovery of Gödel-Dummett logic. *Studia Logica*, 73(1):153–157, 2003.

Maximum Entropy Models for Σ_1 Sentences

Soroush Rafiee Rad
Institute for Logic, Language and Computation, UvA, The Netherlands.
soroush.r.rad@gmail.com

Abstract

In this paper we investigate the most uninformative models of Σ_1 sentences. We will show that the two main approaches for defining the Maximum Entropy models on first order languages are well defined for Σ_1 sentences and that they agree on sets of sentences consisting of only Σ_1 sentences.

Keywords: Maximum Entropy, probabilistic models, existential sentence, Objective Bayesian Epistemology

1 Introduction

The Maximum Entropy model for a sentence ϕ represents the most uninformative model of ϕ. To be more precise, given a consistent sentence ϕ and a formula $\psi(x_1, \ldots, x_n)$ from a first order language L, let M be an structure for L with domain $\{a_1, a_2, \ldots\}$ which we only know to be a model of ϕ. A natural question about this M is to ask how likely it is for M to be also a model of ψ, in other words, what probability should one assign to M being also a model of ψ. When ϕ identifies a unique model N (i.e. $M = N$), this question may be answered by checking the validity of $\psi(a_{i_1}, \ldots, a_{i_n})$ in N. If ϕ admits more than one model, however, knowing that M is a model of ϕ under-determines M and the validity of ψ in M may be uncertain. In this sense ϕ induces an assignment of probabilities to the sentences of the language, where the probability assigned to ψ is intended as the probability that a random model of ϕ is also a model of ψ. This will in turn induce a probability distribution on the set of structures for L with domain $\{a_1, a_2, \ldots, \}$.

We are interested in the least informative of such assignments with respect to M which we shall call the Maximum Entropy model of ϕ, i.e., the Maximum Entropy model of ϕ is identified with the assignment of probabilities that leaves M as unconstrained as possible beyond being a model of ϕ. In this sense it gives a probabilistic description that specifies

---

I would like to thank the referees for their helpful comments. I would also like to thank Jeff Paris for his invaluable guidance and advice in the development of these results.

M to the extent that it is characterised by ϕ while remaining as free as possible beyond that. Note, however, that a Maximum Entropy model is not a model in the sense of a structure for the language, but rather a probability function on the set of sentences of the language that characterises an uncertain (i.e. under-determined) structure. It is important to emphasise at this point that in what follows, we shall say "model" to refer to these probability functions. We shall instead say "term models" to refer to the structures. More generally, given a set of linear constraints K the Maximum Entropy model of K is the probability function over the sentences of L which satisfies the constraints given in K while remaining maximally uninformative beyond that. When considering a set of linear constraints K, we use "models of K" and "solutions for K", interchangeably.

These probability functions have been extensively investigated and applied in various disciplines from statistics [5] and physics, [7] to computer science, pattern recognition [3], computational linguistics [2] as well as economics and finance [6]. Another prototypical example where Maximum Entropy models are of great relevance is formal epistemology and the study of rational belief formation [8, 16, 17]. In this setting the problem of interest is how should an agent in possession of some evidence form rational belief? To be slightly more precise, the question is; given sentences $\phi_1, \ldots, \phi_n$ as the agent's evidence, what would be the credence x she has to assign to some arbitrary sentence ψ such that x represents a rational belief of the agent in the context of her evidence. Equivalently, one can ask which probability function over the sentences of the language best represents the degrees of belief of the agent.

The most popular proposal for formalising the concept of *least informative* is to take Shannon's entropy as the measure for the informational content of a probability function. Given a set of constraints, one approach, see for example [11], is to choose the probability function satisfying the constraints with maximum Shannon entropy as the least informative one. A second approach, followed for example by Williamson [16, 17], uses the relative Shannon entropy instead. To make the idea clear, consider the problem we started with and a case in which there is no information (and thus no restrictions) concerning the structure M. In this case the satisfaction of a sentence ψ in M is maximally uncertain and thus the assignment of probabilities should be maximally equivocal. We shall call this probability function (which we shall shortly define precisely) $P_=$. The second approach for defining the "least informative", requires the assignment of probabilities to satisfy the given constraints and remain informationally as close as possible to $P_=$, where the informational difference between two probability functions is measured by their relative entropy. It is not hard to check that on propositional languages both approaches are well defined and result in the same unique answer [13].

The literature on justification of Maximum Entropy or its underlying principles is extensive and it remains the strongest candidate for the formalisation of the least informative probability function [11, 15, 17]. The major part of this literature is concerned with propositional languages, however, there have been attempts to generalise both these approaches to the first order case. To generalise the first approach, Barnett and Paris, [1], propose to define the Maximum Entropy models on a first order language as the limit of the Maximum Entropy models on finite sub-languages. They showed that for constraint sets from languages with only unary predicates, this limit exists and the resulting probability function does satisfy the constraints. To generalise the second approach one has to move to a more sophisticated notion of informational distance.

This paper further investigates the Maximum Entropy models and the extent to which they can be defined for first order languages; in particular we shall investigate the Maximum Entropy models for existential sentences. The paper unfolds as follows: Section 2 reviews preliminaries and notation, as well as the definition of the Maximum Entropy probability functions over propositional and first order languages; and Section 3 proves the main theorems. We will then conclude with a discussion in Section 4.

2 Preliminaries and Notation

Throughout this paper, we will work with a first order language L with finitely many relation symbols, no function symbols, no equality and countably many constant symbols $a_1, a_2, a_3, \ldots$. Furthermore we assume that these constants exhaust the universe. Let RL, SL and TL denote the sets of relation symbols, sentences and the term models for L respectively, where a *term model* is a structure M for the language L with domain $M = \{a_i \mid i = 1, 2, \ldots\}$ where every constant symbol is interpreted as itself. For more details on the preliminary definitions and results please see [9, 12].

Definition 1. $w : SL \to [0,1]$ *is a probability function if for every* $\theta, \phi, \exists x \psi(x) \in SL$,
P1. If $\models \theta$ then $w(\theta) = 1$.
P2. If $\models \neg(\theta \land \phi)$ then $w(\theta \lor \phi) = w(\theta) + w(\phi)$.
P3. $w(\exists x \psi(x)) = \lim_{n \to \infty} w(\bigvee_{i=1}^{n} \psi(a_i))$.

Definition 2. *Let* $\mathcal{L}$ *be a finite propositional language with propositional variables* $p_1, \ldots, p_n$. *Atoms of* $\mathcal{L}$ *are the sentences* $\{\alpha_i \mid i = 1, \ldots J\}$, *of the form* $\bigwedge_{i=1}^{n} p_i^{\epsilon_i}$ *where* $\epsilon_i \in \{0, 1\}$, $p^1 = p$ *and* $p^0 = \neg p$.

Take a propositional language $\mathcal{L}$. For every sentence $\phi \in S\mathcal{L}$, there is unique set $\Gamma_\phi \subseteq \{\alpha_i \mid i = 1, \ldots, J\}$ such that $\models \phi \leftrightarrow \bigvee_{\alpha_i \in \Gamma_\phi} \alpha_i$. It can be easily checked that $\Gamma_\phi = \{\alpha_j \mid \alpha_j \models \phi\}$. Thus if w is a probability function $w(\phi) = w(\bigvee_{\alpha_i \models \phi} \alpha_i) = \sum_{\alpha_i \models \phi} w(\alpha_i)$ as the α_i's are mutually inconsistent. On the other hand since $\models \bigvee_{i=1}^{J} \alpha_i$ we have $\sum_{i=1}^{J} w(\alpha_i) = 1$. So the probability function w will be uniquely determined by its values on the α_i's, i.e., by the vector $< w(\alpha_1), \ldots, w(\alpha_J) > \in \mathbb{D}^{\mathcal{L}} = \{\vec{x} \in \mathbb{R}^J \mid \vec{x} \geq 0, \sum_{i=1}^{J} x_i = 1\}$. Conversely if $\vec{a} \in \mathbb{D}^{\mathcal{L}}$ we can define a probability function $w' : S\mathcal{L} \to [0, 1]$ such that $< w'(\alpha_1), \ldots, w'(\alpha_J) > = \vec{a}$ by setting $w'(\phi) = \sum_{\alpha_i \models \phi} a_i$.

Now consider a first order language L. Although the atoms of L are not expressible in the language (as they will require infinite conjunctions), the *state descriptions* for the finite sub-languages will play a similar role to that of atoms in the propositional case.

Definition 3. *Let L be a first order language with the finite set of relation symbols RL and let L^k be the sub-language of L with only constant symbols $a_1, ..., a_k$. The state descriptions of L^k are the sentences $\Theta_1^k, ..., \Theta_{n_k}^k$ of the form*

$$\bigwedge_{\substack{i_1,...,i_j \leq k \\ R_i \in RL \;\; j\text{-}ary}} R_i(a_{i_1}, ..., a_{i_j})^{\epsilon_{i_1,...,i_j}}$$

where $\epsilon_{i_1,...,i_j} \in \{0, 1\}$ and $R_i^1 = R_i$ and $R_i^0 = \neg R_i$.

Throughout this paper we will denote the set of state descriptions for L and L^r by Γ and Γ^r respectively. Furthermore, we will write Γ_ϕ (res. Γ_ϕ^r) for the set of state descriptions of L (res. L^r) that are consistent with the sentence ϕ.

For a quantifier free sentence $\theta \in SL$ let k be an upper bound on the i such that a_i appears in θ. Then θ can be thought of as being from the propositional language $\mathcal{L}^k$ with propositional variables $R_i(a_{i_1}, ..., a_{i_j})$ for $i_1, ..., i_j \leq k$, $R_i \in RL$. The sentences Θ_i^k will be the atoms of $\mathcal{L}^k$ and as before $\models \theta \leftrightarrow \bigvee_{\Theta_i^k \models \theta} \Theta_i^k$ and for every probability function w, $w(\theta) = w(\bigvee_{\Theta_i^k \models \theta} \Theta_i^k) = \sum_{\Theta_i^k \models \theta} w(\Theta_i^k)$. Thus to determine $w(\theta)$ we only need to determine the values $w(\Theta_i^k)$ and to require

$$w(\Theta_i^k) \geq 0 \text{ and } \sum_{i=1}^{n_k} w(\Theta_i^k) = 1 \quad (1)$$

$$w(\Theta_i^k) = \sum_{\Theta_j^{k+1} \models \Theta_i^k} w(\Theta_j^{k+1}) \quad (2)$$

to ensure that w satisfies P1 and P2. The following theorem due to Gaifman [4], ensures that this is indeed enough to determine w on all sentences. Let $QFSL$ be the set of quantifier free sentences of L.

Theorem 1. *Let $v : QFSL \to [0, 1]$ satisfy P1 and P2 for $\theta, \phi \in QFSL$. Then v has a unique extension $w : SL \to [0, 1]$ that satisfies P1, P2 and P3. In particular if $w : SL \to [0, 1]$ satisfies P1, P2 and P3 then w is uniquely determined by its restriction to $QFSL$.*

Just as a probability function on the set of sentences of a propositional language is determined by its values on the atoms, a probability function on the set of sentences of a first order language is determined by its values on the state descriptions. We note that the set of state descriptions of L^k is the same as the set of term models for L^k with domain $\{a_1, ..., a_k\}$.

Definition 4. *Define the equivocator, $P_=$, as the probability function that for each k, assigns equal probabilities to the Θ_i^k's (the state descriptions of L^k), i.e., the most non-committal probability function.*

Notice that this determines $P_=$ on all of SL by Theorem 1 and the preceding argument.

Definition 5. *A sentence ϕ from a first order language L is called a Σ_1 sentence iff ϕ is logically equivalent to a sentence of the form $\exists \vec{x} \theta(\vec{x})$ where $\theta(\vec{x})$ is quantifier free.*

Definition 6. *A constraint set K is a finite satisfiable set of linear constraints of the from $\{\sum_{j=1}^{n} a_{ij} w(\theta_j) = b_i \mid i = 1, \ldots, m\}$, where $\theta_j \in SL, a_{ij}, b_i \in \mathbb{R}$ and w is a probability function. Every finite satisfiable set of sentences $K = \{\phi_1, \ldots, \phi_n\}$ is identified with the constraint set $\{w(\phi_1) = 1, \ldots, w(\phi_n) = 1\}$ induced by it and in particular we shall identify every sentence ϕ with the constraint $w(\phi) = 1$.*

We shall next give the definition of Maximum Entropy solutions for a set of linear constraints K as above. Our results in Section 3, however, are concerned only with the constraints that are induced by a sentence. In particular, by the Maximum Entropy model of the sentences ϕ we mean the Maximum Entropy probability function that satisfies the corresponding constraint $w(\phi) = 1$.

Definition 7. *The Shannon entropy of the probability function, W, defined on a set $X = \{x_1, \ldots, x_n\}$ (so $0 \leq W(x_i) \leq 1$ and $\sum_i W(x_i) = 1$), is given by*

$$E(W) = -\sum_{i=1}^{n} W(x_i) \log(W(x_i)).$$

The Shannon entropy is the most commonly used measures for the informational content of a probability function, [14].

Definition 8. *An inference process, N, on L, is a function that on each set of linear constraints K, returns a probability function on SL, $N(K)$, that satisfies K.*

We will write ME for the inference process that on each set of constraints K, returns the maximum entropy probability function that satisfies K, denoted as $ME(K)$. There are two approaches for defining Maximum Entropy probability functions that satisfy a set of constraints. We shall start from a propositional case first and then move to the first order languages. Let $\mathcal{L}$ be a propositional language with atoms $\alpha_1, \ldots, \alpha_J$ and K a set of linear constraints. The first approach is to define $ME(K)$ as the unique probability function over the sentences of the language that satisfies K and for which the Shannon entropy $-\sum_{i=1}^{J} w(\alpha_i) \log(w(\alpha_i))$ is maximised. Since K consists of only linear constraints, the set of probability functions that satisfy K is convex and so is the function $f(x) = -\sum_{i=1}^{J} x_i \log(x_i)$, hence the uniqueness.

An alternative approach is studied by Williamson [16], which we will denote by ME_W. In this approach Maximum Entropy probability functions that satisfy a set of constraints K are defined by minimising the divergence from the probability function $P_=$, which has the maximum Shannon entropy. The information theoretic divergence of a probability function W from the probability function V is given by their relative entropy and defined as:

$$RE(W, V) = \sum_{i=1}^{J} W(\alpha_i) \log(\frac{W(\alpha_i)}{V(\alpha_i)}).^{1}$$

Williamson defines the Maximum Entropy probability function for a set of constraints K, $ME_W(K)$, as the probability function w, that satisfies K and has the minimum relative entropy to $P_=$, i.e. $\sum_{i=1}^{J} w(\alpha_i) \log\left(\frac{w(\alpha_i)}{P_=(\alpha_i)}\right)$, amongst all those probability functions that satisfy K.

Proposition 1. *Let $\mathcal{L}$ be a propositional language and K a set of linear constraints. Then $ME(K)(\phi) = ME_W(K)(\phi)$ for all $\phi \in S\mathcal{L}$.*

Proof. Let $\alpha_1, \ldots, \alpha_J$ be the atoms of $\mathcal{L}$. Notice that

$$RE(w, P_=) = \sum_{i=1}^{J} w(\alpha_i) \log\left(\frac{w(\alpha_i)}{P_=(\alpha_i)}\right) = \sum_{i=1}^{J} w(\alpha_i) \log(w(\alpha_i)) - \sum_{i=1}^{J} w(\alpha_i) \log(P_=(\alpha_i)) =$$

$$\sum_{i=1}^{J} w(\alpha_i) \log(w(\alpha_i)) - \sum_{i=1}^{J} w(\alpha_i) \log(1/J) = -E(w) + \log(J).$$

Let w be a probability function that satisfies K then w minimises $RE(w, P_=)$ if and only if w maximises $E(w)$. Hence $ME_W(K)$ and $ME(K)$ specify the same probability function. ∎

Thus the two approaches agree for constraint sets from a propositional language. The main difficulty for extending these definitions to first order languages is that in the case of a first order language one does not have access to the atomic sentences in order to express the entropy or the relative entropy. In the first order case one has only access to state descriptions over finite sub-languages.

To extend the first approach to a first order language L, Barnett and Paris [1], propose to define the Maximum Entropy probability function that satisfies K as the limit of the Maximum Entropy models of K restricted to finite sub-languages, L^k. These finite sub-languages can essentially be treated as propositional languages where the Maximum Entropy models are well defined for every set of linear constraints. To be more precise let L be a first order

---
[1] Notice that RE is not a distance measure since it is not symmetric, so it is not the distance between W and V but rather the divergence of W from V.

language with relation symbols $RL = \{R_1, \ldots, R_t\}$ and constant symbols $\{a_1, a_2, \ldots\}$, and let K be a set of linear constraints as above. Define $\mathcal{L}^r$ to be the propositional language with propositional variables $R_i(a_{i_1}, \ldots, a_{i_j})$ for $R_i \in RL$ and $a_{i_1}, \ldots, a_{i_j} \in \{a_1, \ldots a_r\}$. If k is the maximum such that a_k appears in K, for $r \geq k$ define $(-)^{(r)} : SL^k \to S\mathcal{L}^r$ as

$$(R_i(a_{i_1}, \ldots, a_{i_n}))^{(r)} = R_i(a_{i_1}, \ldots, a_{i_n})$$
$$(\neg \phi)^{(r)} = \neg(\phi)^{(r)}$$
$$(\phi \vee \psi)^{(r)} = (\phi)^{(r)} \vee (\psi)^{(r)}$$
$$(\exists x \phi(x))^{(r)} = \bigvee_{i=1}^{r} (\phi(a_i))^{(r)}$$

For a set of linear constraints K, let $K^{(r)}$ be the result of replacing every θ appearing in K with $\theta^{(r)}$ and notice that for a state description Θ^k of L^k and $r \geq k$, $(\Theta^k)^{(r)} = \Theta^k$. Barnett and Paris [1], propose to define the Maximum Entropy probability function on first order languages as follows:

Definition 9. *(ME) Let L be a first order language and K a set of linear constraints. For a state description Θ_i^k of L^k, let $ME(K)(\Theta_i^k) = \lim_{r \to \infty} ME(K^{(r)})(\Theta_i^k)$.*
This determines $ME(K)$ on all state descriptions and thus on all quantifier free sentences, which is uniquely extended to all $\psi \in SL$ by Theorem 1.

For the second approach, ME_W, Williamson first defines the r-divergence of a probability function W from a probability function V by

$$RE_r(W, V) = \sum_{i=1}^{J_r} W(\Theta_i^r) \log \left(\frac{W(\Theta_i^r)}{V(\Theta_i^r)} \right)$$

where Θ_i^r's are state descriptions of L^r. Thus the r-divergence of W from V is the divergence of W from V when they are restricted to L^r. Then for probability functions U, V and W, U is closer to V than W if there exists N such that for all $r > N$, $d_r(U, V) < d_r(U, W)$. Williamson [16] defines the Maximum Entropy probability functions on first order languages as:

Definition 10 (ME_W). *Let K be a set of linear constraints as before. The Maximum Entropy model of K, $ME_W(K)$, is the probability function, w, satisfying K such that there is no probability function v that satisfies K and $d_r(v, P_=) < d_r(w, P_=)$ for all r eventually.*

The main questions here are whether or not the Maximum Entropy probability functions, given by Definitions 9 and 10, are well defined for every constraint set K from a first order language, i.e, whether or not the limit in Definition 9, or the closest probability function to $P_=$ as in Definition 10, exist for every K, and when they are well defined, whether or not the resulting probability functions satisfy K. In [1] Barnett and Paris showed that for any set of linear constraints over monadic first order languages, the Maximum Entropy probability function is indeed well defined and that it satisfies the constraints. On the other hand in the general case for constraint sets containing sentences with quantifier complexity of Σ_2, Π_2 or higher the Maximum Entropy probability functions that satisfy the constraints are not always well defined (see [13]). The case of Π_1 sentences has been studied and partially answered by Paris and Rafiee Rad in [10] and in this paper we will focus on knowledge bases consisting of a Σ_1 sentence, i.e., constraint sets of the form $\{w(\exists \vec{x} \phi(\vec{x})) = 1\}$ where $\phi(\vec{x})$ is quantifier free.

3 The Maximum Entropy Models for Σ_1 Sentences

We will now turn to our main result concerning the Maximum Entropy models of sentences with quantifier complexity of Σ_1. We will show that both approaches for defining Maximum Entropy models are well defined for these sentences and agree with each other. As was pointed out before, for our purpose, every Σ_1 sentence $\exists \vec{x} \theta(\vec{x})$ is identified with the constraint set $\{w(\exists \vec{x}\theta(\vec{x})) = 1\}$.

Lemma 2. *Let $\phi \in SL$ be a satisfiable Σ_1 sentence of the form $\exists x_1, ..., x_t \theta(a_1, ..., a_l, \vec{x})$ and let Γ^l_ϕ be the set of state descriptions of L^l that are consistent with ϕ. Then $P_=(\phi \mid \bigvee \Gamma^l_\phi) = 1$.*

Proof. Let $\gamma = \neg\phi = \forall x_1, ..., x_t \neg\theta(\vec{a}, \vec{x})$. Let $\vec{a}$ be all the constants appearing in θ with l the largest such that a_l appears in $\vec{a}$ and let Γ^l be the set of state descriptions of L^l. First notice that for $\Theta^{(l)}_j \in \Gamma^l$ if $\Theta^{(l)}_j \vDash \gamma$ then $\Theta^{(l)}_j \vDash \neg\phi$ and thus $\Theta^{(l)}_j \notin \Gamma^l_\phi$. We show that for every $\Theta^{(l)}_j \in \Gamma^l_\phi$, $P_=(\Theta^{(l)}_j \wedge \gamma) = 0$. If $\Theta^{(l)}_j$ is inconsistent with $\gamma^{(l)}$ then[2] $P_=(\Theta^{(l)}_j \wedge \gamma) = 0$. This is so because if $\Theta^{(l)}_j$ is inconsistent with $\gamma^{(l)}$ then $\Theta^{(l)}_j \vDash \bigvee_{i_1,...,i_t \leq l} \theta(\vec{a}, a_{i_1}, ..., a_{i_t})$ so $\Theta^{(l)}_j \vDash \exists \vec{x} \theta(\vec{a}, \vec{x}) \equiv \neg\gamma$. So $P_=(\Theta^{(l)}_j \wedge \gamma) \leq P_=(\neg\gamma \wedge \gamma) = 0$.

Let $\Gamma^l_{\phi, \gamma^{(l)}}$ be the set of state descriptions in Γ^l_ϕ that are consistent with $\gamma^{(l)}$. For $\Theta^{(l)}_j \in \Gamma^l_{\phi, \gamma^{(l)}}$ let $Q_i(\vec{a}, x_1, ..., x_t)$, $i \in I$ enumerate formulae of the form

$$\Theta^{(l)}_j \wedge \bigwedge_{\substack{y_{i_1},...,y_{i_j} \in \{a_1,...,a_l\} \cup \{x_1,...,x_t\} \\ \{y_{i_1},...,y_{i_j}\} \cap \{x_1,...,x_t\} \neq \emptyset \\ R \in RL, \ j-ary}} \pm R(y_{i_1}, ..., y_{i_j}).$$

Since $\neg\theta(\vec{a}, \vec{x})$ is not a tautology, and since $\Theta^{(l)}_j \nvDash \gamma$ there is some strict subset J of I such that $\vDash \Theta^{(l)}_j \wedge \neg\theta(\vec{a}, \vec{x}) \leftrightarrow \bigvee_{j \in J} Q_j(\vec{a}, \vec{x})$. To see this notice that the sentences $Q_i(\vec{a}, x_1, ..., x_t)$

---

[2]Remember that $\gamma^{(l)} = \bigwedge_{i_1,...,i_t \leq l} \neg\theta(\vec{a}, a_{i_1}, ..., a_{i_t})$

are state descriptions of a language L but with constants $a_1, \ldots, a_l, x_1, \ldots x_t$, which extend the state description $\Theta_j^{(l)} \in \Gamma_{\phi,\gamma^{(l)}}^l$. Then since $\Theta_j^{(l)} \wedge \neg \theta(\vec{a}, \vec{x})$ is a sentence in the language $L^{a_1,\ldots,a_n,x_1,\ldots x_t}$ that implies $\Theta_j^{(l)}$, it will be equivalent to a disjunction of some of these state descriptions. Now, for $i_1 < i_2 < \ldots < i_t < r$ the number of extensions of $Q_i(\vec{a}, a_{i_1}, \ldots, a_{i_t})$ to a state description of L^r is the same for each i so $P_=(Q_i(\vec{a}, a_{i_1}, \ldots, a_{i_t})) = \frac{1}{|T|}$ and for disjoint $\vec{a}^1, \ldots, \vec{a}^r$, $P_=(Q_{n_1}(\vec{a}, \vec{a}^1) \wedge \ldots \wedge Q_{n_r}(\vec{a}, \vec{a}^r)) = \frac{1}{|T|^r}$. So

$$P_=(\Theta_j^{(l)} \wedge \forall x_1, \ldots, x_t \neg \theta(\vec{a}, \vec{x})) \leq P_=(\Theta_j^{(l)} \wedge \bigwedge_{i=1}^{r} \neg \theta(\vec{a}, \vec{a}^i)) = \sum_{n_1,\ldots,n_r \in J} P_=(\bigwedge_{i=1}^{r} Q_{n_i}(\vec{a}, \vec{a}^i)) = \left(\frac{|J|}{|T|}\right)^r.$$

And $\left(\frac{|J|}{|T|}\right)^r \to 0$ as $r \to \infty$. Thus for all $\Theta_j^{(l)} \in \Gamma_\phi^l$, $P_=(\Theta_j^{(l)} \wedge \gamma) = 0$ and thus $P_=(\gamma \mid \Theta_j^{(l)}) = 0$. So for every $\Theta_j^{(l)} \in \Gamma_\phi^l$, $P_=(\phi \mid \Theta_j^{(l)}) = 1$ and thus $P_=(\phi \mid \bigvee \Gamma_\phi^l) = 1$ as required. ∎

Theorem 3. *Let ϕ be a satisfiable Σ_1 sentence of the form $\exists x_1, \ldots, x_t \theta(a_1, \ldots, a_l, \vec{x})$ and let Γ_ϕ^l be the set of state descriptions of L^l that are consistent with ϕ. For $K = \{w(\phi) = 1\}$ and $\psi \in SL$, $ME_W(K)(\psi) = P_=(\psi \mid \bigvee \Gamma_\phi^l)$.*

Proof. First by Lemma 2, $P_=(- \mid \bigvee \Gamma_\phi^l)$ satisfies K. It is also the closest probability function to $P_=$ that satisfies K. To see this notice that if w is a probability function that satisfies K then $w(\phi) = 1$. Thus for all $k \geq l$, both w and $P_=(- \mid \bigvee \Gamma_\phi^l)$ assign probability zero to the state descriptions of L^k that are inconsistent with $\phi^{(k)}$. For those state descriptions that are consistent with $\phi^{(k)}$, $P_=(- \mid \bigvee \Gamma_\phi^l)$ assigns equal probability while w assigns different probability to at least some of them. Thus for $k \geq l$ on each L^k, $P_=(- \mid \bigvee \Gamma_\phi^l)$ has a higher entropy that w and thus has a smaller k-divergence from $P_=$. Hence by definition $P_=(- \mid \bigvee \Gamma_\phi^l)$ is closer than w to $P_=$. ∎

Theorem 3 specifies the Maximum Entropy models for Σ_1 sentences as characterised by ME_W and Definition 10. We shall now turn to the Maximum Entropy models as characterised by ME and the limit in the Definition 9.

Theorem 4. *Let ϕ be the satisfiable Σ_1 sentence $\exists \vec{x} \theta(a_1, \ldots, a_l, \vec{x})$, Γ_ϕ^l be the set of state descriptions of L^l that are consistent with ϕ and $K = \{w(\phi) = 1\}$. Then for $\psi \in SL$, $ME(K)(\psi) = P_=(\psi \mid \bigvee \Gamma_\phi^l)$.*

Proof.
Let $\Lambda = \bigvee \Gamma_\phi^l$. We will show that for quantifier free ψ, $ME(K)(\psi) = P_=(\psi \mid \Lambda)$. This establishes that $ME(K)$ agrees with $P_=(- \mid \Lambda)$ on quantifier free sentences and thus, by Theorem 1, they will agree on all SL, that is, for all $\psi \in SL$, $ME(K)(\psi) = P_=(\psi \mid \Lambda)$.

Let Γ^r be the set of state descriptions of L^r and Γ^r_K be the subset of Γ^r that satisfy $\phi^{(r)}$. For $\Theta^k_i \in \Gamma^k$ define for $r \geq k$, $\Gamma^r_{k,i} = \{\Psi^r_j \in \Gamma^r \mid \Psi^r_j \vDash \Theta^k_i\}$. In other words, $\Gamma^r_{k,i}$ is the set of state description of L^r that extend the state description Θ^k_i of L^k. Notice that $|\Gamma^r_{k,i}| = |\Gamma^r_{k,j}|$ for $\Theta^k_i, \Theta^k_j \in \Gamma^k$ because state descriptions of L^k will all have the same number of extensions to state descriptions of L^{k+1}. Let ${}^K\Gamma^r_{k,i} = \Gamma^r_K \cap \Gamma^r_{k,i}$ be the set of extensions of Θ^k_i to a state description of L^r that satisfies $\phi^{(r)}$. Take Γ^l_ϕ as the set of state descriptions of L^l that are consistent with ϕ, and let $\Gamma^l_{\neg\phi} = \Gamma^l - \Gamma^l_\phi$.

Notice that $ME(K^{(r)})$ assigns probability zero to those state descriptions of L^r that are inconsistent with $\phi^{(r)}$ (so those not in Γ^r_K) since it should assign probability 1 to $\phi^{(r)}$,

$$\Psi^r \in \Gamma^r \setminus \Gamma^r_K, \quad ME(K^{(r)})(\Psi^r) = 0. \tag{3}$$

Next notice also that $ME(K^{(r)})$ assigns equal probability to those state descriptions that are consistent with $\phi^{(r)}$ (i.e to those in Γ^r_K). To see this, suppose not and define the probability function w on SL^r that agrees with $ME(K^{(r)})$ (i.e. assigns zero probability) on those state descriptions that are inconsistent with $\phi^{(r)}$ but divides the full probability measure equally among those in Γ^r_K. Then w satisfies $K^{(r)}$ but it is easy to check that w has strictly higher entropy than $ME(K^{(r)})$, on L^r, which is a contradiction with the choice of $ME(K^{(r)})$ as the Maximum Entropy probability function on L^r that satisfies $K^{(r)}$, so

$$\Psi^r \in \Gamma^r_K, \quad ME(K^{(r)})(\Psi^r) = \frac{1}{|\Gamma^r_K|}. \tag{4}$$

Thus by (3) and (4), for the state description Θ^k_i, $k \geq l$,

$$ME(K^{(r)})(\Theta^k_i) = \sum_{\substack{\Psi^r \in \Gamma^r \\ \Psi^r \vDash \Theta^k_i}} ME(K^{(r)})(\Psi^r) = \sum_{\substack{\Psi^r \in \Gamma^r_K \\ \Psi^r \vDash \Theta^k_i}} ME(K^{(r)})(\Psi^r) = \frac{|{}^K\Gamma^r_{k,i}|}{|\Gamma^r_K|}.$$

The state descriptions in $\Gamma^l_{\neg\phi}$ are inconsistent with ϕ and thus have no extension to a state description of L^r that satisfies $\phi^{(r)}$. Hence Γ^r_K includes only extensions of state descriptions in Γ^l_ϕ and we have $\Gamma^r_K = \bigcup_{\Theta^l_j \in \Gamma^l_\phi} {}^K\Gamma^r_{l,j}$ and since ${}^K\Gamma^r_{l,j}$'s include extensions of different state description of L^l and are thus disjoint,

$$|\Gamma^r_K| = \sum_{\Theta^l_j \in \Gamma^l_\phi} |{}^K\Gamma^r_{l,j}|. \tag{5}$$

On the other hand, for $k \geq l$, $P_=(-\mid \Lambda)$ assigns equal probabilities to all state descriptions of L^k that are consistent with $\Lambda = \bigvee \Gamma^l_\phi$ and zero to those that are not. Thus those with non-zero probability are exactly those state descriptions of L^k that are extensions of some state description in Γ^l_ϕ and the number of these state descriptions is $\sum_{\Theta^l_j \in \Gamma^l_\phi} |\Gamma^k_{l,j}|$. Thus $P_=(\Theta^k_i | \Lambda) = 0$ if Θ^k_i extends a state description in $\Gamma^l_{\neg\phi}$ and $P_=(\Theta^k_i | \Lambda) = \frac{1}{\sum_{\Theta^l_j \in \Gamma^l_\phi} |\Gamma^k_{l,j}|}$ if Θ^k_i extends a state description in Γ^l_ϕ.

To show $ME(K)(\psi) = P_=(\psi | \Lambda)$ for quantifier free ψ, it is enough to show that for each k and each state description $\Theta^k_i \in \Gamma^k$, $ME(K)(\Theta^k_i) = P_=(\Theta^k_i | \Lambda)$. By definition, this is

$$\lim_{r \to \infty} ME(K^{(r)})(\Theta^k_i) = P_=(\Theta^k_i | \Lambda). \tag{6}$$

For $k \geq l$, the state descriptions of L^k are extensions of either a state description in Γ^l_ϕ or a state description in $\Gamma^l_{\neg\phi}$. The state description in $\Gamma^l_{\neg\phi}$ are inconsistent with ϕ and thus have no extension to L^r that satisfies $\phi^{(r)}$, that is

$$^K\Gamma^r_{k,s} = \emptyset \quad \text{for} \quad \Theta^k_s \in \Gamma^l_{\neg\phi},$$

and so $ME(K^{(r)})(\Theta^k_s) = 0$. Hence for those Θ^k_i that extend a state description in $\Gamma^l_{\neg\phi}$,

$$\lim_{r \to \infty} ME(K^{(r)})(\Theta^k_i) = 0 = P_=(\Theta^k_i | \Lambda).$$

For those Θ^k_i that extend a state description in Γ^l_ϕ, we have to show that

$$\lim_{r \to \infty} \frac{|^K\Gamma^r_{k,i}|}{|\Gamma^r_K|} = \frac{1}{\sum_{\Theta^l_j \in \Gamma^l_\phi} |\Gamma^k_{l,j}|}. \tag{7}$$

Using, (5) and the fact that $|\Gamma^k_{l,j}|$ is the same for all $\Theta^l_j \in \Gamma^l_\phi$, to show 7 we will show that[3]

$$\lim_{r \to \infty} \frac{|^K\Gamma^r_{k,i}| \sum_{\Theta^l_j \in \Gamma^l_\phi} |\Gamma^k_{l,j}|}{\sum_{\Theta^l_j \in \Gamma^l_\phi} |^K\Gamma^r_{l,j}|} = \lim_{r \to \infty} \frac{|^K\Gamma^r_{k,i}| |\Gamma^l_\phi| |\Gamma^k_{l,j}|}{\sum_{\Theta^l_j \in \Gamma^l_\phi} |^K\Gamma^r_{l,j}|} = 1. \tag{8}$$

Lemma 5. *Let K, $^K\Gamma^r_{k,i}$ and $\Gamma^r_{k,i}$ be as defined above then* $\lim_{r \to \infty} \frac{|^K\Gamma^r_{k,i}|}{|\Gamma^r_{k,i}|} = 1$.

Proof.
Notice that $|\frac{^K\Gamma^r_{k,i}}{\Gamma^r_{k,i}}|$ is the probability that a random extension of the state description $\Theta^k_i \in \Gamma^k$ to a state description of L^r will satisfy the $K^{(r)}$.[4] Remember that K consists of a Σ_1

---

[3] Notice that $\sum_{\Theta^l_j \in \Gamma^l_\phi} |\Gamma^k_{l,j}| \neq 0$ and does not depend on r.

[4] The denominator is the total number of extensions of $\Theta^k_i \in \Gamma^k$ to a state description of L^r and the nominator is the number of those extensions of $\Theta^k_i \in \Gamma^k$ to a state description of L^r that satisfy $K^{(r)}$.

sentence $\exists x_1, ..., x_t \theta(\vec{a}, x_1, ..., x_t)$, l is the largest that a_l appears in $\theta(\vec{a}, \vec{x})$, and that Θ_i^k extends description in Γ_ϕ^l, say Ψ^l, and let's calculate this probability.

Take $\Theta_i^k \in \Gamma^k$ and let's consider its extensions to state descriptions of L^{k+t}. Let $L^{a_{i_1},...a_{i_n}}$ be language L with only constant symbols $a_{i_1}, ..., a_{i_n}$ and let Δ_i $i = 1, ..., M$ enumerate the state descriptions of $L^{\{a_1,...,a_l\} \cup \{a_{k+1},...,a_{k+t}\}}$ that extend Ψ^l (thus they agree with Θ_i^k when restricted to $a_1, ..., a_l$). Then state descriptions of L^{k+t} that are extension of Θ_i^k can be written in the form $\Theta_{i,m}^{k+t} \equiv \Theta_i^k \wedge \Delta_j \wedge V_h(a_1, ..., a_{k+t})^5$ with $m = 1, ..., |\Gamma_{k,i}^{k+t}|$, $j = 1, ..., M$, and $h = 1, ..., \frac{|\Gamma_{k,i}^{k+t}|}{M}$. At least one of the Δ_j's satisfies $\theta(\vec{a}, a_{k+1}, ..., a_{k+t})$ and will hence satisfies $K^{(k+t)}$. The probability that an arbitrary $\Theta_{i,m}^{k+t}$ satisfies $K^{(k+t)}$ will be the number of $\Theta_{i,m}^{k+t}$'s that satisfies $K^{(k+t)}$ divided by the total number of $\Theta_{i,m}^{k+t}$'s that is *at least*, $\frac{|\Gamma_{k,i}^{k+t}|}{M} \cdot \frac{1}{|\Gamma_{k,i}^{k+t}|} = \frac{1}{M}$, and so the probability that a random $\Theta_{i,m}^{k+t}$ does not satisfy $K^{(k+t)}$ will be *at most* as much as the maximum probability that Δ_j does not satisfy $\theta(\vec{a}, a_{k+1}, ..., a_{k+t})$ that is $1 - \frac{1}{M}$. Now consider the extension of Θ_i^k to a state description of L^{k+pt},

$$\Theta_{i,m}^{k+pt} \equiv \Theta_i^k \wedge \Delta_{j_1}^1 \wedge \Delta_{j_2}^2 \wedge ... \wedge \Delta_{j_p}^p \wedge V'_h(a_1, ..., a_{k+pt})$$

with $m = 1, ..., |\Gamma_{k,i}^{k+pt}|$, $j_1, ..., j_p = 1, ..., M$, $h = 1, ..., \frac{|\Gamma_{k,i}^{k+pt}|}{M^p}$ and where Δ_j^s enumerate the state description of $L^{\{a_1,...,a_l\} \cup \{a_{k+(s-1)t+1},...,a_{k+st}\}}$ that extend Ψ^l. The probability that $\Theta_{i,m}^{k+pt}$ does not satisfy $K^{(k+pt)}$ is at most as high as the probability that $\Delta_j^1 \not\models \theta(\vec{a}, a_{k+1}, ..., a_{k+t}), ..., \Delta_j^p \not\models \theta(\vec{a}, a_{k+(p-1)t+1}, ..., a_{k+pt})$ so $0 \leq 1 - \frac{|{}^K\Gamma_{k,i}^{k+pt}|}{|\Gamma_{k,i}^{k+pt}|} \leq (1 - \frac{1}{M})^p$. Let $p \to \infty$, then $0 \leq \lim_{r \to \infty} 1 - \frac{|{}^K\Gamma_{k,i}^r|}{|\Gamma_{k,i}^r|} \leq \lim_{p \to \infty}(1 - \frac{1}{M})^p = 0$. Hence, we have $\lim_{r \to \infty} 1 - \frac{|{}^K\Gamma_{k,i}^r|}{|\Gamma_{k,i}^r|} = 0$ and $\lim_{r \to \infty} \frac{|{}^K\Gamma_{k,i}^r|}{|\Gamma_{k,i}^r|} = 1$ as required. ∎

All state descriptions of L^k have the same number of extensions to a state description of L^r for $k < r$ thus $|\Gamma_{k,i}^r| = |\Gamma_{k,j}^r|$ for $\Theta_i^k, \Theta_j^k \in \Gamma^k$ and also $|\Gamma_{l,j}^r|$ is the same for all $\Theta_j^l \in \Gamma_\phi^l$. Hence, $|\Gamma_{l,j}^k| \, |\Gamma_{k,i}^r| = |\Gamma_{l,j}^r|^6$ and so,

$$\lim_{r \to \infty} \frac{\sum_{\Theta_j^l \in \Gamma_\phi^l} |{}^K\Gamma_{l,j}^r|}{|\Gamma_{l,j}^k| \, |\Gamma_{k,i}^r|} = \lim_{r \to \infty} \sum_{\Theta_j^l \in \Gamma_\phi^l} \frac{|{}^K\Gamma_{l,j}^r|}{|\Gamma_{l,j}^r|} = \sum_{\Theta_j^l \in \Gamma_\phi^l} \lim_{r \to \infty} \frac{|{}^K\Gamma_{l,j}^r|}{|\Gamma_{l,j}^r|} = |\Gamma_\phi^l|$$

---

[5] $V_h(a_1, ..., a_{k+t})$ enumerate sentence of the form $\bigwedge_{\substack{i_1,...,i_j \leq k+t \\ R \in RL_{j-arey}}} R_i(a_{i_1}, ..., a_{i_j})^{\epsilon_{i_1...i_j}}$ where $\{a_{i_1}, ..., a_{i_j}\}$ intersects both $\{a_{l+1}, ..., a_k\}$ and $\{a_{k+1}, ..., a_{k+t}\}$.

[6] What this says is that the number of extensions of Θ_j^l to a state description of L^k times the number of extensions of a state description of L^k to an state description of L^r (which is the same for all $\Theta_i^k \in \Gamma^k$), is equal to the number of extensions of Θ_j^l to an state description of L^r.

where the last equality follows from Lemma 5. Then

$$\lim_{r\to\infty} \frac{|{}^K\Gamma^r_{k,i}||\Gamma^l_\phi||\Gamma^k_{l,j}|}{\sum_{\Theta^l_j\in\Gamma^l_\phi} |{}^K\Gamma^r_{l,j}|} = |\Gamma^l_\phi| \lim_{r\to\infty} \frac{|{}^K\Gamma^r_{k,i}|}{|\Gamma^r_{k,i}|} \lim_{r\to\infty} \frac{|\Gamma^k_{l,j}||\Gamma^r_{k,i}|}{\sum_{\Theta^l_j\in\Gamma^l_\phi} |{}^K\Gamma^r_{l,j}|} = 1$$

and this establishes 8 as required and completes the proof. ∎

Corollary 1. *For a knowledge base K consisting of a Σ_1 sentence, and a sentence $\psi \in SL$, $ME(K)(\psi) = ME_w(K)(\psi)$.*

4 Discussion

We studied the Maximum Entropy probability functions as the canonical characterisation of some under-determined structure about which we have some partial information. The strongest candidate for this characterisation is the "least informative" probability function that satisfies the given partial information which is in turn formalised in terms of (relative) Shannon Entropy.

For propositional languages, the Maximum Entropy probability function that satisfies a given set of linear constraints is well defined and has been extensively studied. Our goal in this paper was to contribute to the investigation of these probability functions for first order languages. Barnett and Paris had shown in [1] that such probability functions are well defined for constraint sets from a monadic first order language. The case of Π_1 sentences has been investigated and partially answered by Paris and Rafiee Rad in [10] while for the sentences with the quantifier complexity of Σ_2, Π_2 or above these models are not necessarily well defined.

In this paper we have proved that the Maximum Entropy models are well defined for Σ_1 sentences and showed how these models are closely related to $P_=$, the most non-committal probability function. Furthermore, we showed that the two main approaches to defining Maximum Entropy models on first order languages, agree on the Σ_1 sentences.

References

[1] Barnett, O.W. and Paris, J.B., "Maximum Entropy inference with qualified knowledge", in *Logic Journal of the IGPL*, 16(1):85-98, 2008.

[2] Berger, A., Della Pietra, S. & Della Pietra, V., "A maximum Entropy Approach to Natural Language Processing", in *Com. Linguistics*, 22(1):39–71, 1996.

[3] Chen, C. H., "Maximum Entropy Analysis for Pattern Recognition", in *Maximum Entropy and Bayesian Methods*, P. F. Fougere (eds), Kluwer Academic Publisher, 1990.

[4] Gaifman, H. "Concerning measures in first order calculi", in *Israel J. of Mathematics*, 24: 1–18, 1964.

[5] Jaynes, E. T., "Notes on Present Status and Future Prospects" in *Maximum Entropy and Bayesian Methods*, W.T. Grandy & L.H. Schick, (ed), 1–13, 1990.

[6] Jaynes, E. T., "How Should We Use Entropy in Economics?", 1991, manuscript available at: http://www.leibniz.imag.fr/LAPLACE/Jaynes/prob.html.

[7] Kapur, J. N., "Non-Additive Measures of Entropy and Distributions of Statistical Mechanics", in *Ind Jour Pure App Math*, 14(11):1372–1384, 1983.

[8] Landes, J. and Williamson, J. "Objective Bayesianism and Maximum Entropy Principle", in *Entropy*, 15(9):3528–3591, 2013.

[9] Paris, J.B., *The Uncertain Reasoner's Companion*, Cambridge University Press, 1994.

[10] Paris, J.B. and Rad, S.R., "A note on the least informative model of a theory", in *Programs, Proofs, Processes, CiE 2010*, Eds. F. Ferreira, B. Löwe, E. Mayordomo, and L. Mendes Gomes, Springer LNCS 6158, 342–351, 2010.

[11] Paris, J.B. and Vencovská, "In defence of the maximum entropy inference process", in *International Journal of Approximate Reasoning*, 17(1):77–103, 1997.

[12] Paris, J.B. and Vencovská, *Pure Inductive Logic*, Cambridge University Press, 2015.

[13] Rafiee Rad, S., *Inference Processes for First Order Probabilistic Languages*, PhD Thesis, University of Manchester 2009. available at http://www.maths.manchester.ac.uk/~jeff/

[14] Shannon, C. E. & Weaver, W. *The Mathematical Theory of Communication*, University of Illinois Press, 1949.

[15] Williamson, J., "From Bayesian epistemology to inductive logic", in *Journal of Applied Logic*, 2, 2013.

[16] Williamson, J., "Objective Bayesian probabilistic logic", in *Journal of Algorithms in Cognition, Informatics and Logic*, 63:167-183, 2008.

[17] Williamson, J., *In Defence of Objective Bayesianism*, Oxford University Press, 2010.

Elementary Unification in Modal Logic *KD45*

Philippe Balbiani
CNRS, Toulouse University, France.
Philippe.Balbiani@irit.fr

Tinko Tinchev
Sofia University, Bulgaria.
tinko@fmi.uni-sofia.bg

Abstract

KD45 is the least modal logic containing the formulas $\Box x \to \Diamond x$, $\Box x \to \Box\Box x$ and $\Diamond x \to \Box\Diamond x$. It is determined by the class of all serial, transitive and Euclidean frames. The elementary unifiability problem in *KD45* is to determine, given a formula $\varphi(x_1, \ldots, x_n)$, whether there exists formulas $\psi_1, \ldots, \psi_n$ such that $\varphi(\psi_1, \ldots, \psi_n)$ is in *KD45*. It is well-known that the elementary unifiability problem in *KD45* is *NP*-complete. In our paper, we show that every *KD45*-unifiable formula has a projective unifier. As a corollary, we conclude that *KD45* has unitary type for elementary unification.

Keywords: Modal logic *KD45*. Elementary unification. Most general unifier. Projective formula. Unification type.

1 Introduction

Modal logics like *S5* or *KD45* are essential to the design of logical systems that capture elements of reasoning about knowledge [13, 21]. There exists variants of these logics with one or several agents, with or without common knowledge, etc. As in any modal logic, the questions addressed in their setting usually concern their axiomatizability and their decidability. Another desirable question which one should address whenever possible concerns the unifiability of formulas. A formula $\varphi(x_1, \ldots, x_n)$ is unifiable in a modal logic L iff there exists formulas $\psi_1, \ldots, \psi_n$ such that $\varphi(\psi_1, \ldots, \psi_n)$ is in L. See [1, 11, 14, 15] for details.

Special acknowledgement is heartly granted to the referees for the feedback we have obtained from them. Their comments and suggestions have greatly helped us to improve the correctness and the readability of our paper. Philippe Balbiani and Tinko Tinchev were partially supported by the programme RILA (contracts 34269VB and DRILA01/2/2015).

Results about the unifiability problem have been already obtained in many modal logics. Rybakov [22, 23] demonstrated that the unifiability problem in transitive modal logics like *K4* and *G* is decidable. Wolter and Zakharyaschev [24] showed that the unifiability problem is undecidable for any modal logic between *K* and *K4* extended with the universal modality. The notion of projectivity has been introduced by Ghilardi [15] to determine the unification type, finitary, of transitive modal logics like *K4* and *G*. The unification type, nullary, of modal logics like *K*, *KD* and *KT* has been established in [6, 18].

Within the context of description logics, checking subsumption of concepts is not sufficient and new inference capabilities are required. One of them, the unifiability of concept terms, has been introduced by Baader and Narendran [4] for $\mathcal{FL}_0$. Baader and Küsters [2] established the $EXPTIME$-completeness of the unifiability problem in $\mathcal{FL}_{reg}$ whereas Baader and Morawska [3] established the *NPTIME*-completeness of the unifiability problem in $\mathcal{EL}$. Much remains to be done, seeing that the computability of the unifiability problem and the unification types are unknown in multifarious modal logics and description logics.

KD45 is the least modal logic containing the formulas $\Box x \to \Diamond x$, $\Box x \to \Box\Box x$ and $\Diamond x \to \Box\Diamond x$. It is determined by the class of all serial, transitive and Euclidean frames. The elementary unifiability problem in *KD45* is to determine, given a parameter-free formula $\varphi(x_1, \ldots, x_n)$, whether there exists parameter-free formulas $\psi_1, \ldots, \psi_n$ such that $\varphi(\psi_1, \ldots, \psi_n)$ is in *KD45*. It is well-known that the elementary unifiability problem in *KD45* is *NP*-complete. Moreover, as proved by Ghilardi and Sacchetti [16], the unifiability problem in *KD45* is directed and, consequently, *KD45* has either unitary type, or nullary type. See also [7, 19]. The directedness of *KD45* is a consequence of the characterization by Ghilardi and Sacchetti of the normal extensions of *K4* with a directed unifiability problem. This characterization uses advanced notions from algebraic and relational semantics of normal modal logics.

In our paper, we directly show that every *KD45*-unifiable parameter-free formula has a projective unifier. As an immediate corollary, we conclude that *KD45* has unitary type for elementary unification. Section 2 defines the syntax and the semantics of *KD45*. In Section 3, definitions about the elementary unifiability problem in *KD45* are given. Sections 4–6 introduce and study arrows, setarrows and tips which will be our main tools for proving our results. In Section 7, definitions about acceptable agreements as a simplified version of bounded morphisms are given. Section 8 introduces and studies types which are sets of tips. In Sections 9–11, intermediate results about types needed to show that every *KD45*-unifiable parameter-free formula has a projective unifier are proved.

2 Syntax and semantics

Let *VAR* be a countable set of *variables* (with typical members denoted x, y, etc). Let $(x_1, x_2, \ldots)$ be an enumeration of *VAR* without repetitions. The set *FOR* of all *formulas* (with typical members denoted φ, ψ, etc) is inductively defined as follows:

- $\varphi ::= x \mid \bot \mid \neg\varphi \mid (\varphi \vee \psi) \mid \Box\varphi$.

We write $\varphi(x_1, \ldots, x_n)$ to denote a formula whose variables form a subset of $\{x_1, \ldots, x_n\}$. The result of the replacement in $\varphi(x_1, \ldots, x_n)$ of variables $x_1, \ldots, x_n$ in their places with formulas $\psi_1, \ldots, \psi_n$ will be denoted by $\varphi(\psi_1, \ldots, \psi_n)$. We define the other Boolean constructs as usual. We will follow the standard rules for omission of the parentheses. Let φ be a formula. We will write $\Diamond\varphi$ for $\neg\Box\neg\varphi$. We will respectively write $\varphi^\bot$ and $\varphi^\top$ for $\neg\varphi$ and φ. Let Γ be a finite set of formulas. Considering that $\bigvee \emptyset = \bot$ and $\bigwedge \emptyset = \top$, we will write $\nabla\Gamma$ for the conjunction of the following formulas:

- $\Box\bigvee\{\varphi : \varphi$ is a formula in $\Gamma\}$,

- $\bigwedge\{\Diamond\varphi : \varphi$ is a formula in $\Gamma\}$.

A *model* is a function $V : VAR \longrightarrow 2^{\mathbb{N}}$ associating to each variable x a set $V(x)$ of nonnegative integers. We inductively define the *truth* of a formula φ in model V at nonnegative integer s, in symbols $V, s \models \varphi$, as follows:

- $V, s \models x$ iff $s \in V(x)$,

- $V, s \not\models \bot$,

- $V, s \models \neg\varphi$ iff $V, s \not\models \varphi$,

- $V, s \models \varphi \vee \psi$ iff either $V, s \models \varphi$, or $V, s \models \psi$,

- $V, s \models \Box\varphi$ iff for all positive integers t, $V, t \models \varphi$.

As a result, $V, s \models \Diamond\varphi$ iff there exists a positive integer t such that $V, t \models \varphi$. Moreover, $V, s \models \varphi^\bot$ iff $V, s \not\models \varphi$ and $V, s \models \varphi^\top$ iff $V, s \models \varphi$. In other respect, $V, s \models \nabla\Gamma$ iff

- for all positive integers t, there exists $\varphi \in \Gamma$ such that $V, t \models \varphi$,

- for all $\varphi \in \Gamma$, there exists a positive integer t such that $V, t \models \varphi$.

We shall say that a model V is *uniform* iff for all variables x, either $V(x) = \emptyset$, or $V(x) = \mathbb{N}$. We shall say that a formula φ is *satisfiable* iff there exists a model V such that $V, 0 \models \varphi$. We shall say that a formula φ is *valid*, in symbols $\models \varphi$, iff for all models V, $V, 0 \models \varphi$. The following result is well-known and can be proved by using the canonical model construction, the technique of the generated subframe and the bounded morphism lemma [10].

Proposition 1. *For all formulas φ, $\models \varphi$ iff $\varphi \in$ KD45.*

Proof. Left to the reader. □

3 Unification

A *substitution* is a function $\sigma : VAR \longrightarrow FOR$ associating to each variable a formula. We shall say that a substitution σ is *closed* iff for all variables x, $\sigma(x)$ is a variable-free formula. For all formulas $\varphi(x_1, \ldots, x_n)$, let $\sigma(\varphi(x_1, \ldots, x_n))$ be $\varphi(\sigma(x_1), \ldots, \sigma(x_n))$. The *composition* $\sigma \circ \tau$ of the substitutions σ and τ is the substitution associating to each variable x the formula $\tau(\sigma(x))$. We shall say that a substitution σ is *equivalent* to a substitution τ, in symbols $\sigma \simeq \tau$, iff $\models \sigma(x) \leftrightarrow \tau(x)$ for all variables x. We shall say that a substitution σ is more *general* than a substitution τ, in symbols $\sigma \preceq \tau$, iff there exists a substitution υ such that $\sigma \circ \upsilon \simeq \tau$. Note that the notation $\tau \preceq \sigma$ is also used in many papers. We shall say that a formula φ is *unifiable* iff there exists a substitution σ such that $\models \sigma(\varphi)$. In that case, σ is a *unifier* of φ. We shall say that a unifiable formula φ is *projective* iff there exists a unifier σ of φ such that $\models \varphi \wedge \Box\varphi \to (\sigma(x) \leftrightarrow x)$ for all variables x. The following results are well-known [1].

Proposition 2. *Let φ be a formula. If φ is unifiable then φ possesses a closed unifier.*

Proof. Since the set of all valid formulas is closed with respect to the rule of uniform substitution, therefore if σ is a unifier of φ then for all closed substitutions τ, $\sigma \circ \tau$ is a closed unifier of φ. □

Proposition 3. *Let $\varphi(x_1, \ldots, x_n)$ be a $\Box$-free formula. The following conditions are equivalent:*

1. *$\varphi(x_1, \ldots, x_n)$, considered as a Boolean formula, is satisfiable.*

2. *$\varphi(x_1, \ldots, x_n)$, considered as a modal formula, is unifiable.*

Proof. Suppose $\varphi(x_1, \ldots, x_n)$, considered as a Boolean formula, is satisfiable. Hence, there exists $(\psi_1, \ldots, \psi_n)$ in $\{\bot, \top\}^n$ such that $\varphi(\psi_1, \ldots, \psi_n)$ is classically equivalent to $\top$. Thus, $\varphi(\psi_1, \ldots, \psi_n)$ is KD45-equivalent to $\top$. Consequently, $\varphi(x_1, \ldots, x_n)$, considered as a modal formula, is unifiable.
Reciprocally, suppose $\varphi(x_1, \ldots, x_n)$, considered as a modal formula, is unifiable. Let σ be a unifier of $\varphi(x_1, \ldots, x_n)$. Let V be a model. Since σ is a unifier of $\varphi(x_1, \ldots, x_n)$, therefore $V, 0 \models \varphi(\sigma(x_1), \ldots, \sigma(x_n))$. Let $(\psi_1, \ldots, \psi_n)$ in $\{\bot, \top\}^n$ be such that for all $i \in \{1, \ldots, n\}$, if $V, 0 \not\models \sigma(x_i)$ then $\psi_i = \bot$ else $\psi_i = \top$. Since $V, 0 \models \varphi(\sigma(x_1), \ldots, \sigma(x_n))$, therefore $\varphi(\psi_1, \ldots, \psi_n)$ is classically equivalent to $\top$. Hence, $\varphi(x_1, \ldots, x_n)$, considered as a Boolean formula, is satisfiable. □

Proposition 4. *The elementary unifiability problem in KD45 is NP-complete.*

Proof. Remark that every variable-free formula is either *KD45*-equivalent to $\bot$, or *KD45*-equivalent to $\top$. Hence, by Proposition 2, in order to determine if a given formula $\varphi(x_1, \ldots, x_n)$ is unifiable, it suffices to nondeterministically choose $(\psi_1, \ldots, \psi_n)$ in $\{\bot, \top\}^n$ such that $\models \varphi(\psi_1, \ldots, \psi_n)$. Since the validity of a given variable-free formula can be checked in polynomial time, therefore the elementary unifiability problem in *KD45* is in *NP*. As for the *NP*-hardness of the elementary unifiability problem in *KD45*, it follows from Proposition 3. □

Proposition 5. *Let φ be a unifiable formula. If φ is projective then φ possesses a most general unifier.*

Proof. Suppose φ is projective. Let σ be a unifier of φ such that $\models \varphi \wedge \Box\varphi \to (\sigma(x) \leftrightarrow x)$ for all variables x. Let τ be a unifier of φ and x be a variable. Hence, $\models \tau(\varphi)$ and $\models \varphi \wedge \Box\varphi \to (\sigma(x) \leftrightarrow x)$. Thus, $\models \tau(\varphi) \wedge \Box\tau(\varphi)$. Since $\models \varphi \wedge \Box\varphi \to (\sigma(x) \leftrightarrow x)$, therefore $\models \tau(\varphi) \wedge \Box\tau(\varphi) \to ((\sigma \circ \tau)(x) \leftrightarrow \tau(x))$. Since $\models \tau(\varphi) \wedge \Box\tau(\varphi)$, therefore $\models (\sigma \circ \tau)(x) \leftrightarrow \tau(x)$. Since x is an arbitrary variable, therefore $\sigma \circ \tau \simeq \tau$. Consequently, $\sigma \preceq \tau$. □

> From now on, let us fix $n \in \mathbb{N}$.

Formulas of the form $\varphi(x_1, \ldots, x_n)$ will be called *n-formulas*. From now on, they will be denoted $\varphi(\vec{x})$.

4 Arrows

We define $A_n = \{\bot, \top\}^n$. Elements of A_n are n-tuples of bits. They will be called *n-arrows*. They will be denoted α, β, etc. Remark that $Card(A_n) = 2^n$. For all n-arrows $\alpha = (\alpha_1, \ldots, \alpha_n)$, we will write $\tilde{\alpha}(\vec{x})$ for the associated n-formula

- $\tilde{\alpha}(\vec{x}) = x_1^{\alpha_1} \wedge \ldots \wedge x_n^{\alpha_n}$.

The following result says that the n-formula associated to an n-arrow is always satisfiable.

Lemma 6. *Let α be an n-arrow. There exists a model V such that $V, 0 \models \tilde{\alpha}(\vec{x})$.*

Proof. Left to the reader. □

Remark that for all n-arrows $\alpha = (\alpha_1, \ldots, \alpha_n)$ and for all n-tuples $\vec{\psi}$ of formulas, $\tilde{\alpha}(\vec{\psi}) = \psi_1^{\alpha_1} \wedge \ldots \wedge \psi_n^{\alpha_n}$. As a result,

Lemma 7. *Let $\vec{\psi}$ be an n-tuple of formulas, V be a model and s be a nonnegative integer. For all n-arrows α, β, if $V, s \models \tilde{\alpha}(\vec{\psi})$ and $V, s \models \tilde{\beta}(\vec{\psi})$ then $\alpha = \beta$.*

Proof. Let α, β be n-arrows. Suppose $V, s \models \tilde{\alpha}(\vec{\psi})$, $V, s \models \tilde{\beta}(\vec{\psi})$ and $\alpha \neq \beta$. Let $i \in \{1, \ldots, n\}$ be such that either $\alpha_i = \bot$ and $\beta_i = \top$, or $\alpha_i = \top$ and $\beta_i = \bot$. Without loss of generality, assume $\alpha_i = \bot$ and $\beta_i = \top$. Since $V, s \models \tilde{\alpha}(\vec{\psi})$ and $V, s \models \tilde{\beta}(\vec{\psi})$, therefore $V, s \models \neg \psi_i$ and $V, s \models \psi_i$: a contradiction. $\square$

For all n-tuples $\vec{\psi}$ of formulas, for all models V and for all nonnegative integers s, let $\alpha[\vec{\psi}, V, s]$ be the n-arrow such that for all $i \in \{1, \ldots, n\}$,

- if $V, s \not\models \psi_i$ then $\alpha_i[\vec{\psi}, V, s] = \bot$ else $\alpha_i[\vec{\psi}, V, s] = \top$.

As a result, $\tilde{\alpha}[\vec{\psi}, V, s](\vec{x}) = x_1^{\alpha_1[\vec{\psi}, V, s]} \wedge \ldots \wedge x_n^{\alpha_n[\vec{\psi}, V, s]}$ and

Lemma 8. *Let $\vec{\psi}$ be an n-tuple of formulas, V be a model and s be a nonnegative integer. $V, s \models \tilde{\alpha}[\vec{\psi}, V, s](\vec{\psi})$.*

Proof. By definition of $\alpha[\vec{\psi}, V, s]$. $\square$

Moreover,

Lemma 9. *Let $\vec{\psi}$ be an n-tuple of formulas, V be a model and s be a nonnegative integer. $\alpha[\vec{\psi}, V, s]$ is the unique n-arrow α such that $V, s \models \tilde{\alpha}(\vec{\psi})$.*

Proof. By Lemmas 7 and 8. $\square$

The following result will be useful when we study the most general unifiers of unifiable n-formulas.

Lemma 10. *Let V be a model. For all n-arrows α, there exists a model V' such that $V', 0 \models \tilde{\alpha}(\vec{x})$ and for all variables x and for all positive integers s, $s \in V'(x)$ iff $s \in V(x)$.*

Proof. Left to the reader. $\square$

5 Setarrows

Let $S_n = 2^{A_n} \setminus \{\emptyset\}$. Elements of S_n are nonempty sets of n-arrows. They will be called *n-setarrows*. They will be denoted a, b, etc. Remark that $Card(S_n) = 2^{2^n} - 1$. For all n-setarrows $a = \{\alpha^0, \ldots, \alpha^k\}$, we will write $\tilde{a}(\vec{x})$ the associated n-formula

- $\tilde{a}(\vec{x}) = \triangledown\{\widetilde{\alpha^0}(\vec{x}), \ldots, \widetilde{\alpha^k}(\vec{x})\}$.

The following result says that the n-formula associated to an n-setarrow is always satisfiable.

Lemma 11. *Let a be an n-setarrow. There exists a model V such that $V, 0 \models \tilde{a}(\vec{x})$.*

Proof. Left to the reader. □

Remark that for all n-setarrows $a = \{\alpha^0, \ldots, \alpha^k\}$ and for all n-tuples $\vec{\psi}$ of formulas, $\tilde{a}(\vec{\psi}) = \triangledown\{\widetilde{\alpha^0}(\vec{\psi}), \ldots, \widetilde{\alpha^k}(\vec{\psi})\}$. As a result,

Lemma 12. *Let $\vec{\psi}$ be an n-tuple of formulas and V be a model. For all n-setarrows a, b, if $V, 0 \models \tilde{a}(\vec{\psi})$ and $V, 0 \models \tilde{b}(\vec{\psi})$ then $a = b$.*

Proof. Let a, b be n-setarrows. Suppose $V, 0 \models \tilde{a}(\vec{\psi})$, $V, 0 \models \tilde{b}(\vec{\psi})$ and $a \neq b$. Let α be an n-arrow such that either $\alpha \in a$ and $\alpha \notin b$, or $\alpha \notin a$ and $\alpha \in b$. Without loss of generality, assume $\alpha \in a$ and $\alpha \notin b$. Since $V, 0 \models \tilde{a}(\vec{\psi})$, therefore there exists a positive integer s such that $V, s \models \tilde{\alpha}(\vec{\psi})$. Since $V, 0 \models \tilde{b}(\vec{\psi})$ and $\alpha \notin b$, therefore by Lemma 7, for all positive integers s, $V, s \not\models \tilde{\alpha}(\vec{\psi})$: a contradiction. □

For all n-tuples $\vec{\psi}$ of formulas and for all models V, let $a[\vec{\psi}, V]$ be the n-setarrow

- $a[\vec{\psi}, V] = \{\alpha[\vec{\psi}, V, s] : s \text{ is a positive integer}\}$.

As a result, $\tilde{a}[\vec{\psi}, V](\vec{\pi}) = \triangledown\{x_1^{\alpha_1[\vec{\psi}, V, s]} \wedge \ldots \wedge x_n^{\alpha_n[\vec{\psi}, V, s]} : s \text{ is a positive integer}\}$ and

Lemma 13. *Let $\vec{\psi}$ be an n-tuple of formulas and V be a model. $V, 0 \models \tilde{a}[\vec{\psi}, V](\vec{\psi})$.*

Proof. By definition of $a[\vec{\psi}, V]$. □

Moreover,

Lemma 14. *Let $\vec{\psi}$ be an n-tuple of formulas and V be a model. $a[\vec{\psi}, V]$ is the unique n-setarrow a such that $V, 0 \models \tilde{a}(\vec{\psi})$.*

Proof. By Lemmas 12 and 13. □

The following result will be useful when we study the most general unifiers of unifiable n-formulas. It can be proved by induction on $\varphi(\vec{x})$.

Lemma 15. *Let $\vec{\psi}$ be an n-tuple of formulas. Let V, V' be models such that $a[\vec{\psi}, V] = a[\vec{\psi}, V']$. Let $\varphi(\vec{x})$ be an n-formula. For all nonnegative integers s, s', if $\alpha[\vec{\psi}, V, s] = \alpha[\vec{\psi}, V', s']$ then $V, s \models \varphi(\vec{\psi})$ iff $V', s' \models \varphi(\vec{\psi})$.*

Proof. Left to the reader. □

307

6 Tips

Let $\mathcal{P}_n = A_n \times S_n$. Elements of $\mathcal{P}_n$ are couples consisting of an n-arrow component and an n-setarrow component. They will be called n-tips. They will be denoted p, q, etc. Remark that $Card(\mathcal{P}_n) = 2^n \times (2^{2^n} - 1)$. For all n-tips $p = (\alpha, a)$, we will write $\widetilde{p}(\vec{x})$ the associated n-formula

- $\widetilde{p}(\vec{x}) = \widetilde{\alpha}(\vec{x}) \wedge \widetilde{a}(\vec{x})$.

The following result says that the n-formula associated to an n-tip is always satisfiable.

Lemma 16. *Let p be an n-tip. There exists a model V such that $V, 0 \models \widetilde{p}(\vec{x})$.*

Proof. Left to the reader. □

Remark that for all n-tips $p = (\alpha, a)$ and for all n-tuples $\vec{\psi}$ of formulas, $\widetilde{p}(\vec{\psi}) = \widetilde{\alpha}(\vec{\psi}) \wedge \widetilde{a}(\vec{\psi})$. As a result,

Lemma 17. *Let $\vec{\psi}$ be an n-tuple of formulas and V be a model. For all n-tips p, q, if $V, 0 \models \widetilde{p}(\vec{\psi})$ and $V, 0 \models \widetilde{q}(\vec{\psi})$ then $p = q$.*

Proof. By Lemmas 7 and 12. □

For all n-tuples $\vec{\psi}$ of formulas and for all models V, let $p[\vec{\psi}, V]$ be the n-tip

- $p[\vec{\psi}, V] = (\alpha[\vec{\psi}, V, 0], a[\vec{\psi}, V])$.

As a result, $\widetilde{p}[\vec{\psi}, V](\vec{x}) = x_1^{\alpha_1[\vec{\psi}, V, 0]} \wedge \ldots \wedge x_n^{\alpha_n[\vec{\psi}, V, 0]} \wedge \nabla \{x_1^{\alpha_1[\vec{\psi}, V, s]} \wedge \ldots \wedge x_n^{\alpha_n[\vec{\psi}, V, s]} : s$ is a positive integer$\}$ and

Lemma 18. *Let $\vec{\psi}$ be an n-tuple of formulas and V be a model. $V, 0 \models \widetilde{p}[\vec{\psi}, V](\vec{\psi})$.*

Proof. By definition of $p[\vec{\psi}, V]$. □

Moreover,

Lemma 19. *Let $\vec{\psi}$ be an n-tuple of formulas and V be a model. $p[\vec{\psi}, V]$ is the unique n-tip p such that $V, 0 \models \widetilde{p}(\vec{\psi})$.*

Proof. By Lemmas 17 and 18. □

7 Acceptable agreements

In this section, we give definitions of acceptable agreements as a simplified version of bounded morphisms. We shall say that a function $f : \mathbb{N} \longrightarrow \mathbb{N}$ associating to each nonnegative integer a nonnegative integer is *acceptable* iff for all positive integers s, $f(s)$ is a positive integer and $f^{-1}(s)$ contains a positive integer. We shall say that a function $f : \mathbb{N} \longrightarrow \mathbb{N}$ associating to each nonnegative integer a nonnegative integer is an *n-agreement* between models V and V' iff for all $i \in \{1, \ldots, n\}$ and for all nonnegative integers s, $s \in V(x_i)$ iff $f(s) \in V'(x_i)$.

Lemma 20. *Let f be an acceptable n-agreement between models V and V'. Let $\varphi(\vec{x})$ be an n-formula. For all nonnegative integers s, $V, s \models \varphi(\vec{x})$ iff $V', f(s) \models \varphi(\vec{x})$.*

Proof. By induction on $\varphi(\vec{x})$. □

We shall say that a function $f : \mathbb{N} \longrightarrow \mathbb{N}$ associating to each nonnegative integer a nonnegative integer is an *ω-agreement* between models V and V' iff for all variables x and for all nonnegative integers s, $s \in V(x)$ iff $f(s) \in V'(x)$.

Lemma 21. *Let f be an acceptable ω-agreement between models V and V'. Let φ be a formula. For all nonnegative integers s, $V, s \models \varphi$ iff $V', f(s) \models \varphi$.*

Proof. By induction on φ. □

8 Types

Let $\mathcal{T}_n = 2^{A_n \times S_n}$. Elements of $\mathcal{T}_n$ are sets of n-tips. They will be called n-*types*. They will be denoted T, U, etc. Remark that $Card(\mathcal{T}_n) = 2^{2^n \times (2^{2^n} - 1)}$. We shall say that an n-type T is *complete* for an n-setarrow a iff for all n-arrows α, $(\alpha, a) \in T$. We shall say that an n-type T is *empty* for an n-setarrow a iff for all n-arrows α, if $\alpha \in a$ then $(\alpha, a) \notin T$. We shall say that an n-type T is *full* for an n-setarrow a iff for all n-arrows α, if $\alpha \in a$ then $(\alpha, a) \in T$. We shall say that an n-type T is *saturated* iff for all n-arrows α, β and for all n-setarrows a, if $(\alpha, a) \in T$ and $\beta \in a$ then $(\beta, a) \in T$. The following result will be of crucial importance in the remaining sections of our paper.

Proposition 22. *Let T be a saturated n-type. For all n-setarrows a, exactly one of the following conditions holds: (i) T is complete for a; (ii) T is not complete for a and T is empty for a; (iii) T is not complete for a and T is full for a.*

Proof. Left to the reader. □

We shall say that an n-type T is *closed* iff for all n-setarrows a, there exists an n-arrow γ such that if T is not complete for a then either T is empty for a and $(\gamma, \{\gamma\}) \in T$, or T is full for a and $(\gamma, a) \in T$. We shall say that an n-type is *perfect* iff it is saturated and closed.

9 From tuples of formulas to perfect types

Let $\vec{\psi}$ be an n-tuple of formulas. Let $T[\vec{\psi}]$ be the n-type

- $T[\vec{\psi}] = \{p[\vec{\psi}, V] : V \text{ is a model}\}$.

The aim of this section is to demonstrate that $T[\vec{\psi}]$ is perfect.

Lemma 23. $T[\vec{\psi}]$ *is saturated.*

Proof. Let β, γ be n-arrows and b be an n-setarrow such that $(\beta, b) \in T[\vec{\psi}]$ and $\gamma \in b$. Let V be a model such that $\beta = \alpha[\vec{\psi}, V, 0]$ and $b = a[\vec{\psi}, V]$. Recall that $\gamma \in b$. Let s be a positive integer such that $\gamma = \alpha[\vec{\psi}, V, s]$. Let V' be the model such that for all variables x, if $s \notin V(x)$ then $V'(x) = V(x) \setminus \{0\}$ else $V'(x) = V(x) \cup \{0\}$. Let f be the acceptable function such that $f(0) = s$ and for all positive integers t, $f(t) = t$. The reader may easily verify that f is an ω-agreement between V' and V. Since $\gamma = \alpha[\vec{\psi}, V, s]$ and $f(0) = s$, therefore by Lemma 21, $\gamma = \alpha[\vec{\psi}, V', 0]$. Moreover, since $b = a[\vec{\psi}, V]$, therefore by Lemma 21, $b = a[\vec{\psi}, V']$. Hence, $(\gamma, b) \in T[\vec{\psi}]$. Since β, γ are arbitrary n-arrows and b is an arbitrary n-setarrow such that $(\beta, b) \in T[\vec{\psi}]$ and $\gamma \in b$, therefore $T[\vec{\psi}]$ is saturated. □

Lemma 24. *There exists an n-arrow γ such that $(\gamma, \{\gamma\}) \in T[\vec{\psi}]$.*

Proof. Let V be a uniform model. The reader may easily verify that $a[\vec{\psi}, V] = \{\alpha[\vec{\psi}, V, 0]\}$. Hence, $(\alpha[\vec{\psi}, V, 0], \{\alpha[\vec{\psi}, V, 0]\}) \in T[\vec{\psi}]$. □

Lemma 25. $T[\vec{\psi}]$ *is closed.*

Proof. By Lemma 23, $T[\vec{\psi}]$ is saturated. Hence, by Proposition 22, for all n-setarrows a, exactly one of the following conditions holds: (i) $T[\vec{\psi}]$ is complete for a; (ii) $T[\vec{\psi}]$ is not complete for a and $T[\vec{\psi}]$ is empty for a; (iii) $T[\vec{\psi}]$ is not complete for a and $T[\vec{\psi}]$ is full for a. By Lemma 24, let γ be an n-arrow such that $(\gamma, \{\gamma\}) \in T[\vec{\psi}]$. For all n-setarrows a, let $\gamma^{T,a}$ be an arbitrary n-arrow if condition (i) holds, the n-arrow γ if condition (ii) holds and an arbitrary n-arrow in a if condition (iii) holds. The reader may easily verify that for all n-setarrows a, if $T[\vec{\psi}]$ is not complete for a then either $T[\vec{\psi}]$ is empty for a and $(\gamma^{T,a}, \{\gamma^{T,a}\}) \in T[\vec{\psi}]$, or $T[\vec{\psi}]$ is full for a and $(\gamma^{T,a}, a) \in T[\vec{\psi}]$. □

From all this, it follows that

Proposition 26. $T[\vec{\psi}]$ *is perfect.*

Proof. By Lemmas 23 and 25. □

10 From perfect types to tuples of formulas

Let T be a perfect n-type. Hence, T is saturated and closed. Thus, by Proposition 22, for all n-setarrows a, exactly one of the following conditions holds: (i) T is complete for a; (ii) T is not complete for a and T is empty for a; (iii) T is not complete for a and T is full for a. Since T is closed, therefore for all n-setarrows a, let $\gamma^{T,a}$ be an n-arrow such that if T is not complete for a then either T is empty for a and $(\gamma^{T,a}, \{\gamma^{T,a}\}) \in T$, or T is full for a and $(\gamma^{T,a}, a) \in T$. For all n-tips $p = (\alpha, a)$, let $\delta^{T,p}$ be the n-arrow such that if $p \notin T$ then $\delta^{T,p} = \gamma^{T,a}$ else $\delta^{T,p} = \alpha$. Let $\vec{\psi}[T](\vec{x})$ be the n-tuple of n-formulas such that for all $i \in \{1, \ldots, n\}$,

- $\psi_i[T](\vec{x}) = \bigvee\{\widetilde{p}(\vec{x}) \wedge \delta_i^{T,p} : p \text{ is an } n\text{-tip}\}$.

The aim of this section is to demonstrate that $T = T[\vec{\psi}[T](\vec{x})]$.

Lemma 27. *Let p be an n-tip. If $p \in T$ then $\models \widetilde{p}(\vec{x}) \rightarrow \widetilde{p}(\vec{\psi}[T](\vec{x}))$.*

Proof. Suppose $p \in T$. Let β be the n-arrow component of p and b be the n-setarrow component of p. Let V be a model such that $V, 0 \models \widetilde{p}(\vec{x})$. Let $i \in \{1, \ldots, n\}$. Since $V, 0 \models \widetilde{p}(\vec{x})$, therefore by Lemma 17, $V, 0 \models \psi_i[T](\vec{x}) \leftrightarrow \delta_i^{T,p}$. Since $p \in T$, therefore $\delta_i^{T,p} = \beta_i$. Since $V, 0 \models \psi_i[T](\vec{x}) \leftrightarrow \delta_i^{T,p}$, therefore $V, 0 \models \psi_i[T](\vec{x}) \leftrightarrow \beta_i$. Since $V, 0 \models \widetilde{p}(\vec{x})$, therefore $V, 0 \models \widetilde{\beta}(\vec{x})$. Since $V, 0 \models \psi_i[T](\vec{x}) \leftrightarrow \beta_i$, therefore $V, 0 \models \psi_i[T](\vec{x}) \leftrightarrow x_i$. Let s be a positive integer. Since $V, 0 \models \widetilde{p}(\vec{x})$, therefore $V, s \models \widetilde{b}(\vec{x})$. Recall that s is a positive integer. Let α be an n-arrow such that $\alpha \in b$ and $V, s \models \widetilde{\alpha}(\vec{x})$. Let q be the n-tip with n-arrow component α and n-setarrow component b. Since $V, s \models \widetilde{b}(\vec{x})$ and $V, s \models \widetilde{\alpha}(\vec{x})$, therefore $V, s \models \widetilde{q}(\vec{x})$. Hence, by Lemma 17, $V, s \models \psi_i[T](\vec{x}) \leftrightarrow \delta_i^{T,q}$. Since T is saturated, $p \in T$ and $\alpha \in b$, therefore $q \in T$. Thus, $\delta_i^{T,q} = \alpha_i$. Since $V, s \models \psi_i[T](\vec{x}) \leftrightarrow \delta_i^{T,q}$, therefore $V, s \models \psi_i[T](\vec{x}) \leftrightarrow \alpha_i$. Since $V, s \models \widetilde{\alpha}(\vec{x})$, therefore $V, s \models \psi_i[T](\vec{x}) \leftrightarrow x_i$. Since s is an arbitrary positive integer, therefore $V, 0 \models \Box(\psi_i[T](\vec{x}) \leftrightarrow x_i)$. Since $V, 0 \models \psi_i[T](\vec{x}) \leftrightarrow x_i$, therefore $V, 0 \models (\psi_i[T](\vec{x}) \leftrightarrow x_i) \wedge \Box(\psi_i[T](\vec{x}) \leftrightarrow x_i)$. Since i is arbitrary in $\{1, \ldots, n\}$, therefore $V, 0 \models (\psi_1[T](\vec{x}) \leftrightarrow x_1) \wedge \ldots \wedge (\psi_n[T](\vec{x}) \leftrightarrow x_n) \wedge \Box((\psi_1[T](\vec{x}) \leftrightarrow x_1) \wedge \ldots \wedge (\psi_n[T](\vec{x}) \leftrightarrow x_n))$. Since $V, 0 \models \widetilde{p}(\vec{x})$, therefore $V, 0 \models \widetilde{p}(\vec{\psi}[T](\vec{x}))$. Since V is an arbitrary model such that $V, 0 \models \widetilde{p}(\vec{x})$, therefore $\models \widetilde{p}(\vec{x}) \rightarrow \widetilde{p}(\vec{\psi}[T](\vec{x}))$. □

Lemma 28. $T \subseteq T[\vec{\psi}[T](\vec{x})]$.

Proof. Let p be an n-tip such that $p \in T$. By Lemma 16, let V be a model such that $V, 0 \models \widetilde{p}(\vec{x})$. Since $p \in T$, therefore by Lemma 27, $\models \widetilde{p}(\vec{x}) \rightarrow \widetilde{p}(\vec{\psi}[T](\vec{x}))$. Since $V, 0 \models \widetilde{p}(\vec{x})$, therefore $V, 0 \models \widetilde{p}(\vec{\psi}[T](\vec{x}))$. Hence, by Lemma 19, $p = p[\vec{\psi}[T](\vec{x}), V]$. Thus, $p \in T[\vec{\psi}[T](\vec{x})]$. Since p is an arbitrary n-tip such that $p \in T$, therefore $T \subseteq T[\vec{\psi}[T](\vec{x})]$. □

Lemma 29. *Let p be an n-tip with n-setarrow component b. If $p \notin T$ then $\models \widetilde{p}(\vec{x}) \to \widetilde{\gamma^{T,b}}(\vec{\psi}[T](\vec{x}))$.*

Proof. Suppose $p \notin T$. Let V be a model such that $V, 0 \models \widetilde{p}(\vec{x})$. Let $i \in \{1, \ldots, n\}$. Since $V, 0 \models \widetilde{p}(\vec{x})$, therefore by Lemma 17, $V, 0 \models \psi_i[T](\vec{x}) \leftrightarrow \delta_i^{T,p}$. Since $p \notin T$, therefore $\delta_i^{T,p} = \gamma_i^{T,b}$. Since $V, 0 \models \psi_i[T](\vec{x}) \leftrightarrow \delta_i^{T,p}$, therefore $V, 0 \models \psi_i[T](\vec{x}) \leftrightarrow \gamma_i^{T,b}$. Since i is arbitrary in $\{1, \ldots, n\}$, therefore $V, 0 \models \widetilde{\gamma^{T,b}}(\vec{\psi}[T](\vec{x}))$. Since V is an arbitrary model such that $V, 0 \models \widetilde{p}(\vec{x})$, therefore $\models \widetilde{p}(\vec{x}) \to \widetilde{\gamma^{T,b}}(\vec{\psi}[T](\vec{x}))$. □

Lemma 30. *Let b be an n-setarrow. If T is empty for b then $\models \widetilde{b}(\vec{x}) \to \widetilde{\{\gamma^{T,b}\}}(\vec{\psi}[T](\vec{x}))$.*

Proof. Suppose T is empty for b. Let V be a model such that $V, 0 \models \widetilde{b}(\vec{x})$. Let $i \in \{1, \ldots, n\}$. Let s be a positive integer. Since $V, 0 \models \widetilde{b}(\vec{x})$, therefore $V, s \models \widetilde{b}(\vec{x})$. Recall that s is a positive integer. Let α be an n-arrow such that $\alpha \in b$ and $V, s \models \widetilde{\alpha}(\vec{x})$. Let p be the n-tip with n-arrow component α and n-setarrow component b. Since $V, s \models \widetilde{b}(\vec{x})$ and $V, s \models \widetilde{\alpha}(\vec{x})$, therefore $V, s \models \widetilde{p}(\vec{x})$. Hence, by Lemma 17, $V, s \models \psi_i[T](\vec{x}) \leftrightarrow \delta_i^{T,p}$. Since T is empty for b and $\alpha \in b$, therefore $p \notin T$. Thus, $\delta_i^{T,p} = \gamma_i^{T,b}$. Since $V, s \models \psi_i[T](\vec{x}) \leftrightarrow \delta_i^{T,p}$, therefore $V, s \models \psi_i[T](\vec{x}) \leftrightarrow \gamma_i^{T,b}$. Since s is an arbitrary positive integer, therefore $V, 0 \models \Box(\psi_i[T](\vec{x}) \leftrightarrow \gamma_i^{T,b})$. Since i is arbitrary in $\{1, \ldots, n\}$, therefore $V, 0 \models \Box((\psi_1[T](\vec{x}) \leftrightarrow \gamma_1^{T,b}) \wedge \ldots \wedge (\psi_n[T](\vec{x}) \leftrightarrow \gamma_n^{T,b}))$. Consequently, $V, 0 \models \widetilde{\{\gamma^{T,b}\}}(\vec{\psi}[T](\vec{x}))$. Since V is an arbitrary model such that $V, 0 \models \widetilde{b}(\vec{x})$, therefore $\models \widetilde{b}(\vec{x}) \to \widetilde{\{\gamma^{T,b}\}}(\vec{\psi}[T](\vec{x}))$. □

Lemma 31. *Let b be an n-setarrow. If T is full for b then $\models \widetilde{b}(\vec{x}) \to \widetilde{b}(\vec{\psi}[T](\vec{x}))$.*

Proof. Suppose T is full for b. Let V be a model such that $V, 0 \models \widetilde{b}(\vec{x})$. Let $i \in \{1, \ldots, n\}$. Let s be a positive integer. Since $V, 0 \models \widetilde{b}(\vec{x})$, therefore $V, s \models \widetilde{b}(\vec{x})$. Recall that s is a positive integer. Let α be an n-arrow such that $\alpha \in b$ and $V, s \models \widetilde{\alpha}(\vec{x})$. Let p be the n-tip with n-arrow component α and n-setarrow component b. Since $V, s \models \widetilde{b}(\vec{x})$ and $V, s \models \widetilde{\alpha}(\vec{x})$, therefore $V, s \models \widetilde{p}(\vec{x})$. Hence, by Lemma 17, $V, s \models \psi_i[T](\vec{x}) \leftrightarrow \delta_i^{T,p}$. Since T is full for b and $\alpha \in b$, therefore $p \in T$. Thus, $\delta_i^{T,p} = \alpha_i$. Since $V, s \models \psi_i[T](\vec{x}) \leftrightarrow \delta_i^{T,p}$, therefore $V, s \models \psi_i[T](\vec{x}) \leftrightarrow \alpha_i$. Since $V, s \models \widetilde{\alpha}(\vec{x})$, therefore $V, s \models \psi_i[T](\vec{x}) \leftrightarrow x_i$. Since s is an arbitrary positive integer, therefore $V, 0 \models \Box(\psi_i[T](\vec{x}) \leftrightarrow x_i)$. Since i is arbitrary in $\{1, \ldots, n\}$, therefore $V, 0 \models \Box((\psi_1[T](\vec{x}) \leftrightarrow x_1) \wedge \ldots \wedge (\psi_n[T](\vec{x}) \leftrightarrow x_n))$. Since $V, 0 \models \widetilde{b}(\vec{x})$, therefore $V, 0 \models \widetilde{b}(\vec{\psi}[T](\vec{x}))$. Since V is an arbitrary model such that $V, 0 \models \widetilde{b}(\vec{x})$, therefore $\models \widetilde{b}(\vec{x}) \to \widetilde{b}(\vec{\psi}[T](\vec{x}))$. □

Lemma 32. $T[\vec{\psi}[T](\vec{x})] \subseteq T$.

Proof. Let p be an n-tip such that $p \in T[\vec{\psi}[T](\vec{x})]$. Let V be a model such that $p = p[\vec{\psi}[T](\vec{x}), V]$. Hence, by Lemma 18, $V, 0 \models \widetilde{p}(\vec{\psi}[T](\vec{x}))$. Let $q = p[\vec{x}, V]$. Thus, by Lemma 18, $V, 0 \models \widetilde{q}(\vec{x})$. Case "$q \in T$": Hence, by Lemma 27, $\models \widetilde{q}(\vec{x}) \to \widetilde{q}(\vec{\psi}[T](\vec{x}))$. Since $V, 0 \models \widetilde{q}(\vec{x})$, therefore $V, 0 \models \widetilde{q}(\vec{\psi}[T](\vec{x}))$. Since $V, 0 \models \widetilde{p}(\vec{\psi}[T](\vec{x}))$, therefore by Lemma 17, $p = q$. Since $q \in T$, therefore $p \in T$. Case "$q \notin T$": Let a be the n-setarrow component of q. Since $q \notin T$, therefore by Lemma 29, $\models \widetilde{q}(\vec{x}) \to \widetilde{\gamma^{T,a}}(\vec{\psi}[T](\vec{x}))$. Since $V, 0 \models \widetilde{q}(\vec{x})$, therefore $V, 0 \models \widetilde{\gamma^{T,a}}(\vec{\psi}[T](\vec{x}))$. Since $V, 0 \models \widetilde{p}(\vec{\psi}[T](\vec{x}))$, therefore by Lemma 7, $\gamma^{T,a}$ is the n-arrow component of p. Since $q \notin T$, therefore T is not complete for a. Since T is saturated, therefore by Proposition 22, either T is empty for a, or T is full for a. In the former case, $(\gamma^{T,a}, \{\gamma^{T,a}\}) \in T$. Moreover, by Lemma 30, $\models \widetilde{a}(\vec{x}) \to \widetilde{\{\gamma^{T,a}\}}(\vec{\psi}[T](\vec{x}))$. Since $V, 0 \models \widetilde{q}(\vec{x})$, therefore $V, 0 \models \widetilde{\{\gamma^{T,a}\}}(\vec{\psi}[T](\vec{x}))$. Since $V, 0 \models \widetilde{p}(\vec{\psi}[T](\vec{x}))$, therefore by Lemma 12, $\{\gamma^{T,a}\}$ is the n-setarrow component of p. Since $\gamma^{T,a}$ is the n-arrow component of p and $(\gamma^{T,a}, \{\gamma^{T,a}\}) \in T$, therefore $p \in T$. In the latter case, $(\gamma^{T,a}, a) \in T$. Moreover, by Lemma 31, $\models \widetilde{a}(\vec{x}) \to \widetilde{a}(\vec{\psi}[T](\vec{x}))$. Since $V, 0 \models \widetilde{q}(\vec{x})$, therefore $V, 0 \models \widetilde{a}(\vec{\psi}[T](\vec{x}))$. Since $V, 0 \models \widetilde{p}(\vec{\psi}[T](\vec{x}))$, therefore by Lemma 12, a is the n-setarrow component of p. Since $\gamma^{T,a}$ is the n-arrow component of p and $(\gamma^{T,a}, a) \in T$, therefore $p \in T$. Since p is an arbitrary n-tip such that $p \in T[\vec{\psi}[T](\vec{x})]$, therefore $T[\vec{\psi}[T](\vec{x})] \subseteq T$. □

From all this, it follows that

Proposition 33. $T = T[\vec{\psi}[T](\vec{x})]$.

Proof. By Lemmas 28 and 32. □

11 About most general unifiers

Let $\varphi(\vec{x})$ be an n-formula. Let T be the n-type

- $T = \{p : \models \widetilde{p}(\vec{x}) \to \varphi(\vec{x}) \land \Box\varphi(\vec{x})\}$.

Lemma 34. *T is saturated.*

Proof. Let α, β be n-arrows and a be an n-setarrow such that $(\alpha, a) \in T$ and $\beta \in a$. Hence, $\models \widetilde{\alpha}(\vec{x}) \land \widetilde{a}(\vec{x}) \to \varphi(\vec{x}) \land \Box\varphi(\vec{x})$. Let V be a model such that $V, 0 \models \widetilde{\beta}(\vec{x})$ and $V, 0 \models \widetilde{a}(\vec{x})$. By Lemma 10, let V' be a model such that $V', 0 \models \widetilde{\alpha}(\vec{x})$ and for all variables x and for all positive integers s, $s \in V'(x)$ iff $s \in V(x)$. Since $V, 0 \models \widetilde{a}(\vec{x})$, therefore $V', 0 \models \widetilde{a}(\vec{x})$. Since $\models \widetilde{\alpha}(\vec{x}) \land \widetilde{a}(\vec{x}) \to \varphi(\vec{x}) \land \Box\varphi(\vec{x})$ and $V', 0 \models \widetilde{\alpha}(\vec{x})$, therefore $V', 0 \models \varphi(\vec{x}) \land \Box\varphi(\vec{x})$. Moreover, recall that $\beta \in a$. Let s_β be a positive integer such that $V', s_\beta \models \widetilde{\beta}(\vec{x})$. Since $V', 0 \models \varphi(\vec{x}) \land \Box\varphi(\vec{x})$, therefore $V', s_\beta \models \varphi(\vec{x}) \land \Box\varphi(\vec{x})$. Let $f : \mathbb{N} \longrightarrow \mathbb{N}$ be the function associating to each nonnegative integer a nonnegative integer such that $f(0) = s_\beta$ and for all positive integers s, $f(s) = s$. The reader may easily verify that f is an acceptable n-agreement between V and V'. Since $V', s_\beta \models \varphi(\vec{x}) \wedge \Box\varphi(\vec{x})$ and $f(0) = s_\beta$, therefore by Lemma 20, $V, 0 \models \varphi(\vec{x}) \wedge \Box\varphi(\vec{x})$. Since V is an arbitrary model such that $V, 0 \models \widetilde{\beta}(\vec{x})$ and $V, 0 \models \widetilde{a}(\vec{x})$, therefore $\models \widetilde{\beta}(\vec{x}) \wedge \widetilde{a}(\vec{x}) \to \varphi(\vec{x}) \wedge \Box\varphi(\vec{x})$. Thus, $(\beta, a) \in T$. □

Lemma 35. *Let $\vec{\chi}$ be an n-tuple of variable-free formulas. There exists an n-arrow γ such that $\models \widetilde{\gamma}(\vec{\chi})$.*

Proof. Left to the reader. □

> From now on, let us assume $\varphi(\vec{x})$ is unifiable.

The aim of this section is to demonstrate that $\varphi(\vec{x})$ is projective.

Lemma 36. *There exists an n-arrow γ such that $(\gamma, \{\gamma\}) \in T$.*

Proof. Since $\varphi(\vec{x})$ is unifiable, therefore by Proposition 2, let σ be a closed substitution such that $\models \sigma(\varphi(\vec{x}))$. Let $\vec{\chi}$ be the n-tuple of variable-free formulas such that for all $i \in \{1, \ldots, n\}$, $\chi_i = \sigma(x_i)$. Since $\models \sigma(\varphi(\vec{x}))$, therefore $\models \varphi(\vec{\chi})$. Hence, $\models \varphi(\vec{\chi}) \wedge \Box\varphi(\vec{\chi})$. Since $\vec{\chi}$ is an n-tuple of variable-free formulas, therefore by Lemma 35, let γ be an n-arrow such that $\models \widetilde{\gamma}(\vec{\chi})$. Thus, $\models \widetilde{\gamma}(\vec{\chi}) \wedge \Box\widetilde{\gamma}(\vec{\chi})$. Let V be a model such that $V, 0 \models \widetilde{\gamma}(\vec{x})$ and $V, 0 \models \widetilde{\{\gamma\}}(\vec{x})$. Since $\models \widetilde{\gamma}(\vec{\chi}) \wedge \Box\widetilde{\gamma}(\vec{\chi})$, therefore $V, 0 \models \widetilde{\gamma}(\vec{\chi})$ and $V, 0 \models \Box\widetilde{\gamma}(\vec{\chi})$. Let $i \in \{1, \ldots, n\}$. Since $V, 0 \models \widetilde{\gamma}(\vec{x})$ and $V, 0 \models \widetilde{\gamma}(\vec{\chi})$, therefore $V, 0 \models \chi_i \leftrightarrow x_i$. Let s be a positive integer. Since $V, 0 \models \widetilde{\{\gamma\}}(\vec{x})$ and $V, 0 \models \Box\widetilde{\gamma}(\vec{\chi})$, therefore $V, s \models \widetilde{\gamma}(\vec{x})$ and $V, s \models \widetilde{\gamma}(\vec{\chi})$. Hence, $V, s \models \chi_i \leftrightarrow x_i$. Since s is an arbitrary positive integer, therefore $V, 0 \models \Box(\chi_i \leftrightarrow x_i)$. Since $V, 0 \models \chi_i \leftrightarrow x_i$, therefore $V, 0 \models (\chi_i \leftrightarrow x_i) \wedge \Box(\chi_i \leftrightarrow x_i)$. Since i is arbitrary in $\{1, \ldots, n\}$, therefore $V, 0 \models (\chi_1 \leftrightarrow x_1) \wedge \ldots \wedge (\chi_n \leftrightarrow x_n) \wedge \Box((\chi_1 \leftrightarrow x_1) \wedge \ldots \wedge (\chi_n \leftrightarrow x_n))$. Since $\models \varphi(\vec{\chi}) \wedge \Box\varphi(\vec{\chi})$, therefore $V, 0 \models \varphi(\vec{\chi}) \wedge \Box\varphi(\vec{\chi})$. Since $V, 0 \models (\chi_1 \leftrightarrow x_1) \wedge \ldots \wedge (\chi_n \leftrightarrow x_n) \wedge \Box((\chi_1 \leftrightarrow x_1) \wedge \ldots \wedge (\chi_n \leftrightarrow x_n))$, therefore $V, 0 \models \varphi(\vec{x}) \wedge \Box\varphi(\vec{x})$. Since V is an arbitrary model such that $V, 0 \models \widetilde{\gamma}(\vec{x})$ and $V, 0 \models \widetilde{\{\gamma\}}(\vec{x})$, therefore $\models \widetilde{\gamma}(\vec{x}) \wedge \widetilde{\{\gamma\}}(\vec{x}) \to \varphi(\vec{x}) \wedge \Box\varphi(\vec{x})$ Thus, $(\gamma, \{\gamma\}) \in T$. □

Lemma 37. *T is closed.*

Proof. By Lemma 34, T is saturated. Hence, by Proposition 22, for all n-setarrows a, exactly one of the following conditions holds: (i) T is complete for a; (ii) T is not complete for a and T is empty for a; (iii) T is not complete for a and T is full for a. By Lemma 36, let γ be an n-arrow such that $(\gamma, \{\gamma\}) \in T$. For all n-setarrows a, let $\gamma^{T,a}$ be an arbitrary n-arrow if condition (i) holds, the n-arrow γ if condition (ii) holds and an arbitrary n-arrow in a if condition (iii) holds. The reader may easily verify that for all n-setarrows a, if T is not complete for a then either T is empty for a and $(\gamma^{T,a}, \{\gamma^{T,a}\}) \in T$, or T is full for a and $(\gamma^{T,a}, a) \in T$. □

Lemma 38. *T is perfect.*

Proof. By Lemmas 34 and 37. □

By Lemma 37, T is closed. Hence, for all n-setarrows a, let $\gamma^{T,a}$ be an n-arrow such that if T is not complete for a then either T is empty for a and $(\gamma^{T,a}, \{\gamma^{T,a}\}) \in T$, or T is full for a and $(\gamma^{T,a}, a) \in T$. For all n-tips $p = (\alpha, a)$, let $\delta^{T,p}$ be the n-arrow such that if $p \notin T$ then $\delta^{T,p} = \gamma^{T,a}$ else $\delta^{T,p} = \alpha$. Let $\vec{\psi}[T](\vec{x})$ be the n-tuple of n-formulas such that for all $i \in \{1, \ldots, n\}$, $\psi_i[T](\vec{x}) = \bigvee \{\widetilde{p}(\vec{x}) \wedge \delta_i^{T,p} : p \text{ is an } n\text{-tip}\}$. Thus, by Proposition 33 and Lemma 38, $T = T[\vec{\psi}[T](\vec{x})]$.

Lemma 39. $\models \varphi(\vec{\psi}[T](\vec{x}))$.

Proof. Let V be a model. Since $T = T[\vec{\psi}[T](\vec{x})]$, therefore $p[\vec{\psi}[T](\vec{x}), V] \in T$. Hence, $\models \widetilde{p}[\vec{\psi}[T](\vec{x}), V](\vec{x}) \to \varphi(\vec{x}) \wedge \Box\varphi(\vec{x})$. Thus, $\models \widetilde{p}[\vec{\psi}[T](\vec{x}), V](\vec{\psi}[T](\vec{x})) \to \varphi(\vec{\psi}[T](\vec{x})) \wedge \Box\varphi(\vec{\psi}[T](\vec{x}))$. By Lemma 18 it holds $V, 0 \models \widetilde{p}[\vec{\psi}[T](\vec{x}), V](\vec{\psi}[T](\vec{x}))$. Since it holds $\models \widetilde{p}[\vec{\psi}[T](\vec{x}), V](\vec{\psi}[T](\vec{x})) \to \varphi(\vec{\psi}[T](\vec{x})) \wedge \Box\varphi(\vec{\psi}[T](\vec{x}))$, therefore $V, 0 \models \varphi(\vec{\psi}[T](\vec{x})) \wedge \Box\varphi(\vec{\psi}[T](\vec{x}))$. Consequently, $V, 0 \models \varphi(\vec{\psi}[T](\vec{x}))$. Since V is an arbitrary model, therefore $\models \varphi(\vec{\psi}[T](\vec{x}))$. □

Lemma 40. *For all $i \in \{1, \ldots, n\}$, $\models \varphi(\vec{x}) \wedge \Box\varphi(\vec{x}) \to (\psi_i[T](\vec{x}) \leftrightarrow x_i)$.*

Proof. Let $i \in \{1, \ldots, n\}$. Let V be a model such that $V, 0 \models \varphi(\vec{x}) \wedge \Box\varphi(\vec{x})$. Let $q = p[\vec{x}, V]$. By Lemma 18, $V, 0 \models \widetilde{q}(\vec{x})$. Case "$q \in T$": Hence, by Lemma 17, $V, 0 \models \psi_i[T](\vec{x}) \leftrightarrow \delta_i^{T,q}$. Let α be the n-arrow component of q. Since $q \in T$, therefore $\delta_i^{T,q} = \alpha_i$. Since $V, 0 \models \psi_i[T](\vec{x}) \leftrightarrow \delta_i^{T,q}$, therefore $V, 0 \models \psi_i[T](\vec{x}) \leftrightarrow \alpha_i$. Since $V, 0 \models \widetilde{q}(\vec{x})$, therefore $V, 0 \models \alpha_i \leftrightarrow x_i$ Since $V, 0 \models \psi_i[T](\vec{x}) \leftrightarrow \alpha_i$, therefore $V, 0 \models \psi_i[T](\vec{x}) \leftrightarrow x_i$. Case "$q \notin T$": Hence, $\not\models \widetilde{q}(\vec{x}) \to \varphi(\vec{x}) \wedge \Box\varphi(\vec{x})$. Let V' be a model such that $V', 0 \models \widetilde{q}(\vec{x})$ and $V', 0 \not\models \varphi(\vec{x}) \wedge \Box\varphi(\vec{x})$. Since $V, 0 \models \widetilde{q}(\vec{x})$, therefore by Lemma 15, $V, 0 \not\models \varphi(\vec{x}) \wedge \Box\varphi(\vec{x})$: a contradiction. Since V is an arbitrary model such that $V, 0 \models \varphi(\vec{x}) \wedge \Box\varphi(\vec{x})$, therefore $\models \varphi(\vec{x}) \wedge \Box\varphi(\vec{x}) \to (\psi_i[T](\vec{x}) \leftrightarrow x_i)$. □

From all this, it follows that

Proposition 41. *$\varphi(\vec{x})$ is projective.*

Proof. Let σ be the substitution such that for all positive integers i, if $i \in \{1, \ldots, n\}$ then $\sigma(x_i) = \psi_i[T](\vec{x})$ else $\sigma(x_i) = x_i$. By lemma 39, the reader may easily verify that σ is a unifier of $\varphi(\vec{x})$. Moreover, by Lemma 40, $\models \varphi(\vec{x}) \wedge \Box\varphi(\vec{x}) \to (\sigma(x) \leftrightarrow x)$ for all variables x. □

As a corollary, we conclude that

Corollary 42. *KD45 has unitary type for elementary unification, i.e. every unifiable formula possesses a most general unifier.*

Proof. By Propositions 5 and 41. □

12 Conclusion

Much remains to be done. For example, one may consider the unifiability problem when the language is extended by a countable set of parameters (with typical members denoted p, q, etc). In this case, the unifiability problem is said to be non-elementary. It consists to determine, given a formula $\varphi(p_1, \ldots, p_m, x_1, \ldots, x_n)$, whether there exists formulas $\psi_1, \ldots, \psi_n$ such that $\models \varphi(p_1, \ldots, p_m, \psi_1, \ldots, \psi_n)$. Another example, one may also consider the unifiability problem, the elementary one or the non-elementary one, this time in modal logic $K45$ or in modal logic $K5$. More generally, the unifiability problem, the elementary one or the non-elementary one, in modal logics extending $K5$ is of interest, knowing that these modal logics are *coNP*-complete [17]. Other *coNP*-complete modal logics are all proper extensions of $S5 \times S5$ [8, 9] and all finitely axiomatizable tense logics of linear time flows [20]. Thus, one may consider whether our method is applicable to the unifiability problem in these modal logics. A similar question can be asked as well with respect to the linear temporal logic considered by Babenyshev and Rybakov [5]. In other respect, what becomes of the unifiability problem, the elementary one or the non-elementary one, when the language is extended by the universal modality or the difference modality? Finally, considering the tight relationships between unifiability of formulas and admissibility of inference rules as explained in [1, 12, 15], one may ask whether all normal modal logics extending $K5$ are almost structurally complete, i.e. one may ask whether all admissible non-derivable inference rules are passive in these logics.

References

[1] Baader, F., Ghilardi, S.: *Unification in modal and description logics.* Logic Journal of the IGPL **19** (2011) 705–730.

[2] Baader, F., Küsters, R.: *Unification in a description logic with transitive closure of roles.* In Nieuwebhuis, R., Voronkov, A. (editors): *Logic for Programming and Automated Reasoning.* Springer (2001) 217–232.

[3] Baader, F., Morawska, B.: *Unification in the description logic $\mathcal{EL}$.* In Treinen, R. (editor): *Rewriting Techniques and Applications.* Springer (2009) 350–364.

[4] Baader, F., Narendran, P.: *Unification of concept terms in description logics.* Journal of Symbolic Computation **31** (2001) 277–305.

[5] Babenyshev, S., Rybakov, V.: *Unification in linear temporal logic LTL.* Annals of Pure and Applied Logic **162** (2011) 991–1000.

[6] Balbiani, P., Gencer, Ç.: *KD is nullary.* Journal of Applied Non-Classical Logics (to appear).

[7] Balbiani, P., Gencer, Ç.: *Unification in epistemic logics.* Journal of Applied Non-Classical Logics **27** (2017) 91–105.

[8] Bezhanishvili, N., Hodkinson, I.: *All normal extensions of S5-squared are finitely axiomatizable.* Studia Logica **78** (2004) 443–457.

[9] Bezhanishvili, N., Marx, M.: *All proper normal extensions of S5-square have the polynomial size model property.* Studia Logica **73** (2003) 367–382.

[10] Blackburn, P., de Rijke, M., Venema, Y.: *Modal Logic.* Cambridge University Press (2001).

[11] Dzik, W.: *Unification Types in Logic.* Wydawnicto Uniwersytetu Slaskiego (2007).

[12] Dzik, W., Stronkowski, M.: *Almost structural completeness: an algebraic approach.* Annals of Pure and Applied Logic **167** (2016) 525–556.

[13] Fagin, R., Halpern, J., Moses, Y., Vardi, M.: *Reasoning About Knowledge.* MIT Press (1995).

[14] Gencer, Ç., de Jongh, D.: *Unifiability in extensions of K4.* Logic Journal of the IGPL 17 (2009) 159–172.

[15] Ghilardi, S.: *Best solving modal equations.* Annals of Pure and Applied Logic **102** (2000) 183–198.

[16] Ghilardi, S., Sacchetti, L.: *Filtering unification and most general unifiers in modal logic.* The Journal of Symbolic Logic **69** (2004) 879–906.

[17] Halpern, J., Rêgo, L.: *Characterizing the NP-PSPACE gap in the satisfiability problem for modal logic.* Journal of Logic and Computation **17** (2007) 795–806.

[18] Jeřábek, E.: *Blending margins: the modal logic K has nullary unification type.* Journal of Logic and Computation **25** (2015) 1231–1240.

[19] Jeřábek, E.: *Rules with parameters in modal logic I.* Annals of Pure and Applied Logic **166** (2015) 881–933.

[20] Litak, T., Wolter, F.: *All finitely axiomatizable tense logics of linear time flows are coNP-complete.* Studia Logics **81** (2005) 153–165.

[21] Meyer, J.-J., van der Hoek, W.: *Epistemic Logic for AI and Computer Science.* Cambridge University Press (1995).

[22] Rybakov, V.: *A criterion for admissibility of rules in the model system S4 and the intuitionistic logic.* Algebra and Logic **23** (1984) 369–384.

[23] Rybakov, V.: *Admissibility of Logical Inference Rules.* Elsevier (1997).

[24] Wolter, F., Zakharyaschev, M.: *Undecidability of the unification and admissibility problems for modal and description logics.* ACM Transactions on Computational Logic **9** (2008) 25:1–25:20.

Probabilistic Formal Verification of Communication Network-based Fault Detection, Isolation and Service Restoration System in Smart Grid

Syed Atif Naseem
Izmir University of Economics, Izmir, Turkey.
syedatifnaseem@gmail.com

Riaz Uddin
Department of Electrical Engineering, Faculty of Electrical and Computer Engineering, NED University of Engineering and Technology, Karachi, Pakistan.

Osman Hasan
Faculty of Engineering, National University of Science and Technology, Islamabad, Pakistan.

Diaa E. Fawzy
Faculty of Engineering, Izmir University of Economics, Izmir, Turkey.

Abstract

Communication network plays a significant task in distribution system of smart grid when it comes to sending and receiving the bi-directional flows of communication data, information and important control messages between the sending (Intelligent Electrical Device) IED and receiving IED of the components of smart grid in a coupling network (Power and Communication Network). Occurrence of fault in the power network does not affect the communication network because of the introduction of back up battery and power supplies provided to the main router of communication system. This motivated us to study the accuracy of the flow of information in the communication network that gives commands to the power network at the time of fault detection and restoration etc. In this regard, the major contribution of this paper is (i) to develop the Markovain model of the FDIR behavior in distribution network of Smart Grid and (ii) formally verify the model in PRISM model checker tool in order to analyze the system (a) accuracy, (b) efficiency and (c) reliability by developing logical properties in tool. More-over the Markovian model of the (iii) mechanism of sending/receiving of the data packet (IEEE 802.11 DCF) is also develop and integrate it with FDIR in PRISM model checker to investigate the overall system behavior. Another main purpose to construct the probabilistic Markovian model of FDIR along with communication network is to (iv) analyze the frequency of fault occurrence in distribution network in terms of probability and (v) predict the failure probability of different component of distribution network in order to take a corrective action, maintenance. So that, the faulty component can be replaced in advance to avoid the complete failure of system. Moreover, we also (vi) analyze and predict the probability at which the load switches of distribution network work properly by making the faulty component detach itself upon the occurrence of fault. Finally, (vii) predicting the probability to recover the system through particular non-active switch is also analyzed and verified along with the comparison between FDIR model with wireless communication network and FDIR model with ideal communication network (such as Ethernet or Fiber-optics) is also analyzed and discussed.

Keywords: FDIR, DMS, Smart Grid, Formal Verification, Probabilistic Model Checker, PRISM, Wireless Communication Network.

1 Introduction

The conventional electricity networks [1, 2] were developed more than a century ago when the concepts of power generation and consumption off electricity was not much complicated (i.e., without high-level automation and communication inputs etc.). The traditional or existing grids is also called a one way flow of energy where electricity produced at the centralized generation end increases its voltage through step up transformer and sends the energy through the transmission line and upon reaching to consumer end decreases its voltage through step down transformer. It is difficult for the conventional network to make the grid to fulfill the requirement of average variation of demand of electricity in the real time period. Up-grading the traditional electric power grid to the future power grid by accumulating the components (such as voltage sensors, current sensors, fault detectors and two ways digital communication networks etc.) is being done. Therefore, it is possible for the future grid [3-5] giving a concept of bi-directional flow of energy along with communication data [6] and control messages of power network in a coupling network. It also consists of communication technology, sensing and measuring instrument, electric storage, demand response, renewable energies integration and information technologies. In addition, it can store the electrical energy in electrical vehicle system [7, 8] and used it when-ever it is necessary. The renewable energies like bio-mass energy, solar and wind energy are also integrated into the distribution system of future grid to fulfill the requirement of high demand of electricity in the 21st century [9].

A. FDIR with Communication Network

Fault detection, Isolation and service restoration system (FDIR) [10, 11] plays a significant task in the distribution system of smart grid where it finds the exact location of fault whenever any malfunction occurs due to failure of switches or fault current. Usually, the first step is to trip off the main circuit breaker of substation; when protection relay detects the over current exceeding the set value in it. The IED [12] installed on the circuit breaker sends the FDIR start message (FASM) to other IEDS installed at load switches to start the process of FDIR which detects the exact location of fault in order to detach this component from the system and restore the power system through another Tie switch which were present in a circuit as a non-active component [10, 11]. Communication network [13-15] plays a very crucial part in detecting and restoring the power of the substation of smart grid. As fault occurs in distribution network, the communication network of smart grid is also separately energized from the backup power battery [16, 17] and the IEDs of different component present in smart grid starts communicating with each other. In this regard, the control messages of power devices are sent by the circuit breaker IED of the substation to the other connected components. If the IEDs of different component of substation is connected through Ethernet [18] or Fiber optics [13, 19], then there is a small probability of failure of communication network possibly due to less delay [20] (in sending/receiving the control signals/messages) as compared to wireless communication network [21] which not only suffer from large delay of control messages due to network congestion [12], time consuming message process [22] and malicious jamming attacks [23] but also depends on weather conditions which plays the important role in power distribution network in order to avoid the ultimate cascading failures.

B. Existing Analysis Approaches of FDIR with Communication Network

There are variety of procedures reported for the analyses pertaining to FDIR in Smart grid system such as integrated with wired communication networks [18, 19] and integrated with wireless communication network [21], which include either the numerical [24, 25] or simulation based approach [26]. The numerical based approach is basically dependent on a number of iterative methods that produce outcome generally based on the purpose of the number of iterations. In contrast, simulations-based approaches depend on generating the result by taking into account the subset of all possible scenarios of the system. Thus both techniques are scalable and user friendly but have some limitations to generate the accurate, reliable and absolute results of the study. Both these techniques cannot take out the whole bugs while doing the study of the system is due to investigate the models with a subset of all likely combination and rounding off errors of sampling based approach of the study. The above mentioned issues in iterative- or simulation-based method also encourages researchers to use formal methods [27, 28] in the safety critical domain (whose failure may result in loss or severe damage to human/equipment/property) which tests the model and taken out the bugs after rigorous verification of the model through temporal logic specification. Formal methods basically build a Markov model in a mathematical form which is related to the genuine structure and formally verify the accuracy of mathematical model with in a computer through temporal logic specification which in turn increases the probability of finding design errors. A mathematical model is then translated in to the language of model checker and LTL, CTL or PCTL property [29] is fed in to model checker along the translated mathematical model which gives the result true, if the model satisfies the temporal logic property, otherwise false result with counter example will be given by the model checker. Basically formal verification of system [27, 30] can be done in three different ways based on its reason, judgment, self-expression and clarity. Theorem proving, symbolic simulation and model checking are the three methods often used to verify the reactive system, stochastic process and mathematically model of suitable reason [31]. Up till now, a variety of approaches are used to implement FDIR in distribution system but probabilistic model checker i.e., formal verification of FDIR along with wireless communication network has never been done before and to support this claim, a table of related literature is presented below to compare and summarize the work which have been performed on FDIR of the distribution network.

2 Related work and contribution

To connect the IEDS of different component of smart grid with each other, different communication network such as Ethernet, Wifi, PLC etc., are used and in this regards, [20] discusses the coupling network of communication network with power network and analyzes the IEEE test cases (9, 14, 30, 118, 300 Bus case) of different sizes network with the communication network. The main theme in this work is to find the probability of communication network failure on two different timing condition i.e., 3ms and 10ms. It also suggests that the probability of failure of Ethernet communication network is much lower as compared to wireless communication network. The work in [43] proposes the multi-hop wireless network with a

Fault Detection, Isolation & Restoration System (FDIR)		
Technique	Literature Reference	Formal Verification
Compare centralized & decentralized architecture	[32]	No
Restoration Scheme	[33]	No
Decentralized Structure	[34]	No
Different kinds of Agents to restore power system	[35]	No
MAS design for restoration	[36]	No
Integration Of Technique	[37]	No
Restoration Scheme	[38]	No
Restoration Scheme	[39]	No
Substation Restoration Technique	[40]	No
Shortening of restoration Time	[41]	No
Monitoring the Limited current	[42]	No

Table 1: Comparison of FDIR approach with Formal Verification

frequency-reprocess configuration of cellular network and addresses the challenges to send and receive the huge information in future grid applications. It presents the planning of system for analyzing the reporting of network and capability. The Work in [44] describes the wireless smart grid communication system and explain the home area network in which the sensors are installed in the home appliances and form the wireless mesh network. The work also inspects the topologies of networking and wireless data packet simulation result is also shown. In [45], the important issues on smart grid technologies specifically related to the communication network technology and information technology network are discussed and provides the present situation regarding the ability of smart grid communication system. Work in [46] analyzes the reliability and resilience of smart grid communication network by using the IEEE 802.11 communication technology in both infrastructure single-hop and mesh multiple-hop topologies for upgraded meters in a system called Building Area Network (BAN). Another work in [47] proposes the wireless mesh network for a smart distributing grid and then analyzed the security framework under this communication architecture.

Besides the wireless communication network to make a reliable communication link between the IEDS of different element of the substation of future grid, a vast number of work is also presented to show that communication of IEDS through Power line communication is possible and suggests that it is a more reliable medium as compared to other communication technology. In this regard, [48] proposes a narrow band power line communication (PLC) network for outdoor communication component from smart meters to data centers on low voltage or medium voltage power lines in the 3-500 kHz spectrum band. It also presented detailed information on the different types of interference occurred over the power lines which degrade the quality of communication system of the smart grid. The work in [49] develops the iterative algorithm called water filling for PLC system in order to analyze the multichannel modulation techniques. It also describes the different kinds of noises available at the PLC channel and the power spectral density phenomenon is used to represent the intensity of channel noises. In [50], the general model of the broadband PLC network architecture is presented to connect the high voltage lines, substation and low voltage lines of the consumer end. It also proposes the algorithm called recursive to approximate the carrier frequency equalization and its performance is evaluated through the maximum likelihood approach. Another work in [51] presents the overview on PLC system from two different standardization bodies i.e., IEEE and ITU, which presents the similarities and dissimilarities between these two standards. The paper also gives detailed information on physical layer specification and Mac protocol of PLC network of both standards respectively. Work in [52] suggested a solution to integrate two heterogeneous network architectures by combining PLC with back bone of IP based network. It also discussed the critical issues of energy management application by highlighting the reliability, availability, coverage distance, communication delay and security standard of communication network. Literature in [53] presents a solution to integrate the active management system in the network infrastructure when number of distribution and generation setups are involved in the substation of smart grid. It also discussed the standard protocol and technology of different communication network in terms of data rate, bit error rate and installation cost of each wired and wireless medium.

Keeping the above issues in iterative- or simulation-based method in mind, achieving the absolute correctness and system reliability analysis in real world problem, we become motivated to use the formal methods [27, 28] in the safety critical domain (whose failure may result in loss or severe damage to human/equipment/property) which tests the model and takes out the bugs after rigorous verification of the model through temporal logic specification. Up till now, probabilistic formal verification of FDIR along with communication network has never been performed for the study and verification of FDIR classification in distribution network of future/Smart grids. However a number of approaches used to implement FDIR in distribution system but no one performed the probabilistic verification as summarized in the Table 1. On the other hand, above mentioned FDIR approaches in Table

1, section A,B mainly discussed the restoration of fault in distribution network but they did not give any idea on switching and communication failures (possibly in terms of probabilities) of FDIR in distribution network with communication networks (Ethernet or Wifi), so that some preventive actions may be designed for fast isolation and restoration of Smart Grid system. This mainly motivated us to analyze the switching failure of FDIR component along with the failure probability of communication network in order to determine the expected time necessary to recover a system after the switching fault or communication fault has occurred in FDIR of smart grid. Furthermore, the above approaches in Table 1 also did not perform any formal verification on FDIR in their respective system which is important to verify the switching and communication logics among different components in a distribution/smart grid. In order to implement our proposed formal verification notion, (i) a Markovian representation (shown in section VI) of FDIR for an established Tianjin Electric Power Corporation network [10] is integrated along with a mechanism of sending/receiving the data packet (IEEE 802.11 DCF) (considered as smart grid). (ii) This Markovian FDIR model is employed in PRISM [54] model checker tool to formally verify the system accuracy, availability, efficiency and reliability with wireless communication network. Furthermore, several more important studies (contributions) such as (iii) the comparison between FDIR model with wireless communication network and with ideal communication network (such as Ethernet or Fiber-optics) is performed and (iv) the probability of (a) switching and communication failures of FDIR in any distribution network / Smart Grid (b) tripping-off the switch within the limited time period (c) to recover the system automatically within the least possible time after the occurrence of fault is also predicted and discussed in detail.

The rest of the paper is organized as follows: Section III presents a summary on probabilistic model checker (PRISM tool), FDIR behavior and justification of formal model. Section IV explains the exploitation of an established Tianjin Electric Power architecture as Smart Grid. Section V discusses the proposed methodology of modeling FDIR in PRISM. Section VI explains the FDIR model with wireless communication network. Section VII is dedicated to formal verification of FDIR with wireless communication network and Ideal communication network. Section VIII explains the comparison between the FDIR with wireless communication network and FDIR with Ideal communication network. Section XI concludes the paper with future research work.

3 Preliminaries

In this section, the general summary on the probabilistic model checking and prism model checker tool [54-56] along with a brief introduction on FDIR is presented that is later formally verified in this paper. The important interdependent behavior in fault management scenario of communication network with power network is also provided. In this regard, it is required to go through some preliminaries as follows:

A. Probabilistic Model Checking

1) What is Probabilistic Model Checking

A classification that exhibits random behavior, probabilistic model checker is used [55], [56] for the formal study and verification of such system and therefore can be represented as Markov chains [57].However, depending upon its nature, application and usefulness, a system behavior which is probabilistic in nature is represented as DTMC [58], CTMC [59], MDP [60] and PTA [61], [57]. Fig. 1 shows the working example of the probabilistic model where every state transition to the other state is based on the applied probabilities. In DTMC, the present states move to next state by fulfilling the certain condition with the applied probabilities, whereas in CTMC the present state transit to next state does not depend only the probabilities to make such transition but also include the delay before making the transition and move to the next state. These random delays usually are represented as exponential probability distributions [57]. MDPs and PTAs are with non-deterministic transitions whereas DTMCs and CTMC are fully probabilistic transitions.

Once the probabilistic Markov model of the random behavior of system is finalized, the verification and analyzing of such system can be done through the probabilistic temporal logic properties of the model checking tool. There are number of specification language available for probabilistic model checking verification and some of the specification language is mentioned here i.e., PLTL, CTL and LTL etc., [29]. The probabilistic linear temporal logic property along with the Markov model of the random system which is uttered in the form of prism language i.e., Alur's Reactive modules formalism is fed in to the probabilistic model checker tool in order to check all possible execution by reaching each state of the model and satisfying the specification by applying the certain condition through temporal logic property. In addition, PRISM tool [54] supports model checking for every Markov model i.e., for DTMC [58], CTMC [59], MDP [60] and PTAs [61]. It is a generic tool and we found it quite appropriate for our work.

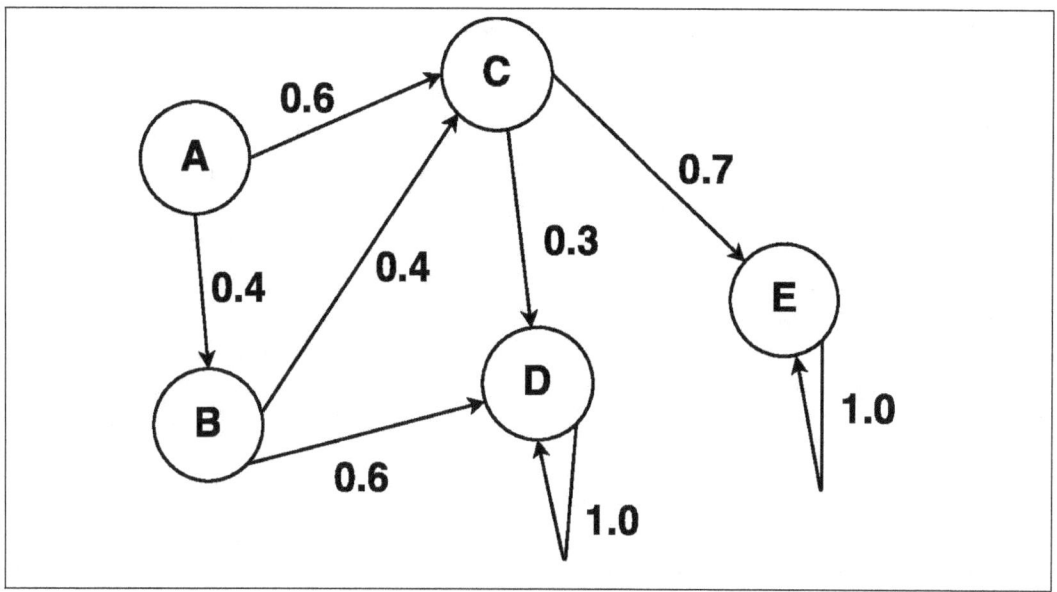

Figure 1: Working of Probabilistic Models

2) PRISM Model Checker

The system which show probabilistic behavior can be analyzed, verified and investigated through PRISM tool [54], [62]. The Prism model checker is basically based on probabilistic modeling technique in which the probabilistic performance of a structure is formulated depends on the reactive modules formalism and then transformed this probabilistic behavior in to prism language [70]. The Prism tool has a built-in Simulation tab which is discrete in nature and it can be used for statistical data analysis. Furthermore, it is designed for the verification of every kind of Markov processes, i.e., CTMC [59], DTMC [63], MDP [60] and PTA [61]. Details of how to fed a Morkovian model in prism tool with command in prism language can be seen in [76].

B. Details of Fault Localization, Isolation and Restoration

1) Fault Detection, Isolation and Supply Restore Behavior in Distribution Network

Whenever the fault occurs in the substation of smart grid due to the malfunction of transformer or the fault current exceeds the set value, the over current relay of substation trips-off the circuit breaker of that particular substation and the IED associated to this circuit breaker transmits the alarm message along with 'FDIR start message' (FASM) to other load switches IEDs which are connected and controlled by the IEDs of the circuit breaker of substation. Tie switch which is present in the substation (but not alive) discard the FASM message by their IEDs. The IEDs of other load switches which are connected to the substation receive the FASM message and starts the process of fault localization and check the fault flag status at the feeder terminal unit (FTU) of load switches. The error flags of load switches are raised by FTU whenever the protection relay sense the faulty current in the substation of distribution network and trips-off the CB of the substation. The IEDs of CB communicates with each IEDs of load switches in order to find the exact location of fault by checking the fault flags set at the local feeder terminal unit of the load switches. The IED of circuit breaker synchronizes itself with each IEDs of load switches by sending and receiving the important control data messages. Once the fault is determined and the fault flag raised at the FTU of any load switch, the fault localization process is completed and the IED of that particular load switch begins the fault isolation process, trips-off the particular load switch and detach this load switch from the rest of the circuit with in a limited time period and send the ISOM message to each IEDs of load components (such as switches, circuit breaker, protection relay) and Tie switches of the feeder of substation in order to restore the power of substation through Tie switch and start the process of the closing preparation of Tie switch. Basically the isolation results message (ISOM) sends the two types of messages i.e., error result of isolation and the plan of restoration of the power supply of the system. After the completion of fault localization process and fault isolation process, the supply restoration process starts and its main purpose is to restore the power supply of substation through Tie switch within the limited time and connect the Tie switch to core feeder or reserve feeder depending upon the lesser energy space between each other. If the non-active switch cannot re-establish the power delivery of substation through main source then it will select the reserve energy feeder from the faultless energy side of the substation off smart grid.

Fault flags play a significant task in defining the state of each IED of the component of smart grid as shown in Fig. 2. Basically there are four possible states to each IED of the component present in the substation i.e., Fault, Restore, Outage and Normal and it is given in Fig. 2. During normal operation of substation, all IEDs of components are in normal state whereas fault flags set IEDs are in outage state. Whenever the fault occurs in the substation, the over current protection relay detects the faulty current in a substation and trips-off the circuit breaker, the fault flags set IED of the associate component which changes its state from normal to faulty state and the other de-energized section i.e. Tie switches IEDs in the distri-

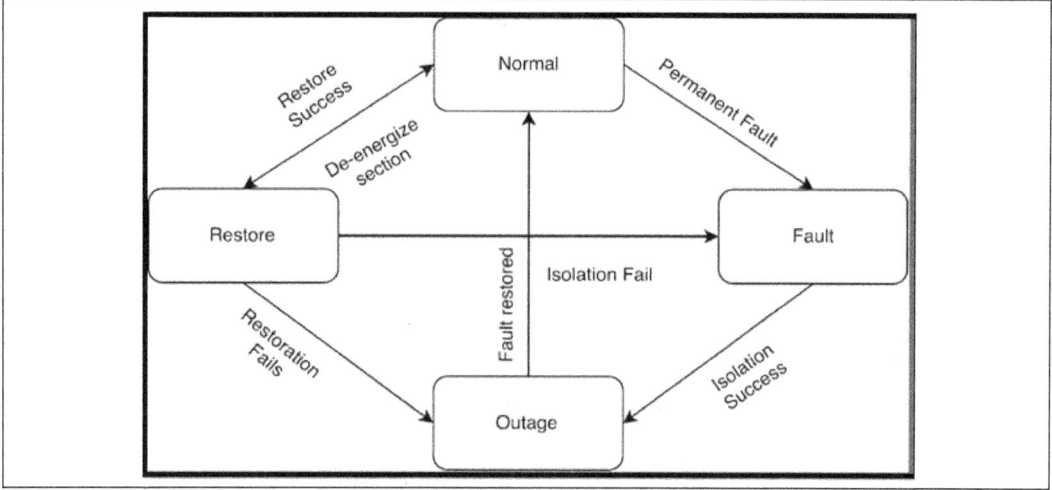

Figure 2: State of Section

bution network turns into the restore section. The faulty state load switch starts the process of FDIR and if the isolation process is successful within the limited time by tripping-off the load switch successfully, then the faulty state status of the load switch changes into outage state but if the isolation process fails and the load switch does not trip-off within a limited time then the faulty section will expand and take more de-energized load switch from restore section and change them into the faulty section. After completing the process of fault localization and isolation process, supply restoration service automatically powers the restore section and tries to connect the Tie switch with the other load switches of substation. If the restore switch is successfully closed within a limited time in the restore section then the state of restore section turns into the normal state but if the restore switch does not close and the process is failed then the restore section is changed into the outage state. When the fault is cleared either automatically or manually in the outage section, then the outage state changes back to normal state.

2) Behavior of coupling networks when fault is occurred in distribution system

It is of the interest to analyze the interdependent behavior of two coupled network i.e., communication network and power network in a fault management scenario and discuss each step of network in brief in order to understand the whole working condition of coupled network when fault occurs on the arrangement as given in Fig. 3.

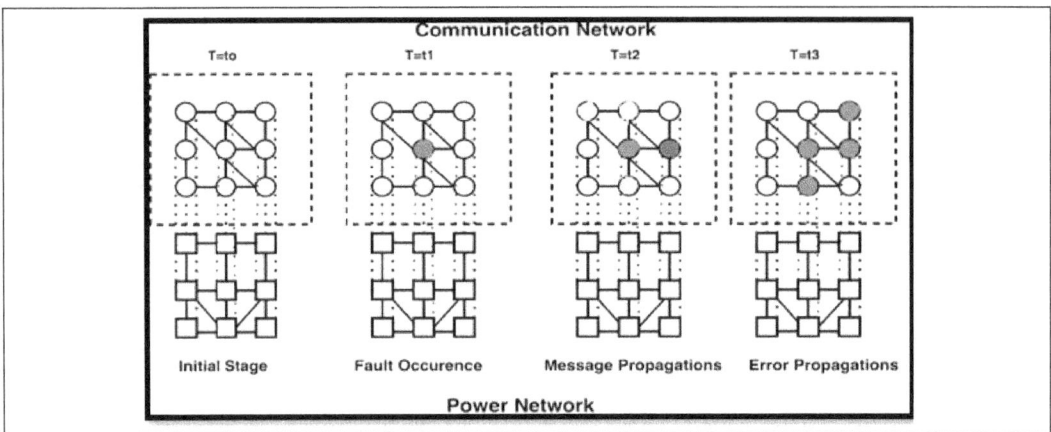

Figure 3: Coupling Network [20]

At the initial stage when time T=to, both coupled network runs normally and the nodes of communication network communicates with each other normally with full of reliability and availability. At T =T1, a fault is occurred in the power network due to faulty current in the substation because of malfunction of transformer and over-current relay trips-off the main circuit breaker. The IED associated to power nodes captures the malfunction state and starts sending the alarm messages to all the nodes connected to it. Since time delay plays a very crucial part in smart grid, at T= t2 the alarm message propagate in all direction and is received at the receiver node in three direction within the possible time delay missing the one direction indicated as purple color node. The reason not to reach the particular node within the time period is because of time consuming message process, malicious jamming attack and network congestion. Without the expected coordination, the three IEDs which received earlier the alarm message will not clear the actual fault and the missed IED node will remain in the same state and will become a fault node to possibly damage other nearby device. At T =t3 more devices can be damaged due to this alarm missed IED node and the number of faulty devices may possibly increase. Based on this situation a reliable communication network is required for the proper operation of IEDS where probability of failure of communication network is very low such as in wired system liked Ethernet and Fiber-optics networks but installation cost is much higher as compare to wireless communication network or PLC network.

3) Justification of the formal model of FDIR with Wireless communication network

The work in [11] overviewed and discussed the fault processing technologies distribution systems in smart grid and mentioned the fault detection, isolation and supply restore system due to short circuit occurred. In addition to this it defines the ground fault processing of the single phase. It describes the principle of the self-healing control of an open-loop/closed-loop of the distribution system and established the simplified model of FDIR system. It defines the standard of distribution automation system of fault processing depending upon centralized intelligence together with the master station, sub-station, feeder terminal unit and the communication system through which data is transferred to all controllers of the load switches. The rule of DAS fault processing program, fault isolation program and service restoration program are also illustrated in the book. In addition, the approach [10] gives the principle and flow chart of the complete process of fault localization, isolation and supply restoration system of the distribution network of Future Grid.

The Literature in [64-67] describe the principle and standard of the IEEE 802.11 DCF. We are now interested to analyze that how the switches, circuit breaker and relays of smart grid communicates with each other wirelessly by sending and receiving the important messages of FDIR algorithm in smart grid after the occurrence of fault and performed their function properly in least possible time with accuracy. Therefore, we developed the Markovian model of the basic access mechanism of the IEEE802.11 DCF (see Fig.11) along with receiving station of wireless communication system and then integrate the model with the overall model of FDIR in order to formally verify the model in prism model checker through temporal properties to predict and analyze the failure probability of the certain component of the substation of smart grid.

4 Tianjin electric power architecture

A radial distribution system of Tianjin Electric Power Corporation [10] is given in Fig. 4 along with IEDs connected to each component of the distribution system. The overview on china's smart grid can be found in literature [68-70] and Tianjin Eco city is one of the pilot project of smart grid where integration of the necessary component of smart grid is demonstrated and accomplished in 2011. In this way, this Tianjin Electric Power architecture along with communication IEDs is considered as Intelligent/Smart Grid. The IEDs of associated component is wired connected (Ethernet) to other IEDs of the component of substation in order to send and receive the control messages of power network. The distribution system of Tianjin Electric Power Architecture is basically consist of four feeders in which the substation A carries the feeder represented as 101 and substation B carries the feeder named as

102 while the substation C carries two feeders named as 103 and 104. Each feeder of the substation has a circuit breaker which is controlled through the over-current protection relay. To communicate the circuit breaker, over current protection relay and load switches of substation, Intelligent Electrical Device (IED) is installed on each element of the future grid. The switches present in the distribution network are load switches which are operated and controlled through IEDs and feeder terminal unit (FTU) also connected with every load switch to define the status of the load switch through various flags. The IED of circuit breaker implements the FDIR process by sending the alarm message along with FASM messages to each IEDs of load switch. When the error occurs in the substation of the distribution network, the protection relays of circuit breaker is energized and trip-off the circuit breaker and sending the FASM message to each load switches of the substation. The IED installed on each load switches starts the process of FDIR by checking the fault flags in each feeder terminal unit of load switches. The switches around faulty section is tripped-off for some time in order to isolate the fault by detaching the faulty load switch from the circuit and restore the substation through tie switches which is located at the non-faulty section of substation. The embedded software of IEDs of each load switch sends the information together with the voltage, current, power, position and fault flags status to the other IEDs of circuit breaker and relays. The IED of each component is also synchronized itself with the neighboring IED in order to send and receive the data and perform its function properly.

5 Proposed methodology

For proper execution of FDIR process in smart grid, some general requirement should be met. The proposed formal verification methodology for fault detection, isolation and recovery system along with wireless communication network is depicted in Fig. 5 and each block of proposed methodology are explain below:

A. Modeling FDIR Algorithm in Prism

The proposed method can be used to verify and study the probabilistic Markov model of FDIR along with wireless communication network in PRISM model checker [54]. The Markov model of FDIR and wireless communication network consists of circuit breaker, protection relay, IEDs, Wireless communication network, Load switches and feeder terminal unit. From the Nordel statistics [71], the realistic values of failure probabilities of components in substation are taken for all kinds of faults occurring (due to Power Transformer, Instrument Transformer, Circuit

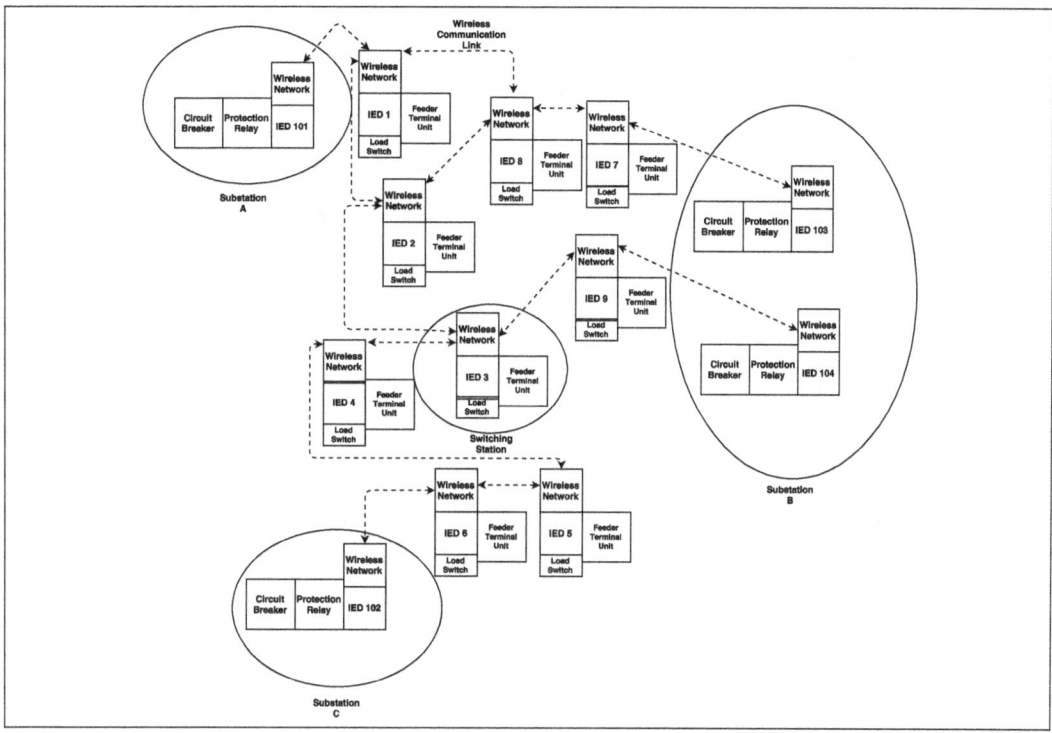

Figure 4: Tianjin Electric Power Network showing Wireless Communication Link

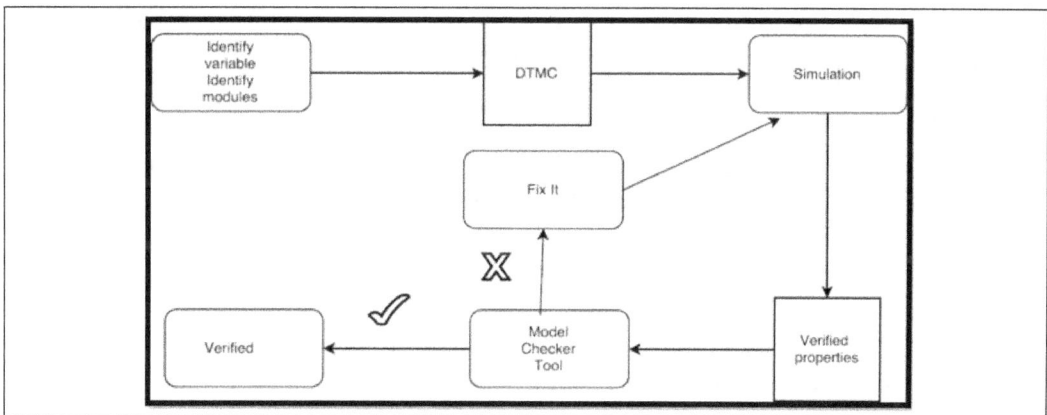

Figure 5: Proposed Methodology

Breaker, Disconnector, Surge Arresters and spark Gap, Bus bar, Control equipment, Common ancillary equipment, and other substation faults) in a substation during the years (2000-2007). In this regard, we have selected the worst case of 0.476 failure probability of all the components of substation of Norway Country among other Nordic Countries to be used for the Tianjin Electric Power System. More specifically, 0.316 failure probability of control equipment (i.e., IEDs) and 0.023 failure probability of load switches (i.e., disconnector and Ancillary equipment) are taken as a reference failure probabilities from [71] to be used in Eq. (1) for the calculation of the overall failure probabilities of components and IEDs which comes out to be 0.6415.

$$PFDIRCC = (PFDIR + PCC) - (PFDIR * PCC), \qquad (1)$$

Where PFDIRCC is the overall failure probability of the system PFDIR is the FDIR component failure probability PCC is the control and communication component (IED) failure probability Similarly, [20] presented and derived the realistic values of probabilities of failures of alarm message transmission to load switch IED within the specified time (e.g., 3ms for 9 bus system). To cater realistic scenario, we have considered our network to be a 9 - bus system whose failure probability is found to be 0.57 (57The following step allows us to model the FDIR process with communication network in PRISM tool.

1) Identifying Modules

The first step to construct the Markov model of any system is to identifying and defines the number of modules present in the whole process. FDIR process along with wireless communication network of the substation basically consists of five modules in which three modules define the FDIR complete process where as two modules will construct the wireless communication network. Each module of the whole process consists of number of states and these states of each module transit to another state by satisfying the condition of augmented probabilities.

2) Identify Variable

Unique Variables are initialized by giving its data type in each modules of FDIR along with wireless communication network which can be used in other modules of the whole system by sharing its data. The variables in each module act as a global variable and therefore can be used in other module of the system.

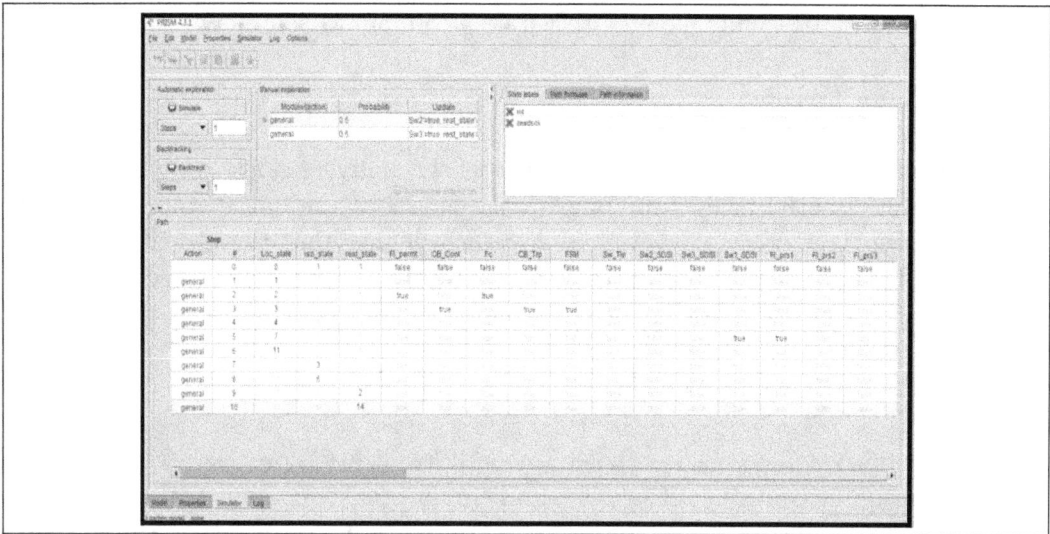

Figure 6: Simulation Result of FDIR Behavior along Wireless Communication Network

B. Functional Verification using Simulation

One of the most significant benefit of using PRISM tool [54] is that it has a built-in simulation tablet which is useful to check and simulate the model first and take out the bugs in the initial state before going to formal verification process which is exhaustive and time consuming. The main concept is to construct the probabilistic model in PRISM language and then compiled it in order to check the errors. Once the error has taken out from the model with the help of built-in simulator of prism model checker tool, then we can verify the model through probabilistic temporal logic property. The simulator of PRISM model checker manually check the model step by step and often detect some critical errors in the initial state which can be corrected in the model at this stage before assigning the temporal logic property in the model as shown in Fig.6.

C. Formal Function Verification

As illustrated in Fig. 5, after completing the process of simulation and now we need to test the model by applying certain condition in the form of probabilistic specification in PRISM model checker and verify the model rigorously and exhaustively. We are interested to develop the following temporal logic property to check and verify the FDIR model along with wireless communication network.

1) Deadlock Freedom

It is of the interest to verify our FDIR model through Deadlock freedom property which is very essential to find the bugs in the model at the initial state. Deadlock is a very important check for any system, which defines that the algorithm is performing well. It basically reaches to every state of the model and checks whether this state of model will move further or remain at the present state.

2) Fault Localization Model Turned On

This property will ensure that when the fault localization process is going on and simultaneously the controller of switches and circuit breaker communicates with each other in order to find the exact location of fault in the substation of Smart Grid, the other process of FDIR will remain idle and switch off and we will verify this by assigning the property in our model.

3) Fault Isolation Model Turned On

When the fault isolation process begins, the others process of FDIR will remains in an idle position. The objective of the fault isolation process is to detach/trip the faulty component of FDIR with in the short possible time and we are interested to verify this through temporal logic property.

4) No Two Process run at same time

To avoid malfunctioning, no two process of FDIR will run at the same time. This is very important as the output of one process is required by the second process to perform its task properly. We need to verify this by assigning the temporal property in the model.

5) Restoration Model should not take more than 60 sec

The proposed model is developed in such a way that it will take a least possible time to restore the power supply of the substation of the distribution network through non faulty zone and we will verify this by assigning the appropriate property in our model.

6) To find the probability that fault will occur in switch no. 1 of the system

As Prism is a probabilistic model checker, we are interested to find the probability at which the fault will occur on switch no. 1 of the distribution system and fault flag 1 is high after the completion of the process of fault localization. We need to develop the probabilistic property in order to get the probability that switch 1 is the faulty switch.

7) To find the probability that fault flag 2 is high in switch no. 2

It is of the interest to find the probability at which the fault flag 2 is high on load switch 2 after the completion of the fault localization process. We will verify this by assigning the probabilistic property in our model.

8) To find the probability that switch no. 2 trip off at the limited time

Find the probability at which any switch trip off with in the limited time whenever it gets the ISOM message is very important. In this regard, we need to verify the FDIR model through this probabilistic property. Therefore, switch no. 2 is taken as a reference to find the probability at which it will be tripped-off from the circuit.

9) To find the probability to recover the system through switch no. 3

It is of the interest to find the probability at which the system is recovered by Tie switch. As this is very essential property, switch no. 3 is dedicated as a recovery or Tie switch in order to find the probability at which it will recover the FDIR model of the smart grid after the occurrence of fault.

10) Finding probability of CB's IED transmitting the data
In the end, the probability is found at which the IED of circuit breaker sends the data to another IED of load switches. We develop the temporal logic property and fed into model checker to find the probability at which IED sends the data.

6 Markov model

To analyze the behavior of FDIR in PRISM [54], it mainly involves three modules i.e. fault detection, isolation and service restoration system and the model of FDIR behavior is selected as DTMC [63, 72-74]. The first step is to initialize the variables in PRISM tool and translate the Finite State Machine (FSM) i.e., Markov chain into PRISM language. After modeling the behavior of FDIR in PRISM tool, we need to compile it and test its functionality. It is good to perform the simulation first in order to find many unpredicted errors or bugs in the models before formal verification. With the purpose of staying away from the 'state space explosion' in formal verification approach [54], we apply the abstraction on the whole system and verify the FDIR behavior on substation A along with three load switches. Other substation of Tianjin electric power corporation is also connected with three load switches where FDIR algorithm is running in order to detect and restore the system. The Probabilistic model of FDIR and wireless communication network along with the brief description of each module is given below:

A. Fault Detection Model

In Markov model of Fig. 7, all variables and constants are initialized in the $Loc\_State' = 1$ as explained in Fig. 8. The fault permit probability (0.476) is initially taken from the Nordel analysis of Norway country as explained in section V.A. The Tianjin distribution network runs smoothly until the permanent fault (Fl_permit) does not occur. There are two possibilities (1) When the fault occurs due to switching failure of distribution network (with the probability $fl\_permt = 0.476$), the FDIR process starts at $Loc\_State' = 2$ using ideal communication medium. (2) When the fault occurs due to IED plus switching failure of distribution network (with the probability $fl\_permt = 0.64158$) the FDIR process starts at $Loc\_State' = 2$ using wireless communication medium. As the fault current exceeds the maximum value in $Loc\_State' = 2$ with probability $=1$, the protection relay sense this faulty over-current and trip the circuit breaker. The circuit breaker IEDs/controller is activated and sends the FASM message along with ACMP message to its connected load switches IEDs in the $Loc\_State' = 3$. To send the data packet with probability 1, it uses the parameters of FHSS i.e., frequency hopping spread spectrum [64-67] as a physical layer to send and receive the data at a transmission bit rate of 2 Mbps. Due to weather condition, delay and noisy environment, 0.43 is the probability [20] that each load switch controller received the FASM and ACMP message correctly without any delay and distortion. Once the load switches receive this signal, it initially checks whether it is a Tie switch or not. If it is not a Tie switch, fault processing start and check the fault flag status (Fl_flg) in the Feeder terminal unit. If Fl_flg status is active then this is the faulty switch otherwise it forward the FASM and ACMP messages to another load switches. If fault location does not find in any load switches, then the FASM message will be discarded and send back to circuit breaker controller ($Loc\_state' = 3$).

B. Fault Isolation Model

After completing the process of fault detection and finding the fault flags in the particular load switch, the fault isolation process starts in order to isolate this faulty load switch from the rest of the network as shown in Fig. 9. The variables and constants are initialized in the $Iso\_State' = 1$ as explained in Fig. 10 of Fault isolation model. In $Iso\_State' = 2, 0.977$ is the probability [71] that the load switch trip off with in the limited time with-out any delay. If the load switch trips-off, the isolation is successful and send the ISOM message to the other load switches indicating the faulty section and starts the process of closing preparation in the restoration section through Tie switch within the limited time. On the other hand if the load switch does not trip-off within the limit time (0.023 probability) [71], the isolation fails by sending the ISOM message and expands its faulty area by including more de-energized switch from the restore section with control switch ID=0. It is necessary for the Tie switch to receive the ISOM message at the time of closing; otherwise it cancels the process of closing preparation of the Tie switch.

C. Supply Restoration Model

In $Rest\_State' = 1$, variable and constants are initialized as defined in the Fig. 12 of the supply restoration model. The Tie Switch receives the ISOM message with a probability of 0.43 [20], starts the reclosing process with probability of 0.977 [71] and issues the RESM message to its neighboring switches. But if it fails to receive the ISOM message (failure probability 0.57) [20] or reclosing (failure probability 0.023) [71], it issues the RESM message by putting Tie switch ID=0 and develops a new restoration scheme with another Tie switch. ISOM message compares the actual load needed with the available power source and sends the restoration policy to all the Tie load switches connected with it.

D. IEEE 802.11 DCF of Basic Access Mechanism

Approaches [64-67] describe the principle and standard of the IEEE 802.11 DCF. We are interested to analyze that how the switches, circuit breaker and relays of smart grid communicates with each other wirelessly by sending and receiving the important messages of FDIR algorithm in smart grid after the occurrence of fault and performed their function properly in least possible time with accuracy. We developed the Markovian model of the basic access mechanism of the IEEE802.11 DCF along

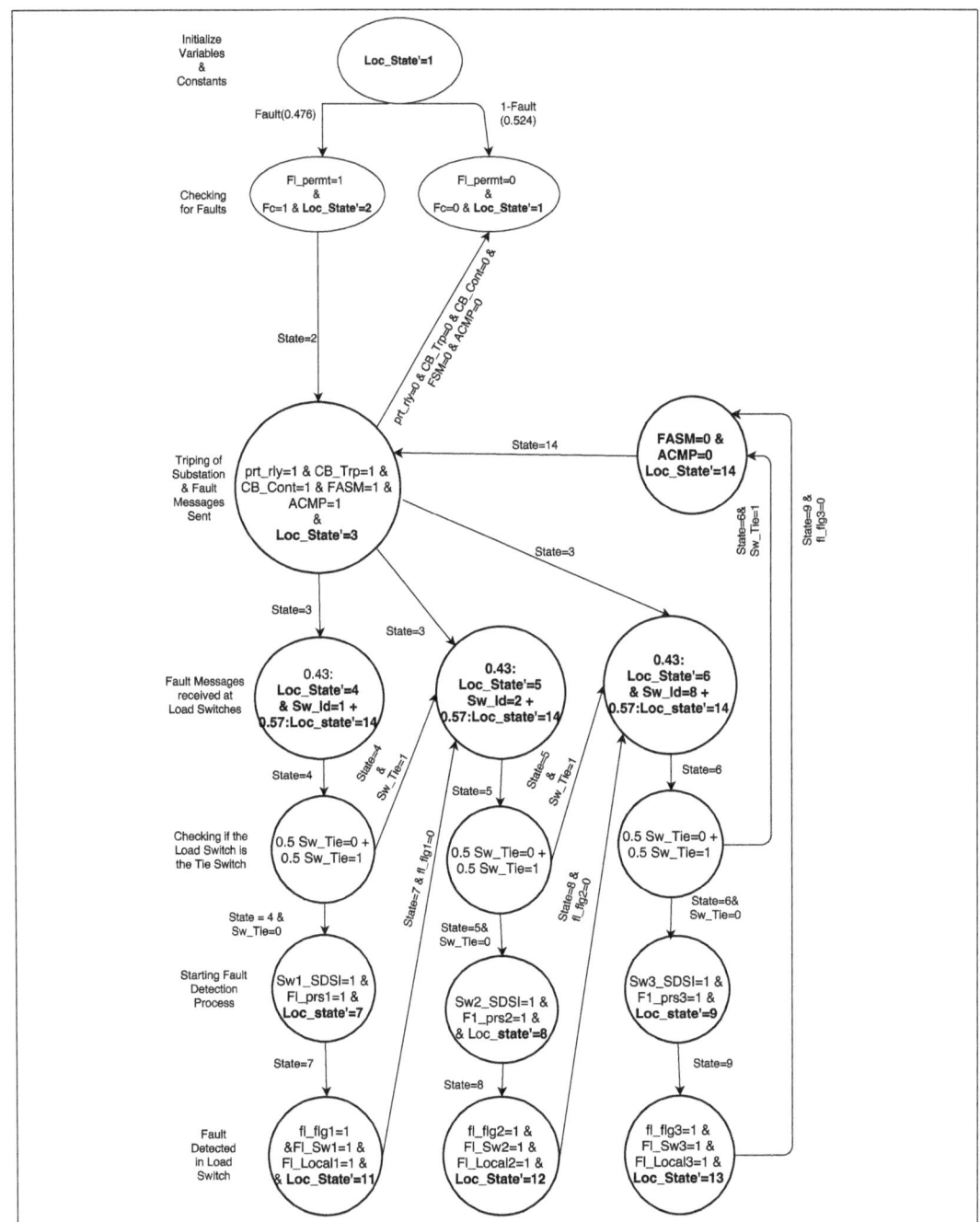

Figure 7: Fault Detection Model

Meaning of Variables and Constants			
Variables/Constants	Meanings	Variables/Constants	Meanings
Fl_permt	Permanent Fault	ACMP	Available Capacity of Main Power Source
FC	Fault Current	Sw_Tie	Tie Switch
CB_Cont	Circuit Breaker Controller	Fl_Prs	Fault Process
CB_Trip	Circuit Breaker Trip	Prt_rly	Protection Relay
FASM	FDIR Message Start	Fl_flg	Fault Flag
Fl_Local	Fault Localization	Fl_Sw	Fault Switch
Sw_SDSI	Switch Data Service Instance	Sw_Id	Load Switch ID
Loc_state	Localization state		

Figure 8: Fault Detection Model Parameters

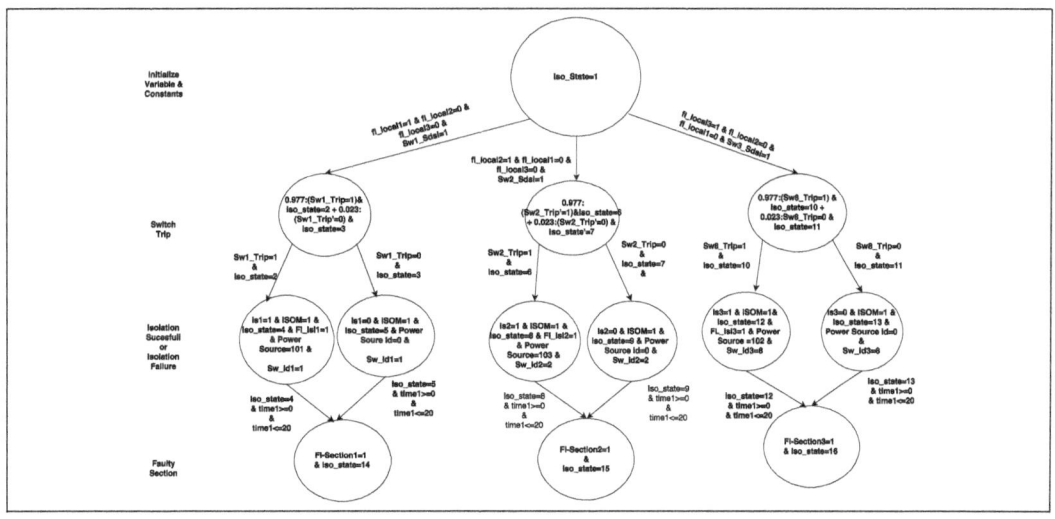

Figure 9: Fault Isolation Model

Meaning of Variables and Constants			
Variables/Constants	Meanings	Variables/Constants	Meanings
Power Source	Power Source ID	ISOM	Isolation Result Message
Iso_state	Isolation state	Sw_Trip	Switch Trip
Fl_Section	Faulty Section	Fl_Isl	Fault Isolation

Figure 10: Fault Isolation Model Parameters

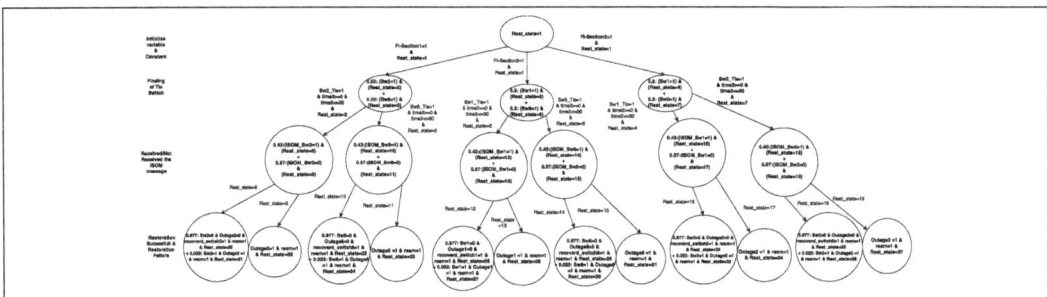

Figure 11: Supply Restoration Model

Meaning of Variables and Constants			
Outage	Outage Switch	resm	Restoration Result Message
recovered_switch	Tie_Switch	Sw_Tie	Switch Tie
rest_state	Restoration State	ISOM_Sw	Tie Switch received the ISOM Message

Figure 12: Supply Restoration Model Parameters

with receiving station of wireless communication system and then integrate the model with the overall model of FDIR in order to formally verify the model in PRISM model checker through temporal properties and analyze the failure probability of the certain component of the substation of smart grid. In our model, we take circuit breaker IED of substation of smart grid along with three load switches IEDs of the distribution network and construct the Discrete Time Markov Chain (DTMC) of these four IEDs in which substation IED sends the fault messages to each load switch connected to it. To send the data packet with probability 1, it uses the parameters of FHSS i.e., frequency hopping spread spectrum as a physical layer to send and receive the data at a transmission bit rate of 2 Mbps. With probability 0.43 at 3ms delay, all the data packets are sent and received by switches and circuit breaker of the smart grid. The probability 0.43 through which circuit breaker IED can send the data, if the channel is open, the sender sends the data and enter into vulnerable period where back-off value (a random number generator) is '0' and if the channel is not free the station enters into random period where it waits for back-off value reaching to '0'in order to send the data. The total instance for sending the data can be varied i.e., non deterministic. If the data packet is sent by the sender is successfully received at the receiving end then it waits for the acknowledgement. After getting the acknowledgement, the sender checks the value of back-off, if it is '0' then the sender sends the another data packet but if back-off value is not '0', it will wait till the back-off value reaches to 0 in order to send the data packet again.

The sender waits 200 μs, if it receives the acknowledgement, it means the data packet sent accurately otherwise the time-out occurs and it enter into the random process where the sender waits for the medium to check a back-off value reaching to '0' in order to resume its transmission by sending again the data packets. The back-off value only reaches to '0' when the channel is idle for certain instances and if the channel is active by sending and receiving the data packets of another station, the back-off decreasing value is stopped and wait for a channel to be free in order to decrease its value and reaches to '0' so that the sender can send and receive the data packets with another station. The new Markovian model of the IEEE 802.11 DCF is developed and is given in fig. 13.

E. Receiving Station Markovian Model

The new Discrete Time Markovain model (DTMC) of the receiving station for the wireless system is developed by considering the three elements in which the state transition occurs from one state to another state with probability 1. Again, 0.43 is the probability at which each receiving station waits for the data packet to receive. If the data packets received accurately then it wait for SIFS and send the

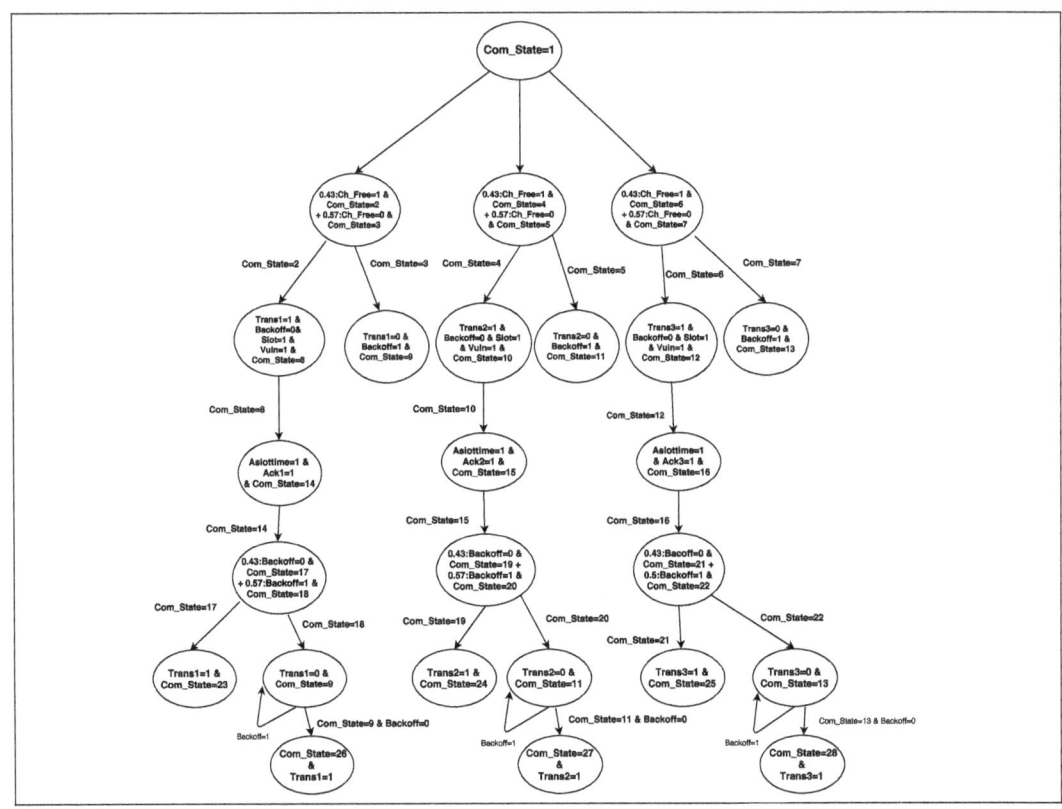

Figure 13: Basic Access Mechanism of IEEE 802.11 DCF

Meaning of Variables and Constants			
Com_State	Communication State	Trans	Transmission of Messages
Backoff	Random generated value	Ack	Acknowledgement
Vuln	Vulnerable state	Ch_Free	Channel Free

Figure 14: Basic Access Mechanism of IEEE 802.11 DCF Parameter Explanation

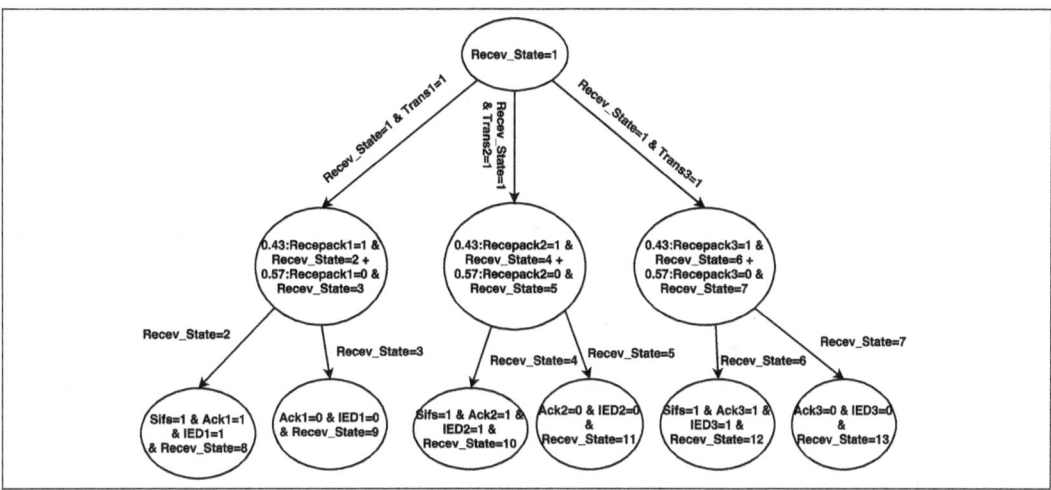

Figure 15: Receiving Station of the Wireless System

Meaning of Variables and Constants			
Recev_State	Receiving Station	Recepack	Receiving Packet
Ack	Acknowledgement	IED	Intelligent Electronic Device
Sifs	Short Interframe Space		

Figure 16: Receiving Station of the Wireless System Parameters

acknowledgment but if the data packet does not receive correctly (0.57 probability), the receiving station do nothing. The probabilistic Markovain model of receiving station is given below in Fig. 15.

F. Integration of FDIR model with Wireless Communication model via Prism Model Checker.

Approach in [75] defines the rule to add two different probabilities of independent events. As communication network has a backup power battery, the power failure of substation will not affect the communication network. Thus, from Eq. (1), the failure probability of communication network (PCC=0.316) can be added up with the failure probability of power network (PFDIR=0.476) by subtracting the joint probability of coupling network (PFDIR * PCC=0.150) with the purpose of getting the whole failure probability (Components + IED) of the Smart Grid system to be (PFDIRCC=0.641584).

7 Formal function verification of FDIR with wireless communication network

In this section, we will put a number of properties in PRISM [54] in order to verify the model formally. In this regard, following important properties fed into the model checker.

A. Property Verification of FDIR with Wireless Communication through PRISM

Property 1: Dead Lock freedom

A desired characteristic of FDIR process is that it never gets stuck in particular position which can be checked by ensuring the deadlock freedom property for the FDIR behavior. Therefore, the deadlock freedom property can be applied for this verification with the purpose that every state of model is reachable.
E[F deadlock]
The temporal operator "E" represents eventually in the future and F represents the value true always if the deadlock occur on the system. The deadlock freedom property checks the whole model in PRISM tool and gives us the result that the algorithm was found to be deadlock free showing result as 'false', meaning that there is no deadlock in the model. In our case study, the dead lock property checks the (0-14) states of fault localization process, (0-16) states of Isolation process and (0-19) states of restoration process and found that all states are deadlock free. Fig. 20 shows the verification result of the property (from PRISM) which is available in Appendix.

Property 2: Fault Localization Model Turned On

When the fault localization process is going on and the controller of circuit breaker communicates with each controller of switches, with the intention to find the exact position of the fault, the isolation process and restoration process should be 'off'. Now, we can verify this by putting the below mentioned property in our model.

$A[FLoc\_state = 0 \& iso\_state = 1 \& rest\_state = 0]$

Where temporal operator "A" represents always and "&" represents conjunction and symbol "F" will always be true when fault current equals to 1 and send the FASM message to each load switch to start the fault localization process. The PRISM tool verified this property by giving the result 'true' i.e., when the process of fault localization is continue, the other two process (Isolation and Restoration) is turned off. The result of this property is shown in Fig. 21 which is available in Appendix.

Property 3: Fault Isolation Model Turned On

When the fault isolation process is going on and the controller of circuit breaker tries to trip off the switch as well as isolate this switch with the other connected switches, relays or component of Smart Grid, it is necessary that the fault localization process and the restoration process should be turned off and we can verify this by assigning the below mentioned property in our model.

$A[Fiso\_state = 1 \& !rest\_state = 1 \& !Loc\_state = 1]$

The prism tool verified this property by giving the result 'true' i.e., whenever switches, relays or any other component of Smart grid trying to isolate itself with the other connected component, the other two process will remain in idle position and will start only once the fault isolation process finishes off. The result of this property is shown in Fig. 22.Available in Appendix.

Property 4 : No Two Process Run At same Time

Another most important property to verify in our model is that the two process will not start at the same time. This can be verified by putting the below mentioned property

$A[FLoc\_state = 1 \& (!(iso\_state = 1 \& rest\_state = 1)) | iso\_state = 1 \&$
$((Loc\_state = 1 \& rest\_state = 1)) | rest\_state = 1 \&$
$((Loc\_state = 1 \& iso\_state = 1))]$

The PRISM tool verified this property by giving the result true and in this way the two processes, fault localization or fault isolation or restoration system will not be started and perform their function at the same time. The result of this property is shown in Fig. 23.

Property 5: Restoration Model should not take more than 60 seconds

The restoration process of the substation of smart grid will start after the fault localization and fault isolation process completed successfully and within 60 s it will complete its process by restoring the substation through the Tie switch.
$A[Frest\_state = 0 \& time2 <= 60]$

The notation represents that "A" always now and in the future, the restoration process starts at restoration state= 1 will take only 60 sec time with probability 1 to recover the system will always be true by symbol "F". The PRISM tool verified this property by giving the result true and in this way restoring model will not take more than 60secs in restoring the power of the Smart Grid. The result of this property is shown in Fig.24.

B. Finding Probabilities of FDIR with Communication Network through PRISM

Property 6: To find the probability that fault will occur in switch no. 1 of the system

It is important to find the probability at which the fault will occur on switch no. 1 of the distribution system after integrating the failure probability of communication system with the FDIR model. The fault flag 1 is high after the completion of the process of fault localization. We developed the probabilistic property in order to get the probability that switch no. 1 is the faulty switch. Result 0.212 is obtained probability by the Prism tool that fault will occur at switch no. 1. The syntax and the result of the temporal logic property is given in Fig 25.
$P =?[trueU(Loc\_state = 11)]$
True always hold probability until loc_state 11 hold probability.

Property 7: Probability of fault flag 2 is high in switch no. 2

It is of the interest to find the probability at which the fault flag 2 is high on load switch no. 2 after the completion of the fault localization process. We developed the probabilistic property and assign on the FDIR model in order to get the probability of fault flag high on switch no. 2. The syntax and result of the probability is shown in Fig. 26
$P =?[trueU(Loc\_state = 12)]$
The PRISM tool results with the 0.349 probability at which the fault flag is high at load Switch no. 2, which can be seen in Fig. 26.

Property 8: Probability that Switch no. 2 trips off within a limited time

It is also required to find the probability at which any switch trips- off within the limited time whenever it gets the ISOM message. We have taken Switch no. 2 as a reference in order to find the probability at which it will trips-off properly. We developed the temporal property and apply on FDIR model. Then, the Prism model checker has given us the probability of 0.427. The syntax and result of the Prism model checker is shown in Fig. 27
$P =?[trueU(iso\_state = 12)\,]$
True always hold probability until the isolation state=12 hold probability.

Property 9: Probability to recover the system through Switch no. 3

Now, finding the probability at which the system is recovered by Tie switch is performed in PRISM. As this is very essential property, we take switch no. 3 as a recovery or Tie switch in order to find the probability at which it will recover in the FDIR model of the smart grid after the occurrence of fault and integrating the communication system failure probability. The PRISM model checker has given us the probability of 0.0918 at which it will restore the system through switch no. 3. The syntax and result of the prism model checker is given below in Fig. 28.
$P =?[trueU(rest\_state = 32)]$ True always hold probability until the restoration state=32 hold probability.

Property 10: To find the probability that IED of CB transmit the Data

Let's find another important probability at which the IED of circuit breaker transmit the data to the other load switches. We develop this property and assign on FDIR model with wireless communication system in order to get the probability of sending the data by IED of circuit breaker. The syntax and result of the property is given below in Fig. 29.
$P =?[trueU(Com\_State = 25)]$

C. Formal Function Verification of FDIR with Ideal communication medium

Literature [20] discusses the communication failure probability in different IEEE test cases and suggest that the communication failure probability of wired system such as Ethernet, Fiber-optics network is very low compared to wireless communication system and PLC network. Due to lower failure probability of Ethernet and fiber-optics communication network, we are integrating the FDIR model with these ideal communication networks by assuming and neglecting the failure probability of wired communication system and then formally verify the FDIR model along with ideal communication system and analyze the failure probability of different component of substation.

Property 1: To find the probability that fault will occur in Switch no. 1 of the system

The probability at which the fault will occur on switch no. 1 of the distribution system can be found, in which fault flag 1 is high after the completion of the process of fault localization. We developed the probabilistic property in order to get the probability that switch no. 1 is the faulty switch 0.197 is the probability given by the Prism tool that fault will occur at switch no. 1 .The syntax and the result of the temporal logic property is given below in Fig. 30.
$P =?[true\ U(Loc\_state = 11)]$

Property 2: Probability that fault flag 2 is high in switch no. 2

Now, we find the probability at which the fault flag 2 is high on load switch no. 2 after the completion of the fault localization process. We develop the probabilistic property and assign on the FDIR model in order to get the probability of fault flag high on switch 2. The syntax and result of the probability is shown below in Fig. 31.
$P =?[trueU(Loc\_state = 12)]$
The Prism tool gives the probability result of 0.3456 at which the fault flag is high at load switch no. 2.

Property 3: Probability that switch no. 2 trip off at the limited time

Now, it is required to find the probability at which any switch trips-off within the limited time whenever it gets the ISOM message. We have taken switch no. 2 as a reference in order to find the probability at which at will trip off properly. We developed the property and apply on FDIR model. The Prism model checker has given us 0.446 probability at which the switch no. 2 trip off properly at the limited time. The syntax and result of the Prism model checker is shown below in Fig. 32.
$P =?[trueU(iso\_state = 12)]$

Property 4 : Probability to recover the system through Switch no. 3

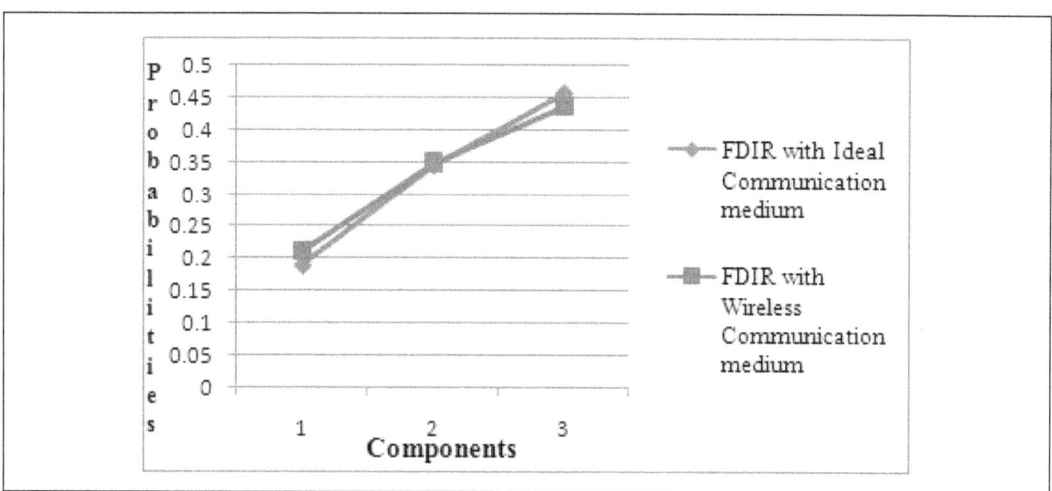

Figure 17: Comparison of Probabilities for Failure of Components

Now, we find the probability at which the system is recovered by Tie switch. As this is very essential property, we take switch no. 3 as a recovery or Tie switch in order to find the probability at which it will recover the FDIR model of the smart grid after the occurrence of fault. The Prism model checker has given us the probability of 0.0959 at which it will restore the system through switch no. 3 as shown in Fig. 33.

8 Comparison of FDIR with ideal communication medium versus FDIR with wireless communication medium

In this section, a valuable comparison done to analyze the failure probabilities of load switches when FDIR connected to ideal communication medium such as Ethernet or Fiber optics medium versus FDIR connected to the wireless communication medium. From graphs, it can be observed that the failure probability of load switches when FIDR connected to ideal communication medium is slightly less as compared to FDIR module integrated in to the wireless communication medium. The failure probability of components in FDIR with ideal communication system ranges from 0.192 to 0.456 where as in wireless communication network with FDIR the ranges extend from 0.212 to 0.437. The comparison graph of two different medium with FDIR module is given below in Fig. 17.

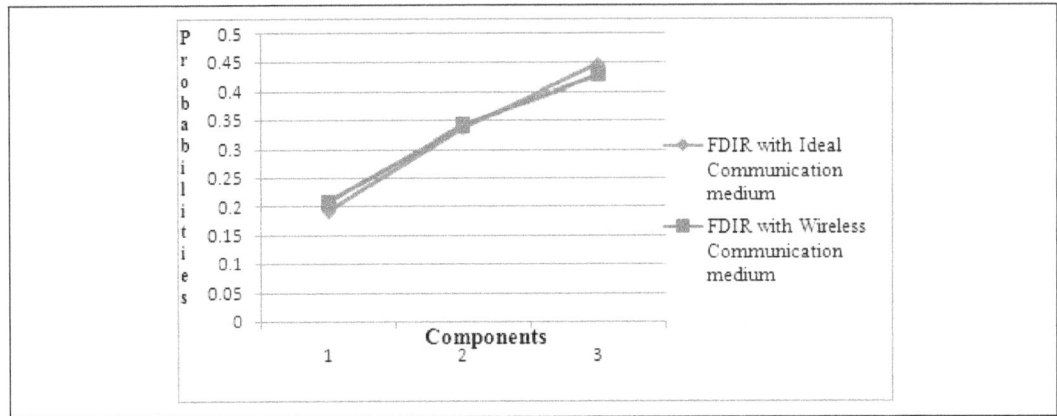

Figure 18: Components Tripping off Probabilities

Figure 18 depicts the probabilities of tripping off load switch within the limited time in order to isolate the load switch from the system. We compare and analyze the probability of tripping off load switches when FDIR module integrated to an ideal communication medium (like Ethernet and Fiber optics medium) versus the FDIR model connected to wireless communication medium. The FDIR with ideal communication system, the Probabilities lies between 0.192 to 0.446 where as in FDIR with wireless communication system, the probabilities lies between the 0.207 to 0.4274. Fig. 18 also shows the comparison result of two different medium integrated with FDIR.

Figure 19 shows the overview on the probabilities of different load switches to recover the system. We compared and analyzed the probabilities of different load switches in order to recover the system when FDIR model integrated with ideal communication medium against the FDIR model connected to wireless communication medium. The probabilities to recover the system in ideal communication system with FDIR ranging from 0.0726 to 0.0959 whereas in wireless communication system with FDIR, the probabilities lies between the 0.0312 to 0.0734. The comparison of two different mediumwith FDIR model can also be seen from Fig. 19 below.

9 Conclusion

The probabilistic Markovian model (DTMC) of FDIR behavior in distribution network of Smart Grid has been successfully developed along with the Markovian model of IEEE 802.11 DCF and integrate it in PRISM model checker in order to verify the whole system and analyze its accuracy, efficiency and reliability by developing

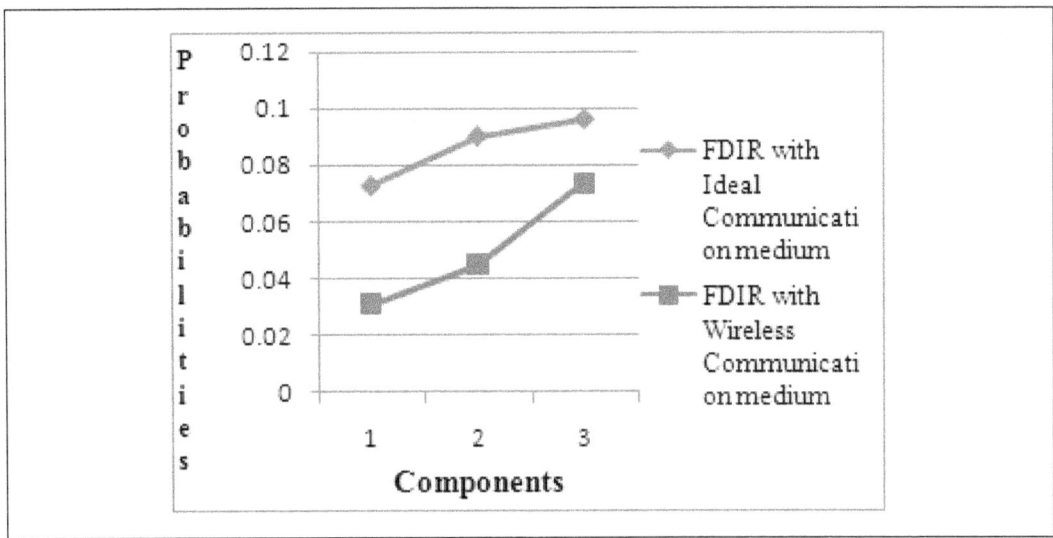

Figure 19: Load Switch Restoration Probabilities

and applying the logical properties in the model. More-over failure probabilities of different component of distribution network in smart grid is predicted when FDIR is connected with wireless communication system and wired communication system. Similarly, we also analyze and predicted the probability at which the load switches of distribution network work properly by making the faulty component detach itself upon the occurrence of fault in wireless communication network as well as wired communication network. More over we also predicted the probability to recover the system through particular non-active switch in wired and wireless communication network. In addition, we also analyze and concluded that restoration process of FDIR will not take more than 60s to restore the power of distribution network of this Smart Grid. Moreover, no malfunction will occur as verified (via PRISM) that two processes will not run at the same time. In the same way, all together ten important probabilistic properties are verified and significant probabilities are predicted to analyze the performance of the Smart Grid Model. Finally, some important comparison results are obtained and discussed when FDIR connected with ideal communication medium as compared to FDIR connected with wireless communication network, which clearly showed ideal communication network has less failure probabilities in Smart Grid.

Appendix

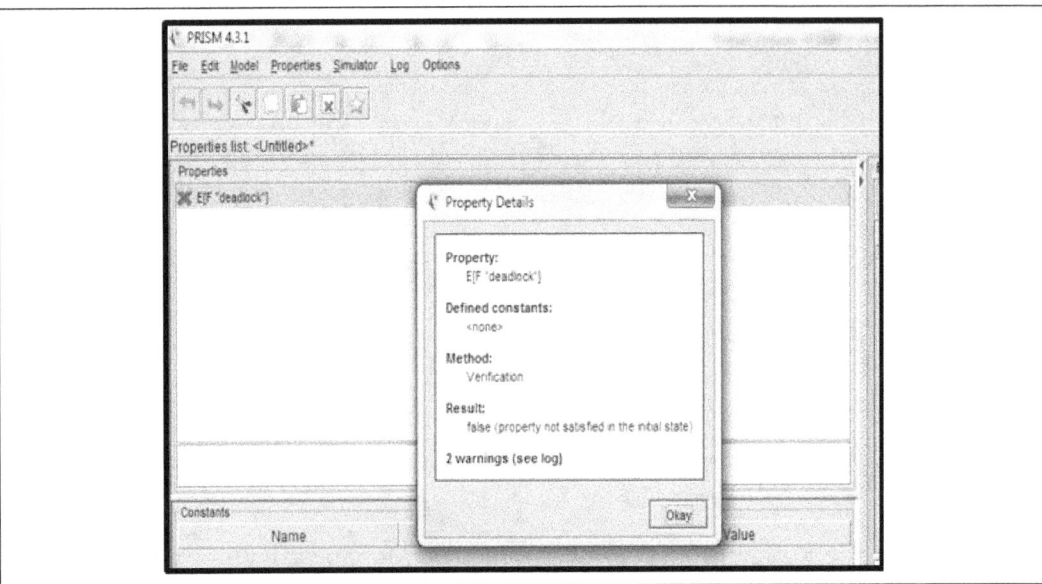

Figure 20: Property of Dead Lock Freedom

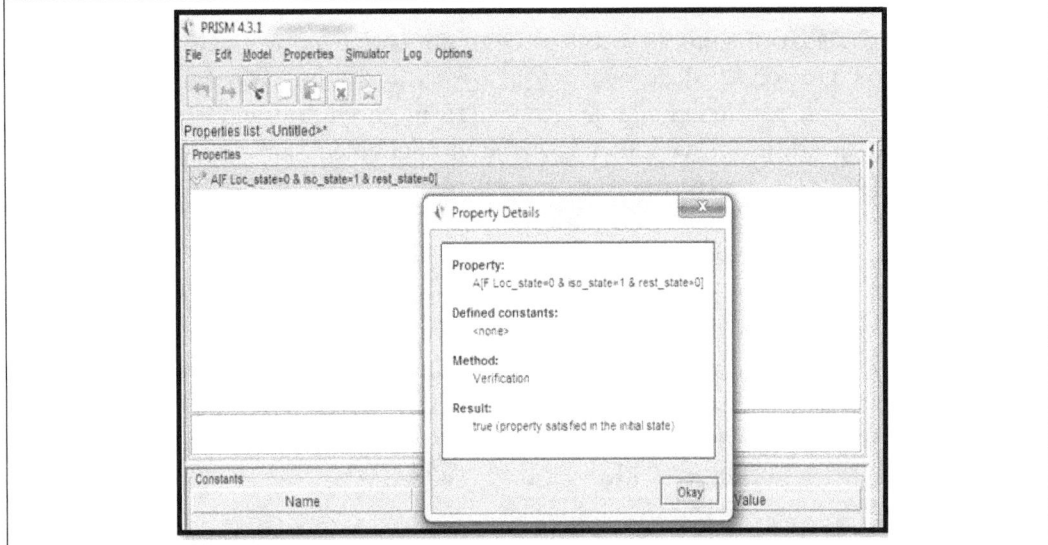

Figure 21: Property of Fault Localization Model

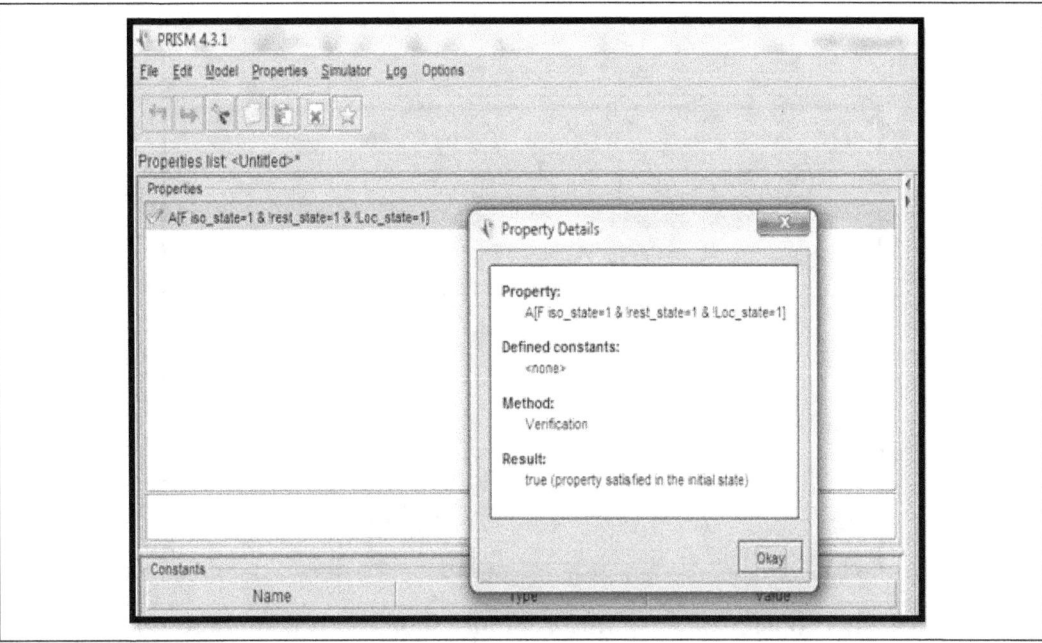

Figure 22: Property of Fault Isolation Model

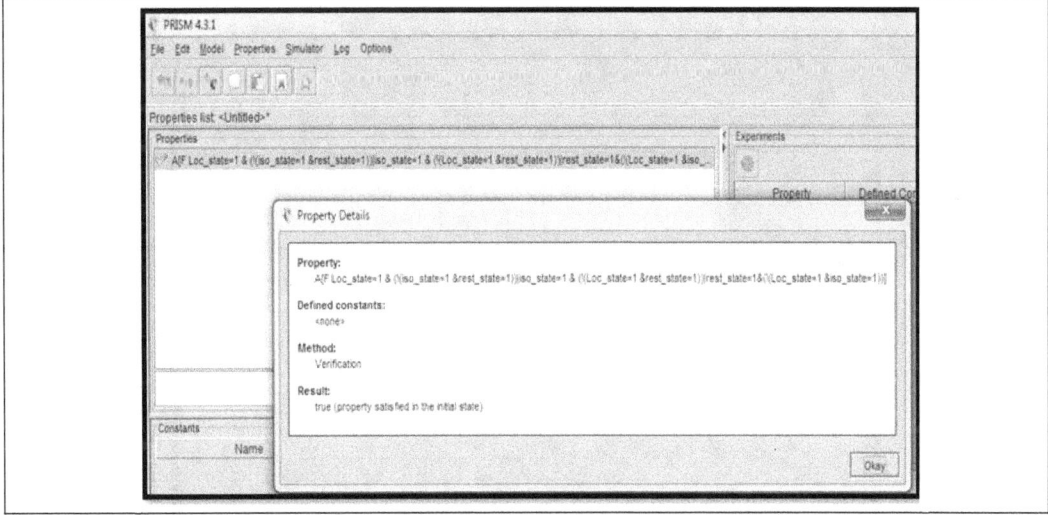

Figure 23: Property of No-Two-Process Run at Same Time

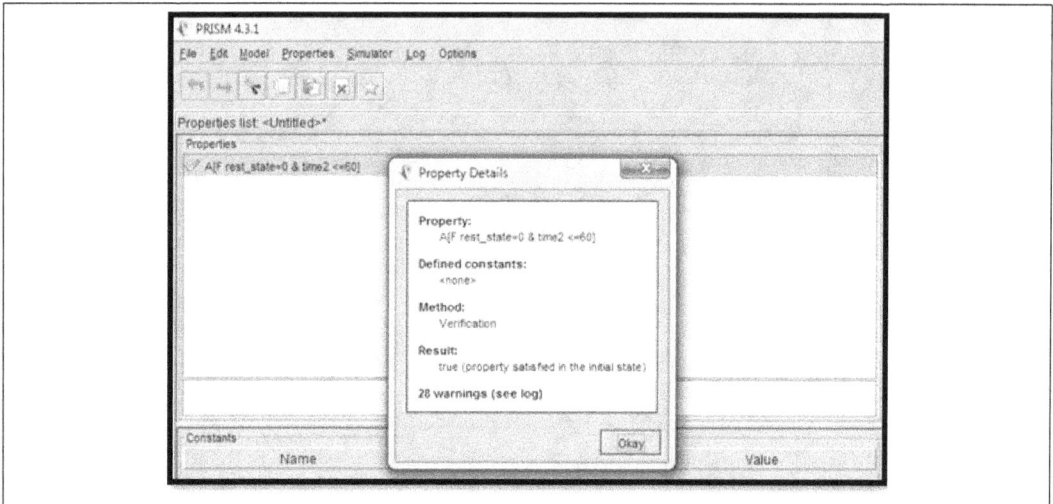

Figure 24: Property of Restoration Model should not take more than 60

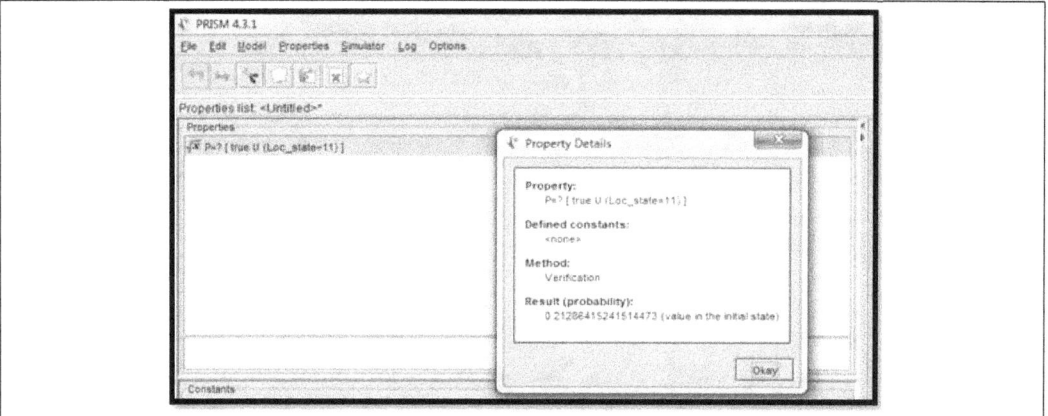

Figure 25: Fault Occurrence Probability of Switch # 1

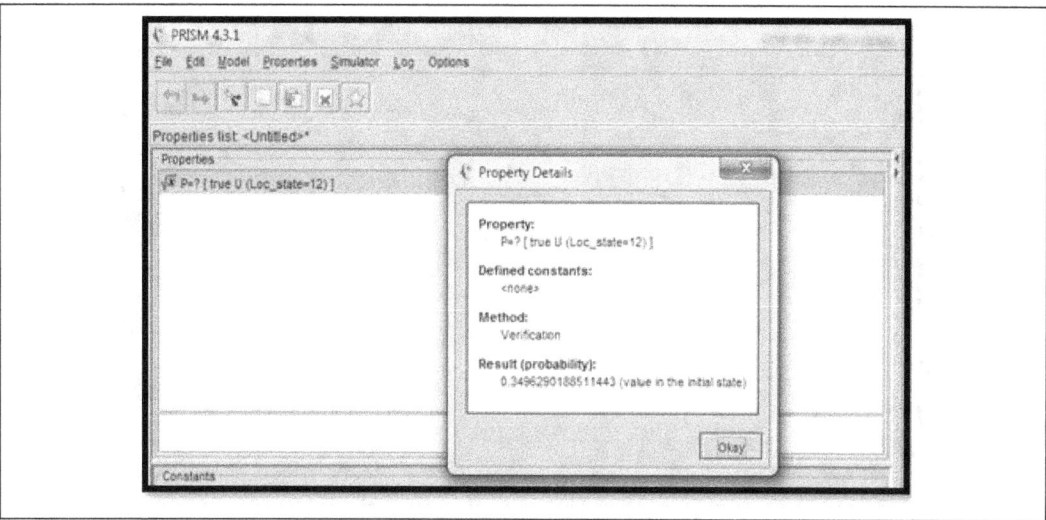

Figure 26: Fault Flag 2 Probability of Switch # 2

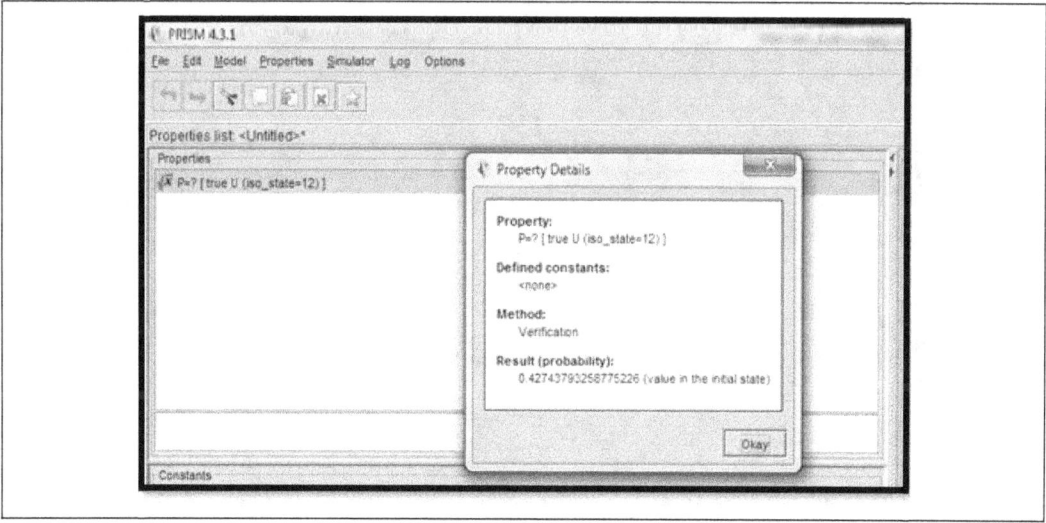

Figure 27: Probability of Switch # 2 Tripping-off within the limited time

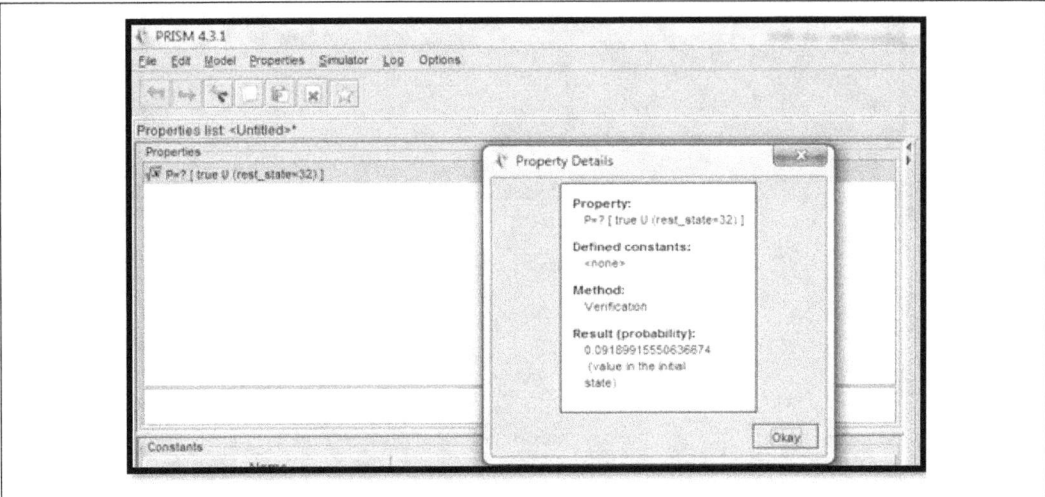

Figure 28: Probability to recover the system through Switch # 3

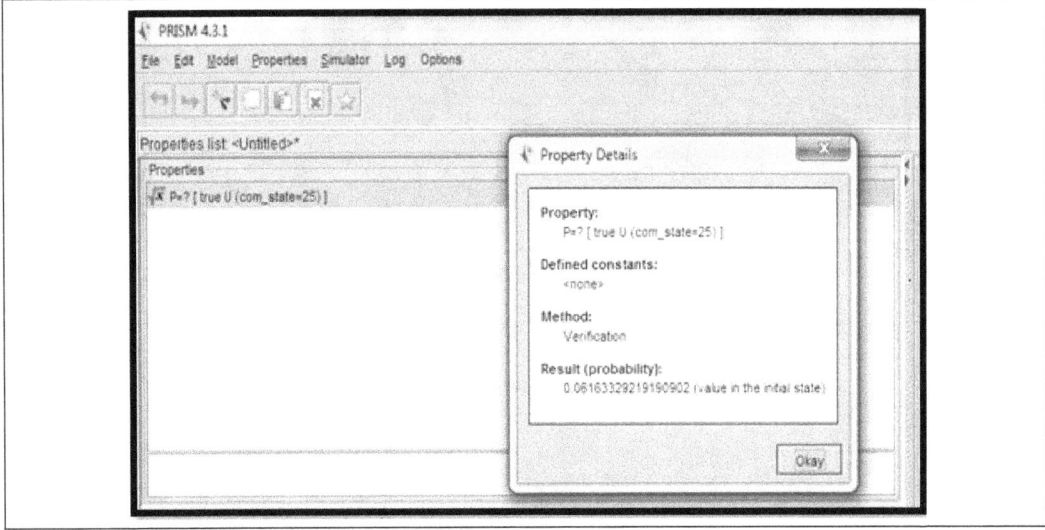

Figure 29: Probability to send the data by IED of CB

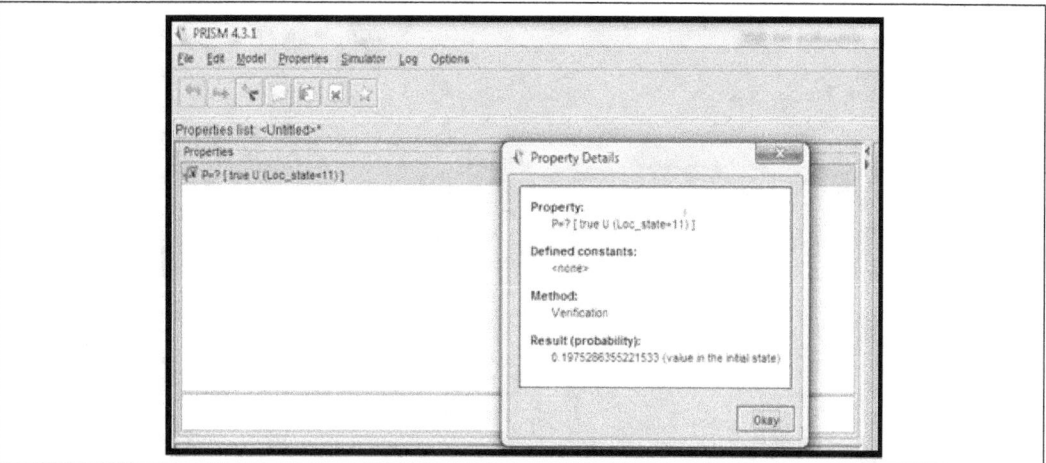

Figure 30: Probability that Switch # 1 is the Faulty Switch

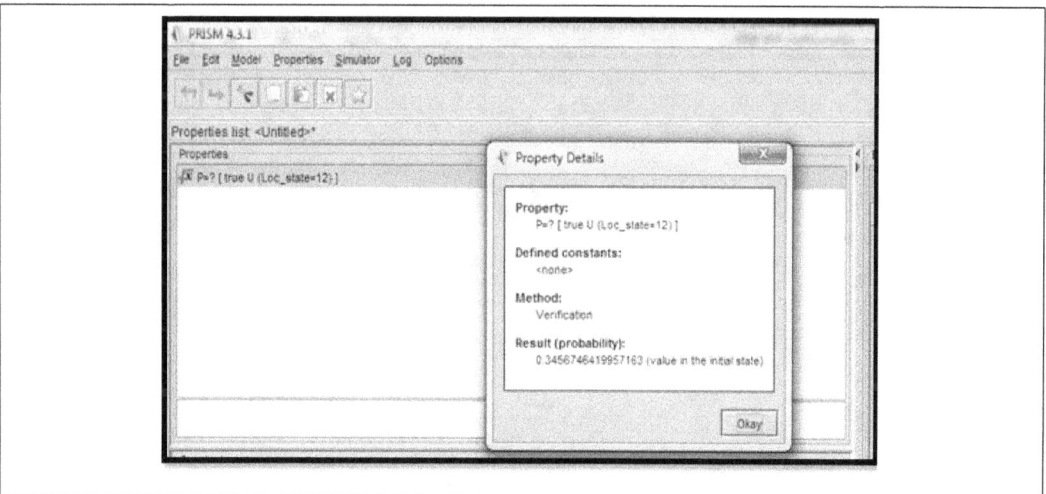

Figure 31: Probability that fault flag 2 is high in switch # 2

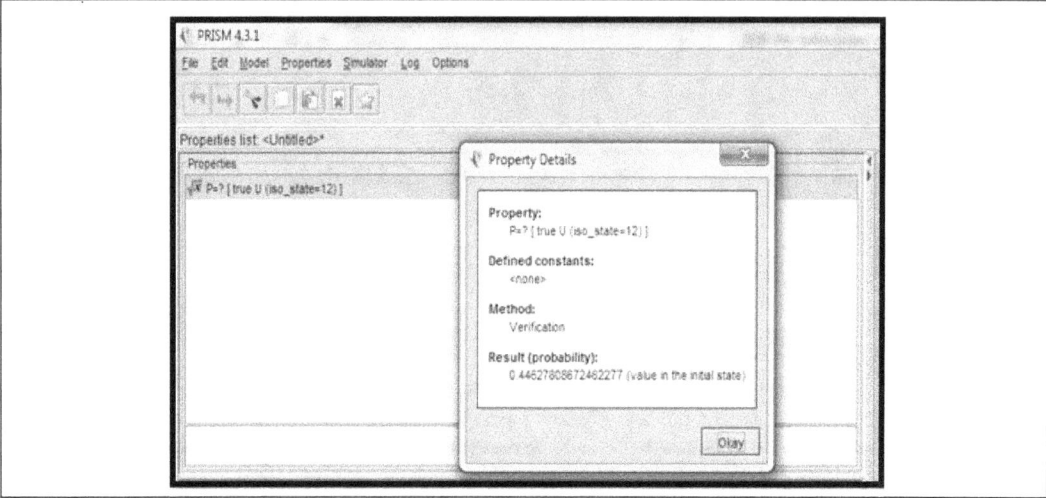

Figure 32: Probability that switch # 2 trip off at the limited time

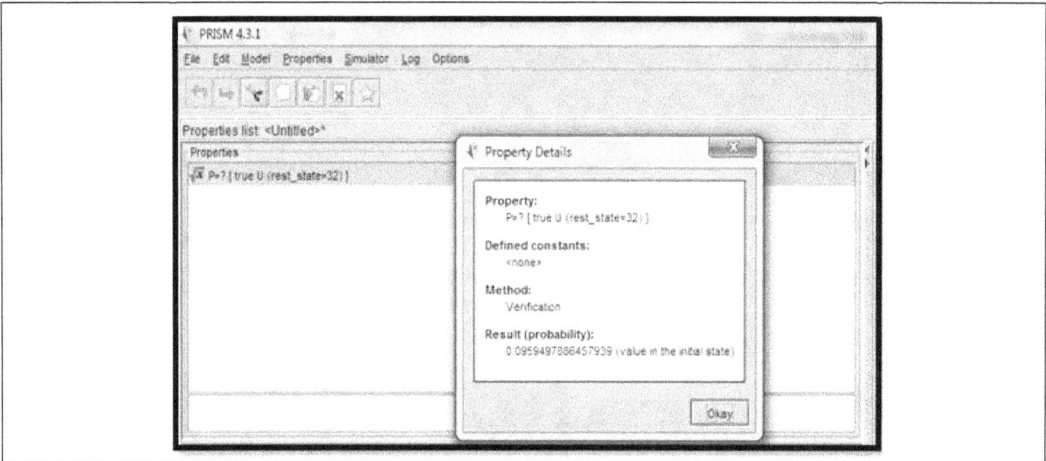

Figure 33: Probability to recover the system through Switch # 3

References

[1] V. Mehta and R. Mehta, "Principles of power system," S. Chand, New Delhi, 2004.

[2] P. Kundur, N. J. Balu, and M. G. Lauby, Power system stability and control vol. 7: McGraw-hill New York, 1994.

[3] E. Platform, "Vision and strategy for europes electricity networks of the future," European SmartGrids Technology Platform, Tech. Rep, 2006.

[4] H. Farhangi, "The path of the smart grid," IEEE power and energy magazine, vol. 8, 2010.

[5] X. Fang, S. Misra, G. Xue, and D. Yang, "Smart grid-The new and improved power grid: A survey," IEEE communications surveys & tutorials, vol. 14, pp. 944-980, 2012.

[6] N. Framework, "Roadmap for Smart Grid Interoperability Standards, Release 1.0, Office of the National Coordinator for Smart Grid Interoperability," Director, 2010.

[7] S. Biradar, R. Patil, and M. Ullegaddi, "Energy storage system in electric vehicle," in Power Quality'98, 1998, pp. 247-255.

[8] M. Hannan, M. Hoque, A. Mohamed, and A. Ayob, "Review of energy storage systems for electric vehicle applications: Issues and challenges," Renewable and Sustainable Energy Reviews, vol. 69, pp. 771-789, 2017.

[9] M. Liserre, T. Sauter, and J. Y. Hung, "Future energy systems: Integrating renewable energy sources into the smart power grid through industrial electronics," IEEE industrial electronics magazine, vol. 4, pp. 18-37, 2010.

[10] W. Ling and D. Liu, "A distributed fault localization, isolation and supply restoration algorithm based on local topology," International Transactions on Electrical Energy Systems, vol. 25, pp. 1113-1129, 2015.

[11] J. G. Liu and X. Zhang, Fault Location and Service Restoration for Electrical Distribution Systems: John Wiley & Sons, 2016.

[12] X. Lu, Z. Lu, W. Wang, and J. Ma, "On network performance evaluation toward the smart grid: A case study of DNP3 over TCP/IP," in Global Telecommunications Conference (GLOBECOM 2011), 2011 IEEE, 2011, pp. 1-6.

[13] W. Wang, Y. Xu, and M. Khanna, "A survey on the communication architectures in smart grid," Computer Networks, vol. 55, pp. 3604-3629, 2011.

[14] Y. Yan, Y. Qian, H. Sharif, and D. Tipper, "A survey on smart grid communication infrastructures: Motivations, requirements and challenges," IEEE communications surveys & tutorials, vol. 15, pp. 5-20, 2013.

[15] E. M. Rogers and D. L. Kincaid, "Communication networks: Toward a new paradigm for research," 1981.

[16] "Cisco, "Connecting cables to cisco 3800 series routers," http://www.cisco.com/en/US/docs/routers/access/3800/hardware/installation/guide/38cable.html#wp10084492009.

[17] "Grass Valley, Trinix - digital video router back-up power supplies," http://www.grassvalley.com/docs/Manuals/routers/trinixnxt/071-8443-02.pdf2009.

[18] J.-D. Decotignie, "Ethernet-based real-time and industrial communications," Proceedings of the IEEE, vol. 93, pp. 1102-1117, 2005.

[19] G. P. Agrawal, Fiber-optic communication systems vol. 222: John Wiley & Sons, 2012.

[20] X. Lu, W. Wang, J. Ma, and L. Sun, "Domino of the smart grid: An empirical study of system behaviors in the interdependent network architecture," in Smart Grid Communications (SmartGridComm), 2013 IEEE International Conference on, 2013, pp. 612-617.

[21] T. S. Rappaport, Wireless communications: principles and practice vol. 2: prentice hall PTR New Jersey, 1996.

[22] X. Lu, W. Wang, Z. Lu, and J. Ma, "From security to vulnerability: Data authentication undermines message delivery in smart grid," in MILITARY COMMUNICATIONS CONFERENCE, 2011-MILCOM 2011, 2011, pp. 1183-1188.

[23] Z. Lu, W. Wang, and C. Wang, "From jammer to gambler: Modeling and detection of jamming attacks against time-critical traffic," in INFOCOM, 2011 Proceedings IEEE, 2011, pp. 1871-1879.

[24] W. H. Tang and A. Ang, Probability Concepts in Engineering: Emphasis on Applications to Civil & Environmental Engineering: Wiley, 2007.

[25] R. Haverkamp, M. Vauclin, J. Touma, P. Wierenga, and G. Vachaud, "A comparison of numerical simulation models for one-dimensional infiltration," Soil Science Society of America Journal, vol. 41, pp. 285-294, 1977.

[26] A.-T. Nguyen, S. Reiter, and P. Rigo, "A review on simulation-based optimization methods applied to building performance analysis," Applied Energy, vol. 113, pp. 1043-1058, 2014.

[27] E. M. Clarke and J. M. Wing, "Formal methods: State of the art and future directions," ACM Computing Surveys (CSUR), vol. 28, pp. 626-643, 1996.

[28] A. Diller, Z: An introduction to formal methods: John Wiley & Sons, Inc., 1990.

[29] C. Baier, J.-P. Katoen, and K. G. Larsen, Principles of model checking: MIT press, 2008.

[30] J. M. Wing, "A specifier's introduction to formal methods," Computer, vol. 23, pp. 8-22, 1990.

[31] O. Hasan and S. Tahar, "Formal verification methods," in Encyclopedia of Information Science and Technology, Third Edition, ed: IGI Global, 2015, pp. 7162-7170.

[32] R. Uluski, "Using distribution automation for a self-healing grid," in Transmission and Distribution Conference and Exposition (T&D), 2012 IEEE PES, 2012, pp. 1-5.

[33] S.-I. Lim, S.-J. Lee, M.-S. Choi, D.-J. Lim, and B.-N. Ha, "Service restoration methodology for multiple fault case in distribution systems," IEEE Transactions on Power Systems, vol. 21, pp. 1638-1644, 2006.

[34] C. P. Nguyen and A. J. Flueck, "Agent based restoration with distributed energy storage support in smart grids," IEEE Transactions on Smart Grid, vol. 3, pp. 1029-1038, 2012.

[35] J. M. Solanki, S. Khushalani, and N. N. Schulz, "A multi-agent solution to distribution systems restoration," IEEE Transactions on Power systems, vol. 22, pp. 1026-1034,

2007.

[36] C. H. Lin, C. S. Chen, T. T. Ku, C. T. Tsai, and C. Y. Ho, "A multiagent?based distribution automation system for service restoration of fault contingencies," European transactions on electrical power, vol. 21, pp. 239-253, 2011.

[37] N. Higgins, V. Vyatkin, N.-K. C. Nair, and K. Schwarz, "Distributed power system automation with IEC 61850, IEC 61499, and intelligent control," IEEE Transactions on Systems, Man, and Cybernetics, Part C (Applications and Reviews), vol. 41, pp. 81-92, 2011.

[38] T. Nagata, Y. Tao, H. Sasaki, and H. Fujita, "A multiagent approach to distribution system restoration," in Power Engineering Society General Meeting, 2003, IEEE, 2003, pp. 655-660.

[39] W. Khamphanchai, S. Pisanupoj, W. Ongsakul, and M. Pipattanasomporn, "A multi-agent based power system restoration approach in distributed smart grid," in Utility Exhibition on Power and Energy Systems: Issues & Prospects for Asia (ICUE), 2011 International Conference and, 2011, pp. 1-7.

[40] I. Lim, Y. Kim, H. Lim, M. Choi, S. Hong, S. Lee, et al., "Distributed restoration system applying multi-agent in distribution automation system," in Power and Energy Society General Meeting-Conversion and Delivery of Electrical Energy in the 21st Century, 2008 IEEE, 2008, pp. 1-7.

[41] I.-H. Lim, T. S. Sidhu, M.-S. Choi, S.-J. Lee, S. Hong, S.-I. Lim, et al., "Design and implementation of multiagent-based distributed restoration system in DAS," IEEE Transactions on Power Delivery, vol. 28, pp. 585-593, 2013.

[42] J. Ghorbani, M. A. Choudhry, and A. Feliachi, "Fault location and isolation using multi agent systems in power distribution systems with distributed generation sources," in Innovative Smart Grid Technologies (ISGT), 2013 IEEE PES, 2013, pp. 1-6.

[43] A. Abdrabou, "A wireless communication architecture for smart grid distribution networks," IEEE Systems Journal, vol. 10, pp. 251-261, 2016.

[44] M. H. U. Ahmed, M. G. R. Alam, R. Kamal, C. S. Hong, and S. Lee, "Smart grid cooperative communication with smart relay," Journal of Communications and Networks, vol. 14, pp. 640-652, 2012.

[45] V. C. Gungor, D. Sahin, T. Kocak, S. Ergut, C. Buccella, C. Cecati, et al., "Smart grid technologies: Communication technologies and standards," IEEE transactions on Industrial informatics, vol. 7, pp. 529-539, 2011.

[46] Y. Tsado, D. Lund, and K. Gamage, "Resilient wireless communication networking for Smart grid BAN," in Energy Conference (ENERGYCON), 2014 IEEE International, 2014, pp. 846-851.

[47] X. Wang and P. Yi, "Security framework for wireless communications in smart distribution grid," IEEE Transactions on Smart Grid, vol. 2, pp. 809-818, 2011.

[48] M. Elgenedy, M. Sayed, M. Mokhtar, M. Abdallah, and N. Al-Dhahir, "Interference mitigation techniques for narrowband powerline smart grid communications," in Smart Grid Communications (SmartGridComm), 2015 IEEE International Conference on,

2015, pp. 368-373.

[49] D. Jiang, "Optimal bit loading algorithm for power-line communication systems subject to individual channel power constraints," in Communication Technology, 2006. ICCT'06. International Conference on, 2006, pp. 1-4.

[50] S. S. Prakash and J. D. S. Lakshmi, "Carrier frequency offset estimation in power line communication networks," in Circuit, Power and Computing Technologies (ICCPCT), 2015 International Conference on, 2015, pp. 1-6.

[51] M. M. Rahman, C. S. Hong, S. Lee, J. Lee, M. A. Razzaque, and J. H. Kim, "Medium access control for power line communications: an overview of the IEEE 1901 and ITU-T G. hn standards," IEEE Communications Magazine, vol. 49, 2011.

[52] T. Sauter and M. Lobashov, "End-to-end communication architecture for smart grids," IEEE Transactions on Industrial Electronics, vol. 58, pp. 1218-1228, 2011.

[53] Q. Yang, J. A. Barria, and T. C. Green, "Communication infrastructures for distributed control of power distribution networks," IEEE Transactions on Industrial Informatics, vol. 7, pp. 316-327, 2011.

[54] http://www.prismmodelchecker.org/.

[55] M. Kwiatkowska, G. Norman, and D. Parker, "PRISM: Probabilistic symbolic model checker," in International Conference on Modelling Techniques and Tools for Computer Performance Evaluation, 2002, pp. 200-204.

[56] H. Oldenkamp, "Probabilistic model checking: A comparison of tools," University of Twente, 2007.

[57] W. R. Gilks, S. Richardson, and D. Spiegelhalter, Markov chain Monte Carlo in practice: CRC press, 1995.

[58] M. Kwiatkowska, G. Norman, and D. Parker, "Stochastic model checking," in SFM, 2007, pp. 220-270.

[59] V. G. Kulkarni, Modeling and analysis of stochastic systems: CRC Press, 2016.

[60] M. L. Puterman, Markov decision processes: discrete stochastic dynamic programming: John Wiley & Sons, 2014.

[61] D. Beauquier, "On probabilistic timed automata," Theoretical Computer Science, vol. 292, pp. 65-84, 2003.

[62] M. Kwiatkowska, G. Norman, and D. Parker, "PRISM: Probabilistic symbolic model checker," Computer performance evaluation: modelling techniques and tools, pp. 113-140, 2002.

[63] A. Ahmed, A. Rashid, and S. Iqbal, "Analysis of Weather Forecasting Model in PRISM," in Frontiers of Information Technology (FIT), 2014 12th International Conference on, 2014, pp. 355-360.

[64] G. Bianchi, "Performance analysis of the IEEE 802.11 distributed coordination function," IEEE Journal on selected areas in communications, vol. 18, pp. 535-547, 2000.

[65] G. Bianchi, "IEEE 802.11-saturation throughput analysis," IEEE communications letters, vol. 2, pp. 318-320, 1998.

[66] H. Wu, Y. Peng, K. Long, S. Cheng, and J. Ma, "Performance of reliable transport pro-

tocol over IEEE 802.11 wireless LAN: analysis and enhancement," in INFOCOM 2002. Twenty-First Annual Joint Conference of the IEEE Computer and Communications Societies. Proceedings. IEEE, 2002, pp. 599-607.

[67] D. Jiunn and R.-S. Chang, "A priority scheme for IEEE 802. 11 DCF access method," IEICE transactions on communications, vol. 82, pp. 96-102, 1999.

[68] http://sites.ieee.org/isgt2014/files/2014/03/Day2_Panel2C_Ni.pdf.

[69] http://www.sciencedirect.com/science/article/pii/S1364032114003761.

[70] Y. Yu, J. Yang, and B. Chen, "The smart grids in China-A review," Energies, vol. 5, pp. 1321-1338, 2012.

[71] http://www.fingrid.fi/fi/asiakkaat/asiakasliitteet/Kayttotoimikunta/2008/19.9.2008/nordel_fault_statistics_2007.pdf.

[72] http://www.me.utexas.edu/~jensen/ORMM/models/unit/markchain/index.html.

[73] http://www.me.utexas.edu/~jensen/ORMM/models/unit/markchain/subunits/example/index.html.

[74] Y. Kwon and G. Agha, "iLTLChecker: a probabilistic model checker for multiple DTMCs," in Quantitative Evaluation of Systems, 2005. Second International Conference on the, 2005, pp. 245-246.

[75] D. C. Montgomery and G. C. Runger, Applied statistics and probability for engineers: John Wiley & Sons, 2010.

[76] R. Alur and A. Thomas, " Reactive modules," Formal Methods in System Design, vol. 15, no. 1, pp. 7-48, 1993.

Elementary-base Cirquent Calculus I: Parallel and Choice Connectives

Giorgi Japaridze
Department of Computing Sciences, Villanova University, USA.
giorgi.japaridze@villanova.edu

Abstract

Cirquent calculus is a proof system manipulating circuit-style constructs rather than formulas. Using it, this article constructs a sound and complete axiomatization **CL16** of the propositional fragment of *computability logic* (the game-semantically conceived logic of computational problems) whose logical vocabulary consists of negation and parallel and choice connectives, and whose atoms represent elementary, i.e. moveless, games.

MSC: primary: 03B47; secondary: 03B70; 03F03; 03F20; 68T15.
Keywords: Proof theory; Cirquent calculus; Resource semantics; Deep inference; Computability logic

1 Introduction

Computability logic, or *CoL* for short, is a long-term project for developing a logic capable of acting as a comprehensive formal theory of computability in the same sense as classical logic is a formal theory of truth (see [24] for a survey). The approach starts by asking what kinds of mathematical objects "computational problems" are in their full generality, and finds that they can be most adequately understood as games played by a machine against its environment, with computability meaning existence of an (algorithmic) winning strategy for the machine. As its next step, CoL tries to identify a collection of the most natural, meaningful and potentially useful operations on games. These operations then form the connectives, quantifiers and other constructs of the logical vocabulary of CoL. Validity of a formula is understood as being "always computable", i.e. computable in virtue of the meanings of its logical operators regardless of how the non-logical atoms are interpreted. The final and most challenging step in developing CoL is finding sound and complete axiomatizations for ever more expressive fragments of this semantically construed logic. The present contribution adds one more brick to this edifice under construction.

Among the main connectives of the language of CoL are *negation* ("*not*") $\neg$, *parallel conjunction* ("*pand*") $\wedge$, *parallel disjunction* ("*por*") $\vee$, *choice conjunction* ("*chand*") $\sqcap$, and *choice disjunction* ("*chor*") $\sqcup$. Where G, H are games, the game-semantical meanings of the above connectives can be briefly characterized as follows. The game $\neg G$ is nothing but G with the roles of the two players interchanged. $G \wedge H$ is a game playing which means playing G and H in parallel, where the machine wins if it wins in both components. $G \vee H$ differs from $G \wedge H$ only in that here winning in just one of the components is sufficient. $G \sqcap H$ is the game where, at the beginning, the environment chooses one of the two components, after which the game continues according to the rules of the chosen component. $G \sqcup H$ is similar, only here it is the machine who makes an initial left-or-right choice. Game operations with similar intuitive characterizations have been studied by Lorenzen [32], Hintikka [11] and Blass [4, 5] in their dialogue/game semantics, with Blass [5] being the first to systematically differentiate between the parallel and choice sorts of operations and pointing out their resemblance with the multiplicative ($\wedge, \vee$) and additive ($\sqcap, \sqcup$) connectives of Girard's [9] linear logic. Many other operators of CoL have no known analogs in the literature. CoL also has two sorts of atoms: *general* atoms stranding for any games, and *elementary* atoms standing for propositions. The latter are understood as games with no moves, automatically won by the machine when true and lost when false. The fragments of CoL with only general atoms [3, 16, 20, 23, 27, 29, 30, 33, 34, 36, 37] are called *general-base*, the fragments with only elementary atoms [14, 17, 28, 31] are called *elementary-base*, and the fragments where both sorts of atoms are present [15, 18, 22, 25, 35] are called *mixed-base*.

All attempts to axiomatize the (whatever-base) full $\{\neg, \wedge, \vee, \sqcap, \sqcup\}$-fragment of CoL within the framework of traditional proof calculi had failed, and it was conjectured [5, 16] that such an axiomatization was impossible to achieve in principle even for the $\{\neg, \wedge, \vee\}$-subfragment. The recent work [8] by Das and Strassburger has positively verified this conjecture. As a way to break the ice, [16] introduced the new sort of a proof calculus called *cirquent calculus*, in which a sound and complete axiomatization of the general-base $\{\neg, \wedge, \vee\}$-fragment of CoL was constructed; this result was later lifted to the mixed-base level in [35]. Rather than being limited to tree-like objects such as formulas, sequents, hypersequents [1] or deep-inference structures [6, 7, 10], cirquent calculus deals with circuit-style constructs dubbed *cirquents*. Cirquents come in a variety of forms and sometimes, as in the present work or in [38, 39], they are written textually rather than graphically, but their essence and main distinguishing feature remains the same: these are syntactic constructs explicitly allowing *sharing* of components between different subcomponents.

Ordinary formulas of CoL are nothing but special cases of cirquents — they are degenerate cirquents where nothing is shared.

Sharing, itself, also takes different forms, such as two ∨-gates sharing a child, or two ⊔-gates sharing the left-or-right choice associated with them without otherwise sharing descendants. Most cirquent calculus systems studied so far [3, 16, 21, 29, 30, 35] only incorporate the first sort of sharing. The idea of the second sort of sharing, dubbed *clustering*, was introduced and motivated in [26]. Among the potential benefits of it outlined in [26] was offering new perspectives on independence-friendly logic [12]. Later work by Wenyan Xu [38, 39] made a significant progress towards materializing such a potential. The present work materializes another benefit offered by clustering: it constructs a sound and complete cirquent calculus axiomatization **CL16** of the full elementary-base {¬, ∧, ∨, ⊓, ⊔}-fragment of CoL. No axiomatizations of any ⊓, ⊔-containing fragments of CoL had been known so far (other than the brute-force constructions of [14, 15, 17, 18, 22, 25], with their deduction mechanisms more resembling games than logical calculi). Generalizing from formulas to cirquents with clustering thus offers not only greater expressiveness, but also makes the otherwise unaxiomatizable CoL or certain fragments of it amenable to being tamed as logical calculi.

2 Games and strategies

As noted, CoL understands computational problems as games played between two players, called the *machine* and the *environment*. The symbolic names for these players are ⊤ and ⊥, respectively. ⊤ is a deterministic mechanical device only capable of following algorithmic strategies, whereas there are no restrictions on the behavior of ⊥. Our sympathies are with ⊤, and by just saying "won" or "lost" without specifying a player, we always mean won or lost by ⊤. ℘ is always a variable ranging over {⊤, ⊥}. ¬℘ means ℘'s adversary, i.e. the player that is not ℘.

A **move** is a finite string over the standard keyboard alphabet. A **labeled move** is a move prefixed with ⊤ or ⊥, with such a prefix (**label**) indicating which player has made the move. A **run** is a (finite or infinite) sequence of labeled moves, and a **position** is a finite run. Runs will be often delimited by "⟨" and "⟩", with ⟨⟩ thus denoting the **empty run**.

Definition 2.1. A **game**[1] is a pair $A = (\mathbf{Lr}^A, \mathbf{Wn}^A)$, where:

---

[1] In CoL, the proper name of the concept defined here is "constant game", with the word "game" reserved for a more general concept; however, since constant games are the only kinds of games we care about in the present paper, we omit the word "constant" and just say "game".

1. $\mathbf{Lr}^A$ is a set of runs satisfying the condition that a finite or infinite run is in $\mathbf{Lr}^A$ iff all of its nonempty finite — not necessarily proper — initial segments are in $\mathbf{Lr}^A$ (notice that this implies $\langle\rangle \in \mathbf{Lr}^A$). The elements of $\mathbf{Lr}^A$ are said to be **legal runs** of A, and all other runs are said to be **illegal**. We say that α is a **legal move** for a player $\wp$ in a position Φ of A iff $\langle \Phi, \wp\alpha\rangle \in \mathbf{Lr}^A$; otherwise α is an **illegal move**. When the last move of the shortest illegal initial segment of Γ is $\wp$-labeled, we say that Γ is a $\wp$-**illegal run** of A; $\wp$-**legal** means "'not $\wp$-illegal".

2. $\mathbf{Wn}^A$ is a function that sends every run Γ to one of the players $\top$ or $\bot$, satisfying the condition that if Γ is a $\wp$-illegal run of A, then $\mathbf{Wn}^A\langle\Gamma\rangle = \neg\wp$.[2] When $\mathbf{Wn}^A\langle\Gamma\rangle = \wp$, we say that Γ is a $\wp$-**won** (or **won by** $\wp$) run of A; otherwise Γ is **lost by** $\wp$. Thus, an illegal run is always lost by the player who has made the first illegal move in it.

It is clear from the above definition that, when defining a particular game A, it would be sufficient to specify what *positions* (finite runs) are legal, and what *legal runs* are won. Such a definition will then uniquely extend to all — including infinite and illegal — runs. We will implicitly rely on this observation in the sequel.

Here is an example of a game, namely, the problem of computing a function f understood as a game. Every legal position of such a game is either $\langle \bot n, \top m\rangle$ or $\langle \bot n\rangle$ or $\langle\rangle$, where n and m are numbers written in (for instance) decimal notation. The empty run $\langle\rangle$ is won by $\top$, every length-1 run $\langle \bot n\rangle$ is won by $\bot$, and a length-2 run $\langle \bot n, \top m\rangle$ is won by $\top$ if and only if $m = f(n)$. The intuitions that determine these arrangements are as follows. A move $\bot n$ is the "input", making which amounts to asking the machine "what is the value of f at n?". A move $\top m$ is the "output", amounting to answering that m is the value of $f(n)$. The machine wins the run $\langle \bot n, \top m\rangle$ if and only if it answered the question correctly, i.e., if $m = f(n)$. The run $\langle \bot n\rangle$ corresponds to the situation where there was an input (n) but the machine failed to generate any output, so the machine loses. And the run $\langle\rangle$ corresponds to the situation where there was no input either, so the machine has nothing to answer for and it therefore wins.

A game is said to be **elementary** iff it has no legal runs other than the (always legal) empty run $\langle\rangle$. That is, an elementary game is a "game" without any (legal) moves, automatically won or lost. There are exactly two such games, for which we use the same symbols $\top$ and $\bot$ as for the two players: the game $\top$ automatically won by player $\top$, and the game $\bot$ automatically won by player $\bot$.[3] Computability logic is a conservative extension of classical logic, understanding classical propositions as elementary games. And, just like classical logic, it sees no difference between any

---

[2] We write $\mathbf{Wn}^A\langle\Gamma\rangle$ for $\mathbf{Wn}^A(\Gamma)$.
[3] Precisely, we have $\mathbf{Wn}^\top\langle\rangle = \top$ and $\mathbf{Wn}^\bot\langle\rangle = \bot$.

two true propositions such as "0 = 0" and "*Snow is white*", and identifies them with the elementary game $\top$; similarly, it treats false propositions such as "0 = 1" or "*Snow is black*" as the elementary game $\bot$.

An **HPM** ("Hard-Play Machine") is a Turing machine with the additional capability of making moves. The adversary can also move at any time, with such moves being the only nondeterministic events from the machine's perspective. Along with the ordinary read/write *work tape*,[4] the machine also has an additional tape called the *run tape*. The latter, at any time, spells the "current position" of the play. The role of this tape is to make the interaction history fully visible to the machine. It is read-only, and its content is automatically updated every time either player makes a move. A more detailed description of the HPM model, if necessary, can be found in [13].

In these terms, a **solution** ($\top$'s winning strategy) for a given game A is understood as an HPM $\mathcal{M}$ such that, no matter how the environment acts during its interaction with $\mathcal{M}$ (what moves it makes and when), the run incrementally spelled on the run tape is a $\top$-won run of A. When this is the case, we write $\mathcal{M} \models A$ and say that $\mathcal{M}$ **wins**, or **solves**, A, and that A is a **computable** game.

There is no need to define $\bot$'s strategies, because all possible behaviors by $\bot$ are accounted for by the different possible nondeterministic updates of the run tape of an HPM.

In the above outline, we described HPMs in a relaxed fashion, without being specific about technical details such as, say, how, exactly, moves are made by the machine, how many moves either player can make at once, what happens if both players attempt to move "simultaneously", etc. As it happens, all reasonable design choices yield the same class of winnable games as long as we consider a certain natural subclass of games called *static*. Intuitively, these are games where the relative speeds of the players are irrelevant because, as Blass has once put it, "it never hurts a player to postpone making moves". Below comes a formal definition of this concept.

For either player $\wp$, we say that a run Υ is a $\wp$-**delay** of a run Γ iff:

- for both players $\wp' \in \{\top, \bot\}$, the subsequence of $\wp'$-labeled moves of Υ is the same as that of Γ, and

- for any $n, k \geq 1$, if the nth $\wp$-labeled move is made later than (is to the right of) the kth $\neg\wp$-labeled move in Γ, then so is it in Υ.

The above conditions mean that in Υ each player has made the same sequence of moves as in Γ, only, in Υ, $\wp$ might have been acting with some delay.

---

[4]In computational-complexity-sensitive treatments, an HPM is allowed to have any (fixed) number of work tapes.

Now, we say that a game A is **static** iff, for either player $\wp$, whenever a run Υ is a $\wp$-delay of a run Γ, we have:

- if Γ is a $\wp$-legal run of A, then so is Υ;
- if Γ is a $\wp$-won run of A, then so is Υ.

All games that we shall see in this paper are static. In fact, they are not merely static, but belong to a special subclass of static games called "enumeration games", where even the order in which the players make their moves is irrelevant, and thus runs can be seen as multisets rather than sequences of labeled moves. Precisely, an **enumeration game** is a game A such that, for any run Γ and any permutation Δ of Γ, Γ is a legal (resp. won) run of A iff so is Δ.

Dealing only with static games, which makes timing technicalities fully irrelevant, allows us to describe and analyze strategies (HPMs) in a relaxed fashion. For instance, imagine HPM $\mathcal{N}$ works by simulating and mimicking the work and actions of another HPM $\mathcal{M}$ in the scenario where $\mathcal{M}$'s imaginary adversary acts in the same way as $\mathcal{N}$'s own adversary. Due to the simulation overhead, $\mathcal{N}$ will generally be much slower than $\mathcal{M}$ in responding to its adversary's moves. Yet, we may safely assume/pretend that the speeds of the two machines do not differ and thus they will be generating identical runs. This is "even more so" when we deal with enumeration games. In what follows we will often implicitly rely on this observation.

3 Syntax

We fix an infinite list of syntactic objects called **elementary game letters**, for which we will be using p, q, r as metavariables. A **positive** (resp. **negative**) **literal** is the expression p (resp. $\neg p$), where p is an elementary game letter. Here p is said to be the **type** of the literal.

We further fix two pairwise disjoint infinite sets $\mathbb{C}(\sqcup)$ and $\mathbb{C}(\sqcap)$ of decimal numerals. The elements of $\mathbb{C}(\sqcup) \cup \mathbb{C}(\sqcap)$ are said to be **clusters**. A cluster c is said to be **disjunctive** if $c \in \mathbb{C}(\sqcup)$, and **conjunctive** if $c \in \mathbb{C}(\sqcap)$.

The symbol $\vee$ (resp. $\wedge$) is said to be **parallel disjunction** (resp. **parallel conjunction**). A **choice disjunction** (resp. **choice conjunction**) is a pair $\sqcup^c$ (resp. $\sqcap^c$), where c is a disjunctive (resp. conjunctive) cluster. A common name for disjunctions and conjunctions of either sort is "**connective**", and the corresponding symbol $\vee, \wedge, \sqcup$ or $\sqcap$ is said to be the **type** of the connective. Given a choice connective $\sqcup^c$ or $\sqcap^c$, c is said to be its **cluster**; in this case we may as well say that the connective **belongs to** — or **is in** — cluster c.

Definition 3.1. A **cirquent** is defined inductively as follows:

- ⊤ and ⊥ are cirquents.

- Each literal is a cirquent.

- If A and B are cirquents, then $(A) \vee (B)$ is a cirquent.

- If A and B are cirquents, then $(A) \wedge (B)$ is a cirquent.

- If A and B are cirquents and c is a conjunctive cluster, then $(A) \sqcap^c (B)$ is a cirquent.

- If A and B are cirquents and c is a disjunctive cluster, then $(A) \sqcup^c (B)$ is a cirquent.

By a **cluster of** a cirquent C we shall mean the cluster c of some choice connective occurring in C. In such a case we may as well say that cluster c **occurs** in C.

Sometimes we may write an expression such as $A_1 \vee \ldots \vee A_n$, where n is a (possibly unspecified) natural number with $n \geq 2$. This is to be understood as any (unspecified) order-respecting $\vee$-combination of the cirquents $A_1, \ldots, A_n$. "Order-respecting" in the sense that A_1 is the leftmost item of the combination, then comes A_2, then A_3, etc. Similarly for $A_1 \wedge \ldots \wedge A_n$. So, for instance, both $(A \wedge B) \wedge C$ and $A \wedge (B \wedge C)$ — and no other cirquent — can be written as $A \wedge B \wedge C$.

Officially, as we see, $\neg$ (**negation**) is only allowed to be applied to elementary game letters. Shall we write $\neg E$ where E is not an elementary game letter, it is to be understood as an abbreviation defined by: $\neg\neg A = A$; $\neg(A \wedge B) = \neg A \vee \neg B$; $\neg(A \vee B) = \neg A \wedge \neg B$; $\neg(A \sqcap^c B) = \neg A \sqcup^c \neg B$; $\neg(A \sqcup^c B) = \neg A \sqcap^c \neg B$. Similarly, $A \rightarrow B$ is an abbreviation of $(\neg A) \vee B$. When writing cirquents, parentheses will usually be omitted if this causes no ambiguity. When doing so, it is our convention that $\neg$ has the highest precedence, then comes $\rightarrow$, then come the choice connectives, and finally the parallel connectives. So, for instance, $\neg A \vee B \rightarrow C \wedge D \sqcap^c E$ means $((\neg(A)) \vee (B)) \rightarrow ((C) \wedge ((D) \sqcap^c (E)))$, i.e., $((A) \wedge (\neg(B))) \vee ((C) \wedge ((D) \sqcap^c (E)))$.

We define the **root** of a cirquent C to be C itself if C is $\top$, $\bot$ or a literal, and $\vee$ (resp. $\wedge$, resp. $\sqcup^c$, resp. $\sqcap^c$) if C is of the form $A \vee B$ (resp. $A \wedge B$, resp. $A \sqcup^c B$, resp. $A \sqcap^c B$). When r is the root of C, we say that C is r-**rooted**.

4 Semantics

We define *LegalRuns* as the set of all runs Γ satisfying the following conditions:

1. Every move of Γ is the string $c.0$ or $c.1$, where c is a cluster.

2. Whenever Γ contains a move $c.i$ where c is a disjunctive cluster, the move is $\top$-labeled.

3. Whenever Γ contains a move $c.i$ where c is a conjunctive cluster, the move is $\bot$-labeled.

4. For any cluster c, Γ contains at most one move of the form $c.i$.

The intuitive meaning of condition 1 is that every move signifies a choice "left" (0) or "right" (1) in some cluster; conditions 2 and 3 say that $\top$ moves (chooses) only in disjunctive clusters and $\bot$ only in conjunctive clusters; and condition 4 says that, in any given cluster, a choice can be made only once.

Given a run $\Gamma \in \textit{LegalRuns}$, we say that a cirquent of the form $A \sqcup^c B$ or $A \sqcap^c B$ is Γ-**resolved** iff Γ contains (exactly) one of the moves $c.0$ or $c.1$; then by the Γ-**resolvent** of the cirquent we mean A if such a move is $c.0$, and B if it is $c.1$. "Γ-unresolved" means "not Γ-resolved". When Γ is clear from the context, we may omit a reference to it and simply say "resolved", "unresolved" or "resolvent".

An **interpretation** is a function $*$ which assigns to each elementary game letter p an element p^* of $\{\top, \bot\}$. Intuitively, such a function tells us whether p, as a proposition, is true or false.

Definition 4.1. Each cirquent C and interpretation $*$ induces a unique game C^*, which we may refer to as "C **under** the interpretation $*$". The set $\mathbf{Lr}^{C^*}$ of legal runs of such a game is nothing a but $\textit{LegalRuns}$. Since $\mathbf{Lr}^{C^*}$ does not depend on C or $*$, subsequently we shall simply say "legal run" rather than "legal run of C^*". The $\mathbf{Wn}^{C^*}$ component of the game C^* is defined by stipulating that a legal run Γ is a won (by the machine) run of C iff one of the following conditions is satisfied:

1. C is $\top$.

2. C is a positive (resp. negative) literal and, where p is the type of that literal, $p^* = \top$ (resp. $p^* = \bot$).

3. C is $A_0 \vee A_1$ (resp. $A_0 \wedge A_1$) and, for at least one (resp. both) $i \in \{0, 1\}$, Γ is a won run of A_i.

4. C is $A_0 \sqcup^c A_1$, it is resolved and, where A_i is the resolvent, Γ is a won run of A_i.

5. C is $A_0 \sqcap^c A_1$ and either it is unresolved, or else, where A_i is the resolvent, Γ is a won run of A_i.

Definition 4.2. Consider a cirquent C.

1. For an interpretation $*$, a **solution** of C under $*$, or simply a solution of C^*, is an HPM $\mathcal{H}$ such that $\mathcal{H} \models C^*$. We say that C is **computable** under $*$, or simply that C^* is computable, iff C^* has a solution.

2. A **logical** (or uniform) **solution** of C is an HPM $\mathcal{H}$ such that, for any interpretation $*$, $\mathcal{H}$ is a solution of C^*. We say that C is **valid** iff it has a logical solution.[5]

Remark 4.3. The cirquents in the present sense can be understood as generalizations of the formulas of system **CL1** of CoL constructed in [14]. Syntactically, the formulas differ from cirquents only in that no clusters are attached to $\sqcup, \sqcap$. Each formula F can be seen as a cirquent C where no two different occurrences of a choice connective belong to the same cluster, i.e., as a cirquent with no sharing of choices associated with $\sqcup, \sqcap$. More specifically, C is a cirquent obtained from F via superscripting each occurrence of $\sqcup$ by a unique disjunctive cluster and each occurrence of $\sqcap$ by a unique conjunctive cluster. Let us call such a C a *cirquentization* of F. We claim without a proof that, given a formula F and a cirquentization C of it, the two are semantically equivalent. Namely, any HPM $\mathcal{F}$ can be transformed into an HPM $\mathcal{C}$ — and vice versa — so that, for any interpretation $*$, we have $\mathcal{F} \models F^*$ iff $\mathcal{C} \models C^*$ (with F^* understood as in [14]). Consequently, F is valid iff C is so.

5 Axiomatics

The system **CL16** introduced in this section is by all accounts a *deep inference* system: for the exception of the Splitting rule, it allows the inference rules to operate on any part of an expression rather than just on the root. One of the most extensively studied deep-inference formalisms is the *calculus of structures* (CoS) [6, 7, 10]. The approach of cirquent calculus and of **CL16** in particular can be seen as a generalization of CoS in that it adds a sharing mechanism to the latter. Namely, even though graphically the cirquents of **CL16** look like formulas, it should be remembered that the indexes attached to $\sqcup, \sqcap$ are not to generate different "versions" of these operators but to indicate the presence (in the case of identical indexes) or absence (in the case of different indexes) of choice-sharing between different occurrences of otherwise *the same* logical operator.

By a **rule of inference** we mean a set $\mathcal{R}$ of pairs $\vec{A} \rightsquigarrow B$, called **applications** of $\mathcal{R}$, where $\vec{A}$ is a tuple consisting of one or two cirquents, called the **premise(s)**, and B is a cirquent, called the **conclusion**. When $\vec{A} \rightsquigarrow B$ is in $\mathcal{R}$, we say that B **follows** from $\vec{A}$ by rule $\mathcal{R}$.

---

[5]In CoL, this sort of validity is called **logical** (or **uniform**) **validity**. There is also another natural sort of validity, called **nonlogical** (or **multiform**) **validity**. Namely, a cirquent (or formula) C is multiformly valid iff, for any interpretation $*$, C^* is computable. Nonlogical validity will not be considered in this paper.

In this section and later we will be using the notation $X[E_1,\ldots,E_n]$ to stand for a cirquent (intuitively "of **structure** X") together with some fixed subcirquents $E_1,\ldots,E_n$. Then, if we later write $X[F_1,\ldots,F_n]$ in the same context, it should be understood as the result of replacing, in $X[E_1,\ldots,E_n]$, all occurrences of $E_1,\ldots,E_n$ by $F_1,\ldots,F_n$, respectively. When this notation is used in the formulation of a rule of inference, our convention is that the context is always set by the conclusion. So, for instance, if we have a (sub)expression $X[E]$ in the conclusion and $X[F]$ in a premise, then $X[F]$ is the result of replacing all occurrences of E by F in $X[E]$ rather than vice versa.

Below is a full list of the rules of inference of our system **CL16**. The first seven rules come in two versions, between which we shall later differentiate by suffixing the name of the rule with "(a)" for the first version and "(b)" for the second version. The last rule takes two premises, while all other rules take a single premise. The rules are written schematically, with A, B, C, D (possibly with indices) acting as variables for subcirquents, a, b, c as variables for clusters, and X, Y as variables for "structures". The names of these rules have been chosen according to the conclusion-to-premises (rather than premises-to-conclusion) intuitions.

Commutativity: $X[B \lor A] \rightsquigarrow X[A \lor B]$ and $X[B \land A] \rightsquigarrow X[A \land B]$.

Associativity: $X[A \lor (B \lor C)] \rightsquigarrow X[(A \lor B) \lor C]$ and $X[A \land (B \land C)] \rightsquigarrow X[(A \land B) \land C)]$.

Identity: $X[A] \rightsquigarrow X[A \lor \bot]$ and $X[A] \rightsquigarrow X[A \land \top]$.

Domination: $X[\top] \rightsquigarrow X[A \lor \top]$ and $X[\bot] \rightsquigarrow X[A \land \bot]$.

Choosing: $X[A_1,\ldots,A_n] \rightsquigarrow X[A_1 \sqcup^c B_1,\ldots,A_n \sqcup^c B_n]$ and $X[B_1,\ldots,B_n] \rightsquigarrow X[A_1 \sqcup^c B_1,\ldots,A_n \sqcup^c B_n]$, where $A_1 \sqcup^c B_1, \ldots, A_n \sqcup^c B_n$ are all $\sqcup^c$-rooted subcirquents of the conclusion.

Cleansing: $X[Y[A] \sqcap^c C] \rightsquigarrow X[Y[A \sqcap^c B] \sqcap^c C]$ and $X[C \sqcap^c Y[B]] \rightsquigarrow X[C \sqcap^c Y[A \sqcap^c B]]$.

Distribution: $X[(A \lor C) \land (B \lor C)] \rightsquigarrow X[(A \land B) \lor C]$ and $X[(A \lor C) \sqcap^c (B \lor C)] \rightsquigarrow X[(A \sqcap^c B) \lor C]$.

Trivialization: $X[\top] \rightsquigarrow X[\neg p \lor p]$, where p is an elementary letter.

Quadrilemma: $X\bigl[\bigl(A \land (C \sqcap^b D)\bigr) \sqcap^a (B \land (C \sqcap^b D))\bigr) \sqcap^c \bigl((((A \sqcap^a B) \land C) \sqcap^b ((A \sqcap^a B) \land D))\bigr)\bigr] \rightsquigarrow X[(A \sqcap^a B) \land (C \sqcap^b D)]$, where c does not occur in the conclusion.

Splitting: $A, B \rightsquigarrow A \sqcap^c B$, where neither A nor B has an occurrence of c.

A **proof** of a cirquent A is a sequence $C_1, \ldots, C_n$ ($n \geq 1$) of cirquents such that $C_1 = \top, C_n = A$ and, for each $i \in \{2, \ldots, n\}$, C_i follows by one of the rules of inference from some earlier cirquents in the sequence. Thus, $\top$ is the only axiom of **CL16**.

Example 5.1. Below is a proof of $p \wedge q \sqcup^c r \to (p \wedge q) \sqcup^d (p \wedge r)$, i.e. of $(\neg p \vee \neg q \sqcap^c \neg r) \vee (p \wedge q) \sqcup^d (p \wedge r)$. For brevity, consecutive applications of Commutativity or Associativity have been combined together in single steps.

1. $\top$ Axiom
2. $\top \wedge \top$ Identity(b): 1
3. $(\neg q \vee \top) \wedge (\neg p \vee \top)$ Domination(a): 2 (twice)
4. $(\neg q \vee (\neg p \vee p)) \wedge (\neg p \vee (\neg q \vee q))$ Trivialization: 3 (twice)
5. $((\neg q \vee \neg p) \vee p) \wedge ((\neg p \vee \neg q) \vee q)$ Associativity(a): 4 (twice)
6. $(p \vee (\neg q \vee \neg p)) \wedge (q \vee (\neg q \vee \neg p))$ Commutativity(a): 5 (three times)
7. $(p \wedge q) \vee (\neg q \vee \neg p)$ Distribution(a): 6
8. $(\neg q \vee \neg p) \vee (p \wedge q)$ Commutativity: 7
9. $(\neg q \vee \neg p) \vee (p \wedge q) \sqcup^d (p \wedge r)$ Choosing(a): 8
10. $(\neg r \vee \top) \wedge (\neg p \vee \top)$ Domination(a): 2 (twice)
11. $(\neg r \vee (\neg p \vee p)) \wedge (\neg p \vee (\neg r \vee r))$ Trivialization: 10 (twice)
12. $((\neg r \vee \neg p) \vee p) \wedge ((\neg p \vee \neg r) \vee r)$ Associativity(a): 11 (twice)
13. $(p \vee (\neg r \vee \neg p)) \wedge (r \vee (\neg p \vee \neg r))$ Commutativity(a): 12 (twice)
14. $(p \wedge r) \vee (\neg r \vee \neg p)$ Distribution(a): 13
15. $(\neg r \vee \neg p) \vee (p \wedge r)$ Commutativity(a): 14
16. $(\neg r \vee \neg p) \vee (p \wedge q) \sqcup^d (p \wedge r)$ Choosing(b): 15
17. $((\neg q \vee \neg p) \vee (p \wedge q) \sqcup^d (p \wedge r)) \sqcap^c ((\neg r \vee \neg p) \vee (p \wedge q) \sqcup^d (p \wedge r))$ Splitting: 9,16
18. $(\neg q \vee \neg p) \sqcap^c (\neg r \vee \neg p) \vee (p \wedge q) \sqcup^d (p \wedge r)$ Distribution(b): 17
19. $(\neg q \sqcap^c \neg r \vee \neg p) \vee (p \wedge q) \sqcup^d (p \wedge r)$ Distribution(b): 18
20. $(\neg p \vee \neg q \sqcap^c \neg r) \vee (p \wedge q) \sqcup^d (p \wedge r)$ Commutativity(a): 19

6 The preservation lemma

Lemma 6.1. *Consider an arbitrary interpretation* $*$.

1. Each application of any of the rules of **CL16** *preserves computability under* $*$ *in the premises-to-conclusion direction, i.e., if all premises are computable under* $*$, *then so is the conclusion.*

*2. Each application of any of the rules of **CL16** other than Choosing also preserves computability under $*$ in the conclusion-to-premises direction, i.e., if the conclusion is computable under $*$, then so are all premises.*

Proof. Consider an arbitrary interpretation $*$. Since $*$ is going to be fixed throughout this proof, for readability we agree to omit explicit references to it. So, for instance, where E is a cirquent, we may write E instead of E^*, or say "... solution of E" instead of "... solution of E under $*$". Throughout this and some later proofs, when trying to show that a given machine $\mathcal{H}$ is a solution of a given game G, we implicitly rely on what is called the "clean environment assumption". According to it, $\mathcal{H}$'s environment never makes moves that are not legal moves of G. Assuming that this condition is satisfied is legitimate, because, if $\mathcal{H}$'s environment makes an illegal move, $\mathcal{H}$ automatically wins.

If $E \leadsto F$ is an application of any of the rules other than Splitting or Choosing, it is not hard to see that E and F are identical as games. So, a solution of E is automatically a solution of F, and vice versa. Let us just look at Cleansing(a) as an illustrative example. Consider an application $X[Y[A] \sqcap^c C] \leadsto X[Y[A \sqcap^c B] \sqcap^c C]$ of this rule. Let Γ be an arbitrary legal run. We want to show that Γ is a won run of E iff it is a won run of F. If c is unresolved in Γ, then the $Y[A \sqcap^c B] \sqcap^c C$ component of the conclusion will be won just like the $Y[A] \sqcap^c C$ component of the premise. Since the two cirquents only differ in that one has $Y[A \sqcap^c B] \sqcap^c C$ where the other has $Y[A] \sqcap^c C$, we find that Γ is a won run of both games or neither. Now assume c is resolved, i.e., Γ contains the move $c.i$ ($i = 0$ or $i = 1$). If $i = 1$, then Γ is a won run of $X[Y[A \sqcap^c B] \sqcap^c C]$ iff it is a won run of $X[C]$ iff it is a won run of $X[Y[A] \sqcap^c C]$. And if $i = 0$, then Γ is a won run of $X[Y[A \sqcap^c B] \sqcap^c C]$ iff it is a won run of $X[Y[A \sqcap^c B]]$ iff it is a won run of $X[Y[A]]$ iff it is a won run of $X[Y[A \sqcap^c C]]$. Thus, in either case, the conclusion is won iff so is the premise.

Consider an application $A, B \leadsto A \sqcap^c B$ of *Splitting*.

For the premises-to-conclusion direction, assume the premises are computable, namely, HPMs $\mathcal{M}_A$ and $\mathcal{M}_B$ are solutions of A and B, respectively. Let $\mathcal{N}$ be an HPM which, at the beginning of the play, waits till the environment makes one of the moves $c.0$ or $c.1$. After that, where $\alpha_1, \ldots, \alpha_n$ are the moves made by the environment before the move $c.0$ (resp. $c.1$) was made, $\mathcal{N}$ starts simulating $\mathcal{M}_A$ (resp. $\mathcal{M}_B$), with $\bot\alpha_1, \ldots, \bot\alpha_n$ on the imaginary run tape of the latter at the very first clock cycle. Whenever $\mathcal{N}$ sees that the simulated machine $\mathcal{M}_A$ (resp. $\mathcal{M}_B$) made a move, $\mathcal{N}$ makes the same move; $\mathcal{N}$ also periodically checks its own run tape to see if the environment has made any new moves in the real play and, if yes, it appends those ($\bot$-prefixed) moves to the imaginary run tape of the simulated machine. In more relaxed and intuitive terms, what we just said about the actions of $\mathcal{N}$ after the environment has moved $c.0$ (resp. $c.1$) can be put as "$\mathcal{N}$ plays exactly like $\mathcal{M}_A$ (resp. $\mathcal{M}_B$) would play in the scenario where, at the very start of the play, the environment made the moves $\alpha_1, \ldots, \alpha_n$". Later, in similar situations, we shall usually describe and analyze HPMs in relaxed terms, without going into technical details of simulation and without even using the word "simulation". Since we exclusively deal with static games, this relaxed approach is safe and valid (see the end of Section 2). Anyway, it is not hard to see that our $\mathcal{N}$ is a solution of $A \sqcap^c B$.

For the conclusion-to-premises direction, assume $\mathcal{N}$ is a solution of $A \sqcap^c B$. Let $\mathcal{M}_A$ (resp. $\mathcal{M}_B$) be an HPM which plays just like $\mathcal{N}$ would in the scenario where, at the very start of the play, $\mathcal{N}$'s adversary made the move $c.0$ (resp. $c.1$). Obviously $\mathcal{M}_A$ and $\mathcal{M}_B$ are solutions of A and B, respectively.

Consider an application $X[A_1, \ldots, A_n] \rightsquigarrow X[A_1 \sqcup^c B_1, \ldots, A_n \sqcup^c B_n]$ of *Choosing(a)*, and assume $\mathcal{M}$ is a solution of the premise. Let $\mathcal{N}$ be an HPM which, at the beginning of the game, makes the move $c.0$, after which it plays exactly as $\mathcal{M}$ would. Obviously $\mathcal{N}$ is a solution of the conclusion. *Choosing(b)* will be handled in a similar way. □

The following is an immediate corollary of Lemma 6.1:

Corollary 6.2. *1. Each application of any of the rules of **CL16** preserves validity in the premise-to-conclusion direction, i.e., if all premises are valid, then so is the conclusion.*

*2. Each application of any of the rules of **CL16** other than Choosing also preserves validity in the conclusion-to-premise direction, i.e., if the conclusion is valid, then so are all premises.*

Remark 6.3. Lemma 6.1 and Corollary 6.2 state the existence of certain solutions. A look back at our proof of those statements reveals that, in fact, this existence is constructive. Namely, in the case of clause (a) of Lemma 6.1, for any given rule, there is a *-independent effective procedure which extracts an HPM $\mathcal{M}$ from the premise(s), the conclusion and HPMs that purportedly solve the premises under *; as long as these purported solutions are indeed solutions, $\mathcal{M}$ is a solution of the conclusion under *. Similarly for clause (b). In the case of clause (a) of Corollary 6.2, for any given rule, there is an effective procedure which extracts an HPM $\mathcal{M}$ from the premise(s), the conclusion and purported logical solutions of the premises; as long as these purported logical solutions are indeed logical solutions, $\mathcal{M}$ is a logical solution of the conclusion. Similarly for clause (b).

7 Soundness and completeness

Below we use the standard notation $^n a$ ("tower of a's of height n") for tertration, defined inductively by $^1 a = a$ and $^{n+1}a = a^{(^n a)}$. So, for instance, $^3 5 = 5^{5^5}$.

Definition 7.1. The **rank** $\overline{C}$ of a cirquent C is the number defined as follows:
1. If C is $\top$, $\bot$ or a literal, then $\overline{C} = 1$.
2. If C is $A \sqcup^c B$ or $A \sqcap^c B$, then $\overline{C} = \overline{A} + \overline{B}$.
3. If C is $A \wedge B$, then $\overline{C} = 5^{\overline{A}+\overline{B}}$.
4. If C is $A \vee B$, then $\overline{C} = {}^{\overline{A}+\overline{B}}5$.

Lemma 7.2. *The rank function is **monotone** in the following sense. Consider a cirquent A with a subcirquent B. Assume B' is a cirquent with $\overline{B'} < \overline{B}$, and A' is the result of replacing an occurrence of B by B' in A. Then $\overline{A'} < \overline{A}$.*

Proof. This is so due to the monotonicity of the functions $x + y$, 5^x and $^x 5$. □

A **surface occurrence** of a subcirquent or a connective in a given cirquent is an occurrence which is not in the scope of a choice connective.

Definition 7.3. We say that a cirquent D is **pure** iff the following conditions are satisfied:
1. D has no surface occurrences of $\bot$ unless D itself is $\bot$.
2. D has no surface occurrence of $\wedge$ which is in the scope of $\vee$.
3. D has no surface occurrence of $\sqcap^c$ (whatever cluster c) which is in the scope of $\vee$.
4. D has no surface occurrence of the form $A_1 \vee \ldots \vee A_n$ such that, for some elementary letter p, both p and $\neg p$ are among $A_1, \ldots, A_n$.
5. D has no surface occurrences of $\top$ unless D itself is $\top$.
6. If D is of the form $A_1 \wedge \ldots \wedge A_n$ $(n \geq 2)$, then at least one A_i $(1 \leq i \leq n)$ is not of the form $B \sqcap^c C$.
7. If D is of the form $A \sqcap^c B$, then neither A nor B contains the cluster c.

Below we describe a procedure which takes a cirquent D and applies to it a series of modifications. Each modification changes the value of D so that the old value of D follows from the new value by one of the single-premise rules (other than Choosing) of **CL16**. The procedure is divided into 7 stages, and the purpose of each stage $i \in \{1, \ldots, 7\}$ is to make D satisfy the corresponding condition i of Definition 7.3.

Procedure Purification applied to a cirquent D: Starting from Stage 1, each of the following 7 stages is a loop that should be iterated until it no longer modifies (the current value of) D; then the procedure goes to the next stage, unless the current stage was Stage 7, in which case the procedure returns (the then-current value of) D and terminates.

Stage 1: If D has a surface occurrence of the form $\bot \vee A$ or $A \vee \bot$, change the latter to A using Identity(a) perhaps in combination with Commutativity(a). Next, if D has a surface occurrence of the form $\bot \wedge A$ or $A \wedge \bot$, change it to $\bot$ using Domination(b) perhaps in combination with Commutativity(b).

Stage 2: If D has a surface occurrence of the form $(A \wedge B) \vee C$ or $C \vee (A \wedge B)$, change it to $(A \vee C) \wedge (B \vee C)$ using Distributivity(a) perhaps in combination with Commutativity(a).

Stage 3: If D has a surface occurrence of the form $(A \sqcap^c B) \vee C$ or $C \vee (A \sqcap^c B)$, change it to $(A \vee C) \sqcap^c (B \vee C)$ using Distributivity(b) perhaps in combination with Commutativity(a).

Stage 4: If D has a surface occurrence of the form $A_1 \vee \ldots \vee A_n$ and, for some elementary letter p, both p and $\neg p$ are among $A_1, \ldots, A_n$, change $A_1 \vee \ldots \vee A_n$ to $\top$ using Trivialization, perhaps in combination with Domination(a), Commutativity(a) and Associativity(a).

Stage 5: If D has a surface occurrence of the form $\top \vee A$ or $A \vee \top$, change it to $\top$ using Domination(a) perhaps in combination with Commutativity(a). Next, if D has a surface occurrence of the form $\top \wedge A$ or $A \wedge \top$, change it to A using Identity(b) perhaps in combination with Commutativity(b).

Stage 6: If D has a surface occurrence of the form $(A \sqcap^a B) \wedge (E \sqcap^b F)$, change it to $\big((A \wedge (E \sqcap^b F)) \sqcap^a (B \wedge (E \sqcap^b F))\big) \sqcap^c \big(((A \sqcap^a B) \wedge E) \sqcap^b ((A \sqcap^a B) \wedge D)\big)$ using Quadrilemma.

Stage 7: If D is of the form $X[E \sqcap^c F] \sqcap^c A$ (resp. $A \sqcap^c X[E \sqcap^c F]$), change it to $X[E] \sqcap^c A$ (resp. $A \sqcap^c X[F]$) using Cleansing.

Lemma 7.4. *Each stage of the Purification procedure strictly reduces the rank of D.*

Proof. Each stage replaces an occurrence of a subcirquent A of D by some cirquent B. In view of Lemma 7.2, in order to show that such a replacement reduces the rank $\overline{D}$ of D, it is sufficient to show that $\overline{B} < \overline{A}$. Keep in mind that the rank of a cirquent is always at least 1.

Stage 1: Each iteration of this stage replaces in D an occurrence of $\bot \vee A$, $A \vee \bot$, $\bot \wedge A$ or $A \wedge \bot$ by A or $\bot$. Of course, both $\overline{A}$ and $\overline{\bot}$ are smaller than $\overline{\bot \vee A}, \overline{A \vee \bot}, \overline{\bot \wedge A}$ and $\overline{A \wedge \bot}$.

Stage 2: Each iteration of this stage replaces in D an occurrence of $(A \wedge B) \vee C$ or $C \vee (A \wedge B)$ by $(A \vee C) \wedge (B \vee C)$. $\overline{(A \wedge B) \vee C}$ (or $\overline{C \vee (A \wedge B)}$) is $[5^{(\overline{A}+\overline{B})}+\overline{C}]_5$

and $\overline{(A \vee C) \wedge (B \vee C)}$ is $5^{[\overline{(A+C)}5 + \overline{(B+C)}5]}$. We want to show that $5^{[\overline{(A+C)}5 + \overline{(B+C)}5]} < [5^{\overline{A+B}} + \overline{C}]5$. We of course have $\overline{A} + \overline{B} + 1 < 5^{\overline{A+B}}$, whence $\overline{A} + \overline{B} + \overline{C} + 1 < 5^{\overline{A+B}} + \overline{C}$, whence $^{[\overline{A}+\overline{B}+\overline{C}+1]}5 < ^{[5^{\overline{A+B}}+\overline{C}]} 5$. We also have

$$5^{[\overline{(A+C)}5 + \overline{(B+C)}5]} = 5^{[\overline{(A+C)}5]} \times 5^{[\overline{(B+C)}5]} = {}^{(\overline{A+C}+1)}5 \times {}^{(\overline{B+C}+1)}5 \leq {}^{[\overline{A}+\overline{B}+\overline{C}+1]}5.$$

Consequently, $5^{[\overline{(A+C)}5 + \overline{(B+C)}5]} < {}^{[5^{\overline{A+B}}+\overline{C}]}5$, as desired.

Stage 3: $\overline{(A \sqcap^c B) \vee C}$ (or $\overline{C \vee (A \sqcap^c B)}$ is ${}^{(\overline{A+B+C})}5$, and $\overline{(A \vee C) \sqcap^c (B \vee C)}$ is ${}^{(\overline{A+C})}5 + {}^{(\overline{B+C})}5$. Taking into account that ranks are always positive, we obviously have ${}^{(\overline{A+C})}5 + {}^{(\overline{B+C})}5 < {}^{(\overline{A+B+C})}5$.

Stage 4: $\overline{\top} = 1 < \overline{A_1 \vee \ldots \vee A_n}$.

Stage 5: Similar to Stage 1.

Stage 6: $\overline{(A \sqcap^a B) \wedge (E \sqcap^b F)}$ is $5^{[\overline{A+B+E+F}]}$, and

$$\overline{\Big((A \wedge (E \sqcap^b F)) \sqcap^a (B \wedge (E \sqcap^b F)) \Big) \sqcap^c \Big(((A \sqcap^a B) \wedge E) \sqcap^b ((A \sqcap^a B) \wedge D) \Big)}$$

is $5^{(\overline{A+E+F})} + 5^{(\overline{B+E+F})} + 5^{(\overline{A+B+E})} + 5^{(\overline{A+B+F})}$. Obviously the latter is smaller than the former.

Stage 7: Each iteration of this stage replaces a subcirquent $E \sqcap^c F$ by E (resp. F). The rank $\overline{E} + \overline{F}$ of $E \sqcap^c F$ is greater than the rank $\overline{E}$ of E (resp. the rank $\overline{F}$ of F). $\square$

Where A is the initial value of D in the Purification procedure and B is its final value (which exists by Lemma 7.4), we call B the **purification** of A.

Lemma 7.5. *For any cirquent A and its purification B, we have:*
1. *If B is provable, then so is A.*
2. *A is valid iff so is B.*
3. *B is pure.*
4. *The rank of B does not exceed the rank of A.*

Proof. Clause 1: When obtaining B from A, each transformation performed during the Purification procedure applies, in the conclusion-to-premise sense, one of the inference rules of **CL16**. Reversing the order of those transformations, we get a derivation of A from B. Appending that derivation to a proof of B (if one exists) yields a proof of A.

Clause 2: Immediate from the two clauses of Lemma 6.2 and the fact that, when obtaining B from A using the Purification procedure, the rule of Choosing is never used.

Clause 3: One by one, Stage 1 eliminates all surface occurrences of $\bot$ in D (unless D itself is $\bot$). So, at the end of the stage, D satisfies condition 1 of Definition 7.3. None of the subsequent steps make D violate that condition, so B, too, satisfies that condition. Similarly, a routine examination of the situation reveals that Stage 2 (resp. 3, ..., resp. 7) of the Purification procedure makes D satisfy condition 2 (resp. 3, ..., resp. 7) of Definition 7.3, and D continues to satisfy that condition throughout the rest of the stages. So, B is pure.

Case 4: Immediate from Lemma 7.4. □

Theorem 7.6. *A cirquent is valid if (soundness) and only if (completeness) it is provable in* **CL16**.

Proof. The soundness part is immediate from clause 1 of Lemma 6.2 and the fact that the axiom ⊤ is valid. The rest of this section is devoted to a proof of the completeness part. Pick an arbitrary cirquent A and assume it is valid. We proceed by induction on the rank of A. Let B be the purification of A.

In view of clauses 2-4 of Lemma 7.5, B is a valid, pure cirquent whose rank does not exceed that of A. We shall implicitly rely on this fact below. By clause 1 of Lemma 7.5, if B is provable, then so is A. Hence, in order to show that A is provable, it suffices to show that B is provable. B cannot be $\bot$ because then, of course, it would not be valid. Similarly, B cannot be a literal because obviously no literal is valid. In view of this observation and B's being pure, it is clear that the following cases cover all possibilities for B.

Case 1: B is ⊤. Then B is an axiom and hence provable.

Case 2: B is $E \sqcup^c F$. Let $\mathcal{H}$ be a logical solution of B. Consider the work of $\mathcal{H}$ in the scenario where the environment does not move until $\mathcal{H}$ makes the move $c.i$, where $i \in \{0,1\}$. Sooner or later $\mathcal{H}$ has to make such a move, for otherwise B would be lost due to being $\sqcup^c$-rooted. Since in the games that we deal with the order of moves is irrelevant, without loss of generality we may assume that the move $c.i$ is made before any other moves. Let B' be the result of replacing in B all subcirquents of the form $X_0 \sqcup^c X_1$ by X_i. Observe that, after the move $c.i$ is made, in any scenario that may follow, $\mathcal{H}$ has to continue and win B'. In other words, $\mathcal{H}$ is a logical solution of (not only B but also) B'. The rank of B' is of course smaller than that of B. Hence, by the induction hypothesis, B' is provable. Then B follows from B' by Choosing.

Case 3: B is $E \sqcap^c F$, and neither E nor F contains the cluster c. By clause 2 of Lemma 6.2, both E and F are valid, because B follows from them by Splitting. The rank of either cirquent is smaller than that of B. Hence, by the induction hypothesis, both E and F are provable. Therefore, by Splitting, so is B.

Case 4: B is $E_1 \vee \ldots \vee E_n$ ($n \geq 2$), where each E_i is either a literal or a cirquent of the form $F \sqcup^c G$; besides, for no elementary letter p do we have that both p and $\neg p$ are among $E_1, \ldots, E_n$. Not all of the cirquents $E_1, \ldots, E_n$ can be literals, for otherwise B would be automatically lost under an interpretation which interprets all those literals as $\bot$, contradicting our assumption that B is valid. With this observation in mind, without loss of generality, we may assume that, for some k with $1 \leq k \leq n$, the first k cirquents $E_1, \ldots, E_k$ are of the form $F_1 \sqcup^{c_1} G_1, \ldots, F_k \sqcup^{c_k} G_k$ and the remaining $n - k$ cirquents $E_{k+1}, \ldots, E_n$ are literals. Let $\mathcal{H}$ be a logical solution of B. Consider the work of $\mathcal{H}$ in the scenario where the environment makes no moves. Note that, at some point, for some $1 \leq j \leq k$, $\mathcal{H}$ should make the move $c_j.i$ ($i \in \{0.1\}$), for otherwise B would be lost under an(y) interpretation which interprets all of the literal cirquents $E_{k+1}, \ldots, E_n$ as $\bot$. Fix such j, i. Let B' be the result of replacing, in B, every subcirquent of the form $X_0 \sqcup^{c_j} X_1$ by X_i. With some analysis left to the reader, $\mathcal{H}$ can be seen to be a logical solution of B'. Thus, B' is valid. The rank of B' is smaller than that of B and hence, by the induction hypothesis, B' is provable. But then so is B, because it follows from B' by Choosing.

Case 5: B is $E_1 \wedge \ldots \wedge E_n$ ($n \geq 2$), where, for some e ($1 \leq e \leq n$), E_e — fix it — is not of the form $F \sqcap^c G$ or $F \wedge G$, nor do we have $E_e \in \{\top, \bot\}$. The validity of B, of course, implies that E_e, as one of its $\wedge$-conjuncts, is also valid. This rules out the possibility that E_e is a literal, because, as we observed earlier, a literal cannot be valid. We are therefore left with one of the following two possible subcases:

Subcase 5.1: E_e is of the form $F \sqcup^c G$. Let $\mathcal{H}$ be a logical solution of B. As in Case 4, consider the work of $\mathcal{H}$ in the scenario where the environment makes no moves. Note that, at some point, $\mathcal{H}$ should make the move $c.0$ or $c.1$, for otherwise B would be lost (under any interpretation). Let us just consider the case of the above move being $c.0$ (the case of it being $c.1$ will be handled in a similar way). Let B_0 be the result of replacing, in B, every subcirquent of the form $X \sqcup^c Y$ (including the conjunct $F \sqcup^c G$) by X. Then, as in Case 4, $\mathcal{H}$ can be seen to be a logical solution of B_0. Thus, B_0 is valid. The rank of B_0 is smaller than that of B and hence, by the induction hypothesis, B_0 is provable. But then so is B, because it follows from B_0 by Choosing(a).

Subcase 5.2: E_e is of the form $F_1 \vee \ldots \vee F_m$, where each F_i ($1 \leq i \leq m$) is either a literal or a cirquent of the form $G \sqcup^c H$, and for no elementary letter p do we have that both p and $\neg p$ are among $F_1, \ldots, F_m$. This case is very similar to Case 4 and, almost literally repeating our reasoning in the latter, we find that B is provable. $\square$

References

[1] A. Avron. *A constructive analysis of RM.* **Journal of Symbolic Logic** 52 (1987), No.4, pp. 939-951.

[2] M. Bauer. *A PSPACE-complete first order fragment of computability logic.* **ACM Transactions on Computational Logic** 15 (2014), No 1, Paper 1.

[3] M. Bauer. *The computational complexity of propositional cirquent calculus.* **Logical Methods is Computer Science** 11 (2015), Issue 1, Paper 12, pp. 1-16.

[4] A. Blass. *Degrees of indeterminacy of games.* **Fundamenta Mathematicae** 77 (1972) 151-166.

[5] A. Blass. *A game semantics for linear logic.* **Annals of Pure and Applied Logic** 56 (1992), pp. 183-220.

[6] K. Brünnler and A. Tiu. *A local system for classical logic.* In: **Lecture Notes in Computer Science** 2250 (2001), pp. 347–361.

[7] P. Bruscoli and A. Guglielmi. *On the proof complexity of deep inference.* **ACM Transactions on Computational Logic** 10 no. 2, Article 14, (2009), pp. 1–34.

[8] A. Das and L. Strassburger. *On linear rewriting systems for Boolean logic and some applications to proof theory.* **Logical Methods in Computer Science** 12 (2016), pp. 1-27.

[9] J.Y. Girard. *Linear logic.* **Theoretical computer science** 50 (1887), pp. 1-102.

[10] A. Guglielmi and L. Strassburger. *Non-commutativity and MELL in the calculus of structures.* In: **Lecture Notes in Computer Science** 2142 (2001), pp. 54-68.

[11] J. Hintikka. **Logic, Language-Games and Information: Kantian Themes in the Philosophy of Logic.** Clarendon Press 1973.

[12] J. Hintikka and G. Sandu. *Game-theoretical semantics.* In: **Handbook of Logic and Language.** J. van Benthem and A ter Meulen, eds. North-Holland 1997, pp. 361-410.

[13] G. Japaridze. *Introduction to computability logic.* **Annals of Pure and Applied Logic** 123 (2003), pp. 1-99.

[14] G. Japaridze. *Propositional computability logic I.* **ACM Transactions on Computational Logic** 7 (2006), pp. 302-330.

[15] G. Japaridze. *Propositional computability logic II.* **ACM Transactions on Computational Logic** 7 (2006), pp. 331-362.

[16] G. Japaridze. *Introduction to cirquent calculus and abstract resource semantics.* **Journal of Logic and Computation** 16 (2006), pp. 489-532.

[17] G. Japaridze. *From truth to computability I.* **Theoretical Computer Science** 357 (2006), pp. 100-135.

[18] G. Japaridze. *From truth to computability II.* **Theoretical Computer Science** 379 (2007), pp. 20-52.

[19] G. Japaridze. *The logic of interactive Turing reduction.* **Journal of Symbolic Logic** 72 (2007), pp. 243-276.

[20] G. Japaridze. *The intuitionistic fragment of computability logic at the propositional*

level. **Annals of Pure and Applied Logic** 147 (2007), pp. 187-227.

[21] G. Japaridze. *Cirquent calculus deepened.* **Journal of Logic and Computation** 18 (2008), pp. 983-1028.

[22] G. Japaridze. *Sequential operators in computability logic.* **Information and Computation** 206 (2008), pp. 1443-1475.

[23] G. Japaridze. *Many concepts and two logics of algorithmic reduction.* **Studia Logica** 91 (2009), pp. 1-24.

[24] G. Japaridze. *In the beginning was game semantics.* In: **Games: Unifying Logic, Language, and Philosophy**. O. Majer, A.-V. Pietarinen and T. Tulenheimo, eds. Springer 2009, pp. 249-350.

[25] G. Japaridze. *Toggling operators in computability logic.* **Theoretical Computer Science** 412 (2011), pp. 971-1004.

[26] G. Japaridze. *From formulas to cirquents in computability logic.* **Logical Methods is Computer Science** 7 (2011), Issue 2 , Paper 1, pp. 1-55.

[27] G. Japaridze. *Separating the basic logics of the basic recurrences.* **Annals of Pure and Applied Logic** 163 (2012), pp. 377-389.

[28] G. Japaridze. *A logical basis for constructive systems.* **Journal of Logic and Computation** 22 (2012), pp. 605-642.

[29] G. Japaridze. *The taming of recurrences in computability logic through cirquent calculus, Part I.* **Archive for Mathematical Logic** 52 (2013), pp. 173-212.

[30] G. Japaridze. *The taming of recurrences in computability logic through cirquent calculus, Part II.* **Archive for Mathematical Logic** 52 (2013), pp. 213-259.

[31] G. Japaridze. *On the system CL12 of computability logic.* **Logical Methods is Computer Science** 11 (2015), Issue 3, paper 1, pp. 1-71.

[32] P. Lorenzen. *Ein dialogisches Konstruktivitätskriterium.* In: **Infinitistic Methods**. In: PWN, Proc. Symp. Foundations of Mathematics, Warsaw, 1961, pp. 193-200.

[33] I. Mezhirov and N. Vereshchagin. *On abstract resource semantics and computability logic.* **Journal of Computer and System Sciences** 76 (2010), pp. 356-372.

[34] M. Qu, J. Luan, D. Zhu and M. Du. *On the toggling-branching recurrence of computability logic.* **Journal of Computer Science and Technology** 28 (2013), pp. 278-284.

[35] W. Xu and S. Liu. *Soundness and completeness of the cirquent calculus system CL6 for computability logic.* **Logic Journal of the IGPL** 20 (2012), pp. 317-330.

[36] W. Xu and S. Liu. *The countable versus uncountable branching recurrences in computability logic.* **Journal of Applied Logic** 10 (2012), pp. 431-446.

[37] W. Xu and S. Liu. *The parallel versus branching recurrences in computability logic.* **Notre Dame Journal of Formal Logic** 54 (2013), pp. 61-78.

[38] W. Xu. *A propositional system induced by Japaridze's approach to IF logic.* **Logic Journal of the IGPL** 22 (2014), pp. 982-991.

[39] W. Xu. *A cirquent calculus system with clustering and ranking.* **Journal of Applied**

Logic 16 (2016), pp. 37-49.

Boolean-valued Models as a Foundation for Locally L^0-Convex Analysis and Conditional Set Theory

Antonio Avilés *
Universidad de Murcia, Dpto. Matemáticas, 30100 Espinardo, Murcia, Spain
avileslo@um.es

José Miguel Zapata †
Universidad de Murcia, Dpto. Matemáticas, 30100 Espinardo, Murcia, Spain
jmzg1@um.es

Abstract

Locally L^0-convex modules were introduced in [D. Filipovic, M. Kupper, N. Vogelpoth. Separation and duality in locally L^0-convex modules. J. Funct. Anal. 256(12), 3996-4029 (2009)] as the analytic basis for the study of multi-period mathematical finance. Later, the algebra of conditional sets was introduced in [S. Drapeau, A. Jamneshan, M. Karliczek, M. Kupper. The algebra of conditional sets and the concepts of conditional topology and compactness. J. Math. Anal. Appl. 437(1), 561-589 (2016)]. By means of Boolean-valued models and its transfer principle we show that any known result on locally convex spaces has a transcription in the frame of locally L^0-convex modules which is also true, and that the formulation in conditional set theory of any theorem of classical set theory is also a theorem. We propose Boolean-valued analysis as an analytic framework for the study of multi-period problems in mathematical finance.

The authors would like to thank an anonymous referee for a careful review of the manuscript and valuable comments, and for pointing out a stem of excellent related works.

*The first author was supported by projects MTM2014-54182-P and MTM2017-86182-P (MINECO,AEI/FEDER, UE) and 19275/PI/14 (Fundación Séneca).

†The second author was supported by the grant associated to the project MTM2014-57838-C2-1-P (MINECO).

Introduction

Boolean-valued models are a tool in mathematical logic that was developed as a way to formalize the method of forcing that Paul Cohen created to solve the first problem in the famous Hilbert's list: it is impossible neither to prove nor to disprove that every infinite set of reals can be bijected either with the natural numbers of with the whole real line [9]. The theory was first formulated by Scott [43] based on some ideas of Solovay, while Vopěnka created independently a similar theory. In this paper we shall see that Boolean-valued models provide a natural framework for certain problems in financial mathematics which involve a multi-period setting, such as representation of dynamic and conditional risk measures and stochastic optimal control. Several recent developments, like the study of *locally L^0-convex modules* and the *algebra of conditional sets* are covered by this theory. The advantage is not only a unified approach for several scattered results in the literature; the important point is that we get at our disposal all the powerful tools of a well developed deep mathematical theory. In particular, the so-called *transfer principle* claims that any known result of set theory has a transcription in the Boolean-valued setting, which is also true.

In order to provide an analytical basis to problems of mathematical finance in a multi-period set-up with a dynamic flow of information, Filipovic et al. [14] considered $L^0 := L^0(\Omega, \Sigma, \mathbb{P})$, the ordered lattice ring of equivalence classes modulo almost sure equality of Σ-measurable random variables, where $(\Omega, \Sigma, \mathbb{P})$ is a probability space that models the market information that is available at some future time. They introduced the topology of almost sure dominance on L^0 and the notion of locally L^0-convex module, and succeeded in developing a randomized version of classical convex analysis. We will show that a locally L^0-convex module can be embedded into a Boolean-valued universe and will systematically study the meaning of several related objects within this framework. Thus, we will show that not only the main results of [14] are consequence of this connection but that any known result of locally convex analysis has a modular transcription which also holds as a consequence of the transfer principle of Boolean-valued models. For instance, we provide randomized versions of celebrated theorems such as the Brouwer's fixed point and the James' compactness theorem.

Conditional set theory was introduced in [11]. The authors need to give a definition of the conditional version of each mathematical concept they want to use (conditional real number, conditional topological space, etc.) and also formulate and prove the conditional version of known results. All this is automatic within the context that we propose: the conditional version of any known result is automatically true. Nevertheless, we should mention that this connection exhibits that conditional set theory provides a practical tool to manage objects from Boolean-valued models and gives intuition to anyone who is not familiarized with the formalisms of Boolean-valued models.

These developments have been applied to mathematical finance. For instance, we find applications to representation of conditional risk measures (see eg [5, 15, 18]), to equilibrium theory (see [2, 7]), to optimal stochastic control (see [28]) and to financial preferences (see [10]). Also, as commented, there is a significant number of related works. For instance, modules endowed with L^0-valued norms have been applied to the study of ultrapowers of Lebesgue-Bochner spaces by Haydon et al. [27]. Guo together with other co-authors have widely studied generalizations of functional analysis results in L^0-modules endowed with the topology of stochastic convergence with respect to a family of L^0-valued seminorms, also called the (ϵ, λ)-topology (see eg [21, 23, 24] and references therein). In this regard, a study of the relations between the (ϵ, λ)-topology and the locally L^0-convex topology induced by L^0-valued seminorms can be found in [22]. Eisele and Taieb [13] extended some functional analysis results to modules over the ring L^∞. A randomized version of finite-dimensional analysis in $(L^0)^d$ is developed in [8] and also a version of the Brouwer fixed point theorem in this context is established in [12]. Other results and counter-examples on locally L^0-convex modules can be found in [48, 46]. Further studies of dual pairs and weak topologies in the context of conditional sets are provided in [41, 47].

In addition, we should highlight that the Boolean-valued model approach shows that the study of locally L^0-convex modules naturally fits the framework of the well-developed theory of lattice-normed spaces (i.e. norms that take values in a vector lattice) and dominated operators, originated in the 1930s by L.V. Kantorovich (see [32]), field that has been widely researched and fruitfully exploited by A. G. Kusraev and S. S. Kutateladze. For a thorough account we refer the reader to eg [36, 39] and their extensive lists of references.

The paper is organized as follows: In the first section we give a short introduction to Boolean-valued models, provide some intuition and recall the basic elements and principles of the Boolean-valued machinery. In Section 2, we explain a precise connection between the framework of locally L^0-convex modules and Boolean-valued locally convex analysis; give a list of basic elements of locally L^0-convex analysis and explain their meanings within Boolean-valued locally convex analysis; and as example of application we derive the main theorems of [14] and modular versions of James' compactness theorem and Brouwer's fixed point theorem. Finally, Section 3 is devoted to provide a precise connection between conditional set theory and Boolean-valued models.

1 Foundations of Boolean-valued models

Let us try to give an intuitive idea of what Boolean-valued analysis is and how it can fit in mathematical finance. We would like to talk about what will happen in a particular moment in the future. This future is uncertain, it is influenced by events that we do not know yet. These possible events that might influence the future will be coded by a complete Booolean algebra $\mathcal{A} = (\mathcal{A}, \vee, \wedge, ^c, 0, 1)$. The simplest case that we can think of is that the future that we are interested in is completely determined by the result of flipping a coin. In that case, the algebra of events is $\mathcal{A}_0 = \{a, a^c, 0, 1\}$ where a is the event "we get head", and its negation a^c is "we get tail". In the algebra of events we also have all the events that we can formulate combining others, and so $a \vee a^c = 1$ is the event "we get either head or tail", which is just the true event, and $a \wedge a^c = 0$ is the event "we get both head and tail" which is just the false event. Another example is that our future depends on a randomly chosen (say with Gaussian probability) real number. In that case, the algebra of events would be the measure algebra: measurable subsets of $\mathbb{R}$ modulo null sets. In that case, for example, the class of $[1, 2]$ is interpreted as the event "the random number happened to fall inside the interval $[1, 2]$".

So let us fix the algebra $\mathcal{A}$ of all the events that we can talk about and influence the future. The next element of our theory are the *names*. The names are the nouns of the language with which we talk about objects in the future despite the uncertainties. In the flipping coin example, suppose that I have five dollars and I bet two dollars that the outcome will be tail. Then I can consider the name $\dot{x}$ that represents the amount of money that I will have in the future. The actual value of $\dot{x}$ is unknown, it could be 3 or 7 depending on the coin. In the very simple flipping coin case, a name can be identified with a pair (r, s) of mathematical objects, one for head and one for tail. In the random real case, names will look more complicated but the idea is similar. Examples of names would be $\dot{y}$ that would take value 1 if the random real is positive or -1 if it is negative, and also $\dot{z}$ the name for the random real itself. A special kind of names are those which do not really depend on the unknown events, and those are represented with a $\vee$ symbol above. For example, $\check{5}$ is a name which represents the number 5, no matter what the coin did or what the random real actually happens to be.

Once we understand the idea of a name, the next step is formulating statements about names and deciding what are the truth values of such statements. Playing with the names given above in the flipping coin case, it makes sense to make the following statements: P1. $\dot{x}$ is a positive real number, P2. $\dot{x} = \check{7}$, P3. $\dot{x} < \check{4}$,

P4. $\dot{x} = \dot{x}^2$. While P1 is clearly true and P4 is clearly false, for P2 and P3 we may say that *it depends on what the coin will do*. In Boolean-valued analysis, these statements are not assigned a binary truth value of true or false. The truth valued of a sentence P, denoted by $[\![P]\!]$ is an element of the Boolean algebra $\mathcal{A}$ that corresponds to the event that describes when this sentence is true. Thus $[\![P3]\!] = a^c$ because I have 8 dollars if and only if the flipping will give a tail, and similarly $[\![P4]\!] = a$. In the random case, for instance, the truth value of the sentence $\check{2}\dot{z} < \check{4}$ is exactly the representative of the interval $(-\infty, 2)$ of the measure algebra, while $[\![\dot{z} > \dot{y}]\!]$ is the representative of $(-1, 0) \cup (1, +\infty)$. In all these examples, we are using names for real numbers, but the idea is more general, we can have names for functions, sets, Banach spaces or any mathematical object we want, and state any kind of properties we wish in formal mathematical language. In the framework of set theory, any mathematical object can be considered as a set and any mathematical statement can be re-stated in terms of the belonging relation $\in$ between sets.

The precise formulation of Boolean-valued analysis requires some familiarity with the basics of set theory and logic, and in particular with first order logic, ordinals and transfinite induction. However, if one understands the key ideas and principles, it is possible to work with Boolean-valued models avoiding the underlying machinery that can be conveniently hidden in a black box. For a detailed description we can refer the reader to [4], [30, Chapter 14], or [39, Chapter 2]. We make now a quick review.

Let us consider a universe of sets V satisfying the axioms of the Zermelo-Fraenkel set theory with the axiom of choice (ZFC), and a first-order language $\mathcal{L}$ which allows the formulation of statements about the elements of V. In the universe V we have all possible mathematical objects (real numbers, topological spaces, etc.) that we can talk about in a context of total certainty. The language $\mathcal{L}$ consists of the elements of V plus a finite list of symbols for logic symbols ($\forall, \wedge, \neg$ and parenthesis), variables (with the symbol x we can express any variables we need as $x, xx, xxx, \ldots$) and the verbs $=$ and $\in$. Though we usually use a much richer language by introducing more and more intricate definitions, in the end any usual mathematical statement can be written using only those mentioned. The elements of the universe V are classified into a transfinite hierarchy: $V_0 \subset V_1 \subset V_2 \subset \cdots V_\omega \subset V_{\omega+1} \subset \cdots$, where $V_0 = \emptyset$, $V_{\alpha+1} = \mathcal{P}(V_\alpha)$ is the family of all sets whose elements come from V_α, and $V_\beta = \bigcup_{\alpha < \beta} V_\alpha$ for limit ordinal β.

Now consider the complete Boolean algebra of events $\mathcal{A} = (\mathcal{A}, \vee, \wedge, {}^c, 0, 1)$ which is an element of V. For given $a, b \in \mathcal{A}$, we will write $a \leq b$ whenever $a \wedge b = a$. For a family $\{a_i\}_{i\in I}$ in $\mathcal{A}$, we denote its supremum by $\bigvee_{i\in I} a_i$ and its infimum by $\bigwedge_{i\in I} a_i$. A family $\{a_i\}_{i\in I}$ in $\mathcal{A}$ is said to be a partition of a if $\bigvee_{i\in I} a_i = a$ and $a_i \wedge a_j = 0$ for all $i \neq j$, $i,j \in I$ (notice that I could be infinite in this definition). For given $a \in \mathcal{A}$ we denote by $p(a)$ the set of all partitions of a.

Given this complete Boolean algebra $\mathcal{A}$, one constructs now $V^{(\mathcal{A})}$, the *Boolean-valued model* of $\mathcal{A}$, whose elements are the *names* that we mentioned earlier, that we interpret as nouns with which we talk about the future. We proceed by induction over the class Ord of ordinals of the universe V. We start by defining $V_0^{(\mathcal{A})} := \emptyset$; if $\alpha + 1$ is the successor of α, we define

$$V_{\alpha+1}^{(\mathcal{A})} := \left\{ x \colon x \text{ is an } \mathcal{A}\text{-valued function with } \mathrm{dom}(x) \subset V_\alpha^{(\mathcal{A})} \right\}.$$

The idea is that for $y \in dom(x)$, y will become an element of x in the future if $x(y)$ happens. If α is a limit ordinal $V_\alpha^{(\mathcal{A})} := \bigcup_{\xi < \alpha} V_\xi^{(\mathcal{A})}$. Finally, let $V^{(\mathcal{A})} := \bigcup_{\alpha \in Ord} V_\alpha^{(\mathcal{A})}$.

Given an element x in $V^{(\mathcal{A})}$ we define its *rank* as the least ordinal α such that x is in $V_{\alpha+1}^{(\mathcal{A})}$.

We consider a first-order language which allows to produce statements about $V^{(\mathcal{A})}$. Namely, let $\mathcal{L}^{(\mathcal{A})}$ be the first-order language which is the extension of $\mathcal{L}$ by adding nouns for each element of $V^{(\mathcal{A})}$. Suppose that φ is any formula of the language $\mathcal{L}^{(\mathcal{A})}$, its *Boolean truth value* $[\![\varphi]\!]$ is defined by induction in the length of φ. If one got the right intuition, all the formulas that follow should look natural. We start by defining the Boolean truth value of the *atomic formulas* $x \in y$ and $x = y$ for x and y in $V^{(\mathcal{A})}$. Namely, proceeding by transfinite recursion we define

$$[\![x \in y]\!] = \bigvee_{t \in \mathrm{dom}(y)} y(t) \wedge [\![t = x]\!],$$

$$[\![x = y]\!] = \bigwedge_{t \in \mathrm{dom}(x)} (x(t) \Rightarrow [\![t \in y]\!]) \wedge \bigwedge_{t \in \mathrm{dom}(y)} (y(t) \Rightarrow [\![t \in x]\!]),$$

where, for $a, b \in \mathcal{A}$, we denote $a \Rightarrow b := a^c \vee b$. For non-atomic formulas we have

$$[\![\exists x \varphi(x)]\!] := \bigvee_{u \in V^{(\mathcal{A})}} [\![\varphi(u)]\!] \quad \text{and} \quad [\![\forall x \varphi(x)]\!] := \bigwedge_{u \in V^{(\mathcal{A})}} [\![\varphi(u)]\!];$$

$$[\![\varphi \wedge \psi]\!] := [\![\varphi]\!] \wedge [\![\psi]\!] \quad \text{and} \quad [\![\neg \varphi]\!] := [\![\varphi]\!]^c.$$

It is well-known that every theorem of ZFC is true in $V^{(\mathcal{A})}$ with the Boolean truth value:

Theorem 1.1. *(Transfer Principle) If φ is a theorem of ZFC, then $[\![\varphi]\!] = 1$.*

Also, it will be important to keep in mind the following results, which will allow to manipulate $V^{(\mathcal{A})}$ and are well-known within Boolean-valued models theory:

Theorem 1.2. *(Maximum Principle) Let $\varphi(x)$ be a formula with one free variable x. Then there exists an element u of $V^{(\mathcal{A})}$ such that $[\![\varphi(u)]\!] = [\![\exists x \varphi(x)]\!]$.*

Theorem 1.3. *(Mixing Principle) Let $\{a_i\} \in p(1)$ and let $\{x_i\}$ be a family in $V^{(\mathcal{A})}$. Then there exists an element x in $V^{(\mathcal{A})}$ such that $[\![x = x_i]\!] \geq a_i$ for all i. Moreover, if y is another element of $V^{(\mathcal{A})}$ which satisfies the same, then $[\![x = y]\!] = 1$.*

Let us say that two names x, y are equivalent, and write $x \sim y$, when $[\![x = y]\!] = 1$. The truth value of a formula is not affected when we change a name by an equivalent one. Given a set x in V we define its canonical name $\check{x}$ in $V^{(\mathcal{A})}$. Namely, we put $\check{\emptyset} := \emptyset$ and for x in $V^{(\mathcal{A})}$ we define $\check{x} \colon \mathcal{D} \to \mathcal{A}$, where $\mathcal{D} := \{\check{y} \colon y \in x\}$ and $\check{x}(\check{y}) := 1$ for $y \in x$. It is not difficult to show that $\check{x}$ is an element of $V^{(\mathcal{A})}$. If $x, y, f \in V^{(\mathcal{A})}$ and we say, for instance, that f is a name for a function $f \colon x \to y$, this means that $[\![\text{``}f\text{ is a function from }x\text{ to }y\text{''}]\!] = 1$. The transfer and maximum principles provide us with names $\mathbb{N}^{(\mathcal{A})}$ and $\mathbb{R}^{(\mathcal{A})}$ for the sets of natural numbers and real numbers, respectively. This means: $[\![\text{``}\mathbb{N}^{(\mathcal{A})}\text{ is the set of natural numbers''}]\!] = 1$.

Let $\overline{V}^{(\mathcal{A})}$ be the subclass of $V^{(\mathcal{A})}$ defined by choosing a representative of the least rank in each class of the equivalence relation $\{(x, y) \colon [\![x = y]\!] = 1\}$.[1] Given a name x with $[\![x \neq \emptyset]\!] = 1$ we define its *descent* by

$$x\!\downarrow = \{y \in \overline{V}^{(\mathcal{A})} \colon [\![y \in x]\!] = 1\}.$$

Notice that, if $x \in V_\alpha^{(\mathcal{A})}$, then any element of the class $x\!\downarrow$ is also in $V_\alpha^{(\mathcal{A})}$. Therefore, we have that $x\!\downarrow$ is a set in V.

The following result will be useful later:

Theorem 1.4. *Let x, y be elements of $V^{(\mathcal{A})}$ with $[\![(x \neq \emptyset) \wedge (y \neq \emptyset)]\!] = 1$, let $f \colon x\!\downarrow \to y\!\downarrow$ be a function such that*

$$[\![u = v]\!] \leq [\![f(u) = f(v)]\!] \quad \text{for all } u, v \in x\!\downarrow.$$

Then there exists g in $V^{(\mathcal{A})}$, which is a name for a function between x and y, such that $[\![f(u) = g(u)]\!] = 1$ for all $u \in x\!\downarrow$.

---
[1] The construction can be done by transfinite induction. We choose a representative of each class $\{(x, y) \colon x, y \in V_{\alpha+1}^{(\mathcal{A})} \colon [\![x = y]\!] = 1, [\![x = z]\!] < 1 \text{ for al } z \in \overline{V}_\alpha^{(\mathcal{A})}\}$ and define $\overline{V}_{\alpha+1}^{(\mathcal{A})}$ the set of all theses representatives. For a limit ordinal α we put $V_\alpha^{(\mathcal{A})} := \bigcup_{\xi < \alpha} \overline{V}_\xi^{(\mathcal{A})}$. The class $\overline{V}^{(\mathcal{A})}$ is frequently defined in literature and is called the *separated universe*, see eg [39, 45].

2 A precise connection between locally L^0-convex analysis and Boolean-valued locally convex analysis

Let $(\Omega, \Sigma, \mathbb{P})$ be a probability space of the universe V and let L^0 denote the set of Σ-measurable random variables, which are identified whenever their difference is $\mathbb{P}$-negligible. We denote by $\mathcal{F}$ the measure algebra, which is defined by identifying events whose symmetric difference has probability 0. Then, $\mathcal{F}$ has structure of complete Boolean algebra which satisfies the *countable chain condition*, that is, all partitions are at most countable. Since $\mathcal{F}$ is a complete Boolean algebra, one can consider the corresponding boolean-valued model $V^{(\mathcal{F})}$.

As shown by Takeuti [44], there exists a canonical bijection ϕ between $\mathbb{R}^{(\mathcal{F})}{\downarrow}$ and L^0. Moreover, the image of $\mathbb{N}^{(\mathcal{F})}$ and $\mathbb{Q}^{(\mathcal{F})}$ under ϕ are precisely $L^0(\mathbb{N})$, the set of (equivalence classes) of $\mathbb{N}$-valued random variables; and $L^0(\mathbb{Q})$, the set of (equivalence class) of $\mathbb{Q}$-valued random variables, respectively. Besides, $\phi(r+s) = \phi(r) + \phi(s)$, $\phi(rs) = \phi(r)\phi(s)$, $\phi(0) = 0$, $\phi(1) = 1$, $[\![r = s]\!] = \bigvee \{A \in \mathcal{F} : 1_A \phi(r) = 1_A \phi(s)\}$, and $[\![r \leq s]\!] = \bigvee \{A \in \mathcal{F} : 1_A \phi(r) \leq 1_A \phi(s)\}$, for all $r, s \in \mathbb{R}^{(\mathcal{F})}{\downarrow}$.

Remark 2.1. *Gordon [19] proved that, in general, if $\mathcal{A}$ is an arbitrary complete Boolean algebra, the descent $\mathbb{R}^{(\mathcal{A})}{\downarrow}$ is a universally complete vector lattice (i.e. every family of pairwise disjoint elements is bounded) such that $\mathcal{A}$ is isomorphic to the Boolean algebra of band projections in $\mathbb{R}^{(\mathcal{A})}{\downarrow}$. Then, as a particular case, we find that $\mathbb{R}^{(\mathcal{F})}{\downarrow}$ is isomorphic to the universally complete vector lattice L^0. Moreover, Takeuti [44] also proved that, in the case that $\mathcal{A}$ is the complete Boolean algebra of orthogonal projections in a Hilbert space, then $\mathbb{R}^{(\mathcal{A})}{\downarrow}$ is isomorphic to the universally complete vector lattice of self-adjoint operators whose spectral resolution takes values in $\mathcal{A}$.*

Suppose that E is an L^0-module; that is, E is a module over the ordered lattice ring L^0. We say that E has the *countable concatenation property* whenever for every sequence $\{x_k\}$ in E and every partition $\{A_k\} \in p(\Omega)$ there exists a unique $x \in E$ (denoted by $x = \sum 1_{A_k} x_k$) such that $1_{A_k} x = 1_{A_k} x_k$ for each $k \in \mathbb{N}$. This property and other related are technical assumptions that are typically assumed in literature cf.[14, 22, 48]. It should be pointed out that not every L^0-module has this property (for instance, see [14, Example 2.12] and [46, Example 1.1]). However, every L^0-module E can be made into an L^0-module with this property by considering the quotient of a suitable equivalence relation on $E^{\mathbb{N}} \times p(\Omega)$ (see eg [41]).

The next result describes the relation between L^0-modules and names for real vector spaces in $V^{(\mathcal{F})}$. Gordon [20] provided an equivalence of categories between the category of names for vector spaces and linear functions in $V^{(\mathcal{A})}$ and the category of unital separated injective K-modules and K-module morphisms, where K is a rationally complete semiprime commutative ring and $\mathcal{A}$ is the Boolean algebra of annihilator ideals (see [39] for terminology). It can be verified that L^0 is a rationally complete semiprime commutative ring whose annihilator ideals coincide with band projections. Besides, it can be checked that the countable concatenation property of an L^0-module E is equivalent to the injectivity of E. Thus Theorem 2.1 below is a particular case of the mentioned equivalence of categories in [20]. However, for the convenience of the reader, we provide a self-contained proof for this particular case.[2]

Theorem 2.1. *For fixed an underlying measure algebra $\mathcal{F}$, there is an equivalence of categories between the category of names for real vector spaces and linear functions in $V^{(\mathcal{F})}$, and the category of L^0-modules with the countable concatenation property and L^0-module morphisms.*

Proof. If we take any E in $V^{(\mathcal{F})}$, which is a name for a real vector space, then $E{\downarrow}$ can be endowed with structure of L^0-module with the countable concatenation property. Indeed, for $x, y \in E{\downarrow}$ and $\eta \in L^0$, we define $x + y := u$ where u is the unique element of $E{\downarrow}$ such that $[\![x + y = u]\!] = \Omega$; we define $\eta x = v$, where v is the unique element of $E{\downarrow}$ such that $[\![\phi^{-1}(\eta)x = v]\!] = \Omega$. It follows by inspection that $E{\downarrow}$ is an L^0-module. In addition, if $\{A_k\} \in p(\Omega)$ and $\{u_k\} \subset E{\downarrow}$, then by the mixing principle (Theorem 1.3) there exists a unique $u \in E{\downarrow}$ such that $A_k \leq [\![u = u_k]\!]$ for all $k \in \mathbb{N}$. Let $\tilde{1}_{A_k} := \phi^{-1}(1_{A_k})$. One has $[\![\tilde{1}_{A_k} = 1]\!] = A_k$ and $[\![\tilde{1}_{A_k} = 0]\!] = A_k^c$ each k. Then

$$[\![\tilde{1}_{A_k} u - \tilde{1}_{A_k} u_k]\!] \geq [\![((u = u_k) \wedge (\tilde{1}_{A_k} = 1)) \vee (\tilde{1}_{A_k} = 0)]\!]$$

$$= ([\![u = u_k]\!] \wedge [\![\tilde{1}_{A_k} = 1]\!]) \vee [\![\tilde{1}_{A_k} = 0]\!] = A_k \vee A_k^c = \Omega \quad \text{for all } k.$$

That is, $1_{A_k} u = 1_{A_k} u_k$ for each k. This shows that $E{\downarrow}$ has also the countable concatenation property.

Now, suppose that f, E, F are elements in $V^{(\mathcal{F})}$ such that E, F are names for real vector spaces and f is a name for a linear function from E to F in $V^{(\mathcal{F})}$. Then, slightly abusing the notation, let $f{\downarrow}$ denote the unique map from $E{\downarrow}$ to $F{\downarrow}$ satisfying $[\![f{\downarrow}(x) = f(x)]\!] = \Omega$ for all $x \in E{\downarrow}$. Then it can be verified that $f{\downarrow} : E{\downarrow} \to F{\downarrow}$ is an L^0-module morphism.

Thus, we define the functor $G(E) := E{\downarrow}$, $G(f) := f{\downarrow}$.

Let us turn to the description of the inverse functor. Let E be an L^0-module with the countable concatenation property. We will show that E can be made into

---

[2] In general, for any Boolean-algebra $\mathcal{A}$, one has that $\mathbb{R}^{(\mathcal{A})}{\downarrow}$ is also a rationally complete semiprime commutative ring whose annihilator ideals coincide with bands, thus the equivalence of categories in [20] also applies. For a proof of this particular case see [39, p. 198].

an element $\tilde{E}$ of $V^{(\mathcal{F})}$ which is a name for a real vector space. Indeed, for each $x \in E$ we define $\bar{x}\colon \mathcal{D}_x \to \mathcal{F}$ with

$$\mathcal{D}_x := \{\check{y}\colon y \in E\}, \text{ and } \bar{x}(\check{y}) := A_{x,y} \text{ for } y \in E,$$

where $A_{x,y} := \bigvee\{B \in \mathcal{F}\colon 1_B(x-y) = 0\}$. Then, let $\tilde{E}\colon \mathcal{D} \to \mathcal{F}$ where $\mathcal{D} := \{\bar{x}\colon x \in E\}$ with $\tilde{E}(\bar{x}) := \Omega$ for each $x \in E$.

One has that $\tilde{E}$ is an element of $V^{(\mathcal{F})}$. Moreover, we claim that $[\![\bar{x} = \bar{y}]\!] = A_{x,y}$ for all $x, y \in E$. Indeed, given $x, y \in E$ one has

$$[\![\bar{x} = \bar{y}]\!] = \bigwedge_{u \in E} (A_{x,u} \Rightarrow [\![\check{u} \in \bar{y}]\!]) \wedge \bigwedge_{v \in E} (A_{y,v} \Rightarrow [\![\check{v} \in \bar{x}]\!]). \tag{1}$$

In addition,

$$[\![\check{u} \in \bar{y}]\!] = \bigvee_{w \in E} (A_{y,w} \wedge [\![\check{u} = \check{w}]\!]) = A_{y,u},$$

since $[\![\check{u} = \check{w}]\!] = \Omega$ if $u = w$, and $[\![\check{u} = \check{w}]\!] = \emptyset$ otherwise. Analogously, we obtain $[\![\check{v} \in \bar{x}]\!] = A_{x,v}$.

Therefore, replacing in (1), one has

$$[\![\bar{x} = \bar{y}]\!] = \bigwedge_{u \in E} \left(A^c_{x,u} \vee A_{y,u}\right) \wedge \bigwedge_{v \in E} \left(A^c_{y,v} \vee A_{x,v}\right).$$

By considering above $u = x$ and $v = y$, it follows $[\![\bar{x} = \bar{y}]\!] \leq A_{x,y}$. Now, if we show that $A_{x,y} \leq A^c_{x,u} \vee A_{y,u}$ for each $u \in E$, we obtain the assertion. Aiming at a contradiction, suppose that $\emptyset < A := A_{x,y} \wedge (A^c_{x,u} \vee A_{y,u})^c$. First, let us show that the supremum that defines $A_{x,y}$ is in fact attained for every $x, y \in E$. That is just to show that $1_{A_{x,y}}(x - y) = 0$. Consider a maximal family $\mathfrak{M}$ of pairwise disjoint elements $B \in \mathcal{F}$ such that $1_B(x-y) = 0$. By maximality, $\mathfrak{M} \in p(A_{x,y})$, and by countable chain condition this partition is countable. The uniqueness in the countable concatenation property yields that $1_{A_{x,y}} x = 1_{A_{x,y}} y$. Finally, since $A \leq A_{x,y} \wedge A_{x,u}$ one has $1_A y = 1_A x = 1_A u$. But $A \leq A^c_{y,u}$, hence $1_A y \neq 1_A u$, which is a contradiction.

For any $x \in E$, let $\tilde{x}$ be the representative in $\overline{V}^{(\mathcal{F})}$ of the name $\bar{x}$. The function $E \to \tilde{E}\!\downarrow$ given by $x \mapsto \tilde{x}$ is a bijection. Indeed, if $\tilde{x} = \tilde{y}$, then $\Omega = [\![\bar{x} = \bar{y}]\!] = A_{x,y}$; since $A_{x,y}$ is attained, it follows that $x = y$. Now, suppose that $z \in \tilde{E}\!\downarrow$. Then

$$\Omega = [\![z \in \tilde{E}]\!] = \bigvee_{x \in E} [\![\bar{x} = z]\!].$$

We can find a partition $\{A_k\} \in p(\Omega)$ such that $A_k \leq [\![\bar{x}_k = z]\!]$ for some $x_k \in E$, each k. The countable concatenation property yields an x so that $1_{A_k} x = 1_{A_k} x_k$

for all k. One has $[\![\bar{x} = \bar{x}_k]\!] = A_{x,x_k} \geq A_k$. Due to the mixing principle we obtain $[\![z = \bar{x}]\!] = \Omega$, and taking representatives in $\overline{V}^{(\mathcal{F})}$, we conclude that $z = \tilde{x}$.

For $x, y \in E$ and $r \in \mathbb{R}^{(\mathcal{F})}\!\downarrow$ we put $[\![\tilde{x} + \tilde{y} = u]\!] = \Omega$ whenever $[\![(x+y)^\sim = u]\!] = \Omega$ and $[\![z = r \cdot \tilde{x}]\!] = \Omega$ if $[\![(\phi(r)x)^\sim = z]\!] = \Omega$. Since the mapping $x \mapsto \tilde{x}$ is bijective and due to Theorem 1.4, the operations are well-defined and $\tilde{E}$ is a name for a vector space in $V^{(\mathcal{F})}$.

Now, suppose that $f : E \to F$ is a morphism between L^0-modules with the countable concatenation property. Then we define the application $g : \tilde{E}\!\downarrow \to \tilde{F}\!\downarrow$, $\tilde{x} \mapsto (f(x))^\sim$, which is well defined as $x \mapsto \tilde{x}$ is one-to-one. Using that f is L^0-linear, we have that for every $x, y \in E$, $[\![\tilde{x} = \tilde{y}]\!] = A_{x,y} \leq A_{f(x),f(y)} = [\![(f(x))^\sim = (f(y))^\sim]\!]$. Then according to Theorem 1.4, there exists $\tilde{f}$ in $V^{(\mathcal{F})}$ such that $[\![\tilde{f} : \tilde{E} \to \tilde{F}]\!] = \Omega$ and $[\![(f(x))^\sim = \tilde{f}(\tilde{x})]\!] = \Omega$ for all $x \in E$. In particular, we have that $[\![\tilde{f}(\tilde{x} + \tilde{y}) = \tilde{f}(\tilde{x}) + \tilde{f}(\tilde{y})]\!] = \Omega$ and $[\![\tilde{f}(r \cdot \tilde{x}) = r \cdot \tilde{f}(\tilde{x})]\!] = \Omega$ for all $x, y \in E$ and $r \in \mathbb{R}^{(\mathcal{F})}\!\downarrow$.

We define the functor $H(E) := \tilde{E}$, $H(f) := \tilde{f}$. Then the functors F and H are inverse equivalences. Indeed, suppose that E is an L^0-module with the countable concatenation property. We have proved that $E \to (\tilde{E})\!\downarrow$, $x \mapsto \tilde{x}$ is a bijection. It is easy to verify that it is in fact an isomorphism of L^0-modules which defines a natural isomorphism between the functor GH and the identity functor.

Also, given a name E for a vector space in $V^{(\mathcal{F})}$, we consider the map $E\!\downarrow \to ((E\!\downarrow)^\sim)\!\downarrow$, $x \mapsto \tilde{x}$. By applying Theorem 1.4, we obtain a name for an isomorphism of vector spaces in $V^{(\mathcal{F})}$; that is, $[\![(E\!\downarrow)^\sim \cong E]\!] = \Omega$. Inspection shows that HG is naturally isomorphic to the functor identity. $\square$

Let us introduce some terminology:

If E is an L^0-module with the countable concatenation property:

- $S \subset E$ is said to be:
    1. L^0-*convex*: if $\eta x + (1-\eta)y \in S$ for all $x, y \in S$ and $\eta \in L^0$ with $0 \leq \eta \leq 1$;
    2. L^0-*absorbing*: if for every $x \in E$ there is $\eta \in L^0$, $\eta > 0$, such that $x \in \eta S$;
    3. L^0-*balanced*: if $\eta x \in S$ whenever $x \in S$ and $\eta \in L^0$ with $|\eta| \leq 1$.

- A non-empty subset $S \subset E$ is said to be *stable under countable concatenations*, or simply *stable*, if for every countable family $\{x_k\} \subset S$ and partition $\{A_k\} \in p(\Omega)$, it holds that $\sum 1_{A_k} x_k \in S$.

- A non-empty collection $\mathscr{C}$ of subsets of E is called *stable* if every $S \in \mathscr{C}$ is stable and for every countable family $\{S_k\} \subset \mathscr{C}$ and partition $\{A_k\} \in p(\Omega)$, it holds that $\sum 1_{A_k} S_k \in \mathscr{C}$.

Filipovic et al. [14] introduced the notion of locally L^0-convex module. Let us recall the following particular case, which was introduced in [41] and is a transcription in the present setting (via the equivalence of categories provided in [41, Theorem 1.2]) of the notion of *conditionally locally topological vector space* introduced in [11].

Definition 2.1. *A topological L^0-module $E[\mathcal{T}]$ with the countable concatenation property is said to be a* stable locally L^0-convex module *if there exists a neighborhood base $\mathcal{U}$ of $0 \in E$ such that:*

(i) $\mathcal{U}$ is a stable collection;

(ii) Every $U \in \mathcal{U}$ is L^0-convex, L^0-absorbing and L^0-balanced.

In this case, $\mathcal{T}$ is called a stable locally L^0-convex topology *on E.*

To our knowledge, the next result is new in literature; it describes the connection between names for locally convex spaces in $V^{(\mathcal{F})}$ and stable locally L^0-convex modules:

Theorem 2.2. *For fixed an underlying measure algebra $\mathcal{F}$, there is an equivalence of categories between the category of names for locally convex spaces and continuous linear functions, and the category of stable locally L^0-convex modules and continuous L^0-module morphims.*

Proof. We consider the same functor G as in Theorem 2.1, but restricted to the category of names for locally convex spaces and continuous linear functions in $V^{(\mathcal{F})}$.

Let $E[\mathcal{T}]$ be a name for a locally convex space in $V^{(\mathcal{F})}$; that is, $\mathcal{T}$ is a name for a locally convex topology in $V^{(\mathcal{F})}$. Let $\mathcal{U}$ be a name for a neighborhood base of the origin, such that

$$[\![\forall U \in \mathcal{U}(\text{``}U \text{ is convex''} \wedge \text{``}U \text{ is absorbing''} \wedge \text{``}U \text{ is balanced''})]\!] = \Omega.$$

We know that $E\!\downarrow$ is an L^0-module with the countable concatenation property. Let $\mathcal{U}\!\Downarrow := \{U\!\downarrow : U \in \mathcal{U}\!\downarrow\}$. Then every $U \in \mathcal{U}\!\Downarrow$ is stable due to the mixing principle. In the same way, again due to the mixing principle, it holds that $\sum 1_{A_k} U_k \in \mathcal{U}\!\Downarrow$ whenever $\{U_k\} \subset \mathcal{U}\!\Downarrow$ and $\{A_k\} \in p(\Omega)$. Therefore, $\mathcal{U}\!\Downarrow$ is a stable collection. Also, it is not difficult to show that each $U \in \mathcal{U}\!\Downarrow$ is L^0-convex, L^0-absorbing and L^0-balanced, and $\mathcal{U}\!\Downarrow$ is a neighborhood base of $0 \in E\!\downarrow$ of a topology $\mathscr{T}$. Therefore $E\!\downarrow[\mathscr{T}]$ is a stable locally L^0-convex module.

Now, let f, E, F be elements of $V^{(\mathcal{F})}$ such that E, F are names for locally convex spaces and f is a name for a continuous linear function between E and F in $V^{(\mathcal{F})}$; that is, $[\![f \in L(E, F)]\!] = \Omega$. Then we can consider $f\!\downarrow$, which is an L^0-module morphism between $E{\downarrow}$ and $F{\downarrow}$ such that $[\![f{\downarrow}(x) = f(x)]\!] = \Omega$ for all $x \in E{\downarrow}$. The function $f{\downarrow}$ is also continuous. To see this, consider $U{\downarrow}$ a basic neighborhood of 0 in F. Then, since f is continuous, there is a name for a basic neighborhood W in E such that $[\![W \subset f^{-1}(U)]\!] = \Omega$. Then $W{\downarrow} \subset f{\downarrow}^{-1}(U{\downarrow})$ and this proves that $f{\downarrow}$ is continuous.

Conversely, let $E[\mathcal{T}]$ be a stable locally L^0-convex module. We consider $\tilde{E}$ as in the proof of Theorem 2.1. Let $\mathscr{U}$ be a neighborhood base of $0 \in E$ as in Definition 2.1. For every $U \in \mathscr{U}$, we can define $\tilde{U}: \mathcal{D}_U \to \mathcal{F}$, where $\mathcal{D}_U := \{\tilde{x}: x \in U\}$ and $\tilde{U}(\tilde{x}) := \Omega$. Note that $\tilde{U}$ is an element of $V^{(\mathcal{F})}$. The map $x \mapsto \tilde{x}$ gives a bijection between U and $\tilde{U}{\downarrow}$. Injectivity was checked in the proof of Theorem 2.1. For surjectivity, if $w \in \tilde{U}{\downarrow}$ then $1 = [\![w \in \tilde{U}]\!] = \bigvee\{[\![w = \tilde{x}]\!]: x \in U\}$. Take a maximal family of pairs A_k, x_k such that $x_k \in U$, the A_k are nonzero and pairwise disjoint and $A_k \leq [\![w = \tilde{x}_k]\!]$. Using that U is stable[3], we obtained the desired preimage of w. Now, we consider $\tilde{\mathscr{U}}: \mathcal{D}_{\mathscr{U}} \to \mathcal{F}$ where $\mathcal{D}_{\mathscr{U}} := \{\tilde{U}: U \in \mathscr{U}\}$ and $\tilde{\mathscr{U}}(\tilde{U}) = \Omega$.

For each $U, V \in \mathscr{U}$, let

$$A_{U,V} := \bigvee \{A \in \mathcal{F}: 1_A U = 1_A V\}.$$

We claim that $A_{U,V} = [\![\tilde{U} = \tilde{V}]\!]$. Indeed,

$$[\![\tilde{U} = \tilde{V}]\!] = \bigwedge_{x \in U} [\![\tilde{x} \in \tilde{V}]\!] \wedge \bigwedge_{y \in V} [\![\tilde{y} \in \tilde{U}]\!] = \bigwedge_{x \in U} \bigvee_{y \in V} A_{x,y} \wedge \bigwedge_{y \in V} \bigvee_{x \in U} A_{x,y}.$$

For every $x \in U$, by using that V is stable similarly as above, one has that $1_{A_{U,V}} x = 1_{A_{U,V}} y_x$ for some $y_x \in V$. Then $A_{U,V} \leq A_{x,y_x}$ for all $x \in U$. Likewise, for every $y \in V$ we can find $x_y \in U$ with $A_{U,V} \leq A_{x_y,y}$. We conclude that $A_{U,V} \leq [\![\tilde{U} = \tilde{V}]\!]$.

By using again that V is stable, for each $x \in U$ one can find $y^x \in V$ such that $\bigvee_{y \in V} A_{x,y} = A_{x,y^x}$. Similarly, for every $y \in V$ one can pick up $x^y \in U$ with $\bigvee_{x \in U} A_{x,y} = A_{x^y,y}$. By using this in the expression above for $[\![\tilde{U} = \tilde{V}]\!]$, we get that always $[\![\tilde{U} = \tilde{V}]\!] \leq A_{x,y^x}$ and $[\![\tilde{U} = \tilde{V}]\!] \leq A_{y,x^y}$, and hence for every $x \in U$ we find that $1_{[\![\tilde{U}=\tilde{V}]\!]} x = 1_{[\![\tilde{U}=\tilde{V}]\!]} y^x$ and for every $y \in V$ we have that $1_{[\![\tilde{U}=\tilde{V}]\!]} x^y = 1_{[\![\tilde{U}=\tilde{V}]\!]} y$. It follows that $1_{[\![\tilde{U}=\tilde{V}]\!]} U = 1_{[\![\tilde{U}=\tilde{V}]\!]} V$, and therefore $[\![\tilde{U} = \tilde{V}]\!] \leq A_{U,V}$.

We proved before that the image of any $U \in \mathscr{U}$ via the mapping $x \mapsto \tilde{x}$ is $\tilde{U}{\downarrow}$. Now we claim that the assignment $U \mapsto \tilde{U}{\downarrow}$ is a bijection from $\mathscr{U}$ to $\tilde{\mathscr{U}}{\Downarrow}$. The assignment is injective because $x \mapsto \tilde{x}$ is injective. It is surjective because if $W{\downarrow} \in \tilde{\mathscr{U}}{\Downarrow}$ with $W \in \tilde{\mathscr{U}}{\downarrow}$, then $\Omega = [\![W \in \mathcal{U}]\!] = \bigvee_{U \in \mathscr{U}} [\![W = \tilde{U}]\!]$, we can take

---

[3] In the sequel, we will omit the details of this usage of the countable concatenation property, that follows always the same scheme through a maximal disjoint family of nonzero elements.

a maximal disjoint family $\{B_k\}$ such that $B_k \leq [\![\tilde{U}_k = W]\!]$, and using the mixing principle and that $\mathscr{U}$ is a stable collection, we get that $V := \sum 1_{A_k} U_k$ is a preimage for W.

Using that $\mathscr{U}$ is a neighborhood base of $0 \in E$, it can be verified that $\tilde{\mathscr{U}}$ is a name for a neighborhood base of the origin of a locally convex topology in $V^{(\mathcal{F})}$.

Now, suppose that $f : E_1[\mathscr{T}_1] \to E_2[\mathscr{T}_2]$ is a continuous L^0-module morphism. Then we know that $\tilde{f} : \tilde{E}_1 \to \tilde{E}_2$ is a name for a linear function. It is in fact a name for a continuous linear functional, because if basic neighborhoods satisfy $f(U) \subset V$, then $[\![\tilde{f}(\tilde{U}) \subset \tilde{V}]\!] = \Omega$.

Let $E[\mathscr{T}]$ be a stable locally L^0-convex module. We know that the map $E \to \tilde{E}\downarrow$, $x \mapsto \tilde{x}$ is an isomorphism of L^0-modules. Moreover, we proved before that $\mathscr{U}$ and $\tilde{\mathscr{U}}\downarrow$ are one-to-one relation via $x \mapsto \tilde{x}$. Consequently, $E \to \tilde{E}\downarrow$, $x \mapsto \tilde{x}$ is also a homeomorphism.

If $E[\mathcal{T}]$ is a name for a locally convex space, then $E[\mathcal{T}]\downarrow$ is a stable locally L^0-convex module and, by the argument above, the map $E[\mathcal{T}]\downarrow \to ((E[\mathcal{T}]\downarrow)^{\sim})\downarrow$ is an isomorphism of L^0-modules which is also a homeomorphism. Then, Theorem 1.4 provides a name for a homeomorphism between $E[\mathcal{T}]$ and $(E[\mathcal{T}]\downarrow)^{\sim}$. A close look shows that all these correspondences are natural transformations. □

The important conclusion of the result above is not the equivalence of categories itself, but that, for any stable locally L^0-convex module $E[\mathscr{T}]$, we can find a tailored name $\tilde{E}[\mathcal{T}]$ for a locally convex space such that $E[\mathscr{T}]$ is isomorphic to the descent $\tilde{E}[\mathcal{T}]\downarrow$. This will allow to reinterpret certain objects related to $E[\mathscr{T}]$ within the Boolean-valued universe.

Henceforth, we will fix a stable locally L^0-convex module $E[\mathscr{T}]$ and its corresponding name $\tilde{E}[\mathcal{T}]$ for a vector space given by the equivalence of categories above. Since $E[\mathscr{T}]$ and $\tilde{E}[\mathcal{T}]\downarrow$ are isomorphic stable locally L^0-convex modules and the properties that we will study are preserved by the isomorphism, for simplicity, we will assume w.l.o.g. that one recovers the initial L^0-module by means of the descent, that is, $E[\mathscr{T}] = \tilde{E}[\mathcal{T}]\downarrow$. For the same reasons, we will assume that $L^0 = \mathbb{R}^{(\mathcal{F})}\downarrow$. Let $\bar{L}^0$ denote the set of equivalence classes of Σ-measurable functions with values in $[-\infty, +\infty]$ and let $\bar{\mathbb{R}}^{(\mathcal{F})}$ be a name for the extended real numbers. Clearly, we can also assume $\bar{L}^0 = \bar{\mathbb{R}}^{(\mathcal{F})}\downarrow$.

Next, we will list different relevant objects related to $E[\mathscr{T}]$. All of them are either introduced in the existing literature of L^0-convex analysis [8, 14, 22, 41] or come from transcriptions in the modular setting of elements of conditional set theory [11, 29, 41]. Also, some of these concepts came earlier from Boolean valued analysis as we will explain later in Remark 2.2. Our purpose is to discuss their meanings within

Boolean-valued analysis, providing a bunch of 'building blocks' for the construction of module analogues of known statements of locally convex analysis, which will be also true due to the transfer principle:

- *Stable subsets*: For a given stable subset S of E we define the name $\tilde{S} : \mathcal{D}_S \to \mathcal{F}$ where $\mathcal{D}_S := \{\tilde{x} : x \in S\}$ and $\tilde{S}(\tilde{x}) := \Omega$. Then, $\tilde{S}$ is a name for a subset of $\tilde{E}$ with $\tilde{S}\downarrow = S$. Conversely, if S_0 is a name with $[\![\emptyset \neq S_0 \subset \tilde{E}]\!] = \Omega$, then $S_0\downarrow$ is a stable subset of E satisfying $[\![(S_0\downarrow)^\sim = S_0]\!] = \Omega$.

  Moreover, S is L^0-convex if, and only if, $[\![``\tilde{S} \text{ is convex''}]\!] = \Omega$; $S$ is $L^0$-absorbing if, and only if, $[\![``\tilde{S} \text{ is absorbing''}]\!] = \Omega$; and S is L^0-balanced if, and only if, $[\![``\tilde{S} \text{ is balanced''}]\!] = \Omega$.

- *Stable collections of subsets*: For a given stable collection $\mathscr{C}$ of subsets of E we define the name $\tilde{\mathscr{C}} : \mathcal{D}_\mathscr{C} \to \mathcal{F}$ with $\mathcal{D}_\mathscr{C} := \left\{\tilde{S} : S \in \mathscr{C}\right\}$ and $\tilde{\mathscr{C}}(\tilde{S}) := \Omega$.

  Conversely, if $\mathcal{C}$ is a name for a non-empty collection of non-empty subsets of $\tilde{E}$, we define $\mathcal{C}\Downarrow := \{S\downarrow : S \in \mathcal{C}\downarrow\}$, which is a stable collection of subsets of E.

  Moreover, if $\mathscr{C}$ is a stable collection of subsets of E, one has that $\mathscr{C} = \tilde{\mathscr{C}}\Downarrow$; and if $\mathcal{C}$ is a name for a non-empty collection of non-empty subsets of $\tilde{E}$, then we have $[\![(\mathcal{C}\Downarrow)^\sim = \mathcal{C}]\!] = \Omega$.

  In particular, we know from the proof of Theorem 2.2 that, if $\mathscr{U}$ is a neighborhood base of $0 \in E$ as in Definition 2.1, then $\tilde{\mathscr{U}}$ is a name for a neighborhood base of the origin and

  $$\tilde{\mathscr{U}}\Downarrow = \mathscr{U}. \qquad (2)$$

  Let $\mathscr{C}$ be a stable collection of subsets of E. We denote by $(\cup \tilde{\mathscr{C}})_\mathcal{F}$, $(\cap \tilde{\mathscr{C}})_\mathcal{F}$ names for the union and intersection of $\tilde{\mathscr{C}}$, respectively. Then, it holds that

  $$\cup \mathscr{C} = (\cup \tilde{\mathscr{C}})_\mathcal{F}\downarrow \quad \text{and} \quad \cap \mathscr{C} = (\cap \tilde{\mathscr{C}})_\mathcal{F}\downarrow. \qquad (3)$$

- *Stable open subsets*: Let $O \subset E$ be stable. It follows from the relation (2), that O is open if and only if, $[\![``\tilde{O} \text{ is open''}]\!] = \Omega$.

  Moreover, for any stable subset S of E, one has $[\![(\text{int}(S))^\sim = \text{int}(\tilde{S})]\!] = \Omega$.

- *Stable closed subsets*: Let $C \subset E$ be stable. Then relation (2) allows to show that C is closed if, and only if, $[\![``\tilde{C} \text{ is closed''}]\!] = \Omega$.

  Moreover, for any stable subset S of E, it holds $[\![(\text{cl}(S))^\sim = \text{cl}(\tilde{S})]\!] = \Omega$.

- *Stable filters*: A *stable filter* on E is a filter $\mathscr{F}$, which admits a filter base $\mathscr{B}$ which is a stable collection of subsets of E.

  If $\mathscr{F}$ is a stable filter with base $\mathscr{B}$, where $\mathscr{B}$ is a stable collection, then it can be verified that $[\![``\tilde{\mathscr{B}} \text{ is a filter base}"]\!] = \Omega$. Conversely, if $\mathcal{B}$ is a name for a filter base, then $\mathcal{B}{\Downarrow}$ is the base of some stable filter.

- *Stably compact subsets*: A stable subset S of E is said to be *stably compact*, if every stable filter base $\mathscr{B}$ on S has a cluster point in S.

  We have that $K \subset E$ is stably compact if, and only if, $[\![``\tilde{K} \text{ is compact}"]\!] = \Omega$. This follows because, due to (2), a stable filter base $\mathscr{B}$ has a cluster point if, and only if, $[\![``\tilde{\mathscr{B}} \text{ has a cluster point}"]\!] = \Omega$; and a name for a filter base $\mathcal{B}$ satisfies $[\![``\mathcal{B} \text{ has a cluster point}"]\!] = \Omega$ if, and only if, $\mathcal{B}{\Downarrow}$ has a cluster point.

  We will say that a stable subset K of E is *relatively stably compact*, if $\mathrm{cl}(K)$ is stably compact. Notice that $K \subset E$ is relatively stably compact if, and only if, $[\![``\tilde{K} \text{ is relatively compact}"]\!] = \Omega$.

  Just mention that it was proven in [29, Proposition 5.2] that, when the underlying probability space is atomless, then any Hausdorff stable locally L^0-convex module is anti-compact; that is, the only compact subsets are the finite subsets. This means that the conventional compactness is not interesting because does not allow to establish any meaningful theorem. On the other hand, the transfer principle brings a huge range of theorems involving stable compactness, which shows that stable compactness is by far much richer than classical compactness.

- *Stable functions*: Suppose that $S_1, S_2 \subset E$ are stable. A function $f: S_1 \to S_2$ is said to be *stable* if $f(\sum 1_{A_k} x_k) = \sum 1_{A_k} f(x_k)$ for all $\{x_k\} \subset S_1$ and $\{A_k\} \in p(\Omega)$.

  Since $[\![x = y]\!] \leq [\![f(x) = f(y)]\!]$ for all $x, y \in S_1$, Theorem 1.4 yields a name for a function $\tilde{f}$ between $\tilde{S}_1$ and $\tilde{S}_2$ such that $\tilde{f}{\downarrow} = f$. Moreover, it can be verified that f is continuous if, and only if, $[\![``\tilde{f} \text{ is continuous}"]\!] = \Omega$.

  A function $f: E \to \bar{L}^0$ has the *local property*, if $1_A f(x) = 1_A f(1_A x)$ for all $A \in \mathcal{F}$. If f has the local property, once again, Theorem 1.4 allows to define a name $\tilde{f}$ for a function from $\tilde{E}$ to $\bar{\mathbb{R}}^{(\mathcal{F})}$ so that $\tilde{f}{\downarrow} = f$.

  A function $f: E \to \bar{L}^0$ is:

  1. L^0-*convex*: if $f(\eta x + (1-\eta)y) \leq \eta f(x) + (1-\eta)f(y)$ for all $\eta \in L^0$ with $0 \leq \eta \leq 1$ and $x, y \in E$;

2. *proper*: if $f(x) > -\infty$ for all $x \in E$ and there is some $x_0 \in E$ with $f(x_0) \in L^0$;

3. *lower semi-continuous*: if the sublevel $V_f(\eta) := \{x \in E \colon f(x) \leq \eta\}$ is closed for every $\eta \in \bar{L}^0$.

The *domain* of f is defined by $\text{dom}(f) := \{x \in E \colon f(x) \in L^0\}$.

When f has the local property, one has that f is L^0-convex if, and only if $[\![\text{``}\tilde{f} \text{ is convex''} ]\!] = \Omega$; and $f$ is proper if, and only if, $[\![ \text{``}\tilde{f} \text{ is proper''}]\!] = \Omega$. Further, it can be verified that $V_f(\eta)$ is a stable set for each $\eta \in L^0$ such that $V_f(\eta) \neq \emptyset$. Thus, we can conclude that, f is lower semi-continuous if, and only if, $[\![\text{``}\tilde{f} \text{ is lower semi-continuous''}]\!] = \Omega$.

Finally, just mention that if f is L^0-convex, then f has automatically the local property (see [14, Theorem 3.2]), hence in the statements we will not have to require the latter property whenever f is L^0-convex.

- *Topological dual*: We consider $E^* := E^*[\mathcal{T}]$ the set of all continuous L^0-module morphisms $\mu \colon E \to L^0$. Then, we can consider the name $F \colon \mathcal{D}_{E^*} \to \mathcal{F}$ with $\mathcal{D}_{E^*} := \{\tilde{\mu} \colon \mu \in E^*\}$ and $F(\tilde{\mu}) := \Omega$. Then we have $[\![F = \tilde{E}^*[\mathcal{T}]]\!] = \Omega$, where $\tilde{E}^*[\mathcal{T}]$ denotes a name for the topological dual of $\tilde{E}[\mathcal{T}]$. Moreover, note that we have the relation $E^*[\mathcal{T}] = \{\mu\!\downarrow \colon \mu \in \tilde{E}^*[\mathcal{T}]\!\downarrow\}$.

- *Stable sequences*: A net $\chi = \{x_\mathfrak{n}\}_{\mathfrak{n} \in L^0(\mathbb{N})}$ in E is called a *stable sequence* whenever $x_\mathfrak{n} = \sum_{k \in \mathbb{N}} 1_{\{\mathfrak{n}=k\}} x_k$ for all $\mathfrak{n} \in L^0(\mathbb{N})$. Then, again, Theorem 1.4 provides us with a name $\tilde{\chi}$ for a function from $\mathbb{N}^{(\mathcal{F})}$ to $\tilde{E}$; that is, a name for a sequence in $\tilde{E}$. Besides, we have $\tilde{\chi}\!\downarrow = \chi$.

Bearing in mind relation (2), it can be verified that the net χ converges to $x \in E$ if, and only if, $[\![\text{``}\tilde{\chi} \text{ converges to } \tilde{x} \in \tilde{E}\text{''}]\!] = \Omega$.

A stable sequence $\kappa = \{y_\mathfrak{n}\}_{\mathfrak{n} \in L^0(\mathbb{N})} \subset E$ is called a *stable subsequence* of $\chi = \{x_\mathfrak{n}\}_{\mathfrak{n} \in L^0(\mathbb{N})}$ if there exists a stable sequence $\{\mathfrak{n}_\mathfrak{m}\}_{\mathfrak{m} \in L^0(\mathbb{N})} \subset L^0(\mathbb{N})$, with $\mathfrak{n}_\mathfrak{m} < \mathfrak{n}_{\mathfrak{m}'}$ whenever $\mathfrak{m} < \mathfrak{m}'$, such that $y_\mathfrak{m} = x_{\mathfrak{n}_\mathfrak{m}}$ for all $\mathfrak{m} \in L^0(\mathbb{N})$. In this case, it can be verified that $[\![\text{``}\tilde{\kappa} \text{ is a subsequence of } \tilde{\chi}\text{''}]\!] = \Omega$.

- L^0-*norms*: An L^0-*norm* on E is a function $\|\cdot\| \colon E \to L^0$ such that for all $x, y \in E$ and $\eta \in L^0$ satisfies:

  (i) $\|x\| \geq 0$, with $\|x\| = 0$ if and only if $x = 0$;
  (ii) $\|\eta x\| = |\eta|\|x\|$;
  (iii) $\|x + y\| \leq \|x\| + \|y\|$.

In this case $(E, \|\cdot\|)$ is called an L^0-*normed module*.

The collection of sets $B_\varepsilon := \{x \in E \colon \|x\| < \varepsilon\}$, where $\varepsilon \in L^0$ with $\varepsilon > 0$, is a neighborhood base of $0 \in E$ for a stable locally convex topology $\mathcal{T}$.

Due to (ii), $\|\cdot\|$ has the local property. Then we can define $\|\cdot\|^\sim$, which is a name for a norm on $\tilde{E}$. Furthermore, one has that $[\![``\|\cdot\|^\sim \text{ induces } \mathcal{T}"]\!] = \Omega$.

- *Stable completeness*: Suppose that $(E, \|\cdot\|)$ is an L^0-normed module. A stable sequence $\{x_\mathfrak{n}\}_{\mathfrak{n} \in L^0(\mathbb{N})} \subset E$ is said to be *Cauchy* if for every $\varepsilon \in L^0$, $\varepsilon > 0$, there exists $\mathfrak{n}_0 \in L^0(\mathbb{N})$ such that $\|x_\mathfrak{n} - x_{\mathfrak{n}'}\| \leq \varepsilon$ for all $\mathfrak{n}, \mathfrak{n}' \in L^0(\mathbb{N})$ with $\mathfrak{n}, \mathfrak{n}' \geq \mathfrak{n}_0$. We say that $(E, \|\cdot\|)$ is *stably complete*, if every Cauchy stable sequence is convergent.

  Then, one has that $(E, \|\cdot\|)$ is stably complete if, and only if,

  $$[\![``(\tilde{E}, \|\cdot\|^\sim) \text{ is a Banach space}"]\!] = \Omega.$$

- *Stable weak topologies*: The collection of sets

  $$U_{\{F_k\}, \{A_k\}, \varepsilon} := \{x \in E \colon \sum 1_{A_k} \operatorname*{ess.\,sup}_{\mu \in F_k} |\mu(x)| < \varepsilon\},$$

  where $\{A_k\} \in p(\Omega)$, $\{F_k\}$ is a countable collection of non-empty finite subsets of E^* and $\varepsilon \in L^0$ with $\varepsilon > 0$, is a neighborhood base of $0 \in E$ for a stable locally L^0-convex topology, which is called the *stable weak topology* and is denoted by $\sigma_s(E, E^*)$.

  Then the corresponding name for a locally convex topology provided by the equivalence of categories in Theorem 2.2 is precisely a name for the weak topology of $\tilde{E}[\mathcal{T}]$.

  Analogously, we can define the *stable weak-$*$ topology* $\sigma_s(E^*, E)$.

Remark 2.2. *As mentioned previously, some of the notions listed above were introduced earlier in literature of Boolean-valued analysis under different nomenclature. Stable compactness was formulated in [29] as a transcription of the notion of conditional compactness introduced in [11]. However, stable compactness was first time studied by Kusraev [33] giving rise to the notion of* cyclic compact set. *Later, the notion of* mix-compactness *was introduced by Gutman and Lisovskaya [26]. It turns out that* cyclic compactness *and* mix-compactness *are equivalent notions (see [38, Theorem 2.12.C.5]). These types of compactness have been fruitfully exploited, see for instance results in [35, Sections 1.3 and 1.4], [36, Section 8.5] and the analogues of the boundedness and uniform boundedness principles obtained in [26].*

The notion of stable completeness *is a transcription of the notion of* conditional completeness *introduced in [11]. Descents of complete spaces and Banach spaces were studied earlier by Kusraev [34], originating the notion of Banach-Kantorovich space, which are descents of real Banach spaces as proven in [34] (for further details see [36, Section 8.3] and [39, Section 5.4]).*

Finally, the stably weak and stably weak-∗ topologies *defined above are transcriptions of the notion of* conditional initial topology *induced by* conditional dual pairs *introduced in [11] applied to the pairing* $\langle E, E^* \rangle$. *Descents of dual pairs, which give rise to dual systems with* $\mathbb{R}^{(\mathcal{A})}\!\!\downarrow$*-bilinear forms, were studied earlier in [35]. In particular, [35, Theorem 3.3.10(b)] is related to Theorem 2.2. This type of pairings covers the* stably weak and stably weak-∗ topologies *defined above in the more general framework of modules over universally complete vector lattices.*

Once we have the 'building blocks', let us see some examples to exhibit how they can be assembled to give rise to different statements. Of course, this list is not exhaustive and we can create many other pieces for our puzzle.

Let us start by the main theorems of [14]. For instance, we will see that Theorems 2.8, 3.7 and 3.8 in [14] follow from the transfer principle of Boolean-valued models. Although, these results apply to the more general structure of locally L^0-convex module, they are proved under the assumption that the locally L^0-convex topology is induced by a family of L^0-seminorms (see [14, Definition 2.3]), which is closed under finite suprema and with the so-called *countable concatenation property*.[4] It is not difficult to prove that these properties amount to the existence of a neighborhood base $\mathscr{U}$ of $0 \in E$ as in Definition 2.1. Thus, these results implicitly apply to stable locally L^0-convex modules.

We have the following:[5]

Theorem 2.3. *Let* $E[\mathscr{T}]$ *be a stable locally* L^0*-convex module, and suppose that* S_1, S_2 *are stable* L^0*-convex subsets of* E *with* S_1 *stably compact and* S_2 *closed. If*

$$1_A S_1 \cap 1_A S_2 = \emptyset \quad \text{for all } A \in \mathcal{F} \text{ with } A > \emptyset,$$

then there exists a continuous L^0*-module morphism* $\mu : E \to L^0$ *and* $\varepsilon \in L^0$, $\varepsilon > 0$, *such that*

$$\mu(x) > \mu(y) + \varepsilon \quad \text{for all } x \in S_1, y \in S_2.$$

---

[4]Here, we refer to the countable concatenation property for families of L^0-seminorms, which has not to be missed up with the algebraic countable concatenation property introduced at the beginning of the section.

[5]This statement is more general than [14, Theorem 2.8] as the latter applies to the particular case in which S_1 is a singleton. This statement is also a transcription of [11, Theorem 5.5(ii)] as shown in [29].

Remember the classical separation theorem: If C, K are non-empty convex subsets with C closed, K compact, and C and K have empty intersection, then there is a lineal functional that separates C from K. What we have above is just a reformulation of the statement [[separation theorem]] $= \Omega$, so no proof needed.

In literature, there is a long tradition of studying *conjugates* and *subgradients* of functions taking values in different types of ordered lattice rings such as Kantorovich spaces (see eg [37, chap. 4]), and addressing versions of the classical Fenchel-Moreau theorem in these settings (see eg [37, Theorem 4.3.10(1)] and [35, Theorem 1.2.11]). More recently, Filipovic et al [14] worked with versions of conjugates and subgradients for $\bar{L}^0$-valued functionals defined on L^0-modules. Namely, the *conjugate* of a function $f : E \to \bar{L}^0$ is defined by

$$f^* : E^* \to \bar{L}^0, \quad f^*(\mu) := \operatorname*{ess.\,sup}_{x \in E} (\mu(x) - f(x)),$$

and its *biconjugate* is defined by

$$f^{**} : E \to \bar{L}^0, \quad f^{**}(x) := \operatorname*{ess.\,sup}_{\mu \in E^*} (\mu(x) - f^*(\mu)).$$

An element $\mu \in E^*$ is a *subgradient* of $f : E \to \bar{L}^0$ at $x_0 \in \operatorname{dom}(f)$, if

$$\mu(x - x_0) \leq f(x) - f(x_0) \quad \text{for all } x \in E.$$

The set $\partial f(x_0)$ stands for the set of all subgradients of f at x_0.

The notion of L^0-barrel was introduced in [14]. Namely, a subset S of E is an L^0-*barrel* if it is L^0-convex, L^0-absorbing, L^0-balanced and closed. We will say that a topological L^0-module is *stably barreled* if every stable L^0-barrel is a neighborhood of $0 \in E$.

[14, Theorem 3.8] is a module analogue of the classical Fenchel-Moreau theorem. We have the following statement, which does not need a proof as it follows from its conventional version [3, Theorem 2.22] by means of the transfer principle by just noting that $[\![\tilde{f}^{**} = (f^{**})^\sim]\!] = \Omega$:

Theorem 2.4. *Let $E[\mathscr{T}]$ be a stable locally L^0-convex module and let $f : E \to \bar{L}^0$ be proper lower semi-continuous and L^0-convex. Then $f^{**} = f$.*

Concerning subgradients, we have the following result, which is a generalization of [14, Theorem 3.7] and follows from the transfer principle applied to the so-called Fenchel-Rockafellar theorem, see eg [6, Theorem 1]:

Theorem 2.5. *Let $E[\mathcal{T}]$ be a stable locally L^0-convex module which is stably barreled. Let $f : E \to \bar{L}^0$ be a proper lower semicontinuous L^0-convex function. Then,*

$$\partial f(x) \neq \emptyset \quad \text{for all } x \in \text{int}(\text{dom}(f)).$$

The notion of L^0-*barreled* topological L^0-module was introduced in [14]; namely, $E[\mathcal{T}]$ is L^0-barreled if every L^0-barrel is a neighborhood of $0 \in E$. Thus, the notion of stably barreled topological L^0-module is more general. This was already pointed out in [25], where the statement above was already proven by using the techniques introduced in [14].

Let us see more examples of application of our method. Next, we provide module analogues of the classical James' compactness theorem and also a version of the important Brouwer fixed point theorem.

The following statement is a modular version of a non-linear variation of classical James' compactness theorem, which plays an important role in the study of robust representation of risk measures (see eg [31, Theorem A.1] and [40, Theorem 2]). The statement we present follows from the transfer principle applied to its most general version [42, Theorem 2.4].

Theorem 2.6. *Let $(E, \|\cdot\|)$ be a stably complete L^0-normed module and let $f : E \to \bar{L}^0$ be a proper function with the local property. If for every $\mu \in E^*$ there is an x_0 such that $\mu(x_0) - f(x_0) = f^*(\mu)$, then the set $V_f(\eta) = \{x \in E : f(x) \leq \eta\}$ is relatively stably compact w.r.t. $\sigma_s(E, E^*)$ for every $\eta \in L^0$ with $V_f(\eta) \neq \emptyset$.*

Of course, we also have a modular version of the celebrated James' compactness theorem, which is a consequence of the statement above, and also follows from the transfer principle applied to its classical version:

Theorem 2.7. *Let $(E, \|\cdot\|)$ be a stably complete L^0-normed module and let $K \subset E$ be stable, L^0-convex and L^0-norm bounded (i.e. $\text{ess.sup}_{x \in K} \|x\| < \infty$). Then, K is stably compact w.r.t. $\sigma_s(E, E^*)$ if, and only if, each $\mu \in E^*$ there exists $x_0 \in K$ such that $\mu(x_0) := \text{ess.sup}_{x \in K} \mu(x)$.*

A version of the Brouwer Fixed Point Theorem for $(L^0)^d$ was provided in [12], which corresponds to the finite-dimensional case in our context. Next, we will state a Brouwer fixed point theorem for Hausdorff[6] stable locally L^0-convex modules, which is a direct application of the transfer principle to the so-called Schauder-Tychonov Theorem.

---

[6] In view of (2) and (3), It is not difficult to show that $E[\mathcal{T}]$ is Hausdorff if, and only if, $\bigcap \mathscr{U} = \{0\}$, if, and only if, $[\![\bigcap \tilde{\mathscr{U}} = \{0\}]\!] = \Omega$ and if, and only if, $[\![\text{``}\tilde{E}[\mathcal{T}] \text{ is Hasdorff''}]\!] = \Omega$

Theorem 2.8. *If S is an L^0-convex stably compact subset of a Hausdorff stable locally L^0-convex module $E[\mathcal{T}]$, then any stable continuous function $f : S \to S$ has a fixed point in S.*

Obviously, all these Theorems are just some examples: we can state a version of any theorem T on locally convex spaces and it immediately renders a version for locally L^0-modules of the form $[\![T]\!] = \Omega$.

Finally, let us turn to the discussion of an example of financial application:

The notion of *convex risk measure* was independently introduced by Föllmer and Schied [16] and Fritelli and Gianin [17] as an extension of the notion of *coherent risk measure* introduced in Artzner et al. [1]. Let $\mathscr{X}$ be an ordered vector space with $\mathbb{R} \subset \mathscr{X}$ which models all the financial positions in a financial market. A convex risk measure is a proper convex function $\rho : \mathscr{X} \to \bar{\mathbb{R}}$ which satisfies the following conditions for all $x, y \in \mathscr{X}$:

- *Monotonicity*: if $x \leq y$, then $\rho(y) \leq \rho(x)$;

- *Cash invariance*: $\rho(x + r) = \rho(x) - r$, for all $r \in \mathbb{R}$.

Now, suppose that $(\Omega, \Sigma, \mathbb{P})$ models the market events at some future date $t > 0$. In this case, from a modelling point of view, the risk of any financial position is contingent on the information encoded in the measure algebra $\mathcal{F}$. For instance, the risk measurably depends on the decisions taken by the risk manager in virtue of the market eventualities arisen at time t. Therefore, in this case, the different financial positions can be modelled by an ordered L^0-module $\mathscr{X}$ with $L^0 \subset \mathscr{X}$. Filipovic et al. [15] proposed the following definition: a *conditional convex risk measure* is a proper L^0-convex function $\rho : \mathscr{X} \to \bar{L}^0$ which satisfies the following conditions for all $x, y \in \mathscr{X}$:

- *Monotonicity*: if $x \leq y$, then $\rho(y) \leq \rho(x)$;

- *Cash invariance*: $\rho(x + \eta) = \rho(x) - \eta$, for all $\eta \in L^0$.

Since a conditional convex risk measure $\rho : \mathscr{X} \to \bar{L}^0$ is L^0-convex, in particular, it has the local property, and Theorem 1.4 defines a name for a function $\tilde{\rho}$ from $\tilde{\mathscr{X}}$ to $\bar{\mathbb{R}}^{(\mathcal{F})}$. Moreover, it can be verified that

$$[\![\text{``}\tilde{\rho} \text{ is convex, monotone and cash-invariant''}]\!] = \Omega.$$

We conclude that a conditional convex risk measure ρ can be identified with a name $\tilde{\rho}$ for a convex risk measure within $V^{(\mathcal{F})}$. Thus, the machinery of Boolean-value models and its transfer principle can be applied.

From a modelling point of view, we have that, in the same manner the available market information is encoded in $\mathcal{F}$, the financial strategy followed by the risk manager in order to maximize or hedge future payments can be analytically expressed in terms of the formal language $\mathcal{L}^{(\mathcal{F})}$, which consistently depends on the information of $\mathcal{F}$. Thus the Boolean-valued analysis makes available to us a powerful technology to incorporate trading rules based on equilibrium prices or risk constraints in the mathematical analysis of certain problems of mathematical finance involving a multi-period setting.

3 A precise connection between Conditional set theory and Boolean-valued models

In [11] it was introduced the notion of conditional set:

Definition 3.1. *[11, Definition 2.1] Let X be a non-empty set and let $\mathcal{A}$ be a complete Boolean algebra. A conditional set of X and $\mathcal{A}$ is a set $\mathbf{X}$ such that there exists a surjection $(x, a) \mapsto x|a$ from $X \times \mathcal{A}$ onto $\mathbf{X}$ satisfying:*

(C1) if $x, y \in X$ and $a, b \in \mathcal{A}$ with $x|a = y|b$, then $a = b$;

(C2) (Consistency) if $x, y \in X$ and $a, b \in \mathcal{A}$ with $a \leq b$, then $x|b = y|b$ implies $x|a = y|a$;

(C3) (Stability) if $\{a_i\}_{i \in I} \in p(1)$ and $\{x_i\}_{i \in I} \subset X$, then there exists a unique $x \in X$ such that $x|a_i = x_i|a_i$ for all $i \in I$.

The unique element $x \in X$ provided by C3, is called the *concatenation* of the family $\{x_i\}$ along the partition $\{a_i\}$, and is denoted by $\sum x_i|a_i$.

Let $\mathbf{X}, \mathbf{Y}$ be conditional sets. According to [11, Definition 2.1] a function $f : X \to Y$ is said to be *stable* if

$$f\left(\sum x_i|a_i\right) = \sum f(x_i)|a_i, \quad \text{for } \{a_i\} \in p(1), \{x_i\} \subset X.$$

If $f : X \to Y$ is a stable function, it is simply to verify that

$$\mathbf{G_f} := \{(x|a, f(x)|a) \ x \in X, a \in \mathcal{A}\}$$

is a conditional set of the graph of f and $\mathcal{A}$. $\mathbf{G_f}$ is called the *conditional graph of a conditional function* $\mathbf{f} : \mathbf{X} \to \mathbf{Y}$ (see [11, Definition 2.1]).

A conditional function $\mathbf{f} : \mathbf{X} \to \mathbf{Y}$ is *conditionally injective* if $x|a \neq x'|a$ for all $a > 0$ implies that $f(x)|a \neq f(x')|a$ for all $a > 0$; it is *conditionally surjective* whenever f is surjective; and it is a *conditional bijection* if it is conditionally injective and surjective.

Then the following result gives the relation between conditional sets of the universe V and the boolean-valued universe $V^{(\mathcal{A})}$.

Theorem 3.1. *For fixed a Boolean algebra $\mathcal{A}$, there is an equivalence of categories between the category of conditional sets of $\mathcal{A}$ whose morphisms are conditional functions, and the category of elements x of $V^{(\mathcal{A})}$ such that $[\![x \neq \emptyset]\!] = 1$ whose morphisms are names for functions in $V^{(\mathcal{A})}$.*

Proof. First, suppose that x is an element of $V^{(\mathcal{A})}$ with $[\![x \neq \emptyset]\!] = 1$. Then we consider the equivalence relation on $(x{\downarrow}) \times \mathcal{A}$ given by

$$(u, a) \sim (v, b) \quad \text{whenever } a = b,\ [\![u = v]\!] \geq a.$$

Let us denote by $x|a$ the class of (x, a) and let $\mathbf{x}{\downarrow}$ be the corresponding quotient set. Then $\mathbf{x}{\downarrow}$ is a conditional set of $x{\downarrow}$ and $\mathcal{A}$. Indeed, (C1) and (C2) from Definition 3.1 are trivially satisfied. Further, (C3) follows from the mixing principle (Theorem 1.3).

Suppose that f, X, Y are in $V^{(\mathcal{A})}$ and $[\![f : X \to Y]\!] = 1$. Then $f{\downarrow}$ is a function from $X{\downarrow}$ to $Y{\downarrow}$ such that $[\![f{\downarrow}(u) = f(u)]\!] = 1$ for all $u \in X{\downarrow}$. Now, we claim that $f{\downarrow} : \mathbf{X}{\downarrow} \to \mathbf{Y}{\downarrow}$ is a stable function of the conditional sets $\mathbf{X}{\downarrow}, \mathbf{Y}{\downarrow}$. Indeed, given $\{a_i\} \in p(1)$ and $\{u_i\} \subset X$ we take $u := \sum u_i | a_i \in x{\downarrow}$. We have that $[\![f{\downarrow}(u) = f{\downarrow}(u_i)]\!] = [\![f(u) = f(u_i)]\!] \geq [\![u = u_i]\!] \geq a_i$, and thus $f{\downarrow}(u)|a_i = f{\downarrow}(u_i)|a_i$ each i. This shows that $f{\downarrow}(u) = \sum f{\downarrow}(u_i)|a_i$, hence $f{\downarrow}$ is stable. We can consider the corresponding conditional function $\mathbf{f}{\downarrow}$.

Thereby, we define the functor $G(x) := \mathbf{x}{\downarrow}$, $G(f) := \mathbf{f}{\downarrow}$. Let us construct the inverse functor. Suppose now that $\mathbf{X}$ is a conditional set of X and $\mathcal{A}$. We will construct from X an element $\tilde{X}$ of $V^{(\mathcal{A})}$. Indeed, for every $u \in X$ we define $\tilde{u} : \mathcal{D}_u \to \mathcal{A}$ where $\mathcal{D}_u := \{\check{v} : v \in X\}$, and $\tilde{u}(\check{v}) = a_{u,v}$ with $a_{u,v} := \bigvee \{b \in \mathcal{A} : u|a = v|a\}$ for each $v \in X$. Notice that $u|a_{u,v} = v|a_{u,v}$. The proof is similar to others we have done before: take a maximal disjoint family of elements b such that $u|b = v|b$ and then use uniqueness of (C3) of Definition 3.1.

Let $\tilde{X} : \mathcal{D} \to \mathcal{A}$ where

$$\mathcal{D} = \{\tilde{u} : u \in X\} \text{ and } \tilde{X}(\tilde{u}) = 1 \text{ for each } u \in X.$$

One has that $\tilde{X}$ is an element of $V^{(\mathcal{A})}$. Moreover, we claim that $[\![\tilde{u} = \tilde{v}]\!] = a_{u,v}$

for all $u, v \in X$. Indeed,

$$[\![\bar{u} = \bar{v}]\!] = \bigwedge_{t \in X} \left(a_{u,t} \Rightarrow [\![\check{t} \in \bar{v}]\!]\right) \wedge \bigwedge_{s \in X} \left(a_{v,s} \Rightarrow [\![\check{s} \in \bar{u}]\!]\right). \tag{4}$$

In addition,

$$[\![\check{t} \in \bar{v}]\!] = \bigvee_{w \in X} a_{v,w} \wedge [\![\check{t} = \check{w}]\!] = a_{v,t},$$

because $[\![\check{t} = \check{w}]\!] = 1$ if $t = w$, $[\![\check{t} = \check{w}]\!] = 0$ otherwise. Similarly, one has $[\![\check{s} \in \bar{u}]\!] = a_{u,s}$.

Therefore, replacing in (4), one has

$$[\![\bar{u} = \bar{v}]\!] = \bigwedge_{t \in X} (a^c_{u,t} \vee a_{v,t}) \wedge \bigwedge_{s \in X} (a^c_{v,s} \vee a_{u,s}).$$

By considering above $t = u$ and $s = v$, we obtain

$$[\![\bar{u} = \bar{v}]\!] \leq a_{u,v}$$

For the converse inequality, suppose by contradiction that $0 < a := a_{u,v} \wedge (a^c_{u,t} \vee a_{v,t})^c$ for some t. Since $a \leq a_{u,v}, a_{u,t}$ one has $v|a = u|a = t|a$ by (C2). But $a \leq a^c_{v,t}$ implies that $v|a \neq t|a$, which is a contradiction.

For any $u \in X$, let $\tilde{u}$ denote the canonical representative of $\bar{u}$ in $\overline{V}^{(\mathcal{A})}$. We claim that the map $X \to \tilde{X}\downarrow$ given by $u \to \tilde{u}$ is one-to-one. Indeed, if $\tilde{u} = \tilde{v}$, then $1 = [\![\bar{u} = \bar{v}]\!] = a_{u,v}$, hence $u = v$. On the other hand, given $w \in \tilde{X}\downarrow$, one has

$$1 = [\![w \in \tilde{X}]\!] = \bigvee_{u \in X} [\![\bar{u} = w]\!].$$

As we have done before, we can find by maximality a partition $\{a_i\} \in p(1)$ so that $a_i \leq [\![\bar{u}_i = w]\!]$ for some $u_i \in X$, each i. Then, (C3) of Definition 3.1 provides us with $u \in X$ such that $u|a_i = u_i|a_i$ for all i. We have that $[\![\bar{u} = \bar{u}_i]\!] = a_{u,u_i} \geq a_i$. Hence $a_i \leq [\![\bar{u} = \bar{u}_i]\!] \wedge [\![\bar{u}_i = w]\!]$ for all i, and so $[\![w = \bar{u}]\!] = 1$ and thus $\tilde{u} = w$.

Now suppose that $\mathbf{f} : \mathbf{X} \to \mathbf{Y}$ is a conditional function between the conditional sets $\mathbf{X}, \mathbf{Y}$. We consider the stable function $f : X \to Y$. Let $g : \tilde{X}\downarrow \to \tilde{Y}\downarrow$ be with $g(\tilde{x}) := (f(x))^\sim$, which is well defined since the map $x \mapsto \tilde{x}$ is one-to-one. Given $x, y \in X$, using that f is stable we can show that $[\![\tilde{x} = \tilde{y}]\!] = a_{x,y} \leq a_{f(x),f(y)} = [\![g(\tilde{x}) = g(\tilde{y})]\!]$. Due to Theorem 1.4, we can find $\tilde{f}$ in $V^{(\mathcal{A})}$ with $[\![\tilde{f} : \tilde{X} \to \tilde{Y}]\!] = 1$ and such that $[\![g(\tilde{x}) = \tilde{f}(x)]\!] = 1$ for all $x \in X$.

Thereby, we take the functor $H(\mathbf{X}) := \tilde{X}$ and $H(\mathbf{f}) := \tilde{f}$. We will show that G and H are inverse equivalences. Suppose that x is an element of $V^{(\mathcal{A})}$ with

$[\![x \neq \emptyset]\!] = 1$. We consider the map $x{\downarrow} \to ((x{\downarrow})^{\sim}){\downarrow}$, $u \mapsto \tilde{u}$. Due to Theorem 1.4 it defines a name for a bijection between x and $(x{\downarrow})^{\sim}$. It follows by inspection that there is a natural isomorphism between HG and the identity functor.

If **X** is a conditional set, then we can consider the mapping $\mathbf{X} \mapsto (\tilde{X}){\downarrow}$, $x \mapsto \tilde{x}$. This is a stable bijection, which defines a conditional bijection between the conditional sets **X** and $\tilde{\mathbf{X}}{\downarrow}$. This also gives a natural isomorphism between GH and the identity functor. $\square$

Remark 3.1. *The Boolean-valued part of the proof of Theorem 3.1 is covered by the well-known theorem from Boolean-valued analysis stating the equivalence of the category of names for non-empty sets and names for functions and the category of non-empty mix-complete Boolean sets and contractive functions (see Kusraev and Kutateladze [39, Theorem 3.5.10]). Thus, Theorem 3.1 actually establishes that the category of conditional sets of $\mathcal{A}$ and conditional functions is equivalent to the category of non-empty mix-complete Boolean sets over $\mathcal{A}$ and contractive functions.*

One more time, the important message is not the equivalence of categories provided above, but that for any conditional set **X** we build a tailored name $\tilde{X}$ for a set that induces a conditional set $\tilde{\mathbf{X}}{\downarrow}$ which is essentially **X**.

Let us fix a conditional set **X**. For the forthcoming discussion, we will suppose w.l.o.g. that $\mathbf{X} = \tilde{\mathbf{X}}{\downarrow}$.

Next, we will briefly explain how the main elements of the framework of conditional sets are connected to Boolean-valued analysis. A comprehensive introduction to conditional set theory is given in [11], thus for each unexplained notion we will give an exact reference to its definition in [11]:

- *Conditional subsets*: A non-empty subset S of X is *stable* if $\sum x_i|a_i \in S$ whenever $\{x_i\} \subset S$ and $\{a_i\} \in p(1)$. A *conditional subset* of **X** is a conditional set $\mathbf{S} := \{x|a : x \in S,\, a \in \mathcal{A}\}$, where S is a stable subset of X. For short, we will write $\mathbf{S} \sqsubset \mathbf{X}$.

  Suppose that $\mathbf{S} \sqsubset \mathbf{X}$. We define $\hat{S} : \mathcal{D}_\mathbf{S} \to \mathcal{A}$ with $\mathcal{D}_\mathbf{S} := \{\tilde{x} : x \in S\}$ and $\hat{S}(\tilde{x}) := 1$. Then it can be verified that $\hat{S}$ is a name with $[\![\hat{S} \subset \tilde{X}]\!] = 1$ and $\hat{S}{\downarrow} = \mathbf{S}$.

  Now, suppose that S_0 is a name with $[\![\emptyset \neq S_0 \subset \tilde{X}]\!] = 1$. Then $\mathbf{S}_0{\downarrow} \sqsubset \mathbf{X}$ and $[\![S_0 = (S_0{\downarrow})^{\wedge}]\!] = 1$.

- *Conditional power set*: Let $P(\mathbf{X})$ be the collection of all stable subsets of E. For $S \in P(\mathbf{X})$ and $a \in \mathcal{A}$, we define $\mathbf{S}|a := \{x|b : x \in S,\, b \leq a\}$. The set $\mathbf{P}(\mathbf{X}) := \{\mathbf{S}|a : S \text{ is stable},\, a \in \mathcal{A}\}$ is a conditional set which is called *conditional power set*.

Suppose that $\mathbf{C} \sqsubset \mathbf{P}(\mathbf{X})$. Let $\hat{C} : \mathcal{D}_{\mathbf{C}} \to \mathcal{A}$ with $\mathcal{D}_{\mathbf{C}} := \{\hat{S} : S \in C\}$ and $\hat{C}(\hat{S}) := 1$. Then $\hat{C}$ is a name for a set of subsets of $\tilde{X}$.

Now, given a name C_0 for a non-empty collection of non-empty sets of $\tilde{X}$, we define $C_0 \Downarrow := \{S\downarrow : S \in C_0\downarrow\}$. Then $C_0 \Downarrow$ is a stable set of subsets of X and we can consider the corresponding conditional set $\mathbf{C_0} \Downarrow \sqsubset \mathbf{P}(\mathbf{X})$.

Moreover, if $\mathbf{C} \sqsubset \mathbf{P}(\mathbf{X})$ one has that $\hat{\mathbf{C}} \Downarrow = \mathbf{C}$ and if C_0 is a name for a non-empty collection of non-empty sets of $\tilde{X}$ one has $[\![C_0 = (C_0\Downarrow)^\wedge]\!] = 1$.

In particular, if $\mathbf{C} = \mathbf{P}(\mathbf{X})$, then $\hat{C}$ is a name for the collection of all non-empty subsets of $\tilde{X}$ in $V^{(\mathcal{A})}$.

- *Conditional step functions*: If E is a non-empty set, consider the *conditional set of step functions*, let us say $\mathbf{E_s}$, see [11, Examples 2.3(5)]. Then, the name $\tilde{E}_s$ is precisely the canonical name $\check{E}$ of E in $V^{(\mathcal{A})}$.

  The *conditional natural numbers* $\mathbf{N}$ and the *conditional rational numbers* $\mathbf{Q}$ are introduced in [11] as a particular case of the step functions. It is known that $[\![\mathbb{N}^{(\mathcal{A})} = \check{\mathbb{N}}]\!] = \Omega$ and $[\![\mathbb{Q}^{(\mathcal{A})} = \check{\mathbb{Q}}]\!] = \Omega$, see eg [44]. Thus, it is satisfied that $\tilde{N}$ and $\tilde{Q}$ are names for the natural numbers and the rational numbers of $V^{(\mathcal{A})}$, respectively.

- *Conditional real numbers*: In [11] a conditional set $\mathbf{R}$ which is called *conditional real numbers* is defined, see [11, Definition 4.3]. Then it can be verified that $\tilde{R}$ is a name for the real numbers of $V^{(\mathcal{A})}$.

- *Conditional topologies*: Suppose that $\mathcal{T}$ is a *conditional topology* on $\mathbf{X}$, see [11, Definition 3.1]. Then $\hat{\mathcal{T}}$ is a name for the set of non-empty open sets of a topology on $\tilde{X}$.

  If $\mathcal{T}_0$ is a name for the set of non-empty open sets of a topology on $\tilde{X}$ then $\mathcal{T}_0 \Downarrow$ is a conditional topology.

  Moreover, $\mathbf{O}$ is a *conditional open* subset if and only if $\hat{O}$ is a name for an open set. $\mathbf{C}$ is a *conditional closed* subset if and only if $\hat{C}$ is a name for a closed set. $\mathbf{S}$ is a *conditionally compact subset* (see [11, Definition 3.24]) if and only if $\hat{S}$ is a name for a compact subset.

  Furthermore, $\mathcal{T}$ is *conditionally Hausdorff* (see [11, Section 3]) if and only if $[\![\text{``}\mathcal{T}_0 \text{ is Hausdorff''}]\!] = 1$.

- *Conditional functions:* Given a conditional function $\mathbf{f} : \mathbf{S_1} \to \mathbf{S_2}$, where $\mathbf{S_1}, \mathbf{S_2}$ are conditional subsets of $\mathbf{X}$, then we have a stable function $g : S_1 \to S_2$.

Theorem 1.4 allows to define a name $\hat{f}$ for a function from S_1 to S_2 with $\hat{f}\downarrow = f$.

Conversely, if f is name for a function between non-empty subsets of $\tilde{X}$, then $f\downarrow$ is a stable function between stable subsets of X and it defines a conditional function $\mathbf{f}\downarrow$ between conditional subsets of $\mathbf{X}$.

The same applies to *conditional families, conditional nets* and *conditional sequences*, see [11, Definition 2.20].

Bearing in mind the construction given in the proof of Theorem 3.1, the following is easy to check: $\mathbf{X}$ is a *conditional metric space*, see [11, Definition 4.5], if and only if $\tilde{X}$ is a name for a metric space; $\mathbf{X}$ is a *conditional locally convex space*, see [11, Definition 5.4], if and only if $\tilde{X}$ is a name for a locally convex space; $\mathbf{X}$ is a *conditional normed space*, see [11, Definition 5.11], if and only if $\tilde{X}$ is a name for a normed space; $\mathbf{X}$ is a *conditional Banach space*, see [11, Section 5], if and only if $\tilde{X}$ is a name for a Banach space.

Again, we see that all these objects are some of the building blocks for the main results provided in [11]. Clearly, names for more and more conditional versions of classical objects can be defined by using the same logic.

As an instance of application, we can provide a conditional version of the Schauder–Tychonov fixed point theorem:

Proposition 3.1. *Let $\mathbf{X}$ be a conditional locally convex space which is conditionally Hausdorff. If $\mathbf{C}$ is a conditionally compact conditional subset of $\mathbf{X}$ and $\mathbf{f}: \mathbf{C} \to \mathbf{C}$ is a conditionally continuous conditional function, then there exists $\mathbf{x}$ in $\mathbf{C}$ such that $\mathbf{f}(\mathbf{x}) = \mathbf{x}$.*

We can consider the names $\tilde{X}$, $\hat{C}$ and $\hat{f}$ as described above. If T denotes the statement of the Schauder-Tychonov Theorem, then the statement above, let us say $\mathbf{T}$, is nothing else but a reformulation of the statement '$[\![T]\!] = 1$', which holds due to the transfer principle of Boolean-valued models. Thus $\mathbf{T}$ is also a theorem. Of course, this is just an example. In general, this method can be systematically applied to the different theorems of [11].

References

[1] P. Artzner, F. Delbaen, J. M. Eber, and D. Heath. Coherent measures of risk. *Mathematical Finance*, 9:203–228, 1999.

[2] J. Backhoff and U. Horst. Conditional analysis and a Principal-Agent problem. *SIAM Journal on Financial Mathematics*, 7(1):477–507, 2016.

[3] V. Barbu and T. Precupanu. *Convexity and optimization in Banach spaces.* Springer Science & Business Media, 2012.

[4] J. L. Bell. *Set Theory: Boolean-Valued Models and Independence Proofs.* Oxford Logic Guides. Clarendon Press, 2005.

[5] T. R. Bielecki, I. Cialenco, S. Drapeau, and M. Karliczek. Dynamic assessment indices. *Stochastics*, 88(1):1–44, 2016.

[6] J. M. Borwein and Q. J. Zhu. Variational methods in convex analysis. *Journal of Global Optimization*, 35(2):197–213, 2006.

[7] P. Cheridito, U. Horst, M. Kupper, and T. Pirvu. Equilibrium pricing in incomplete markets under translation invariant preferences. *Mathematics of Operations Research*, 41(1):174 – 195, 2016.

[8] P. Cheridito, M. Kupper, and N. Vogelpoth. Conditional analysis on $\mathbb{R}^d$. *Set Optimization and Applications, Proceedings in Mathematics & Statistics*, 151:179 – 211, 2015.

[9] P. J. Cohen. Set theory and the continuum hypothesis. w.a. benjamin. *Inc., New York*, 1966.

[10] S. Drapeau and A. Jamneshan. Conditional preferences and their numerical representations. *Journal of Mathematical Economics*, 63:106–118, 2016.

[11] S. Drapeau, A. Jamneshan, M. Karliczek, and M. Kupper. The algebra of conditional sets, and the concepts of conditional topology and compactness. *Journal of Mathematical Analysis and Applications*, 437(1):561– 589, 2016.

[12] S. Drapeau, M. Karliczek, M. Kupper, and M. Streckfuss. Brouwer fixed point theorem in $(L^0)^d$. *Fixed Point Theory and Applications*, 301(1), 2013.

[13] K-T. Eisele and S. Taieb. Weak topologies for modules over rings of bounded random variables. *Journal of Mathematical Analysis and Applications*, 421(2):1334–1357, 2015.

[14] D. Filipović, M. Kupper, and N. Vogelpoth. Separation and duality in locally L^0-convex modules. *Journal of Functional Analysis*, 256:3996 – 4029, 2009.

[15] D. Filipović, M. Kupper, and N. Vogelpoth. Approaches to conditional risk. *SIAM Journal of Financial Mathematics*, 3(1):402 – 432, 2012.

[16] H. Föllmer and A. Schied. Convex measures of risk and trading constraints. *Finance and stochastics*, 6(4):429–447, 2002.

[17] M. Frittelli and E. R. Gianin. Putting order in risk measures. *Journal of Banking & Finance*, 26(7):1473–1486, 2002.

[18] M. Frittelli and M. Maggis. Dual representation of quasi-convex conditional maps. *SIAM Journal on Financial Mathematics*, 2(1):357–382, 2011.

[19] E. I. Gordon. K-spaces in Boolean-valued models of set theory. *Dokl. Akad. Nauk SSSR*, 258(4):777–780, 1981.

[20] E. I. Gordon. Rationally complete semiprime commutative rings in boolean valued models of set theory. *Gor kii, VINITI*, (3286-83), 1983.

[21] T. Guo. The relation of Banach-Alaoglu theorem and Banach-Bourbaki-Kakutani-Šmulian theorem in complete random normed modules to stratification structure. *Sci-*

ence in *China Series A Mathematics*, 51:1651–1663, 2008.

[22] T. Guo. Relations between some basic results derived from two kinds of topologies for a random locally convex module. *Journal of Functional Analysis*, 258:3024–3047, 2010.

[23] T. Guo. On Some Basic Theorems of Continuous Module Homomorphisms between Random Normed Modules. *Journal of Function Spaces and Applications*, pages 1–13, 2013.

[24] T. Guo and X. Chen. Random duality. *Science in China Series A: Mathematics*, 52(10):2084–2098, 2009.

[25] T. Guo, S. Zhao, and X. Zeng. Random convex analysis (I): separation and Fenchel-Moreau duality in random locally convex modules. *arXiv preprint arXiv:1503.08695*, 2015.

[26] A. E. Gutman and S. A. Lisovskaya. The boundedness principle for lattice-normed spaces. *Siberian Mathematical Journal*, 50(5):830–837, 2009.

[27] R. Haydon, M. Levy, and Y. Raynaud. *Randomly normed spaces*. Hermann, 1991.

[28] A. Jamneshan, M. Kupper, and J. M. Zapata. Parameter-dependent stochastic optimal control in finite discrete time. *arXiv preprint arXiv:1705.02374*, 2017.

[29] A. Jamneshan and J. M. Zapata. On compactness in L^0-modules. *arXiv preprint arXiv:1711.09785*, 2017.

[30] T. Jech. *Set theory*. Springer Science & Business Media, 2013.

[31] E. Jouini, W. Schachermayer, and N. Touzi. Law invariant risk measures have the fatou property. *Advances in mathematical economics*, pages 49–71, 2006.

[32] L. V. Kantorovich. *To the general theory of operations in semiordered spaces*, volume 1. Dokl. Akad. Nauk SSSR (Russian), 1936.

[33] A. G. Kusraev. Boolean valued analysis of duality between universally complete modules. *Dokl. Akad. Nauk SSSR*, 267(5):1049–1052, 1982.

[34] A. G. Kusraev. Banach-kantorovich spaces. *Siberian Mathematical Journal*, 26(2):254–259, 1985.

[35] A. G. Kusraev. Vector duality and its applications, 1985.

[36] A. G. Kusraev. Dominated operators. In *Dominated Operators*, pages 141–186. Springer, 2000.

[37] A. G. Kusraev and S. S. Kutateladze. *Subdifferentials: Theory and applications*, volume 323. Springer Science & Business Media, 2012.

[38] A. G. Kusraev and S. S. Kutateladze. Boolean valued analysis: Selected topics. *Vladikavkaz: SMI VSC RAS*, 1000(6), 2014.

[39] A. G. Kusraev and S. S. Kutateladze. *Boolean Valued Analysis*. Mathematics and Its Applications. Springer Netherlands, 2012.

[40] J. Orihuela and M. Ruiz-Galán. A coercive james's weak compactness theorem and nonlinear variational problems. *Nonlinear Analysis: Theory, Methods & Applications*, 75(2):598–611, 2012.

[41] J. Orihuela and J. M. Zapata. Stability in locally L^0-convex modules and a condi-

tional version of James' compactness theorem. *Journal of Mathematical Analysis and Applications*, 452(2):1101 – 1127, 2017.

[42] J. Saint-Raymond. Weak compactness and variational characterization of the convexity. *Mediterranean journal of mathematics*, 10(2):927–940, 2013.

[43] D. Scott. A proof of the independence of the continuum hypothesis. *Theory of Computing Systems*, 1(2):89–111, 1967.

[44] G. Takeuti. *Two Applications of Logic to Mathematics*. Publications of the Mathematical Society of Japan. Princeton University Press, 2015.

[45] D. A. Vladimirov. *Boolean algebras in analysis*, volume 540. Springer Science & Business Media, 2013.

[46] J. M. Zapata. Randomized versions of Mazur lemma and Krein-Šmulian theorem. *Journal of Convex Analysis*, 25(3), 2018 (To appear).

[47] J. M. Zapata. Versions of Eberlein-Šmulian and Amir-Lindenstrauss theorems in the framework of conditional sets. *Applicable Analysis and Discrete Mathematics*, 10(2):231–261, 2016.

[48] J. M. Zapata. On the Characterization of Locally L^0-Convex Topologies Induced by a Family of L^0-Seminorms. *Journal of Convex Analysis*, 24(2):383–391, 2017.

ABOUT RELATIONSHIPS BETWEEN TWO INDIVIDUALS

ROBERT DEMOLOMBE
IRIT, Toulouse University, France.
robert.demolombe@orange.fr

Abstract

If an internet user wants to access information about two given individuals, he can submit a request with the names of these individuals. However, the occurrences of these two names do not guarantee that the obtained information expresses a relationship between these individuals. The aim of this paper is to propose a clear definition of sentences which express a relationship between two individuals. We first present an informal analysis, based on examples, of this notion of relationship in the context of atomic sentences, or complex sentences which combines logical connectives or quantifiers. In the next section, we give formal definitions, assuming that sentences are expressed in First Order Logic. We define the notion of "path" between individuals, the notion of link between individuals and the notion of relationship between individuals. A Theorem shows how the relationships which are implicitly expressed in complex formulas can be represented in equivalent formulas expressed with "basic relationships".

In the conclusion we suggest possible extensions where the language involves equality, function symbols or modal operators.

1 Introduction

At the beginning of the seventies, requests to retrieve information using Relational Data Base Systems had the form: *what are the individuals, or tuples of individuals, which fulfill some properties?*, where the properties were formulated either with Relational Algebra [11, 10, 1] or First Order Predicate Calculus [6, 8, 12]. Now, to retrieve information, requests expressed by internet users have the form: *what are the documents which "match" a combination of key words or of short sentences?* However, there are no general formal definition of this concept of "matching".

Key words may denote topics or individuals. Roughly speaking, in the case of individuals the returned documents are documents which contain occurrences of the individual names and, in the case of topics, the returned documents are those which are *about* these topics.

To have more precise definitions of what should be returned it can be assumed that the content of the documents is represented by sentences expressed in a formal language

In [3, 4] a formal characterization of sentences which are about some topics has been proposed, where the notion of "aboutness" may be understood as a kind of matching (see [9, 7] for a more philosophical analysis of aboutness). That could be used, for instance, to retrieve the information about the topics "employment" and "climate change".

In [5] is proposed a formal definition of all the sentences which express information about a given individual. For instance, all the sentences which express information about the individual named "Alan Turing".

Here, we propose a formal definition of all the sentences which express relationships between two individuals. For instance, the sentences which express a relationship between the individuals named "Alan Turing" and "Albert Einstein". This allows to retrieve more specific information than information where "Alan Turing" and "Albert Einstein" occur.

In section 2 we give an informal justification of our definition of relationship between two individuals. A formal definition is presented in section 3. After the analysis of related works in section 4 several possible extensions are presented in the conclusion.

2 Intuitive analysis

The fact that the names of two individuals occur in the same sentence does not guarantee that this sentence expresses a relationship between these individuals.

Let's see, for instance, the sentence: *Romeo is a man and Juliet is a woman*. It does not express a relationship between Romeo and Juliet. Indeed, in semiformal terms, the information expressed by:

f_1: (Romeo is a man) $\wedge$ (Juliet is a woman)

can be represented as well by the two independent sentences:

f_2: (Romeo is a man)

and

f_3: (Juliet is a woman)

At the opposite, the atomic sentence: *Romeo loves Juliet*:

f_4: Romeo loves Juliet

expresses a relationship between Romeo and Juliet since the information expressed by this sentence cannot be expressed by two independent sentences.

The conclusion shown with this example can be extended to any atomic formula where a predicate is used to represent a relationship between two individuals. Now, the question is: "are there other ways than atomic formulas to express this kind of relationships?".

Let's see first atomic formulas which are combined by the logical connectives: negation, conjunction and disjunction.

For the negation it seems to be clear that we should accept that a sentence like: *Romeo does not love Juliet*, in semiformal terms:

f_5: ¬ (Romeo loves Juliet)

informs about a relationship between Romeo and Juliet. This conclusion can be extended to the negation of any formula which expresses a relationship between two individuals.

For the conjunction, we also should accept that if two sentences express relationships between two individuals, then any conjunction of these sentences express relationships between these individuals. For instance: *Romeo loves Juliet and Juliet loves Romeo*, which is represented in semiformal terms by:

f_6: (Romeo loves Juliet) ∧ (Juliet loves Romeo)

In addition, we do not see any objection to accept that the conjunction: *Romeo loves Juliet and Juliet is a woman*, which is represented by:

f_7: (Romeo loves Juliet) ∧ (Juliet is a woman)

expresses a relationship between Romeo and Juliet, even if the formula *Juliet is a woman* is not about this relationship.

However, we can have some doubts to accept a conjunction of the kind: *Romeo loves Juliet and Romeo does not love Juliet*:

f_8: (Romeo loves Juliet) ∧ ¬ (Romeo loves Juliet)

because it is an inconsistent formula. Nevertheless, even if it does not express information about a relationship between Romeo and Juliet it is about Romeo and Juliet in the sense of aboutness presented in [4]. That is the reason why we have accepted that f_8 expresses a relationship between Romeo and Juliet.

For the disjunction, we should accept that the disjunction of two sentences which express relationships between two individuals expresses a relationship between these individuals. For instance, the sentence: *Romeo loves Juliet or Juliet loves Romeo*, represented by:

f_9: (Romeo loves Juliet) ∨ (Juliet loves Romeo)

expresses a relationship between Romeo and Juliet. We also can accept that a sentence expresses a relationship between Romeo and Juliet in the cases where only one term of the disjunction expresses this relationship. For instance, the sentence: *If we are on Sunday, Romeo meets Juliet*, represented by:

f_{10}: ¬ (on Sunday) ∨ (Romeo meets Juliet)

can express information about the fact that Romeo meets Juliet if we know that we are on Sunday.

For the same reason as we have accepted that inconsistent sentences may be about a relationship between Romeo and Juliet, we have accepted that some tautologies may be about a relationship between Romeo and Juliet. For instance the sentence: *Romeo loves Juliet or Romeo does not love Juliet*, represented by:

f_{11}: (Romeo loves Juliet) ∨ ¬ (Romeo loves Juliet)

It may be that the relationship between two individuals in a sentence is not represented by the composition of atomic formulas where these relationships explicitly appear. The link between the individuals may be represented by a kind of "path" which relates an individual to another individual which is itself related to another individual and so on ... For instance, in the sentence: *Romeo leaves in Verona and Juliet leaves in Verona*, represented by:

f_{12}: (Romeo leaves in Verona) ∧ (Juliet leaves in Verona)

there is a path from Romeo to Verona and from Verona to Juliet (the ordering in the past is irrelevant). In general, we call a "path" a conjunction of positive or negative atomic formulas. For instance, in: *Juliet knows Paris and Paris does not know Mercutio and Romeo knows Mercutio*, which is represented by:

f_{13}: (Juliet knows Paris) ∧ ¬ (Paris knows Mercutio) ∧ (Romeo knows Mercutio)

we have a path from Juliet to Romeo via Paris and Mercutio.

In some sentences we may have what we call a "link" between two individuals where the paths between these individuals is not explicit. For instance, in the sentence: *Romeo is in Verona or Juliet is in Verona*, represented by:

f_{14}: (Romeo is in verona) ∨ (Juliet is in Verona)

there is no explicit path between Romeo and Juliet in the sense that we have defined before. Nevertheless, it is clear that there is a link between Romeo and Juliet. This link can be made explicit if we observe that if f_{14} is true there is at least one the atomic sentences: *Romeo leaves in Verona*, or *Juliet leaves in Verona* which is true, and f_{14} is logically equivalent to f_{15}:

f_{15}: ((Romeo is in verona) ∧ (Juliet is in Verona)) ∨
((Romeo is in verona) ∧ ¬ (Juliet is in Verona)) ∨
(¬ (Romeo is in verona) ∧ (Juliet is in Verona))

In f_{15} we have a disjunction of three sentences such that each one expresses a path between Romeo and Juliet.

In the following we shall call "basic relationship" between two individuals a relationship which is explicitly represented by a path, that is a conjunction of literals.

For quantifiers we may have quantified variables which are involved in the definition of a path. For instance, for the universal quantifier, the sentence: *Romeo does everything that Juliet wants he does*, represented by:

f_{16}: $\forall x$ ((Juliet wants x) $\rightarrow$ (Romeo does x))

In a similar way, for existential existential quantifiers we may have sentences like: *Romeo and Juliet leaves in the same city*, represented by:

f_{17}: $\exists x$ ((Romeo leaves in x) $\wedge$ (Juliet leaves in x))

It is worth noting that relationships may be "hidden" inside complex sentences. For instance, in the sentence: *Juliet loves Romeo or Marutio and she does not love Paris*, represented by:

f_{18}: ((Juliet loves Romeo) $\vee$ (Juliet loves Marutio)) $\wedge$ $\neg$ (Juliet loves Paris)

it is not easy to perceive that there is a relationship between Romeo and Paris.

In the next section, formal syntactical criteria are defined to characterize the fact that a formula contains some kinds of implicit relationships between two individuals.

3 Formalization

In this section, after to define the formal language which is used to represent the sentences, we define the notion of *Link*. The intuitive interpretation of $Link_{a,b}\phi$ is that in the sentence ϕ there are predicates and logical connectives or quantifiers that define a relationship between the individuals denoted by a and b.

Since it may be difficult to intuitively see this relationship, we have defined the notion of *Path* which is much more simple to percieve. A *Path* is just a set of literals where the predicate names could be interpreted as the edges of a graph and the names of the constants or of the variables could be interpreted as the nodes. According to this intuitive interpretation, $Path_{t_1,t_2}(Lit)$ holds in the set of literals Lit if there is a path in this graph from t_1 to t_2. In addition it is required that Lit is minimal, in the sense that it does not contain literals which could be removed without eliminating the existence of this path in Lit.

Latter on is defined the notion of basic relationship. The main difference between a path and a basic relationship is that a path is a set of literals which are implicitly connected by conjunctions, while a basic relationship is a formula which is a conjunction of these literals. The notion of path is justified by technical motivations. Indeed, a path allows to have a unique representation for several formulas which express the same basic relationship. For instance, in the set of literals: $\{p(a,x), \neg q(x,c), r(c,a)\}$, there is a path from a to b which corresponds to formulas like: $r(c,a) \wedge \neg q(x,c) \wedge r(a,x)$ or: $\neg q(x,c) \wedge r(c,a) \wedge p(a,x)$.

Definition 1. *Language L.*
The language L' is a First Order Language defined as follows.
Let P be a set of predicate symbols, C a set of constant symbols and V a set of variable symbols, the set A of atomic formulas is the set of formulas of the form: $p(t_1, t_2, \ldots, t_n)$ such that p is in P and the t_i are either in C or in V.
The formulas ϕ in L' are defined by:
$\phi := Atom \mid \neg\phi \mid \phi \wedge \phi \mid \forall x \phi$
where $Atom$ is in A and x is in V.
The formulas in the language L are the formulas ϕ in L' such that, if ϕ contains a sub-formula of the form $\forall x \psi(x)$, there is no other sub-formula in ϕ of the form $\forall x \theta(x)$.

Definition 2. *Relationship between two individuals.*
Let ϕ be a formula in L and t_i and t_j which denote two different symbols either in C or in V, the formula ϕ expresses one, or several relationships between t_i and t_j iff ϕ satisfies the property $Link_{t_i,t_j}\phi$, where $Link_{t_i,t_j}\phi$ is recursively defined as follows:

- If ϕ is an atomic formula of the form $p(t_1, t_2, \ldots, t_n)$, then we have $Link_{t_i,t_j}\phi$ iff there exist t_k and t_l in ϕ such that t_i is the symbol t_k and t_j is the symbol t_l[1].

- If $\phi = \neg\phi_1$, then we have $Link_{t_i,t_j}\phi$ iff we have $Link_{t_i,t_j}\phi_1$.

- If $\phi = \phi_1 \wedge \phi_2$, then we have $Link_{t_i,t_j}\phi$ iff we have $Link_{t_i,t_j}\phi_1$ or $Link_{t_i,t_j}\phi_2$ or there exists t in C or in V such that we have $Link_{t_i,t}\phi_1$ and $Link_{t,t_j}\phi_2$.

- If $\phi = \forall x \phi_1$, then we have $Link_{t_i,t_j}\phi$ iff we have $Link_{t_i,t_j}\phi_1$.

The language L can be extended to disjunctions and existential quantifiers. According to the standard definitions we have:
$\phi_1 \vee \phi_2 \stackrel{def}{=} \neg(\neg\phi_1 \wedge \neg\phi_2)$ and $\exists x \phi \stackrel{def}{=} \neg\forall x \neg\phi$
Then, we can easily show that, according to these definitions, the property $Link_{t_i,t_j}\phi$ is extended as follows:

- If $\phi = \phi_1 \vee \phi_2$, then we have $Link_{t_i,t_j}\phi$ iff we have $Link_{t_i,t_j}\phi_1$ or $Link_{t_i,t_j}\phi_2$ or there exists t in C or in V such that we have $Link_{t_i,t}\phi_1$ and $Link_{t,t_j}\phi_2$.

- If $\phi = \exists x \phi_1$, then we have $Link_{t_i,t_j}\phi$ iff we have $Link_{t_i,t_j}\phi_1$.

---
[1]Here and in the following the symbols t_i, t_j, t_k and t_l are in the metalanguage. They are used to denote constant symbols or variable symbols in the language L.

Examples. The following examples show how to derive properties of the kind: $Link_{a,b}\phi$.

Let $G_1 = \neg\exists x(p(a,x) \land \neg q(x,b))$. We have:
(1) $Link_{a,x}p(a,x)$ and (2) $Link_{x,b}q(x,b)$
From (2) we have: (3) $Link_{x,b}\neg q(x,b)$.
From (1) and (3) we have: (4) $Link_{a,b}(p(a,x) \land \neg q(x,b))$.
From (4) we have: (5) $Link_{a,b}\exists x(p(a,x) \land \neg q(x,b))$.
From (5) we have: $Link_{a,b}G_1$.
Let $G_2 = \neg((p(a,c) \lor p(a,d)) \land (p(c,b) \lor r(d,b)))$.
We have:
(1) $Link_{a,c}p(a,c)$ and (2) $Link_{a,d}p(a,d)$ and (3) $Link_{c,b}p(c,b)$ and
(4) $Link_{d,b}r(d,b)$.
From (1) we have: (5) $Link_{a,c}(p(a,c) \lor p(a,d))$.
From (3) we have: (6) $Link_{c,b}(p(c,b) \lor r(d,b))$.
From (5) and (6) we have: (7) $Link_{a,b}((p(a,c) \lor p(a,d)) \land (p(c,b) \lor r(d,b)))$.
From (7) we have: $Link_{a,b}G_2$.
Let $G3 = (p(a,b) \lor p(a,d)) \land r(d,b)$.
We have: (1) $Link_{a,b}p(a,b)$ and (2) $Link_{a,d}p(a,d)$ and (3) $Link_{d,b}r(d,b)$.
From (1) we have: (4) $Link_{a,b}(p(a,b) \lor p(a,d))$.
Then, we have: $Link_{a,b}G_3$.
From (2) we also have: (5) $Link_{a,d}(p(a,b) \lor p(a,d))$.
From (5) and (3) we also have: $Link_{a,b}G_3$.

Lemma 1. *We have: $Link_{a,b}\phi$ iff we have: $Link_{b,a}\phi$.*

Proof. The proof follows from the definition of $Link_{a,b}\phi$.

Lemma 2. *We can have $\vdash \phi_1 \leftrightarrow \phi_2$ and $Link_{a,b}\phi_1$ and not $Link_{a,b}\phi_2$.*

Proof. Example: $\phi_1 = (p(a,c) \land q(c,b)) \lor (p(a,c) \land \neg q(c,b))$ and $\phi_2 = p(a,c)$.

Definition 3. *Path in a set of literals.*

A literal l in L is a formula ϕ in L such that ϕ is an atomic formula or ϕ is the negation of an atomic formula.

Let Lit be a set of literals $\{l_1, l_2, \ldots, l_n\}$, there is a path between t_1 and t_2 in Lit, where t_1 and t_2 are either in C or in V, iff we have $Path_{t_1,t_2}(Lit)$, where $Path_{t_1,t_2}(Lit)$ is defined as follows.

We have $Path_{t_1,t_2}(Lit)$ iff 1) there exists a literal l_i in Lit such that t_1 and t_2 are in l_i, or 2) there exists t in C or in V such that $Path_{t_1,t}(Lit)$ and $Path_{t_2,t}(Lit)$, and 3) Lit is minimal, in the sense that if we remove a literal, then for the new set of literals Lit', we do not have $Path_{t_1,t_2}(Lit')$

Examples: $Lit_1 = \{p(a,b)\}$, $Lit_2 = \{p(a,c), \neg q(c,d), r(d,b)\}$, $Lit_3 = \{p(a,c), \neg q(c,d)\}$. We have: $Path_{a,b}(Lit_1)$, $Path_{a,b}(Lit_2)$ and $Path_{a,d}(Lit_3)$. However, we do not have $Path_{a,d}(Lit_2)$ because Lit_2 is not minimal for the path $<a,d>$.

Definition 4. *Basic relationship between two individuals.*

Let Lit be a set of literals.

The formula ϕ is a basic relationship between the constant symbols a and b iff ϕ is a conjunction of all the literals in Lit and we have $Path_{a,b}(Lit)$.

Examples. If $\phi_1 = p(a,c) \wedge \neg q(c,d) \wedge r(d,b)$ and $\phi_2 = r(d,b) \wedge p(a,c) \wedge q(c,d)$, then ϕ_1 and ϕ_2 are basic relationships between a and b.

Lemma 3. *If ϕ is in L and ϕ is a basic relationship between a and b, then there is a relationship in ϕ between a and b in the sense that we have: $Link_{a,b}\phi$.*

Proof. The proof is by induction on the number of literals in ϕ.

Lemma 4. *Let ϕ be a formula in L of the form $\phi = l_1 \vee l_2 \vee \ldots \vee l_n$, where the l_is are literals in L. The formula ϕ is logically equivalent to*
$\phi' = \bigvee (l'_1 \wedge l'_2 \wedge \ldots \wedge l'_n)$ *where the disjunction $\bigvee$ is extended to all the conjunctions of the l'_is such that l'_i is either l_i or $\neg l_i$ (if l_i is a negative literal of the form $\neg A_i$, then $\neg l_i$ is replaced by A_i) and such that there exist at least one l'_i such that $l'_i = l_i$*
[2].

Example. Let ϕ be the formula $A_1 \vee \neg A_2 \vee A_3$. ϕ is logically equivalent to $\phi' = (A_1 \wedge \neg A_2 \wedge A_3) \vee (\neg A_1 \wedge \neg A_2 \wedge A_3) \vee (A_1 \wedge A_2 \wedge A_3) \vee (A_1 \wedge \neg A_2 \wedge \neg A_3) \vee (\neg A_1 \wedge A_2 \wedge A_3) \vee (\neg A_1 \wedge \neg A_2 \wedge \neg A_3) \vee (A_1 \wedge A_2 \wedge \neg A_3)$.

Lemma 5. *If $r_{a,b} = l_1 \wedge l_2 \wedge \ldots \wedge l_n$ is a basic relationship between a and b, then its negation $\neg r_{a,b}$ is logically equivalent to a disjunction of basic relationships between a and b of the form: $\bigvee(l''_1 \wedge l''_2 \wedge \ldots \wedge l''_n)$, where the l''_is are obtained from the $\neg l_i$s in the same way as the l'_is are obtained from the l_is in the Lemma 4.*

Example. Let $r_{a,b} = \neg A_1 \wedge A_2 \wedge A_3$ be a basic relationship between a and b. We have $\neg r_{a,b}$ logically equivalent to:
$(A_1 \wedge \neg A_2 \wedge \neg A_3) \vee (A_1 \wedge \neg A_2 \wedge \neg A_3) \vee (A_1 \wedge A_2 \wedge \neg A_3) \vee (\neg A_1 \wedge \neg A_2 \wedge \neg A_3) \vee (A_1 \wedge A_2 \wedge A_3) \vee (\neg A_1 \wedge \neg A_2 \wedge A_3) \vee (\neg A_1 \wedge A_2 \wedge \neg A_3)$.

Theorem 1. *If ϕ is a formula of the language L extended with the disjunction $\vee$ and the existential quantifier $\exists$, such that we have $Link_{a,b}\phi$, then there exists a set of basic relationships between a and b, denoted by $r_{a,b}$, such that we have:*

---

[2]The formula $\phi' = \bigvee(l'_1 \wedge l'_2 \wedge \ldots \wedge l'_n)$ is a shorthand for the: formula
$\phi'' = (l'_{1,1} \wedge l'_{1,2} \wedge \ldots l'_{1,n_1}) \vee (l'_{2,1} \wedge l'_{2,2} \wedge \ldots l'_{2,n_2}) \vee \ldots \vee (l'_{m,1} \wedge l'_{m,2} \wedge \ldots l'_{m,n_m}))$.

$\vdash \phi \leftrightarrow QX(\bigwedge(\bigvee(r_{a,b} \wedge C) \vee D))$
where $QX(\bigwedge(\bigvee(r_{a,b} \wedge C) \vee D))$ is in prenex normal form and QX is a list of quantifiers of the form: $q_1 x_1 q_2 x_2 \ldots q_n x_n$ where the q_i may be $\forall$ or $\exists$ and $x_1, \ldots, x_n$ are the variables in $\bigwedge(\bigvee(r_{a,b} \wedge C) \vee D)$ and C and D are formulas in L.
$\bigwedge(\bigvee(r_{a,b} \wedge C) \vee D)$ is a simplified notation for formulas of the kind:
$((r_{a,b}^{1,1} \wedge C_1^1) \vee (r_{a,b}^{1,2} \wedge C_1^2) \vee \ldots \vee (r_{a,b}^{1,i} \wedge C_1^i) \vee \ldots \vee (r_{a,b}^{1,n_1} \wedge C_1^{n_1}) \vee D_1) \wedge$
$\ldots$
$((r_{a,b}^{i,1} \wedge C_i^1) \vee (r_{a,b}^{i,2} \wedge C_i^2) \vee \ldots \vee (r_{a,b}^{i,j} \wedge C_i^j) \vee \ldots \vee (r_{a,b}^{i,n_i} \wedge C_i^{n_i}) \vee D_i) \wedge$
$\ldots$
$((r_{a,b}^{n,1} \wedge C_n^1) \vee (r_{a,b}^{n,2} \wedge C_n^2) \vee \ldots \vee (r_{a,b}^{n,k} \wedge C_n^k) \vee \ldots \vee (r_{a,b}^{n,n_n} \wedge C_n^{n_n}) \vee D_n)$
where the $r_{a,b}^{i,j}$s are basic relationships between a and b.

Proof. The proof is by induction on the complexity degree $deg(\phi)$ of ϕ.
Case $deg(\phi) = 0$. In that case ϕ is an atomic formula and if we have $Link_{a,b}\phi$, then ϕ is a basic relationship between a and b.
Induction assumption: (H) Theorem 1 holds for every formula ϕ such that $deg(\phi) \leq n$.
Case $deg(\phi) = n + 1$.
   Case of negation
   If $\phi = \neg \phi_1$ $Link_{a,b}\phi$ entails $Link_{a,b}\phi_1$. From (H) we have:
$\phi_1 \leftrightarrow QX(\bigwedge(\bigvee(r_{a,b} \wedge C) \vee D))$
   and
$\phi \leftrightarrow \neg QX(\bigwedge(\bigvee(r_{a,b} \wedge C) \vee D))$
   If the negation is distributed on the quantifiers we have:
$\phi \leftrightarrow Q'X \neg (\bigwedge(\bigvee(r_{a,b} \wedge C) \vee D))$
   If the negation is distributed on the conjunctions and after that it is distributed on the disjunctions, we have:
$\phi \leftrightarrow Q'X(\bigvee(\bigwedge(\neg(r_{a,b} \wedge C) \wedge \neg D))$
   If the disjunctions are distributed on the conjunctions we have:
$\phi \leftrightarrow Q'X(\bigwedge(\bigvee(\neg(r_{a,b} \wedge C) \wedge \neg D))$
   The formula $F_1 = \neg(r_{a,b} \wedge C) \wedge \neg D$ is equivalent to $F_2 = (\neg r_{a,b} \vee \neg C) \wedge \neg D$ which is equivalent to: $F_3 = (\neg r_{a,b} \wedge \neg D) \vee (\neg C \wedge \neg D)$.
   From Lemma 5 $\neg r_{a,b}$ is equivalent to a disjunction of basic relationships between a and b which is denoted by: $\bigvee r'_{a,b}$. Then, F_3 is equivalent to: $F_4 = \bigvee(r'_{a,b} \wedge \neg D) \vee (\neg C \wedge \neg D)$ which is of the kind: $\bigvee(r'_{a,b} \wedge C') \vee D'$.
   Therefore, we have:
$\phi \leftrightarrow Q'X(\bigwedge(\bigvee(r'_{a,b} \wedge C') \vee D'))$

Case of conjunction

If $\phi = \phi_1 \wedge \phi_2$ $Link_{a,b}\phi$ entails $Link_{a,b}\phi_1$ or $Link_{a,b}\phi_2$ or there exists t in C or in V such that we have $Link_{a,t}\phi_1$ and $Link_{t,b}\phi_2$.

Subcase 1) If we have $Link_{a,b}\phi_1$ and not $Link_{a,b}\phi_2$, from (H) we have:
$\phi_1 \leftrightarrow QX(\bigwedge(\bigvee(r_{a,b} \wedge C_1) \vee D_1))$
then, we have:
$\phi \leftrightarrow QX(\bigwedge(\bigvee(r_{a,b} \wedge C_1) \vee D_1)) \wedge \phi_2$

If ϕ_2 is distributed on all the terms of the conjunction and then on the disjunctions, we have:
$\phi \leftrightarrow QX(\bigwedge(\bigvee(r_{a,b} \wedge C_1 \wedge \phi_2) \vee (D_1 \wedge \phi_2)))$
which is of the kind:
$\phi \leftrightarrow QX(\bigwedge(\bigvee(r_{a,b} \wedge C) \vee D))$

Subcase 2) If we have not $Link_{a,b}\phi_1$ and $Link_{a,b}\phi_2$, the proof is very similar to the proof in 1).

Subcase 3) If we have $Link_{a,b}\phi_1$ and $Link_{a,b}\phi_2$, from (H) we have:
$\phi_1 \leftrightarrow Q_1X_1(\bigwedge(\bigvee(r_{a,b} \wedge C_1) \vee D_1))$
and
$\phi_2 \leftrightarrow Q_2X_2(\bigwedge(\bigvee(r_{a,b} \wedge C_2) \vee D_2))$

The variable names in ϕ_2 are changed in order to have different variable names in ϕ_1 and ϕ_2. That leads to:
$\phi_2 \leftrightarrow Q'_2X'_2(\bigwedge(\bigvee(r'_{a,b} \wedge C'_2) \vee D'_2))$
where the variable names in $(r'_{a,b} \wedge C'_2) \vee D'_2$ are changed according to the changes in $Q'_2X'_2$.

Then, we have (see [2] section 3.3):
$\phi \leftrightarrow Q_1X_1Q'_2X'_2((\bigwedge(\bigvee(r_{a,b} \wedge C_1) \vee D_1)) \wedge (\bigwedge(\bigvee(r'_{a,b} \wedge C'_2) \vee D'_2)))$
Therefore, if we aggregate the conjunctions we have:
$\phi \leftrightarrow QX(\bigwedge(\bigvee(r_{a,b} \wedge C) \vee D))$

Subcase 4) If we have not $Link_{a,b}\phi_1$ and not $Link_{a,b}\phi_2$, there exists t in C or in V such that we have $Link_{a,t}\phi_1$ and $Link_{t,b}\phi_2$, and from (H) we have:
$\phi_1 \leftrightarrow Q_1X_1(\bigwedge(\bigvee(r_{a,t} \wedge C_1) \vee D_1))$
and
$\phi_2 \leftrightarrow Q_2X_2(\bigwedge(\bigvee(r_{t,b} \wedge C_2) \vee D_2))$
For the same reason as in 3) we have:
$\phi \leftrightarrow Q_1X_1Q'_2X'_2(\bigwedge((\bigvee(r_{a,t} \wedge C_1) \vee D_1) \wedge (\bigwedge(\bigvee(r'_{t,b} \wedge C'_2) \vee D'_2)))$
then, we have:
$\phi \leftrightarrow Q_1X_1Q'_2X'_2(\bigwedge((\bigvee(r_{a,t} \wedge C_1) \vee D_1) \wedge (\bigvee(r'_{t,b} \wedge C'_2) \vee D'_2)))$
In the formula $(\bigvee(r_{a,t} \wedge C_1) \vee D_1) \wedge (\bigvee(r'_{t,b} \wedge C'_2) \vee D'_2)$ if we distribute the conjunction on the disjunctions, we have a disjunction of terms of the following forms:

$\bigvee(r_{a,t} \wedge C_1) \wedge \bigvee(r'_{t,b} \wedge C'_2)$ or $\bigvee(r_{a,t} \wedge C_1) \wedge D'_2$ or $D_1 \wedge \bigvee(r'_{t,b} \wedge C'_2)$ or $D_1 \wedge D'_2$
which are the forms:
$r_{a,t} \wedge C_1 \wedge r'_{t,b} \wedge C'_2$ or $r_{a,t} \wedge C_1 \wedge D'_2$ or $D_1 \wedge (r'_{t,b} \wedge C'_2)$ or $D_1 \wedge D'_2$

From the definition of basic relationships between a and b, $r_{a,t} \wedge C_1 \wedge r'_{t,b} \wedge C'_2$ is of the kind $r_{a,b} \wedge C$, while the other terms are not necessarily about basic relationships between a and b and they are denoted by D.

Therefore, we have:
$\phi \leftrightarrow QX(\bigwedge(\bigvee(r_{a,b} \wedge C) \vee D))$

Case of disjunction

If $\phi = \phi_1 \vee \phi_2$ $Link_{a,b}\phi$ entails $Link_{a,b}\phi_1$ or $Link_{a,b}\phi_2$ or there exists t in C or in V such that we have $Link_{a,t}\phi_1$ and $Link_{t,b}\phi_2$.

Subcase 1) If we have $Link_{a,b}\phi_1$ and not $Link_{a,b}\phi_2$, from (H) we have:
$\phi_1 \leftrightarrow QX(\bigwedge(\bigvee(r_{a,b} \wedge C_1) \vee D_1))$
then, we have:
$\phi \leftrightarrow QX(\bigwedge(\bigvee(r_{a,b} \wedge C_1) \vee D_1)) \vee \phi_2$
if the disjunction of ϕ_2 is distributed on the conjunctions, we have:
$\phi \leftrightarrow QX(\bigwedge(\bigvee(r_{a,b} \wedge C_1) \vee D_1 \vee \phi_2))$
Therefore, we have:
$\phi \leftrightarrow QX(\bigwedge(\bigvee(r_{a,b} \wedge C) \vee D))$

Subcase 2) If we have not $Link_{a,b}\phi_1$ and $Link_{a,b}\phi_2$, the proof is very similar to the proof in 1).

Subcase 3) If we have $Link_{a,b}\phi_1$ and $Link_{a,b}\phi_2$, from (H) we have:
$\phi_1 \leftrightarrow Q_1 X_1(\bigwedge(\bigvee(r_{a,b} \wedge C_1) \vee D_1))$
and
$\phi_2 \leftrightarrow Q_2 X_2(\bigwedge(\bigvee(r_{a,b} \wedge C_2) \vee D_2))$
Then, if the variables in X_2 a renamed, we have:
$\phi \leftrightarrow Q_1 X_1 Q'_2 X'_2((\bigwedge(\bigvee(r_{a,b} \wedge C_1) \vee D_1)) \vee (\bigwedge(\bigvee(r'_{a,b} \wedge C'_2) \vee D'_2)))$
The formula in the scope of the quantifiers is of the form:
$F_1 = (\delta_1 \wedge \delta_2 \wedge \ldots \wedge \delta_i \wedge \ldots \wedge \delta_n) \vee (\gamma_1 \wedge \gamma_2 \wedge \ldots \wedge \gamma_j \wedge \ldots \wedge \gamma_m)$
If the disjunction is distributed on the conjunctions, we have:
$F_2 = (\delta_1 \vee \gamma_1) \wedge (\delta_1 \vee \gamma_2) \wedge \ldots \wedge (\delta_i \vee \gamma_j) \wedge \ldots \wedge (\delta_n \vee \gamma_m)$
In this formula the terms of the kind $(\delta_i \vee \gamma_j)$ have the form:
$(\bigvee(r_{a,b} \wedge C_i) \vee D_i) \vee (\bigvee(r_{a,b} \wedge C_j) \vee D_j)$
which are of the form: $\bigvee(r_{a,b} \wedge C) \vee D$.
Then F_2 is equivalent to:
$F_3 = \bigwedge(\bigvee(r_{a,b} \wedge C) \vee D)$
Therefore, we have:
$\phi \leftrightarrow QX(\bigwedge(\bigvee(r_{a,b} \wedge C) \vee D))$

Subcase 4) If we have not $Link_{a,b}\phi_1$ and not $Link_{a,b}\phi_2$, there exists t in C or in V such that we have $Link_{a,t}\phi_1$ and $Link_{t,b}\phi_2$, and from (H) we have:

In the same way as in case 3) we have:
$\phi \leftrightarrow Q_1 X_1 Q_2' X_2'((\bigwedge(\bigvee(r_{a,t} \wedge C_1) \vee D_1)) \vee (\bigwedge(\bigvee(r'_{t,b} \wedge C_2') \vee D_2')))$

The terms which are named $(\delta_i \vee \gamma_j)$ in case 3) have in case 4) the form:
$(\bigvee(r_{a,t} \wedge C_i) \vee D_i) \vee (\bigvee(r_{t,b} \wedge C_j) \vee D_j)$
which is equivalent to:
$F_1 = \bigvee((r_{a,t} \wedge C_i) \vee (r_{t,b} \wedge C_j)) \vee D_i \vee D_j$
the formula: $(r_{a,t} \wedge C_i) \vee (r_{t,b} \wedge C_j)$ is equivalent to :
$F_2 = ((r_{a,t} \wedge C_i) \wedge (r_{t,b} \wedge C_j)) \vee ((r_{a,t} \wedge C_i) \wedge \neg(r_{t,b} \wedge C_j)) \vee (\neg(r_{a,t} \wedge C_i) \wedge (r_{t,b} \wedge C_j))$
the term $((r_{a,t} \wedge C_i) \wedge \neg(r_{t,b} \wedge C_j))$ in F_2 is equivalent to: $((r_{a,t} \wedge C_i) \wedge (\neg(r_{t,b} \vee \neg C_j))$
which is equivalent to:
$F_3 = ((r_{a,t} \wedge \neg r_{t,b} \wedge C_i) \vee (r_{t,b} \wedge C_i \wedge \neg C_j)$

From Lemma 5 $\neg r_{t,b}$ is equivalent to a disjunction of the form: $\bigvee r'_{t,b}$. Then, $r_{a,t} \wedge \neg r_{t,b}$ is equivalent to: $\bigvee(r_{a,t} \wedge r'_{t,b})$.

Then, F_3 is equivalent to a formula of the form F_4:
$F_4 = \bigvee(r_{a,t} \wedge r'_{t,b}) \wedge C_i \vee E_i$

Since $r_{a,t} \wedge r'_{t,b}$ is equivalent to a basic relationship $r_{a,b}$ between a and b, F_4 is equivalent to F_5:
$F_5 = \bigvee(r_{a,b} \wedge C_i \vee E_i)$

It can be shown in a similar way that $(\neg(r_{a,t} \wedge C_i) \wedge (r_{t,b} \wedge C_j))$ is equivalent to a formula F_5' of the form:
$F_5' = \bigvee(r_{a,b} \wedge C_i') \vee E_i'$

We also have $(r_{a,t} \wedge C_i) \wedge (r_{t,b} \wedge C_j)$ which is equivalent to $r_{a,b} \wedge C_i \wedge C_j$. Therefore, F_2 is equivalent to F_6:
$F_6 = (r_{a,b} \wedge C_i \wedge C_j) \vee (\bigvee(r_{a,b} \wedge C_i') \vee E_i') \vee (\bigvee(r_{a,b} \wedge C_i'') \vee E_i'')$

The formula F_6 is of the form of F_7:
$F_7 = \bigvee(r_{a,b} \wedge C) \vee D$

Since each term $(\delta_i \vee \gamma_j)$ of the conjunction is equivalent to a formula of the form of F_7 we have:
$\phi \leftrightarrow QX(\bigwedge(\bigvee((r_{a,b} \wedge C) \vee D))$

QED.

It is worth noting that the proof of Theorem 1 is constructive in the sense that it shows what are the $r_{a,b}$s in a formula from the $r_{a,b}$s in the subformulas. Then, it can be used a basis to exhibit the $r_{a,b}$s which hold in a formula.

For instance, if we call $T(\phi)$ the transformations which exhibits the $r_{a,b}$s, and $\phi = \neg \phi_1$, from the fact that we have:
$T(\phi_1) = QX(\bigwedge(\bigvee((r_{a,b} \wedge C) \vee D))$
we can infer that:

$T(\phi) = Q'X(\bigwedge(\bigvee((r'_{a,b} \wedge \neg D) \vee (\neg C \wedge \neg D)))$

where we have $\forall x_i$ (respectively $\exists x_i$) in $Q'X$ if we have $\exists x_i$ (respectively $\forall x_i$) in QX and the $r'_{a,b}$s are defined from $\neg r_{a,b}$ as it is shown in Lemma 5.

Examples. Let's see the examples presented after Definition 2.

For $G_1 = \neg \exists x(p(a,x) \wedge \neg q(x,b))$, G_1 is logically equivalent to:
$\forall x(\neg p(a,x) \vee q(x,b))$
which is equivalent to:
$\forall x((\neg p(a,x) \wedge q(x,b)) \vee (p(a,x) \wedge q(x,b)) \vee (\neg p(a,x) \wedge \neg q(x,b))$

If we use the notations:
$r^1_{a,b} \stackrel{def}{=} \neg p(a,x) \wedge q(x,b)$
$r^2_{a,b} \stackrel{def}{=} p(a,x) \wedge q(x,b)$
$r^3_{a,b} \stackrel{def}{=} \neg p(a,x) \wedge \neg q(x,b)$

G_1 is equivalent to: $\forall x(r^1_{a,b} \vee r^2_{a,b} \vee r^3_{a,b})$.

If we use the transformation $T(\phi)$, to compute $T(\phi)$ for the formula: $\phi = \neg \phi_1$, where $\phi_1 = \exists x(p(a,x) \wedge \neg q(x,b))$ we have:
$T(\phi_1) = \exists x(r_{a,b})$,
where $r_{a,b} = p(a,x) \wedge \neg q(x,b)$

If we unify $T(\phi_1)$ with the general definition:
$T(\phi_1) = QX(\bigwedge(\bigvee((r_{a,b} \wedge C) \vee D))$
we have only one term in the conjunctions and one term in the disjunctions, that is:
$QX = \exists x$, $C = True$, $D = False$ and $\bigvee(r_{a,b} \wedge C) = r_{a,b}$

The generic form of $T(\phi)$ is $T(\phi) = Q'X(\bigwedge(\bigvee((r'_{a,b} \wedge \neg D) \vee (\neg C \wedge \neg D)))$.
Then, for this example we have: $Q'X = \forall x$ and $T(\phi) = \forall x(\bigvee(r'_{a,b}))$.
where $\bigvee(r'_{a,b}) = \neg r_{a,b} = r^1_{a,b} \vee r^2_{a,b} \vee r^3_{a,b}$

Therefore, as it has been shown before, we have:
$T(\phi) = \forall x(r^1_{a,b} \vee r^2_{a,b} \vee r^3_{a,b})$

For $G_2 = \neg((p(a,c) \vee p(a,d)) \wedge (p(c,b) \vee r(d,b)))$, G_2 is equivalent to:
$\neg((p(a,c) \vee p(a,d)) \vee \neg(p(c,b) \vee r(d,b))$
which is equivalent to:
$(\neg(p(a,c) \wedge \neg p(a,d)) \vee (\neg p(c,b) \wedge \neg r(d,b)))$
which is equivalent to:
$((\neg p(a,c) \wedge \neg p(c,b)) \vee (\neg p(a,c) \wedge \neg r(d,b)) \vee (\neg p(a,d) \wedge \neg p(c,b)) \vee (\neg p(a,d) \wedge \neg r(d,b))$

If we use the notations:
$r^1_{a,b} \stackrel{def}{=} \neg p(a,c) \wedge \neg p(c,b)$
$r^2_{a,b} \stackrel{def}{=} \neg p(a,d) \wedge \neg r(d,b)$

G_2 is equivalent to: $r^1_{a,b} \vee r^2_{a,b} \vee (\neg p(a,c) \wedge \neg r(d,b)) \vee (\neg p(a,d) \wedge \neg p(c,b))$.

For $G3 = (p(a,b) \lor p(a,d)) \land r(d,b)$, G_3 is equivalent to:
$(p(a,b) \land r(d,b)) \lor (p(a,d) \land r(d,b))$

If we use the notations:
$r^1_{a,b} \stackrel{def}{=} p(a,b)$
$r^2_{a,b} \stackrel{def}{=} p(a,d) \land r(d,b)$

G_3 is equivalent to: $(r^1_{a,b} \land r(d,b)) \lor r^2_{a,b}$.

4 Related works

In [4] are defined the relationships between sentences and topics and it is pointed out that, even if some sentences do not express information about the state of the world, because they represent tautologies, they are not necessarily about the same topic. For instance the tautology: *Romeo loves Juliet or Romeo does not love Juliet*, is about the topic "love", while the tautology: *Verona is in Italia or Verona is not in Italia*, is about the topic "geography".

For the same reason we have accepted that a sentence may express a given relationship between two individuals even if it is a tautology.

In [5] is defined, in the semantics, the sentences which express information about a given individual named, for instance, by a. Roughly speaking, this definition is based on the notion of variants of a given model with regard to the individual a. These variants are all the models where at least one tuple in a relation which contain an interpretation of a has a different truth value than the truth value it has in the original model. Then, a sentence which informs about a is a sentence whose truth value changes in at least one of these variants.

The main difference with the work presented in this paper is that a sentence, like: *Romeo is a man and Juliet is a woman* informs about the individual Romeo and also informs about the individual Juliet while it does not express information about a relationship between Romeo and Juliet. Another significant difference is that according to this definition, a tautology does not express information about an individual since its truth value does not change in the variants of a given model.

5 Conclusion

After an informal introduction of the notion of relationship between individuals we have given a formal definition which is based on the notions of path in a conjunction of literals, the notion of link, which is a property of a complex formula, and the notion of relationship which is based on the notion of link. The Theorem 1 shows how the relationships involved in a complex formula can be expressed in terms of basic relationships. Since the proof of the Theorem 1 is constructive, it could be used to define an algorithm to compute these basic relationships.

We have shown in section 3 that the definition of relationship is based on the syntax of a formula and may not be preserved by logical equivalence. A further work might be to give a definition based on the semantics. The idea would be to consider variants of a given model (see section 4) in terms of a path as defined in the Definition 3, that is, models such that the tuples involved in the path are assigned different truth values than in the initial model.

There are several possible extensions of this work. The first one is to introduce equality in order to show that in a sentence like: *Venus is Juliet and Romeo loves Juliet* (in semiformal terms: (Venus = Juliet) $\wedge$ (Romeo loves Juliet)), there is a relationship between Romeo and Venus. The second one might be to introduce function symbols to represent sentences like: *Romeo hates Juliet's father* (formally: Romeo hates (father(Juliet))). A less intuitive extension could be to introduce modal operators for the representation of sentences of the kind: *Juliet believes that Romeo knows that Marutio is an enemy* (formally: $Believes_{Juliet}$ ($Knows_{Romeo}$ (enemy Marutio))), where Juliet and Romeo are not arguments of predicates but indexes of modal operators. More complex cases are formulas like: *Venus is Juliet and Juliet believes that Romeo loves Venus* (formally: (Venus = Juliet) $\wedge$ $Believes_{Juliet}$ (Romeo loves Venus)) which shows that equality is interpreted *de re*, versus the formula: *Romeo believes that Juliet is Venus and Romeo loves Venus* (formally: $Believes_{Romeo}$ ((Juliet = Venus) $\wedge$ (Romeo loves Venus))) where equality is interpreted *de dicto*.

Acknowledgements. Reviewer's comments have been very helpful to improve the quality of the paper.

References

[1] D. Chamberlin, A. Gilbert, and R. Yost. A history of system R and SQL/Data system. In *Very Large Data Bases, 7th International Conference*, 1981.

[2] C-L. Chang and R. Lee. *Symbolic Logic and Mechanical Theorem Proving*. Academic Press, 1973.

[3] R. Demolombe and A.J.I. Jones. Reasoning about Topics: towards a formal theory. In *American Association for Artificial Intelligence Fall Symposium*, 1995.

[4] R. Demolombe and A.J.I. Jones. On sentences of the kind "sentence "p" is about topic "t": some steps toward a formal-logical analysis. In H-J. Ohlbach and U. Reyle, editor, *Logic, Language and Reasoning. Essays in Honor of Dov Gabbay*. Kluwer Academic Press, 1999.

[5] R. Demolombe and L. Fariñas del Cerro. Information about a given entity: from semantics towards automated deduction. *Journal of Logic and Computation*, 20(6), 2010.

[6] R. Demolombe, M. Lemaitre, and J-M. Nicolas. The language of syntex2, an experimental relational like dbms. In *Proceedings of Jerusalem Conference on Information Technology. Jerusalem*, 1978.

[7] N. Goodman. About. *Mind*, LXX(277), 1961.

[8] J. L. Kuhns. Interrogating a relational data file. Technical Report R-511-PR, Rand Corporation, 1970.

[9] H. Putnam. Formalization of the concept "About". *Philosophy of Science*, XXV:125–130, 1958.

[10] J. Ullman. Implementation of Logical Query Languages for Databases. *ACM Transactions On Database Systems*, 10(3), 1985.

[11] J. D. Ullman. *Principles of Database Systems*. Computer Science Press, 1980.

[12] M. Zloof. Query-by-example. In *Proc. of AFIPS Vol4*, 1975.

On the Lattice of the Subvarieties of Monadic $MV(C)$-algebras

Antonio Di Nola
University of Salerno and IIASS, Vietri, Italy.
`adinola@unisa.it`

Revaz Grigolia
Tbilisi Sate University, Georgia.
`revaz.grigolia@tsu.ge, revaz.grigolia359@gmail.com`

Giacomo Lenzi (corresponding)
University of Salerno, Italy.
`gilenzi@unisa.it`

Abstract

The description of the lattice $\mathcal{L}$ of subvarieties of the variety **MMV(C)** generated by monadic MV-algebras, the MV-reduct of which are the algebras from the variety of MV-algebras generated by perfect MV-algebras, is given.

1 Introduction

The finitely valued propositional calculi, which have been described by Łukasiewicz and Tarski in [15], are extended to the corresponding predicate calculi. The predicate Łukasiewicz (infinitely valued) logic QL is defined in the following standard way [14, 17]: the existential (universal) quantifier is interpreted as supremum (infimum) in a complete MV-algebra. Then the valid formulas of predicate calculus are defined as all formulas having value 1 for any assignment. The functional description of the predicate calculus is given by Rutledge in [17]. Scarpellini in [18] has proved that the set of valid formulas is not recursively enumerable. We also refer the reader to the papers [19, 20, 10] concerning the Łukasiewicz predicate calculus.

Monadic MV-algebras were introduced and studied by Rutledge in [17] as an algebraic model for the predicate calculus QL of Łukasiewicz infinite-valued logic, in which only a single individual variable occurs. Rutledge followed P.R. Halmos' study of monadic Boolean algebras. In view of the incompleteness of the predicate calculus the result of Rutledge in [17], showing the completeness of the monadic predicate calculus, has been of great interest.

Let L denote a first-order language based on $\cdot, +, \rightarrow, \neg, \exists$ and let L_m denote a propositional language based on $\cdot, +, \rightarrow, \neg, \exists$. Let $Form(L)$ and $Form(L_m)$ be the set of all formulas of L and L_m, respectively. We fix a variable x in L, associate with each propositional letter p in L_m a unique monadic predicate $p^*(x)$ in L and define by induction a translation $\Psi : Form(L_m) \rightarrow Form(L)$ by putting:

- $\Psi(p) = p^*(x)$ if p is propositional variable,
- $\Psi(\alpha \circ \beta) = \Psi(\alpha) \circ \Psi(\beta)$, where $\circ = \cdot, +, \rightarrow$,
- $\Psi(\exists \alpha) = \exists x \Psi(\alpha)$.

Through this translation Ψ, we can identify the formulas of L_m with monadic formulas of L containing the variable x. Moreover, it is routine to check that $\Psi(MLPC) \subseteq QL$, where $MLPC$ is the monadic Lukasiewicz propositional calculus [8].

For a detailed consideration of Łukasiewicz predicate calculus we refer to [2, 14, 15].

Recall that the variety **MMV(C)** is a subvariety of the variety **MMV** of all monadic MV-algebras defined by the identity $2(x^2) = (2x)^2$ [8, 11], where C is Chang's algebra introduced in [7]. The paper is devoted to the description of a lattice of subvarieties of the variety **MMV(C)**. It is highlighted the subvarieties generated by the subdirectly irreducible $MMV(C)$-algebras the MV-reduct of which is isomorphic to C^m and its subalgebras.

2 Preliminaries on Monadic MV-algebras

The characterization of monadic MV-algebras as pair of MV-algebras, where one of them is a special kind of subalgebra (m-relatively complete subalgebra), is given in [8, 4]. MV-algebras were introduced by Chang in [7] as an algebraic model for infinitely valued Łukasiewicz logic.

An MV-algebra is an algebra $A = (A, \oplus, \odot, ^*, 0, 1)$ where $(A, \oplus, 0)$ is an abelian monoid, and the following identities hold for all $x, y \in A$: $x \oplus 1 = 1$, $x^{**} = x$, $0^* = 1$, $x \oplus x^* = 1$, $(x^* \oplus y)^* \oplus y = (y^* \oplus x)^* \oplus x$, $x \odot y = (x^* \oplus y^*)^*$.

Every MV-algebra has an underlying ordered structure defined by

$$x \leq y \text{ iff } x^* \oplus y = 1.$$

Thus $(A, \leq, 0, 1)$ is a bounded distributive lattice. Moreover, the following property holds in any MV-algebra:

$$x \odot y \leq x \wedge y \leq x \vee y \leq x \oplus y.$$

We introduce some abbreviations: (i) $0x = 0$, $(m+1)x = mx \oplus x$, (ii) $x^0 = 1$, $x^{m+1} = x^m \odot x$.

The operation $x \Rightarrow y = x^* \oplus y = sup\{z : x \odot z \leq y\}$ is named implication. In other words the operation $\Rightarrow$ is adjoint to the operation $\odot$, i. e. $x \odot z \leq y$ iff $z \leq x \Rightarrow y$.

The unit interval of real numbers $[0, 1]$ endowed with the following operations: $x \oplus y = \min(1, x+y), x \odot y = \max(0, x+y-1), x^* = 1-x$, becomes an MV-algebra. It is well known that the MV-algebra $S = ([0,1], \oplus, \odot, ^*, 0, 1)$ generates the variety **MV** of all MV-algebras, i. e. $\mathcal{V}(S) =$ **MV**.

Let $\mathbb{Q}$ denote the set of rational numbers; then $[0,1] \cap \mathbb{Q}$ is another MV-algebra.

There are MV-algebras which are not semisimple, i.e. the intersection of their maximal ideals (the radical of A, notation $Rad(A)$) is different from $\{0\}$. Non-zero elements from the radical of A are called infinitesimals. It is worth to stress that to the existence of infinitesimals in some MV-algebras is due the remarkable difference of behaviour between Boolean algebras and MV-algebras.

Perfect MV-algebras are those MV-algebras generated by their infinitesimal elements or, equivalently, generated by their radical [3]. They generate the smallest non locally finite subvariety of the variety **MV** of all MV algebras.

The class of perfect MV-algebras does not form a variety and contains non-simple subdirectly irreducible MV-algebras. It is worth stressing that the variety generated by all perfect MV-algebras, denoted by **MV(C)**, is also generated by a single MV-chain, actually the MV-algebra C, which have been defined by Chang in [7]. We name $MV(C)$-*algebras* all the algebras from the variety generated by C. Let L_P be the logic corresponding to the variety generated by perfect algebras which coincides with the set of all Łukasiewicz formulas that are valid in all perfect MV-chains, or equivalently that are valid in the MV-algebra C. Actually, L_P is the logic obtained by adding to the axioms of Łukasiewicz sentential calculus the following axiom: $(x \veebar x) \& (x \veebar x) \leftrightarrow (x \& x) \veebar (x \& x)$ (where $\veebar$ is strong disjunction, & strong conjunction in Łukasiewicz sentential calculus), see [3]. Notice that the Lindenbaum algebra of L_P is an $MV(C)$-algebra. The perfect algebra C has relevant properties. Indeed C generates the smallest variety of MV-algebras containing non-Boolean non-semisimple algebras. It is also subalgebra of any non-boolean perfect MV-algebra.

The importance of the class of $MV(C)$-algebras and the logic L_P can be perceived by looking further at the role that infinitesimals play in MV-algebras and

Łukasiewicz logic. Indeed the pure first order Łukasiewicz predicate logic is not complete with respect to the canonical set of truth values $[0,1]$, see [18], [2]. The Lindenbaum algebra of the first order Łukasiewicz logic is not semisimple and the valid but unprovable formulas are precisely the formulas whose negations determine the radical of the Lindenbaum algebra, that is the co-infinitesimals of such algebra. Hence, the valid but unprovable formulas generate the perfect skeleton of the Lindenbaum algebra. So, perfect MV-algebras, the variety generated by them and their logic are intimately related with a crucial phenomenon of the first order Łukasiewicz logic.

An algebra $A = (A, \oplus, \odot, {}^*, \exists, 0, 1)$ is said to be *a monadic MV-algebra (MMV-algebra* for short) if $A = (A, \oplus, \odot, {}^*, 0, 1)$ is an MV-algebra and in addition $\exists$ satisfies the following identities:

E1. $x \leq \exists x$,

E2. $\exists(x \vee y) = \exists x \vee \exists y$,

E3. $\exists(\exists x)^* = (\exists x)^*$,

E4. $\exists(\exists x \oplus \exists y) = \exists x \oplus \exists y$,

E5. $\exists(x \odot x) = \exists x \odot \exists x$,

E6. $\exists(x \oplus x) = \exists x \oplus \exists x$.

Sometimes we shall denote a monadic MV-algebra $A = (A, \oplus, \odot, {}^*, \exists, 0, 1)$ by $(A, \exists)$, for brevity. We can define a unary operation $\forall x = (\exists x^*)^*$ corresponding to the universal quantifier.

Theorem 1. *In any MMV-algebra holds the identity*
E7. $\exists(x \odot \exists y) = \exists x \odot \exists y$.

Proof. It is clear that $\exists(x \odot \exists y) \leq \exists x \odot \exists y$. On the other hand we have $x \odot \exists y \leq \exists(x \odot \exists y) \Rightarrow x \leq \exists y \Rightarrow \exists(x \odot \exists y) \Rightarrow \exists x \leq \exists y \Rightarrow \exists(x \odot \exists y) \Rightarrow \exists x \odot \exists y \leq \exists(x \odot \exists y)$. So, $\exists(x \odot \exists y) = \exists x \odot \exists y$. □

Let A_1 and A_2 be any MMV-algebras. A mapping $h : A_1 \to A_2$ is an MMV-homomorphism if h is an MV-homomorphism and for every $x \in A_1$ $h(\exists x) = \exists h(x)$. Denote by **MMV** the variety and the category of MMV-algebras and MMV-homomorphisms.

As it is well known, MV-algebras form a category that is equivalent to the category of abelian lattice ordered groups (ℓ-groups, for short) with strong unit [16].

Let us denote by Γ the functor implementing this equivalence. If G is an ℓ-group, then for any element $u \in G$, $u > 0$ we let $[0, u] = \{x \in G : 0 \leq x \leq u\}$ and for each $x, y \in [0, u]$ $x \oplus y = u \wedge (x + y)$ and $x^* = u - x$.

Notations. (i) $C_0 = \Gamma(Z, 1)$.

(ii) $C_1 = C \cong \Gamma(Z \times_{lex} Z, (1, 0))$ with generator $(0, 1) = c_1(= c)$, where C is the MV-algebra introduced by Chang in [7] which is important in this paper, because C generates the variety generated by perfect MV-algebras, and $\times_{lex}$ is the lexicographic product.

(iii) $C_m = \Gamma(Z \times_{lex} \cdots \times_{lex} Z, (1, 0, ..., 0))$ with generators $c_1(= (0, 0, ..., 1))$, ..., $c_m(= (0, 1, ..., 0))$, where the number of factors Z is equal to $m + 1$.

(iv) $R^*(A) = Rad(A) \cup \neg Rad(A)$, where $\neg Rad(A) = \{x^* : x \in Rad(A)\}$, where $Rad(A)$ is the intersection of all maximal ideals of the MV-algebra A.

Let $(A, \oplus, \odot, ^*, \exists, 0, 1)$ be a monadic MV-algebra. Let $\exists A = \{x \in A : x = \exists x\}$. By [8], $(\exists A, \oplus, \odot, ^*, 0, 1)$ is an MV-subalgebra of the MV-algebra $(A, \oplus, \odot, ^*, 0, 1)$.

A subalgebra A_0 of an MV-algebra A is said to be relatively complete if for every $a \in A$ the set $\{b \in A_0 : a \leq b\}$ has a least element.

Let $(A, \oplus, \odot, ^*, \exists, 0, 1)$ be a monadic MV-algebra. By [17], the MV-algebra $\exists A$ is a relatively complete subalgebra of the MV-algebra $(A, \oplus, \odot, ^*, 0, 1)$, and $\exists a = inf\{b \in \exists A : a \leq b\}$.

A subalgebra A_0 of an MV-algebra A is said to be *m-relatively complete* [8], if A_0 is relatively complete and two additional conditions hold:

(#) $(\forall a \in A)(\forall x \in A_0)(\exists v \in A_0)(x \geq a \odot a \Rightarrow v \geq a \, \& \, v \odot v \leq x)$,

(##) $(\forall a \in A)(\forall x \in A_0)(\exists v \in A_0)(x \geq a \oplus a \Rightarrow v \geq a \, \& \, v \oplus v \leq x)$.

Notice that two-element Boolean subalgebra of the standard MV-algebra $S = ([0, 1], \oplus, \odot, ^*, 0, 1)$ is relatively complete, but not m-relatively complete.

Proposition 2. *[8] Let A be monadic MV-algebra. Then $\exists A$ is an m-relatively complete subalgebra of the monadic MV-algebra $(A, \exists)$.*

Proposition 3. *(1) [8] If A_0 is m-relatively complete totally ordered MV-subalgebra of the MV-algebra A, then A_0 is a maximal totally ordered subalgebra of A.*

(2) [8, 17] If $(A, \exists)$ is a totally ordered monadic MV-algebra, then $A = \exists A$.

(3) [8, 17] $(A, \exists)$ is a subdirectly irreducible monadic MV-algebra if and only if $\exists A$ is totally ordered.

(4) [8, 17] Any monadic MV-algebra $(A, \exists)$ is isomorphic to a subdirect product of monadic MV-algebras $(A_i, \exists)$ such that $\exists A_i$ is totally ordered.

3 Properties of subvarieties of monadic $MV(C)$-algebras

From the variety of monadic MV-algebras **MMV** [8] select the subvariety **MMV(C)** which is defined by the following equation [11]:

$$(Perf) \quad 2(x^2) = (2x)^2,$$

that is **MMV(C)** = **MMV** + $(Perf)$. The main object of our interest is the variety **MMV(C)**.

According to axiom **E5** of monadic MV-algebras m-relatively complete subalgebra of C coincides with C but not its two-element Boolean subalgebra. Indeed, if $\exists C = \{0, 1\}$, then for any $x \in RadC$ $\exists(x \odot x) = 0$, but $\exists x \odot \exists x = 1$. In other words, $(C, \exists)$ is monadic $MMV(C)$-algebra if $\exists x = x$. Let we have C^n for some non-negative integer n. Then $(C^n, \exists)$ will be $MMV(C)$-algebra, where $\exists(a_1, ..., a_n) = (a_m, a_m, ..., a_m)$ with $a_m = max\{a_1, ..., a_n\}$ and $\forall(a_1, ..., a_n) = (a_m, a_m, ..., a_m)$ with $a_m = min\{a_1, ..., a_n\}$. Notice, that $(C^n, \exists)$ is subdirectly irreducible [8].

We name a monadic MV-algebra $(A, \exists)$ *perfect* if MV-algebra reduct of the algebra is a perfect MV-algebra. Let us denote by **MMV(C)**^{0} the subvariety of **MMV(C)** containing the $MMV(C)$-algebras with trivial monadic operator $\exists = \exists_{id}$, where $\exists_{id}(x) = x$.

Let $Alt_m^C = \forall(2x_1^2) \vee \forall(2x_1^2 \to 2x_2^2) \vee \cdots \vee \forall(2x_1^2 \wedge 2x_2^2 \wedge \cdots \wedge 2x_m^2 \to 2x_{m+1}^2)$ for $0 < m \in \omega$. Let **MMV(C)**^{m} be the subvariety of **MMV(C)** defined by the identity $Alt_m^C = 1$.

Theorem 4. *The identity $Alt_m^C = 1$, for $0 < m \in \omega$, is true in finitely generated subdirectly irreducible algebra $A \in$ **MMV(C)** if and only if A contains as a maximal homomorphic image the monadic Boolean algebra $(2^k, \exists)$ for $k \leq m \in \omega$.*

Proof. The identity Alt_m^C is the instance of the Segerberg's formula $Alt_m = \forall x_1 \vee \forall(x_1 \to x_2) \vee \cdots \vee \forall(x_1 \wedge x_2 \wedge \cdots \wedge x_m \to x_{m+1})$ for modal logic $S5$ [21], the algebraic models of which are monadic Boolean algebras $(2^m, \exists)$ for $0 < m \in \omega$. The identity $Alt_m = 1$ is true in subdirectly irreducible monadic Boolean algebras $(2^k, \exists)$, where $1 \leq k \leq m$, and the algebras of this type generate the variety of monadic Boolean algebras. Moreover, the variety of monadic Boolean algebras is a subvariety of monadic MV-algebras, and the variety **MMV(C)** as well. Let $(A, \exists)$ be a subdirectly irreducible $MMV(C)$-algebra. Then $B(A) = \{2x^2 : x \in A\}$ is a monadic Boolean skeleton of the monadic MV-algebra $(A, \exists) \in$ **MMV(C)** which at the same time is a subalgebra of $(A, \exists)$ and, moreover, is a homomorphic image by the maximal monadic filter [11]. Notice that for any subdirectly irreducible algebra $(A, \exists) \in$ **MMV(C)** and the maximal monadic filter $F \subset A$ the factor algebra $(A, \exists)/F$ is a monadic Boolean algebra, where monadic filter F is a MV-filter of A additionally satisfying the condition: if $x \in F$, then $\forall x \in F$ [8]. So, the identity $Alt_m^C = 1$ is true in the subdirectly irreducible $MMV(C)$-algebras that contains as a maximal homomorphic image the monadic Boolean algebra $(2^k, \exists)$, where $1 \leq k \leq m$. □

From this theorem we immediately obtain

Corollary 5. *There is no a variety* **V** *between the varieties* **MMV(C)**^{m} *and* **MMV(C)**$^{m+1}$ *which is distinct from* **MMV(C)**^{m} *for* $0 < m \in \omega$.

Proof. The proof immediately follows from the fact that there is no variety between the variety of monadic Boolean algebras generated by $(2^m, \exists)$ and the variety of monadic Boolean algebras generated by $(2^{m+1}, \exists)$. □

Corollary 6. *Let A be non-Boolean subdirectly irreducible algebra from* **MMV(C)**$^{m+1}$ − **MMV(C)**^{m}. *Then A generates* **MMV(C)**$^{m+1}$.

Proof. Let A be non-Boolean subdirectly irreducible algebra from **MMV(C)**$^{m+1}$ − **MMV(C)**^{m}. Therefore the identity $Alt_{m+1}^C = 1$ is true in A. Then, according to Corollary 5, $\mathcal{V}(A) = $ **MMV(C)**$^{m+1}$, because $A \notin$ **MMV(C)**^{m}. □

Theorem 7. $\mathcal{V}(\bigcup_{k \in \omega}$ **MMV(C)**$^k) = $ **MMV(C)**.

Proof. Let $F_{\mathbf{MMV(C)}}(\omega)$ be ω-generated free $MMV(C)$-algebra with free generators $g_1, g_2, g_3, \ldots$. Then $g_1, \ldots, g_k \in F_{\mathbf{MMV(C)}}(\omega)$ generates the subalgebra of $F_{\mathbf{MMV(C)}}(\omega)$ which is k-generated free algebra $F_{\mathbf{MMV(C)}}(k)$ in the variety **MMV(C)**.

Since the variety of monadic Boolean algebras **MB** is a subvariety, we have that there exists a homomorphism $h : F_{\mathbf{MMV(C)}}(\omega) \to F_{\mathbf{MB}}(\omega)$ such that $h(g_1), h(g_2), h(g_3), \ldots$ are free generators of $F_{\mathbf{MB}}(\omega)$. Therefore $h(g_1), \ldots, h(g_k)$ are the free generators of k-generated free monadic Boolean algebra $F_{\mathbf{MB}}(k)$. So, $Alt_k^C = 1$ is true in $F_{\mathbf{MB}}(k)$ and, hence, $Alt_k^C = 1$ is true in **MMV(C)**^{m} for $m \leq k$.

At the same time there exists a homomorphism $h_m : F_{\mathbf{MMV(C)}}(k) \to F_{\mathbf{MMV(C)}^m}(k)$, where $F_{\mathbf{MMV(C)}^m}(k)$ is k-generated free algebra in the variety **MMV(C)**^{m}.

Notice that in $F_{\mathbf{MMV(C)}}(k)$ is true $Alt_k^C = 1$, because k-generated $MMV(C)$-algebra contains as a maximal homomorphic image the monadic Boolean algebra $(2^k, \exists)$. So, $F_{\mathbf{MMV(C)}}(k) \cong F_{\mathbf{MMV(C)}}^k(k)$ for $1 \leq k \in \omega$. Thus we have a direct system $(F_{\mathbf{MMV(C)}}(k))_{k \in Z^+}$ with natural embedding sending the generator to the generator. Therefore, $\mathcal{V}(\bigcup_{k \in \omega}$ **MMV(C)**$^k) = $ **MMV(C)**. □

Theorem 8. *Let us suppose that a subdirectly irreducible algebra $A \in \mathbf{MMV(C)}$, which is not monadic Boolean algebra, does not satisfy $Alt_C^m = 1$ for any positive integer m. Then A generates $\mathbf{MMV(C)}$.*

Proof. Let $A \in \mathbf{MMV(C)}$ be a subdirectly irreducible algebra, which is not monadic Boolean algebra, and does not satisfy $Alt_C^m = 1$ for any positive integer m. It means that A is not finitely generated (Theorem 4, because maximal homomorphic image of A is not isomorphic to monadic Boolean algebra $(\mathbf{2}^m, \exists)$ for any $m \in Z^+$). Let A' be ω-generated subalgebra of the algebra A which is not monadic Boolean algebra. Let $a_1, a_2, \ldots$ be generators of A'. Let A'_i be the subalgebra of A' generated by $a_1, a_2, \ldots, a_i \in A'$. Then we have a directed family $(A'_i)_{i \in Z^+}$ by subalgebra embedding, where Z^+ is the set of positive integers. It is clear that A' is generated by $\bigcup_{i \in Z^+} A'_i$. Then there exists cofinal subset J of Z^+ such that A'_j is not monadic Boolean algebra (we exclude from the directed set of the subalgebras that are monadic Boolean algebras) and $A'_j \in \mathbf{MMV(C)}^j$, i. e. $Alt_j^C = 1$ is true in A'_j because A'_j is finitely generated. Therefore, according to Corollary 6, A'_j generates the variety $\mathbf{MMV(C)}^j$. From here we conclude that A' generates the variety $\mathbf{MMV(C)}$, since $\mathcal{V}(\bigcup_{j \in J} \mathbf{MMV(C)}^j) = \mathbf{MMV(C)}$ (Theorem 7). □

So we have the following diagram:

$$\mathbf{MMV(C)}^1 \subset \mathbf{MMV(C)}^2 \subset \cdots \subset \mathbf{MMV(C)}^m \subset \quad \cdots \quad \mathbf{MMV(C)}$$

Fig. 1

Let us consider the identity $(\exists x)^2 \wedge (\exists x^*)^2 = 0$. It holds

Lemma 9. *The identity $(\exists x)^2 \wedge (\exists x^*)^2 = 0$ is satisfied in the subdirectly irreducible $MMV(C)$-algebra $(A, \exists)$ if and only if the MV-algebra reduct of that is a perfect MV-algebra.*

Proof. Let the MV-reduct of the algebra $(A, \exists)$ be perfect MV-algebra. Then any element $x \in A$ belongs to either radical of A or co-radical of A. If x belongs to radical of A, then $\exists x$ also belongs to the radical, the co-radical of A, and, hence, $(\exists x)^2 = 0$. If x belongs to co-radical of A, then x^* and $\exists x^*$ belong to radical of A, and, hence, $(\exists x^*)^2 = 0$.

Now suppose that the MV-algebra reduct of the algebra $(A, \exists)$ is not a perfect MV-algebra. It means that $(A, \exists)$ contains as a subalgebra the Boolean algebra $\mathbf{2}^k$ for some $1 < k \in \omega$. Let $b \in \mathbf{2}^k$ such that b is different from the greatest and the least element of A. So, representing the element b as a sequence of 1 and 0, one of the components should be 1. Therefore, $\exists b = \exists b^* = 1$. So, the identity $(\exists x)^2 \wedge (\exists x^*)^2 = 0$ does not hold in $(A, \exists)$. □

From the variety **MMV(C)** we can pick out the subvariety **MMV(C)**$_{perf}$ by the identity $(\exists x)^2 \wedge (\exists \neg x)^2 = 0$ which is generated by $MMV(C)$-algebras the MV-algebra reduct of which are perfect MV-algebras. Notice that this variety coincides with the variety **MMV(C)**^{1}.

Let $t(x) = (x \vee x^*) \oplus (\forall (x \vee x^*))^*$. Let us give an analysis for the polynomial $t(x)$. It is clear that for any subdirectly irreducible $MMV(C)$-algebra $(A, \exists)$ $t(x)$ belongs to the co-radical of A for every $x \in A$. Moreover, let us consider the algebra $(C^m, \exists)$ and non-Boolean element $(a_1, ..., a_m) \in Rad^*A$ and suppose $a_1 \leq ... \leq a_m$. Then $(\forall (a_1, ..., a_m))^* = (a_1^*, ..., a_1^*)$, and, hence, $t(a_1, ..., a_m) = (1, a_2 \oplus a_1^*, ..., a_m \oplus a_1^*)$. As we see we have got the element the one component of which is equal to 1. Observe that $t(t(a_1, ..., a_m)) = t^2(a_1, ..., a_m) = (1, 1, (a_3 \oplus a_1^*) \oplus (a_2 \oplus a_1^*)^*, ..., (a_m \oplus a_1^*) \oplus (a_2 \oplus a_1^*)^*)$ and $t^m(a_1, ..., a_m) = (1, 1, ..., 1)$. Notice that if $(a_1, ..., a_m) \in Rad^*A$ is a Boolean element, then $t(a_1, ..., a_m) = 1$.

Lemma 10. *The identity $t^m = 1$ is true in $(C^k, \exists)$ for $1 < k \leq m$ and $t^m = 1$ does not hold in $(C^k, \exists)$ for $k > m$.*

Corollary 11. *The identity $t^m = 1$ is true in $(R^*(C^{k_1}) \times ... \times R^*(C^{k_n}), \exists)$ for $1 < k \leq m$, where $k = k_1 + ... + k_n$, and $t^m = 1$ does not hold in $(R^*(C^{k_1}) \times ... \times R^*(C^{k_n}), \exists)$ for $k > m$. Moreover, $Alt_n^C = 1$ is true in $(R^*(C^{k_1}) \times ... \times R^*(C^{k_n}), \exists)$ for $n \leq m$ and is not true for $n > m$.*

Proof. Notice that $(R^*(C^{k_1}) \times ... \times R^*(C^{k_n}), \exists)$ is a subalgebra of $(C^m, \exists)$, where $k_1 + ... + k_n = m$. Therefore we conclude that the identity $t^m = 1$ is true in $(R^*(C^{k_1}) \times ... \times R^*(C^{k_n}), \exists)$.

Since $(R^*(C^{k_1}) \times ... \times R^*(C^{k_n}), \exists)$ has $(2^n, \exists)$ as a maximal homomorphic image, we have that $Alt_n^C = 1$ is true in $(R^*(C^{k_1}) \times ... \times R^*(C^{k_n}), \exists)$ for $n \leq m$ and is not true for $n > m$. □

4 Generating algebras for MMV(C)

Recall that given any class **K** of similar algebras, Jónsson's lemma states that if the variety $HSP(\mathbf{K})$ generated by **K** is congruence-distributive, its subdirectly irreducibles are in $HSP_U(\mathbf{K})$, that is, they are quotients of subalgebras of ultraproducts of members of **K**. (If **K** is a finite set of finite algebras, the ultraproduct operation is redundant.)

Notice that if A is any MV-algebra, then A is subdirectly irreducible iff A is a chain. Similarly, a monadic MV-algebra A is subdirectly irreducible iff $\exists A$ is a chain and A has a minimal nonzero monadic ideal. Moreover, the lattice of congruences of MMV-algebra A is isomorphic to the lattice of congruences of the algebra $\exists A$ (which is really an MV-algebra with trivial operator $\exists$, i.e. $\exists x = x$).

In [11] it is shown that the variety generated by C_1 contains any perfect algebra. So, C_1 and C_n generate the same variety. Here we give similar results for **MMV(C)**.

Let $(\mathbb{C}^k, \exists) = (C_1^k, =, \oplus, \odot, \neg, \exists, 0, 1)$ be the model for $MMV(C)$-theory and $Th((\mathbb{C}^k, \exists))$ the set of all true sentences in $(\mathbb{C}^k, \exists)$.

Let $At(x) = (\forall y)(y \leq x \Rightarrow (y = 0 \vee y = x))$ that means that x is an atom and $A(x_1, ..., x_k) = \wedge_{i \neq j}(x_i \neq x_j) \wedge (\wedge_{i=1}^k At(x_i)) \wedge (\exists(\vee_{i=1}^k x_i) = \vee_{i=1}^k x_i = \wedge_{i=1}^k \exists x_i) \wedge (\exists x_{k+1})(At(x_{k+1}) \Rightarrow (\vee_{i=1}^k (x_i = x_{k+1})) \wedge (\forall y)((y = \exists y) \wedge (y \leq \vee_{i=1}^k x_i)) \Rightarrow (y = 0 \vee y = \vee_{i=1}^k x_i)) \wedge (\wedge_{i=1}^k (\forall x_i = 0)))$.

Notice that in $(\mathbb{C}^k, \exists)$ are true the following sentences:

(i) $(\exists x_1 \exists x_2)A(x_1, ..., x_k)$, which means that there exist only k atoms in $(\mathbb{C}^k, \exists)$, that we denote by $a_1, ..., a_k$ such that $\exists a_1 = \exists a_2 = ... = \exists a_k = \vee_{i=1}^k a_i$ and $\forall a_1 = ... = \forall a_k = 0$;

(ii) $(\forall x \forall y)(((x = \exists x) \wedge (y = \exists y) \wedge (x \leq y) \vee (y \leq x))$, that means that $\exists(\mathbb{C}^k, \exists)$ is a chain;

(iii) $(\exists y_1 \exists y_2 .... \exists y_k)\Phi(y_1, ..., y_k) = (\exists y_1 \exists y_2 .... \exists y_k)(\wedge_{i=1}^k (y_i = 2y_i^2) \wedge \wedge_{i \neq j}(y_i \neq y_j) \wedge \wedge_{i \neq j}(\exists y_i = \exists y_j))$, that means that maximal homomorphic image is isomorphic to $(2^k, \exists)$ and not to $(2^{k+1}, \exists)$, $(2^{k-1}, \exists)$;

(iv) $(\forall x)(t^k(x) = 1)$.

Observe that the both formulas $(\exists y_1 \exists y_2 .... \exists y_k)\Phi(y_1, ..., y_k)$ and $(\forall x)(t^k = 1)$ are true in $(\mathbb{C}^k, \exists)$ and not true in $(\mathbb{C}^{k+1}, \exists)$, $(\mathbb{C}^{k-1}, \exists)$.

We add to the signature the new constants $a_1, ..., a_k$ such that $At(a_1), ..., At(a_k)$ are true in $(\mathbb{C}^k, \exists)$, c_1 and c_2, and let $\Lambda_n^\exists = A(a_1, ..., a_k) \wedge (c_1 = \vee_{i=1}^k a_i) \wedge (c_2 = \exists c_2) \wedge (c_2^2 = 0) \wedge (c_2 \neq nc_1)$ $(n \in Z^+)$ and let us consider a theory

$$T = Th((\mathbb{C}^k, \exists)) \cup \{\Lambda_1^\exists, \Lambda_2^\exists, \Lambda_3^\exists, ...\}.$$

So, in this theory we have terms $nc_1, n = 1, 2, 3, ...$

Proposition 12. *Every finite subtheory $T_0 \subseteq T$ is satisfiable.*

Proof. T_0 contains a finite number of axioms of the kind $\Lambda_{n_1}^\exists, \Lambda_{n_2}^\exists, ..., \Lambda_{n_k}^\exists$. Let c_2 be interpreted in the model $(\mathbb{C}^k, \exists)$ as any mc_1 such that $m > max\{n_1, ..., n_k\}$. □

According to the theorem of compactness there exists a model $(M, \exists) \models T$, that contains atoms, that we denote by $a_1^M, ..., a_k^M$, the elements of the kind $\exists x$ form a chain that contains an atom (let c_1^M be the atom of the chain). The model $(M, \exists)$ has the following properties:

1) $(\mathbb{C}^k, \exists)$ is embedded into $(M, \exists)$: $\varepsilon(c_1) = c_1^M$, $\varepsilon(a_1) = a_1^M$, ... , $\varepsilon(a_k) = a_k^M$;
2) $(M, \exists) \models Th((\mathbb{C}^k, \exists))$;
3) $(M, \exists) \not\cong (\mathbb{C}^k, \exists)$, in particular $c_2^M \in M$ such that $c_2^M \geq nc_1^M$ for every natural number n;
4) $(M, \exists) \cong ((\exists M)^k, \exists)$ because the formulas $(\exists y_1 \exists y_2 .... \exists y_k)\Phi(y_1, ..., y_k)$ and $(\forall x)(t^k = 1)$ are true in the model.

Now we take the elements $a_1^M, ..., a_k^M, c_1^M, c_2^M \in M$ and generate by these elements the subalgebra $(D, \exists)$, which is isomorphic to $(C_2^k, \exists)$.

As a consequence we have

Theorem 13. *The variety generated by $(C_1^k, \exists)$ coincides with the variety generated by $(C_n^k, \exists)$ for any $n \in Z^+$.*

5 Subvarieties of monadic $MV(C)$-algebras

Let m be a positive integer. Then a *partition* of m is a nonincreasing sequence of positive integers $(m_1, m_2, ..., m_n)$ whose sum is m. Each m_i is called a *part* of the partition. We let the function $p(m)$ denote the number of partitions of the integer m. For example, for the number 4: $1+1+1+1 = 2+1+1 = 2+2 = 3+1 = 4$. Let $\hat{p}(m)$ be the set of all partitions of the number m. So, $\hat{p}(4) = \{(1,1,1,1),(2,1,1),(2,2),(3,1)\}$. We define the function $p(m,n)$ to be the number of partitions of m whose largest part is n (or equivalently, the number of partitions of m with n parts). Let $\hat{p}(m,n)$ be the set of all partitions of the number m with n parts. For example $\hat{p}(4,2) = \{(2,2),(3,1)\}$, $\hat{p}(5,2) = \{(4,1),(3,2)\}$, $\hat{p}(5,1) = \{(5)\}$.

Let $\mathbf{MMV(C)}_{(m_1,...,m_n)}^n$ be the subvariety of $\mathbf{MMV(C)}$ generated by the algebra $(R^*(C^{m_1}) \times ... \times R^*(C^{m_n}), \exists)$, where $m_1 + ... + m_n = m$, i. e. $(m_1, ..., m_n) \in \hat{p}(m,n)$.

Lemma 14. *If $n_1 \neq n_2$, then $\mathbf{MMV(C)}_{(m_1,...,m_{n_1})}^{n_1} \neq \mathbf{MMV(C)}_{(m_1,...,m_{n_2})}^{n_2}$.*

Proof. Let us suppose that $n_1 < n_2$. Then $Alt_{n_1}^C = 1$ is true in $\mathbf{MMV(C)}_{(m_1,...,m_{n_1})}^{n_1}$ and is not true in $\mathbf{MMV(C)}_{(m_1,...,m_{n_2})}^{n_2}$. □

Lemma 15. *If $m \neq k$, then $\mathbf{MMV(C)}_{(m_1,...,m_n)}^n \neq \mathbf{MMV(C)}_{(k_1,...,k_n)}^n$, where $m_1 + ... + m_n = m$ and $k_1 + ... + k_n = k$.*

Proof. Let us suppose that $m < k$. Then $t^m(x) = 1$ is true in $\mathbf{MMV(C)}_{(m_1,...,m_n)}^n$ and is not true in $\mathbf{MMV(C)}_{(k_1,...,k_n)}^n$. □

Lemma 16. *Let $\lambda_1, \lambda_2 \in \hat{p}(m,n)$ such that $\lambda_1 \neq \lambda_2$. Then $\mathbf{MMV(C)}_{\lambda_1}^n \neq \mathbf{MMV(C)}_{\lambda_2}^n$.*

447

Proof. Let us suppose that $\lambda_1 = (m_1, ..., m_n)$, $\lambda_2 = (m'_1, ..., m'_n)$ and $\lambda_1 \neq \lambda_2$. Notice that $\mathbf{MMV(C)}^n_{\lambda_1}$ is generated by the subdirectly irreducible algebra $(R^*(C^{m_1}) \times ... \times R^*(C^{m_n}), \exists)$ and $\mathbf{MMV(C)}^n_{\lambda_2}$ is generated by the subdirectly irreducible algebra $(R^*(C^{m'_1}) \times ... \times R^*(C^{m'_n}), \exists)$. Notice that both algebras have maximal homomorphic images which are isomorphic to $(2^n, \exists)$. Observe that an algebra $R^*(C^k)$ is a subalgebra of $R^*(C^m)$ for any $k \leq m$. So, $(R^*(C^{m_1}) \times ... \times R^*(C^k) \times ... \times R^*(C^{m_n}), \exists)$ is a subalgebra of $(R^*(C^{m_1}) \times ... \times R^*(C^{m_i}) \times ... \times R^*(C^{m_n}), \exists)$ for any $k \leq m_i$ ($i \leq n$). Taking into account that $m_1 + ... + m_n = m'_1 + ... + m'_n = m$ we have that $(R^*(C^{m_1}) \times ... \times R^*(C^{m_n}), \exists)$ and $(R^*(C^{m'_1}) \times ... \times R^*(C^{m'_n}), \exists)$ are not subalgebras of each other. Notice also that $Alt^C_n = 1$ and $t^m(x) = 1$ are true in $(R^*(C^{m_1}) \times ... \times R^*(C^{m_n}), \exists)$ and $(R^*(C^{m'_1}) \times ... \times R^*(C^{m'_n}), \exists)$.

Now we will give a first order universal formula that is true in $(R^*(C^{m_1}) \times ... \times R^*(C^{m_n}), \exists)$ and false in $(R^*(C^{m'_1}) \times ... \times R^*(C^{m'_n}), \exists)$.

$\phi^m_{(m_1,...,m_n)} = (\forall x_1, ..., \forall x_m \forall y_1, ..., \forall y_n, \forall x)((\wedge_{i=1}^m At(x_i) \Rightarrow (\exists \vee_{i=1}^m x_i = \vee_{i=1}^m x_i = \wedge_{i=1}^m \exists x_i)) \wedge ((\wedge_{i=1}^n (y_i \oplus y_i = y_i) \wedge (\vee_{i=1}^n y_i = 1) \wedge (\wedge_{i \neq j}(y_i \wedge y_j = 0)) \wedge (\wedge_{i=1}^n (2x^2 \leq y_i \Rightarrow (2x^2 = 0 \vee 2x^2 = y_i)))) \Rightarrow (\vee_{\varphi \in \Phi}((\vee_{i=1}^{m_1} x_i \to y_{\varphi(1)} = 1) \wedge (\vee_{i=m_1+1}^{m_2} x_i \to y_{\varphi(2)} = 1) \wedge ... \wedge (\vee_{i=m_{n-1}+1}^{m_n} x_i \to y_{\varphi(n)} = 1))),$

where $n \leq m$, $At(x)$ means that x is an atom and Φ is the set of all bijections from $\{1, 2, ..., n\}$ to $\{1, 2, ..., n\}$.

Observe that $(R^*(C^{m_1}) \times ... \times R^*(C^{m_n}), \exists)$ (and $(R^*(C^{m'_1}) \times ... \times R^*(C^{m'_n}), \exists)$ as well) contains m atoms. Moreover, in the subformula

$((\wedge_{i=1}^n (y_i \oplus y_i = y_i) \wedge (\vee_{i=1}^n y_i = 1) \wedge (\wedge_{i \neq j}(y_i \wedge y_j = 0)) \wedge (\wedge_{i=1}^n (2x^2 \leq y_i \Rightarrow (2x^2 = 0 \vee 2x^2 = y_i))))$

the elements $y_1, ..., y_n$ are interpreted as atoms of Boolean algebra which is isomorphic to 2^n; and the subformula

$\vee_{\varphi \in \Phi}((\vee_{i=1}^{m_1} x_i \to y_{\varphi(1)} = 1) \wedge (\vee_{i=m_1+1}^{m_2} x_i \to y_{\varphi(2)} = 1) \wedge ... \wedge (\vee_{i=m_{n-1}+1}^{m_n} x_i \to y_{\varphi(n)} = 1))$

is true in $(R^*(C^{m_1}) \times ... \times R^*(C^{m_n}), \exists)$, where only one member of the disjunction, say $(\vee_{i=1}^{m_1} x_i \to y_{\varphi(1)} = 1) \wedge (\vee_{i=m_1+1}^{m_2} x_i \to y_{\varphi(2)} = 1) \wedge ... \wedge (\vee_{i=m_{n-1}+1}^{m_n} x_i \to y_{\varphi(n)} = 1))$ for some $\varphi \in \Phi$, is true in $(R^*(C^{m_1}) \times ... \times R^*(C^{m_n}), \exists)$, and, at the same time, any member of the disjunction is not true in $(R^*(C^{m'_1}) \times ... \times R^*(C^{m'_n}), \exists)$.

So, the formula $\phi^m_{(m_1,...,m_n)}$ is true in $(R^*(C^{m_1}) \times ... \times R^*(C^{m_n}), \exists)$ and not true in $(R^*(C^{m'_1}) \times ... \times R^*(C^{m'_n}), \exists)$. From here we conclude that any homomorphic image of any subalgebra of any ultrapower of $(R^*(C^{m_1}) \times ... \times R^*(C^{m_n}), \exists)$ is not isomorphic to $(R^*(C^{m'_1}) \times ... \times R^*(C^{m'_n}), \exists)$. Consequently, $\mathbf{MMV(C)}^n_{\lambda_1} \neq \mathbf{MMV(C)}^n_{\lambda_2}$. □

Let us consider the algebras $(R^*(C^3) \times R^*(C^2), \exists) \in \mathbf{MMV(C)}^2_{(3,2)}$ and $(R^*(C^4) \times C), \exists) \in \mathbf{MMV(C)}^2_{(4,1)}$. Notice that in this case $\lambda_1 = (3,2)$, $\lambda_2 = (4,1)$. In this case we have

$\phi^2_{(3,2)} = (\forall x_1, \forall x_2, ..., \forall x_5 \forall y_1, \forall y_2)((\wedge^5_{i=1} At(x_i) \Rightarrow (\exists \vee^5_{i=1} x_i = \vee^5_{i=1} x_i = \wedge^5_{i=1} \exists x_i)) \wedge ((\wedge^2_{i=1}(y_i \oplus y_i = y_i) \wedge (\vee^2_{i=1} y_i = 1) \wedge (\wedge_{i \neq j}(y_i \wedge y_j = 0)) \wedge (2x^2 \leq y_1 \Rightarrow 2x^2 = 0 \vee 2x^2 = y_1) \wedge (2x^2 \leq y_1 \Rightarrow 2x^2 = 0 \vee 2x^2 = y_1)) \Rightarrow ((((x_1 \vee x_2 \vee x_3) \rightarrow y_1 = 1)) \wedge ((x_4 \vee x_5 \rightarrow y_2 = 1)) \vee (((x_1 \vee x_2 \vee x_3) \rightarrow y_2 = 1)) \wedge ((x_1 \vee x_2) \rightarrow y_1 = 1))$.

Observe that Boolean skeletons of $(R^*(C^3) \times R^*(C^2), \exists)$ and $(R^*(C^4) \times C), \exists)$ are isomorphic to 2^2. More precisely, the elements of the Boolean skeleton of the algebra $(R^*(C^3) \times R^*(C^2), \exists)$ are $(1,1,1,1,1), (1,1,1,0,0), (0,0,0,1,1)$, $(0,0,0,0,0)$; and the elements of the Boolean skeleton of the algebra $(R^*(C^4) \times C), \exists)$ are $(1,1,1,1,1), (1,1,1,1,0), (0,0,0,0,1), (0,0,0,0,0)$. So, if we interpret y_1 as $(1,1,1,0,0)$, y_2 as $(0,0,0,1,1)$, x_1 as $(c,0,0,0,0)$, x_2 as $(0,c,0,0,0)$, x_3 as $(0,0,c,0,0)$, x_4 as $(0,0,0,c,0)$ and x_5 as $(0,0,0,0,c)$, then the only disjunction $(((x_1 \vee x_2 \vee x_3) \rightarrow y_1 = 1)) \wedge ((x_4 \vee x_5) \rightarrow y_2 = 1))$ is true in $(R^*(C^3) \times R^*(C^2), \exists)$.

Observe that for every MV-algebra A, and for every $MV(C)$-algebra as well, there exists the Belluce lattice $\beta(A)$ which is distributive lattice, the spectral space of which coincides with the spectral space of the MV-algebra A [1].

Q-distributive lattices was introduced by Cignoli in [6]. A Q-distributive lattice is an algebra $(A, \vee, \wedge, \exists, 0, 1)$ such that $(A, \vee, \wedge, 0, 1)$ is a bounded distributive lattice and $\exists$ is a quantifier on A that satisfies the following identities: (Q_0) $\exists 0 = 0$; (Q_1) $a \wedge \exists a = a$; (Q_2) $\exists (a \wedge \exists b) = \exists a \wedge \exists b$; (Q_3) $\exists (a \vee b) = \exists a \vee \exists b$.

If we have monadic $MV(C)$-algebra $(A, \exists)$, then we can obtain Q-distributive lattice $(\beta(A), \vee, \wedge, \exists, 0, 1)$. For Q-distributive lattice $(\beta(A), \vee, \wedge, \exists, 0, 1)$ it is constructed its dual object $(\mathcal{P}(\beta(A)), R, E)$ [6], where $(\mathcal{P}(\beta(A))$ is the set of all prime filters ordered by inclusion and $E \subset \mathcal{P}(\beta(A))^2$ is an equivalence relation on $\mathcal{P}(\beta(A))$ corresponding to the monadic operator $\exists$.

The dual objects (a) $(\mathcal{P}(\beta((R^*(C^3) \times R^*(C^2)))), R, E)$ and (b) $(\mathcal{P}(\beta(((R^*(C^4) \times C), \exists))), R, E)$ are depicted in Fig. 2. Notice that the spectral spaces represented in Fig. 2 correspond to the spectral spaces corresponding to the monadic Gödel algebras [5], where the equivalent elements are inside of ovals.

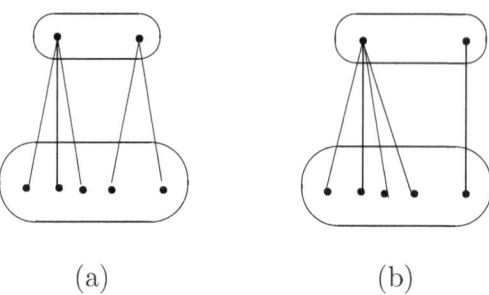

Fig. 2 Dual objects of (a) $(R^*(C^3) \times R^*(C^2), \exists))$, (b) $(R^*(C^4) \times C, \exists)$

Notice that $\mathbf{MMV}(\mathbf{C})^1_{(m)}$ coincides with $\mathbf{MMV}(\mathbf{C})^1_m$. Notice also that any variety $\mathbf{MMV}(\mathbf{C})^n_{(m_1,...,m_n)}$ is a subvariety of $\mathbf{MMV}(\mathbf{C})^m_m$, $m = m_1... + m_n$.

We have subvarieties $\mathbf{MMV}(\mathbf{C})^n_m = \mathbf{MMV}(\mathbf{C})^n + t^m = 1$ where $n \leq m$. Notice that $\mathbf{MMV}(\mathbf{C})^1_1$ coincides with the variety of monadic MV-algebras with trivial monadic operator $\exists x = x$.

It is easy to prove the following

Theorem 17. *1)* $\mathbf{MMV}(\mathbf{C})^n_{(m_1,...,m_n)}$ *is a subvariety of* $\mathbf{MMV}(\mathbf{C})^n_{m_1+...+m_n}$;

2) $\mathbf{MMV}(\mathbf{C})^n_{(m_1,...,m_n)}$ *is a subvariety of* $\mathbf{MMV}(\mathbf{C})^n_{(m'_1,...,m'_n)}$ *iff* $m_i \leq m'_i$, $i = 1,...,n$;

3) $\mathbf{MMV}(\mathbf{C})^n_m$ *is a subvariety of* $\mathbf{MMV}(\mathbf{C})^n_{m+1}$;

4) $\mathbf{MMV}(\mathbf{C})^n_m$ *is a subvariety of* $\mathbf{MMV}(\mathbf{C})^{n+1}_m$ *if* $n < m$.

Let $\mathbf{B}$ be the variety of monadic Boolean algebras and $\mathbf{B}_m$ the subvariety of $\mathbf{B}$ generated by $(2^m, \exists)$ where $1 \leq m < \omega$.

In the following diagram is represented a lattice the elements of which are $\mathbf{MMV}(\mathbf{C})^n$, $\mathbf{MMV}(\mathbf{C})^n_m$, $\mathbf{MB}_m$, $n \leq m, n, m \in Z^+$.

This diagram represents a lattice of subvarieties of $\mathbf{MMV}(\mathbf{C})$.

According to the results above we can define generating set of algebras for the subvarieties.

$\mathbf{MMV(C)}^1 = \mathcal{V}(\{(R^*(C^m), \exists) : 1 < m \in \omega\}), \dots,$
$\mathbf{MMV(C)}^n = \mathcal{V}(\{((R^*(C^m))^n, \exists)), 1 < m \in \omega\}),$
$\mathbf{MMV(C)}_m^n = \mathcal{V}(((R^*(C^{m_1})) \times \dots \times R^*(C^{m_n}), \exists)),$
$\mathbf{MB}_m = \mathcal{V}((2^m, \exists)), \ m \in \omega,$
$\mathbf{MB} = \mathcal{V}(\{(2^m, \exists)), \ m \in \omega\}),$
where $R^*(A) = RadA \cup (\neg RadA)$.

$\mathbf{MMV(C)}^1 \subset \mathbf{MMV(C)}^2 \subset \mathbf{MMV(C)}^3 \subset \dots \subset \mathbf{MMV(C)}^m \subset \dots \mathbf{MMV}(C)$
$\vdots \qquad \vdots \qquad \vdots \qquad \vdots$
$\cup \qquad \cup \qquad \cup \qquad \cup$
$\mathbf{MMV(C)}_m^1 \subset \mathbf{MMV(C)}_m^2 \subset \mathbf{MMV(C)}_m^3 \dots \subset \mathbf{MMV(C)}_m^m$
$\cup \qquad \cup \qquad \cup$
$\vdots \qquad \vdots \qquad \vdots$
$\cup \qquad \cup \qquad \cup$
$\mathbf{MMV(C)}_3^1 \subset \mathbf{MMV(C)}_3^2 \subset \mathbf{MMV(C)}_3^3$
$\cup \qquad \cup$
$\mathbf{MMV(C)}_2^1 \subset \mathbf{MMV(C)}_2^2$
$\cup$
$\mathbf{MMV(C)}_1^1$
$\cup \qquad \cup \qquad \cup \qquad \cup \qquad \cup$
$\mathbf{MB}_1 \ \subset \ \mathbf{MB}_2 \ \subset \ \mathbf{MB}_3 \ \subset \ \dots \ \subset \ \mathbf{MB}_m \ \subset \dots \ \mathbf{MB}$

Fig. 3

We do not know whether Fig. 3 represents all subvarieties of the variety $\mathbf{MMV(C)}$ are represented. Let $S_k R^*(C_k^m)$ be the set of all subalgebras of $R^*(C_k^m)$ having a totally ordered subalgebra isomorphic to C_k.

Conjecture 1. *Let SI^f be the set of all subdirectly irreducible $MMV(C)$-algebras having finite spectral space. SI^f coincides with the set of $MMV(C)$-algebras of the type $(A_1 \times \dots \times A_n, \exists)$, where $A_i \in S_k R^*(C_k^m)$ for $i = 1, \dots, n$, $k \in Z^+$.*

Conjecture 2. *Any proper subvariety of the variety $\mathbf{MMV(C)}$ is generated by the finite number of algebras $(A_1 \times \dots \times A_n, \exists)$, where $A_i \in S_k R^*(C_k^m)$ for $i = 1, \dots, n$, $k \in Z^+$.*

6 Problems

Now we formulate some problems.

1. To give axiomatization of all subvarieties of the variety **MMV(C)**.

The solution the next two problems will help to represent the lattice of all subvarieties of the variety **MMV(C)** and fill the gaps between subvarieties.

2. To show that the varieties generated by $(R^*(C^3) \times R^*(C^2), \exists)$ and $(R^*(C_n^3) \times R^*(C_n^2), \exists)$ $(n > 1)$, the dual object of which is depicted in Fig. 4, coincide.

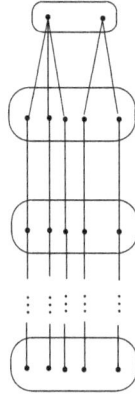

Fig. 4

3. A dual object of some subalgebra $A \in S_2(R^*(C_2^8))$ of the algebra $(C_2^8, \exists)$ is depicted in Fig. 5. To show that the varieties generated by $(R^*(C^3) \times R^*(C^2), \exists)$ and A does not coincide.

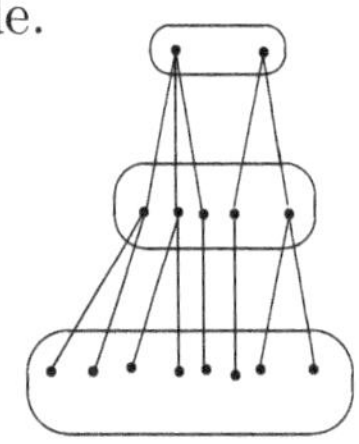

Fig. 5

Acknowledgements

We express our gratitude to the referees for her/his suggestions to improve the readability of this paper.

References

[1] L. P. Belluce, *Semisimple algebras of infinite-valued logic and bold fuzzy set theory*, Canad. J. Math. 38 (1986) 1356âĂŞ1379.

[2] L. P. Belluce, C.C. Chang, A weak completeness theorem for infinite valued 2rst-order logic, J. Symbolic Logic 28 (1963) 43–50.

[3] L. P. Belluce, A. Di Nola, B. Gerla, *Perfect MV-algebras and their Logic*, Applied Categorical Structures Volume 15, Numbers 1-2 (2007), 135-151.

[4] L.P. Belluce, R. Grigolia and A. Lettieri, Representations of monadic MV- algebras, Studia Logica, vol. 81, Issue October 15th, 2005, pp. 125-144.

[5] G. Bezhanishvili, R. Grigolia, "Locally tabular extensions of MIPC", Proceedings of Uppsala Symposium ", Advances in Modal Logic'98" , vol. 2, Csli Publications, Stanford, California, 101-120 (2001)

[6] R. Cignoli, Quantifiers on distributive lattices, Discrete Mathematics, 96(1991), 183-197.

[7] C. C. Chang, *Algebraic Analysis of Many-Valued Logics*, Trans. Amer. Math. Soc., 88(1958), 467-490.

[8] A. Di Nola, R. Grigolia, On Monadic MV-algebras, APAL, Vol. 128, Issues 1-3 (August 2004), pp. 125-139.

[9] A. Di Nola , R. Grigolia, *Profinite MV-spaces*, Discrete Mathematics , Vol. 283, Issues 1-3 (6 June 2004), pp. 61-69.

[10] G. Georgescu, A. Iurgulescu, I. Leustean, Monadic and Closure MV-Algebras, Multi. Val. Logic 3 (1998) 235–257.

[11] A. Di Nola, A. Lettieri, *Perfect MV-algebras are Categorically Equivalent to Abelian ℓ-Groups*, Studia Logica, 53(1994), 417-432.

[12] P.R. Halmos, Algebraic Logic I. Monadic Boolean algebras, Compositio Math. 12 (1955), 217-249.

[13] P.R. Halmos, Algebraic Logic (Chelsea, New York, 1962).

[14] L.S. Hay, An axiomatization of the infinitely many-valued calculus, M.S. Thesis, Cornell University, 1958.

[15] J. Łukasiewicz, A. Tarski, Unntersuchungen Ouber den Aussagenkalkul, Comptes Rendus des seances de la Societe des Sciences et des Lettres de Varsovie 23 (cl iii) (1930) 30–50.

[16] D. Mundici, *Interpretation of AF C^*-Algebras in Lukasiewicz Sentential Calculus*, J. Funct. Analysis **65**, (1986), 15-63.

[17] J.D. Rutledge, A preliminary investigation of the infinitely many-valued predicate calculus, Ph.D. Thesis, Cornell University, 1959.

[18] B. Scarpellini, Die Nichtaxiomatisierbarkeit des unendlichwertigen PrOadikatenkalkulus von Łukasiewicz, J. Symbolic Logic 27 (1962) 159–170.

[19] D. Schwartz, Theorie der polyadischen MV-Algebren endlicher Ordnung, Math. Nachr. 78 (1977) 131–138.

[20] D. Schwartz, Polyadic MV-algebras, Zeit. f. math. Logik und Grundlagen d. Math. 26 (1980) 561–564.

[21] K. Segerberg, An essay in classical modal logic, Uppsala, 1971.

www.ingramcontent.com/pod-product-compliance
Lightning Source LLC
Chambersburg PA
CBHW080919180426
43192CB00040B/2468